Engineering M...

Engineering mechanics is a foundation subject. A s... ...quired during the analysis of complex problems in several core engineering disciplines. This textbook, adapted to meet the syllabi requirements of most universities, begins with an introduction to systems of units and the representation, interaction and concatenation of forces in cartesian space. It shows ways to articulate forces and force systems, work out resultants, and equilibrium conditions through free body diagrams. Forces and moments are related to real applications by introducing beams and loading of beams. Besides applied and incident forces, friction and its related aspects are introduced. This is followed up by an analysis of trusses; the concepts of centroid and center of gravity, moment of inertia, bending moment etc. Moving beyond the mechanics of static bodies, the text takes up problems related to particles and bodies in motion with separate treatments for constant, uniform and variable accelerations. Kinematics and the concept of virtual work conclude this comprehensive introduction to engineering mechanics. The book's strength lies in its ability to relate abstract engineering concepts to real life situations.

Arshad Noor Siddiquee is a Professor at the Department of Mechanical Engineering, Jamia Millia Islamia, New Delhi. He served as Assistant Director at All India Council of Technical Education (AICTE) from 2005 to 2007. He has published more than 100 articles in national and international journals. His research interests include materials structure property correlation, welding engineering, machining, optimization of design and process parameters using fuzzy modelling.

Zahid A. Khan is a Professor at the Department of Mechanical Engineering, Jamia Millia Islamia, New Delhi. He has published more than 90 articles in national and international journals. His research interests include optimization of design and manufacturing processes parameters, artificial neural network (ANN), fuzzy modelling and environmental ergonomics.

Pankul Goel is Associate Professor at the Department of Mechanical Engineering, IMS Engineering College, Ghaziabad. He has more than 15 years experience in teaching and has also spent five years in the industry. He has published number of articles in national and international journals. His research areas include machining, friction stir welding, optimization of design and process parameters.

Engineering Mechanics

Problems and Solutions

Arshad Noor Siddiquee

Zahid A. Khan

Pankul Goel

CAMBRIDGE
UNIVERSITY PRESS

University Printing House, Cambridge CB2 8BS, United Kingdom

One Liberty Plaza, 20th Floor, New York, NY 10006, USA

477 Williamstown Road, Port Melbourne, vic 3207, Australia

314 to 321, 3rd Floor, Plot No.3, Splendor Forum, Jasola District Centre, New Delhi 110025, India

79 Anson Road, #06–04/06, Singapore 079906

Cambridge University Press is part of the University of Cambridge.

It furthers the University's mission by disseminating knowledge in the pursuit of education, learning and research at the highest international levels of excellence.

www.cambridge.org
Information on this title: www.cambridge.org/9781108411622

© Cambridge University Press 2018

This publication is in copyright. Subject to statutory exception and to the provisions of relevant collective licensing agreements, no reproduction of any part may take place without the written permission of Cambridge University Press.

First published 2018

Printed in India by Rajkamal Electric Press, Kundli, Haryana.

A catalogue record for this publication is available from the British Library

ISBN 978-1-108-41162-2 Paperback

Additional resources for this publication at www.cambridge.org/9781108411622

Cambridge University Press has no responsibility for the persistence or accuracy of URLs for external or third-party internet websites referred to in this publication, and does not guarantee that any content on such websites is, or will remain, accurate or appropriate.

Dedicated
to
Family members

Contents

Preface *xiii*

1. **Introduction** **1-33**
   1.1 Introduction to Engineering Mechanics 1
   1.2 Basic Idealizations: Particle, Continuum and Rigid Body 2
   1.3 Units 2
       1.3.1 Types of units 2
       1.3.2 Systems of units 3
   1.4 Scalar and Vector Quantities 4
   1.5 Force and its Characteristics 4
   1.6 Force System 4
       1.6.1 Classification of force system 4
   1.7 Laws of Mechanics 8
       1.7.1 Laws of motion 8
       1.7.2 The gravitational law of attraction 8
       1.7.3 Laws of forces 9
   1.8 Vector Algebra 12
       1.8.1 Vectors' representation 12
       1.8.2 Classification of vectors 13
       1.8.3 Vector operations 16
       1.8.4 Vectorial representation of component of force 20
       1.8.5 Vectorial representation of force passing through two points in space 21
   Theoretical Problems *31*
   Numerical Problems *31*
   Multiple Choice Questions *32*

2.	**Two Dimensional Concurrent Force Systems**		**34-95**
	2.1	Resolution of Force and Force Systems	34
	2.2	Resultant of Two Dimensional Concurrent Forces	37
	2.3	Principle of Transmissibility of Forces	41
	2.4	Free Body Diagrams	42
	2.5	Equations of Equilibrium Conditions	43
	2.6	Lami's Theorem	44
	Numerical Problems		*90*
	Multiple Choice Questions		*94*
3.	**Two Dimensional Non-concurrent Force Systems**		**96-148**
	3.1	Introduction	96
	3.2	Moment	97
	3.3	Couple	98
	3.4	Moment of Couple	98
	3.5	Transfer of a Force to Parallel Position	99
	3.6	Graphical Presentation of Moment	100
	3.7	Varignon's Theorem	101
	3.8	Equations of Equilibrium Conditions	102
	3.9	Types of Supports and their Reactions on Beams	103
	3.10	Types of Beams	105
	3.11	Types of Loading on Beams	105
	Numerical Problems		*144*
	Multiple Choice Questions		*147*
4.	**Friction**		**149-184**
	4.1	Introduction	149
	4.2	Coulomb's Laws of Dry Friction	150
	4.3	Static Friction, Limiting Friction, Kinetic Friction	150
	4.4	Angle of Friction	151
	4.5	Angle of Repose	152
	4.6	Cone of Friction	153
	Numerical Problems		*180*
	Multiple Choice Questions		*183*
5.	**Application of Friction**		**185-244**
	5.1	Ladder Friction	185
	5.2	Wedge Friction	196
	5.3	Screw Friction	206
	5.4	Belt Friction	216
	5.5	Band Brakes	229
	Theoretical Problems		*239*
	Numerical Problems		*239*
	Multiple Choice Questions		*243*

Contents

6.	**Analysis of Trusses**		245–322
	6.1	Introduction	245
	6.2	Classification of Trusses	245
	6.3	Assumptions for the Analysis of Perfect Truss	248
	6.4	Analysis of Forces in the Members of the Truss	248
		6.4.1 Method of joint	248
		6.4.2 Method of section	301
	Theoretical Problems		*315*
	Numerical Problems		*316*
	Multiple Choice Questions		*321*
7.	**Centroid and Centre of Gravity**		323–386
	7.1	Introduction	323
	7.2	Centre of Gravity, Centroid of Line, Plane Area and Volume	323
	7.3	Centroid of L, C, T and I-Sections	326
	7.4	Importance of Axis of Symmetry in Centroid and Centre of Gravity	330
	7.5	Centroid of a Triangle	336
	7.6	Centroid of a Quarter Circle and Semicircle	337
	7.7	Centroid of Composite Sections and Bodies	351
	7.8	Centre of Gravity of Cone and Hemisphere	375
	Numerical Problems		*380*
	Multiple Choice Questions		*385*
8.	**Moment of Inertia**		387–445
	8.1	Moment of Inertia of Plane Area and Mass	387
	8.2	Radius of Gyration	388
	8.3	Parallel Axis Theorem and its Significance	389
	8.4	Perpendicular Axis Theorem	391
	8.5	Moment of Inertia of a Rectangle	392
	8.6	Moment of Inertia of a Triangle	393
	8.7	Moment of Inertia of a Circle, a Quarter Circle and a Semicircle	395
	8.8	Moment of Inertia of Composite Sections and Bodies	400
	8.9	Mass Moment of Inertia of Prismatic Bar, Rectangular Plate, Circular Disc, Solid Cone and Sphere about Axis of Symmetry	430
	Numerical Problems		*438*
	Multiple Choice Questions		*444*
9.	**Shear Force and Bending Moment Diagrams**		446–497
	9.1	Beams	446
	9.2	Types of Beams	446
	9.3	Types of Loads and Beams	447
	9.4	Shear Force and Bending Moment	448
	9.5	Shear Force Diagram (SFD) and Bending Moment Diagram (BMD)	449
	9.6	Sign Convention of Shear Force and Bending Moment in SFD and BMD	449

9.7	Relationship between Load Intensity (w), Shear Force (S) and Bending Moment (M)		450
9.8	Point of Contraflexure or Inflexion		452
9.9	Characteristics of SFD and BMD		452
	Theoretical Problems		493
	Numerical Problems		494
	Multiple Choice Questions		496

10. Kinematics: Rectilinear Motion of Particles — 498-528

10.1	Introduction	498
10.2	Displacement, Velocity and Acceleration	498
10.3	Rectilinear Motion	500
10.4	Rectilinear Motion in Horizontal Direction (X-axis)	500
	10.4.1 Motion with variable acceleration	500
	10.4.2 Motion with uniform acceleration	508
10.5	Graphical Method for Motion Curves	514
10.6	Rectilinear Motion in Vertical Direction (Y-axis)	520
	Theoretical Problems	525
	Numerical Problems	525
	Multiple Choice Questions	527

11. Kinematics: Curvilinear Motion of Particles — 529-559

11.1	Introduction	529
11.2	Rectangular Coordinates	529
11.3	Tangential and Normal Components of Acceleration	531
11.4	Projectile	539
	Theoretical Problems	556
	Numerical Problems	556
	Multiple Choice Questions	557

12 Kinetics of Particles — 560-590

12.1	Introduction	560
12.2	Laws of Motion	560
12.3	D'Alembert's Principle	561
	Theoretical Problems	587
	Numerical Problems	587
	Multiple Choice Questions	589

13. Work and Energy — 591-630

13.1	Introduction	591
13.2	Work Done by a Force	591
13.3	Work Done by a Variable Force	592
13.4	Energy	593
13.5	Work–Energy Principle	593

	13.6	Power	594
	13.7	Principle of Conservation of Energy	595
	Theoretical Problems		*625*
	Numerical Problems		*626*
	Multiple Choice Questions		*629*
14.	**Impulse and Momentum**		**631-660**
	14.1	Introduction	631
	14.2	Principle of Impulse and Momentum	631
	14.3	Principle of Conservation of Momentum	632
	14.4	Collisions of Elastic Bodies	633
		14.4.1 Direct central impact	633
		14.4.2 Oblique/Indirect central impact	634
		14.4.3 Coefficient of restitution	634
	Theoretical Problems		*658*
	Numerical Problems		*658*
	Multiple Choice Questions		*659*
15.	**Kinematics of Rigid Bodies**		**661-707**
	15.1	Introduction	661
	15.2	Rotational Motion	661
	15.3	Angular Displacement, Angular Velocity and Angular Acceleration	661
	15.4	Relationship between Linear and Angular Velocity	663
	15.5	Relationship between Linear, Normal and Angular Acceleration	663
	15.6	Equations of Angular Motion	664
	15.7	General Plane Motion	673
	15.8	Instantaneous Centre	675
	15.9	Relative Velocity	694
	Theoretical Problems		*704*
	Numerical Problems		*704*
	Multiple Choice Questions		*706*
16.	**Kinetics of Rigid Bodies**		**708-741**
	16.1	Introduction	708
	16.2	Kinetics of Rotary Motion	708
		16.2.1 Moment of momentum	708
		16.2.2 Torque and angular momentum	709
	16.3	Kinetic Energy of a Body in Translatory and Rotary Motion	709
	16.4	Principle of Conservation of Energy	710
	16.5	Principle of Work and Energy	711
	Theoretical Problems		*739*
	Numerical Problems		*739*
	Multiple Choice Questions		*740*

17.	**Virtual Work**	**742-782**
	17.1 Introduction	742
	17.2 Principle of Virtual Work	742
	17.3 Work Done by Forces	743
	17.4 Work Done by Moments	744
	Theoretical Problems	778
	Numerical Problems	778
	Multiple Choice Questions	782
Index		**783-788**

Preface

Proper design and analysis of the elements of a structure, machine or installation is critical to its rigidity, safety, cost and reliability. With methods and materials of fabrication, construction and manufacturing changing rapidly, a lot of attention to design and analysis is required so that the system is robust yet cost effective. Further, 'principles of mechanics' is a foundation subject and its importance to several engineering disciplines cannot be overemphasized. A sound understanding of this subject is extremely useful during the analysis of complex problems in core engineering subjects. The subject deals with a variety of materials, in different geometries and loading configurations subjected to various types of loads. During this course, the learner needs to be presented cases similar to those encountered in real engineering problem-situations. This helps the student develop faculties to choose the right approach to analyze problems, situations and arrive at a correct solution.

Due to the importance of engineering mechanics and its applications in many engineering disciplines, it is a part of common engineering curricula. The contents of this volume have been developed to match the syllabi of major universities. This treatise is organized to provide the basic concepts in the initial chapters; advancing to subsequent application of these concepts to a variety of situations encountered in engineering problem solving. The architecture of the volume makes it a self-sufficient introduction to the subject. The book begins with an introduction to basic building blocks of this subject, such as units, system of units of force and force systems. Representation, interaction and concatenation of forces in the Cartesian space are also dealt with. Concept of planar forces is developed through concurrent and non-concurrent forces in chapters 2 and 3. These chapters demonstrate articulation of forces and force systems to resolve forces, to work out resultants, and equilibrium conditions through free body diagrams. The concept and articulation of moments and couples is developed through forces and force systems. Forces and moments are further related to real applications by introducing beams and loading of beams.

Apart from applied and incident forces as dealt with in the beginning, friction and related aspects are introduced in chapters 4 and 5. Forces induced due to friction, their estimation and articulation is presented in these chapters. Frictional problems in common cases such as in wedge, screw, belt friction, etc. are discussed in these chapters. Structural problems related to trusses are dealt with in chapter 6. Analysis of trusses subjected to various loading conditions is demonstrated through joint and section methods. Centroid and centre of gravity is introduced in chapter 7. Work out of centre and centroids of various structural sections and common geometric shapes have been articulated in this chapter, so that the learner can appreciate it during real problem solving. The centroid is further developed and related to area and mass moment of inertia. Moment of inertia is treated in chapter 8. Beginning with simple shapes and sections, the centroid is developed for more complex cases like composite sections and bodies. Analysis of beams for bending through shear force and bending moment is presented in depth, separately, in chapter 9.

After dealing with problems that are static in nature, the analysis of bodies in motion is covered in chapters 10 through 17. Problems relating to dynamics of particles and bodies in motions along rectilinear and curvilinear paths have been presented in chapters 10 and 11. Cases are further structured through treatment of constant, uniform and variable accelerations separately. Concepts are developed through treatment of situations analytically and graphically both. Kinetics of particles and bodies is elaborated through laws of motion and applicable theorems detailed in chapter 12. Chapter 13 covers, in depth, analysis on work and energy.

The abstract concepts of impulse and momentum are related to real application situations such as elastic collision. These aspects are covered in detail in chapter 14. Kinetics and kinematics is covered in chapters 15 and 16. Analysis of motions of bodies and particles in rotation is detailed in chapter 15. Motion and acceleration of bodies in rotation is elaborated through equations of motion in rotation. Analogy has also been drawn by relating linear motion with rotational motion. Chapter 16 deals with kinetics of rigid bodies and develops the concepts of moment of momentum, torque and angular momentum and relates them to the work and energy. The concluding chapter 17 details the concept of virtual work.

This treatise is rich in illustration: a large number of diagrams and worked out examples have been provided. End of chapter exercises have been included so that the reader can apply the knowledge gained in each chapter to solve un-encountered problems.

Chapter 1

Introduction

1.1 Introduction to Engineering Mechanics

Engineering Mechanics is basically a branch of mechanics, associated with the study of effect of forces acting on rigid bodies. Such forces may keep a body or bodies in rest or in motion. If a body remains in rest condition, it means the net effect of all forces acting is zero; this condition of the body is termed as static or equilibrium. However, if the body moves, it means an unbalanced force is acting and causing the motion; this condition of body is termed as dynamics. Thus, engineering mechanics is mainly concerned with the effect of forces on rigid body. However, mechanics is broadly classified into various categories depending upon type (nature) of bodies influenced by forces. The bodies may be solid (rigid or deformable) or fluid. A pictorial chart describing mechanics is given below:

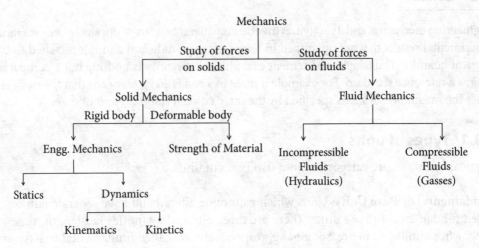

Fig. 1.1 Pictorial chart describing Mechanics

1.2 Basic Idealizations: Particle, Continuum and Rigid Body

A body has its distinct mass which is continuously distributed all over its entire volume. Sometimes the group of various components, i.e., a system; is termed as body. For example, a locomotive is modelled as a body for analysis of its motion but the locomotive is nothing but a group of various engineering components like piston, crank, connecting rod, chassis and wheels.

The different idealizations of bodies are:

Particle
A particle is defined as an individual body whose size does not affect the analysis of forces acting on it. In other words, it is nothing but a point of concentrated mass. For example a helicopter running at high altitude observed form ground, a football viewed by spectators, etc.

Continuum
Atoms and molecules are the main constituent of any matter. The analysis of forces on individual atom and molecule of a body is too complex, perhaps impossible. Thus matter is assumed to be continuously distributed and due to this average behaviour of a body can be measured. This assumption of matter distribution is termed as continuum. The continuum is valid for rigid body as well as for deformable body.

Rigid Body
In nature, there is no perfect rigid body. A rigid body is one that does not deform under the action of forces. Thus its change in volume is negligible in comparison to original volume of the body and each point remains at constant distance from other points in the body. For example, strings and belts are assumed to stretch less and treated as rigid bodies under application.

1.3 Units

Engineering mechanics and its activities involve considerable study of physical phenomena and experimental results that are quantified in terms of magnitude and a unit is attached to each physical quantity. Thus any measurement of a physical quantity is nothing but a comparison against a reference standard. For example if mass of a rod is 20 kg, it means that the mass is 20 times the magnitude of kg, as specified by the international standard unit of mass.

1.3.1 Types of units

Physical quantities are categorized into two types of units:

Fundamental or Base Units: Units which cannot be altered and have separate entities are called fundamental or base units. There are three physical quantities i.e., length, time and mass, whose units i.e., metre, sec and kg, respectively; are called fundamental or base units and extensively used in mechanics.

Introduction

Derived Units: These are units which are derived or dependent on fundamental or base units. For example, the units of force, work, power, density, area, volume, velocity and acceleration; are derived units.

1.3.2 Systems of units

There are four systems of units which are as follows:

M.K.S. (Metre-Kilogram-Second) system

C.G.S. (Centimetre-Gram-Second) system

F.P.S. (Foot-Pound-Second) system

S.I. (International System) system

The units of length, mass and time in the first three systems are defined by individual name of the system. Different countries have adopted one of the first three systems based on their choice and convention followed by the scientific community. However, that has led to some confusion and inconvenience, particularly when conversion of one unit becomes necessary in order to apply in another country based on the system the latter follows. Thus to avoid confusion, a universal standard for units was framed, known as the **SI units** at the Eleventh General Conference of Weights and Measures held during 1960 in Paris. SI units in French and English is referred as Syste´me Internationale d' Unite´s and International System of Units, respectively. This system is now widely used all over the world. In India SI units are being used since 1957 after a statutory decision.

Table 1.1 SI Units

Types of Units	Physical Quantity	Unit	Symbol
Base Units	amount of substance	mole	mol
	current	ampere	A
	length	metre	m
	luminous intensity	candela	cd
	mass	kilogram	kg
	temperature	kelvin	k
	time	second	s
Supplementary Units	plane angle	radian	rad
	solid angle	steradian	sr
Derived units with distinct name	force	newton	N
	frequency	hertz	Hz
	power	watt	W
	pressure, stress	pascal	Pa
	work, heat, energy	joule	J

Derived units in terms of Base units	acceleration	metre/sec	m/s²
	activity(radioactive)	1/sec	s⁻¹
	area	square metre	m²
	concentration	mole/cubic metre	mol/m³
	luminance	candela/square metre	cd/m²
	mass density	kilogram/cubic metre	kg/m³
	specific volume	cubic metre/kilogram	m³/kg
	velocity, speed	metre/sec	m/s
	volume	cubic metre	m³

1.4 Scalar and Vector Quantities

All quantities involved in engineering mechanics are classified into scalar and vector quantities.

Scalar Quantities are those that can be described completely by its magnitude only, like mass, length, volume, power, temperature and time, etc. For example if a person asks a shopkeeper to pack 2 kg potatoes, it is providing sufficient information for the required task. Such quantities are called scalar quantities.

Vector Quantities are those which cannot be described completely by its magnitude only, like force, weight, moment, couple, displacement, velocity, acceleration and momentum, etc. For example, if a person is applying 20 Nm couple on cap of bottle, it does not specify whether person is tightening or loosing cap but when direction is stated clockwise or anticlockwise it serves the full information. Such quantities which are defined by both magnitude as well as direction are called vector quantities.

1.5 Force and its Characteristics

A force is a vector quantity that causes interaction between bodies. It changes or tends to change the position of a body whether the body is in rest or in motion. A force can cause a body to push, pull or twist. A force is specified by four characteristics, i.e., magnitude, point of application, line of action, and sense.

1.6 Force System

Force System is a collection or pattern or group of various forces acting on a rigid body. It is of two types and its classification depends upon number of forces acting in planes.

1.6.1 Classification of force system

The force system is classified into two categories
   (a) Coplanar Force System   (b) Non-Coplanar Force System

1.6.1.1 Co-planar force system

Coplanar force system is the one where a number of forces work in a single plane or common plane. It is further divided into:

(i) **Concurrent coplanar force system:** If a number of forces work through a common point in a common plane, then the force system is called concurrent coplanar force system.

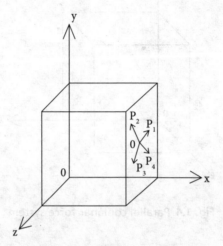

Fig. 1.2 Concurrent coplanar force system

(ii) **Collinear coplanar force system:** If a number of forces have single line of action in a common plane, then the force system is called collinear coplanar force system.

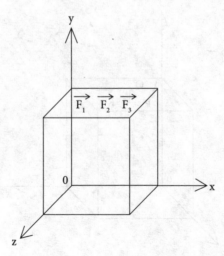

Fig. 1.3 Collinear coplanar force system

(iii) **Parallel coplanar force system:** If a number of forces have parallel line of action in a common plane, then the force system is called parallel coplanar force system.

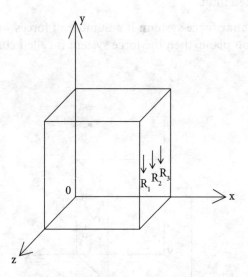

Fig. 1.4 Parallel coplanar force system

(iv) **Non-parallel coplanar force system:** If a number of forces do not have parallel line of action in a common plane, then the force system is known as non-parallel coplanar force system.

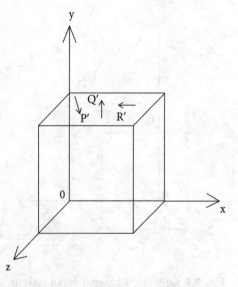

Fig. 1.5 Non parallel coplanar force system

Introduction

1.6.1.2 Non-coplanar force system

It is the one where a number of forces work in different planes. It is further divided into:

(i) **Concurrent non-coplanar force system:** If a number of forces work through a common point in different planes, then the force system is called concurrent non-coplanar force system.

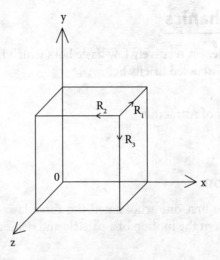

Fig. 1.6 Concurrent non-coplanar force system

(ii) **Non-parallel non-coplanar force system:** When a number of forces are having different line of action in three different planes, then the force system is called non-parallel non-coplanar force system.

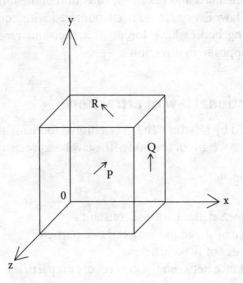

Fig. 1.7 Non-parallel non-coplanar force system

(iii) **Parallel non-coplanar force system:** When a number of forces are having parallel line of action in two different planes, then the force system is known as parallel non-coplanar force system. This force system can exist in two, two-dimensional planes only. For example: the compressive forces in four legs of a table represent this force system.

1.7 Laws of Mechanics

The whole mechanics relies on relatively few basic laws which lay down the foundation of mechanics. The laws are discussed briefly below:

- Laws of Motion
- The Gravitational Law of Attraction
- Laws of Forces

1.7.1 Laws of motion

Sir Isaac Newton was the first one who stated the three laws of motion in his treatise, **Principia.** These laws govern the motion of a particle and demonstrate their validity. These laws are as follows:

(i) **Newton's First law:** A particle always continues to remain at rest or in uniform motion in a straight line in the absence of applied force. This law is also known as the law of inertia.
(ii) **Newton's Second law:** If a particle is subjected to force, the magnitude of the acceleration will be directly proportional to the magnitude of the force and inversely proportional to the mass and lies in the direction of the force.
(iii) **Newton's Third law:** Every action is encountered with equal and opposite reaction, or two interacting bodies have forces of action and reaction collinearly equal in magnitude but opposite in direction.

1.7.2 The gravitational law of attraction

This law was formulated by Newton. The gravitational force on a body, computed by using this formula, reflects the weight of the body. The law is expressed by equation

$$F = G \frac{m_1 m_2}{r^2}$$

where G = universal gravitational constant
F = force of attraction between two particles
m_1, m_2 = masses of two particles
r = distance between the centres of two particles

Introduction

1.7.3 Laws of forces

1.7.3.1 Parallelogram law

This law states that if two forces acting a point are represented as per magnitude and direction by two adjacent sides of a parallelogram, then the diagonal of such parallelogram will represent their resultant force in magnitude and direction. The parallelogram law can determine the resultant by two ways, one graphically and second analytically.

Consider a body where two forces P and Q are acting at a point O as shown in Fig. 1.8.

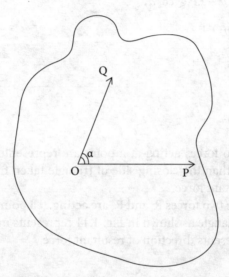

Fig. 1.8 Two concurrent forces acting on rigid body

In the graphical method, as shown in Fig. 1.9, if these forces are represented according to a suitable scale based on their magnitude and direction (α) by sides of parallelogram OA and OB respectively, then length of diagonal OC represents magnitude of the resultant and $\angle$COA (θ) represent its direction which can be measured by using protector.

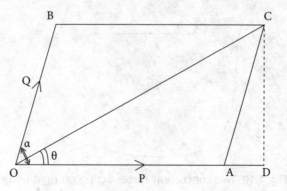

Fig. 1.9 Two concurrent forces represented by parallelogram

In analytical method it can be determined from Fig. 1.9 as follows:

$$OC^2 = OD^2 + CD^2$$
$$= (OA + AD)^2 + CD^2$$
$$= OA^2 + AD^2 + 2.OA.OD + CD^2$$
$$= P^2 + Q^2 \cos^2 \alpha + 2.P.Q.\cos \alpha + Q^2 \sin^2 \alpha$$
$$R^2 = P^2 + Q^2 + 2.P.Q.\cos \alpha$$

Thus resultant, $R = \sqrt{P^2 + Q^2 + 2.P.Q.\cos \alpha}$

and direction, $\tan \theta = CD/OD$

$$\theta = \tan^{-1}\left(\frac{Q \sin \alpha}{P + Q \cos \alpha}\right)$$

1.7.3.2 Triangle law

This law states that if two forces acting at a point are represented as per magnitude and direction taken in order than the closing side of triangle taken from starting point to last point represents the resultant force.

Consider a body where two forces R_1 and R_2 are acting at a point O as shown in Fig. 1.10. The closing side AC of triangle as shown in Fig. 1.11 represents magnitude of the resultant force, R_3 and $\angle CAB$ represents direction of resultant force.

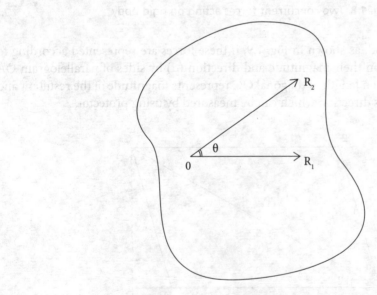

Fig. 1.10 Two concurrent forces acting on rigid body

Introduction

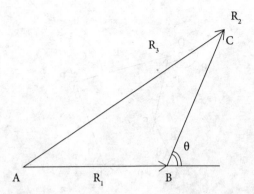

Fig. 1.11 Two concurrent forces representing triangle law

1.7.3.3 Polygon law

This law states that if a number of forces acting at a point are represented as per their magnitude and direction in the correct order, then the closing side from the starting point of first force to the last point of the last force represents the resultant force.

Consider a body where four forces P_1, P_2, P_3 and P_4 are acting at O as shown in Fig. 1.12.

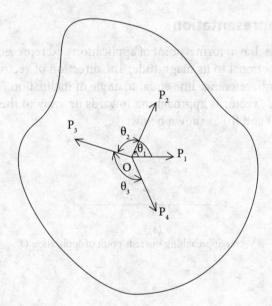

Fig. 1.12 Concurrent forces acting on rigid body

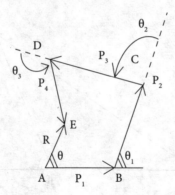

Fig. 1.13 Determination of resultant using polygon law

The length of the closing side AE of polygon as shown in Fig. 1.13 represents magnitude of the resultant force (R) and ∠EAB (θ) represents the direction of the resultant force.

1.8 Vector Algebra

This section includes representation of vectors, classification of vectors and various mathematical operations like addition, subtraction and multiplication of vectors.

1.8.1 Vectors' representation

Graphically, a vector is drawn form a point of application and represented by a line segment whose length is proportional to its magnitude. The direction of vector is represented by its line of action made with reference line equal to angle of inclination. The arrowhead shows sensing of a vector i.e., vector is approaching towards or away to the point of application. Consider Figs. 1.14(a) and (b) as shown below:

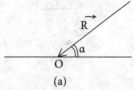

(a)
Vector approaching towards point of application O.

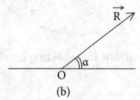

(b)
Vector acting away from point of application O.

Fig. 1.14 Vector representation

Introduction

However, a vector is either written by thick bold uppercase letter (**R**) or by an uppercase letter with an arrow over it ($\vec{R}$). Here an uppercase letter without arrow or bold feature represents the magnitude of vector

i.e., for vector $\vec{R} = |\vec{R}| = (\vec{R})$

and (R) represents the magnitude of vector.

1.8.2 Classification of vectors

Vectors are broadly classified under three categories, like sliding, fixed and free vectors. However depending upon their magnitude they are further classified as unit vector and null vector. The classification is described below:

Sliding Vector is a vector which can move along its line of action without altering its magnitude, direction and sensing. The point of application can lie anywhere along its line of action. For example, consider Fig. 1.15 where force P acting at point A of a rigid body can be transferred about new point of application B, C and so on, along its line of action without changing any effect.

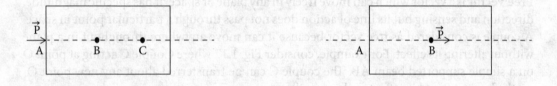

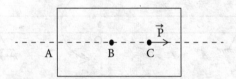

Fig. 1.15 Sliding Vector

Fixed Vector is vector whose point of application remains fixed, i.e., it cannot move without altering the conditions of the problem. Such vectors have specific magnitude, direction and line of action always passes through a particular point in space. The moment value of a force about a point represents fixed vector. For example, consider Fig. 1.16 where force P is acting at point A of a cantilever beam OA. The moment value of a force P about point O remains fixed vector. The force P cannot be transferred about new point of application B, C as its effect will change. Consider Fig. 1.16:

M_o when force P acts at A = P × 7
M_o when force P acts at B = P × 4
M_o when force P acts at C = P × 2

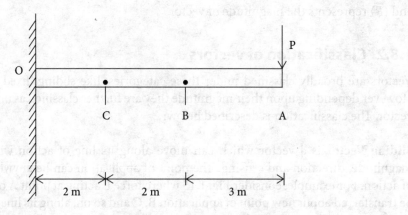

Fig. 1.16 Fixed Vector

Free Vector is a vector which can move freely in any plane or space. It has specific magnitude, direction and sensing but its line of action does not pass through a particular point in space. A couple is considered as free vector because it can move anywhere throughout in a plane without altering its effect. For example, consider Fig. 1.17 where Couple C acting at point O on a simple supported beam AB. The couple C can be transferred about any new point O_1 on the beam AB and its effect on beam will remain same.

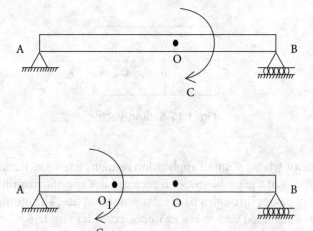

Fig. 1.17 Free Vector

Introduction

Unit Vector is one whose magnitude is equal to one i.e., $\hat{n} = 1$. It is a dimensionless quantity. It is shown by $\hat{n}$ where a hat (^) is placed over letter n. The other vectors can be shown in terms of unit vector by product of their magnitude with unit vector along their direction as shown below:

$$\vec{O} = |\vec{O}|\hat{n} = O\hat{n}$$

The unit vector along the direction of any vector may be obtained as

$$\hat{n} = \frac{\vec{O}}{|\vec{O}|}$$

The main importance of unit vector is to denote a direction in space. The unit vectors along the rectangular coordinate axis i.e., X, Y and Z are represented by $\hat{i}, \hat{j}$ and $\hat{k}$ as shown in Fig. 1.18

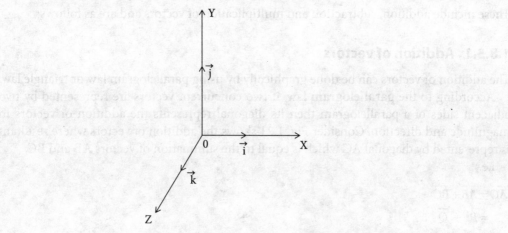

Fig. 1.18 Unit vector along rectangular coordinate axis

Null Vector is one whose magnitude is zero and it is similar to zero in scalar.

Vectors' equality is said when two vectors have same magnitude and direction. However, the point of application may be different like O_1 and O_2 as shown in Fig. 1.19.

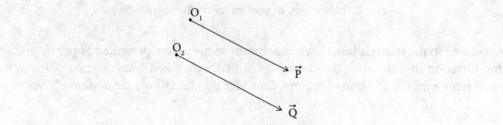

Fig. 1.19 Equal Vectors

Negative Vector is said about a vector when it is reversed from its direction as shown in Fig. 1.20.

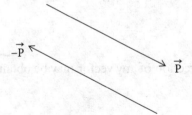

Fig. 1.20 Negative Vector

1.8.3 Vector operations

These include addition, subtraction and multiplication of vectors and are as follows:

1.8.3.1 Addition of vectors

The addition of vectors can be done graphically by using parallelogram law or triangle law.

According to the **parallelogram law,** if two concurrent vectors are represented by two adjacent sides of a parallelogram then its diagonal represents the addition of vectors in magnitude and direction. Consider Fig. 1.21 shows the addition of vectors where resultant is represented by diagonal AC which is equal to the summation of vectors AB and BC.

i.e.,

$$\overrightarrow{AC} = \overrightarrow{AB} + \overrightarrow{BC}$$
$$= \vec{P} + \vec{Q}$$

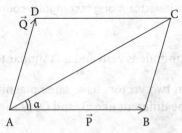

Fig. 1.21 Addition of vectors by parallelogram law

According to the **triangle law,** if two concurrent vectors are represented as per magnitude and direction in order than the closing side of triangle taken from starting point to last point represents the addition of vectors. Consider Fig. 1.22 shows the addition of vectors.

$$\overrightarrow{AC} = \overrightarrow{AB} + \overrightarrow{BC}$$
$$= \vec{P} + \vec{Q}$$

Introduction

Where α shows direction of addition of vectors

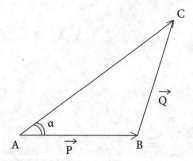

Fig. 1.22 Addition of vectors by triangle law

1.8.3.2 Subtraction of vectors

Vectors can be subtracted by addition of one vector with reversing the second vector (negative vector). This is illustrated as shown in Fig. 1.23.

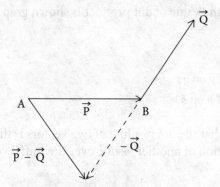

Fig. 1.23 Subtraction of vectors

1.8.3.3 Multiplication of vectors

It is of two types
   (i) Dot or scalar product   (ii) Cross or vector product

Dot or Scalar Product

The dot product of two vectors is equal to the product of their magnitudes and the cosine of the angle between them. If $\vec{p}$ and $\vec{q}$ are two vectors then dot product will be

$$\vec{p}.\vec{q} = pq\cos\alpha$$

which is called a scalar quantity because p and q are scalar quantities and $\cos\alpha$ is a pure numeric value.

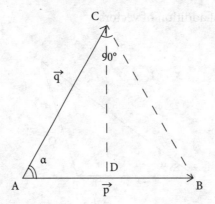

Fig. 1.24 Projection of Vectors in dot product

It can be observed from Fig. 1.24,

AC = projection of vector p on vector q = AB$\cos\alpha$ = $p\cos\alpha$

Similarly,

AD = projection of vector q on vector p = AC$\cos\alpha$ = $q\cos\alpha$

Consider Fig. 1.24 where the value of dot product is shown graphically:

$\vec{p}.\vec{q} = pq\cos\alpha$

$= q(p\cos\alpha)$

$= q(AC$ i.e projection of $\vec{p}$ on $\vec{q})$

or $= p(AD$ i.e projection of $\vec{p}$ on $\vec{q})$

Thus it can be concluded that the dot product of two vectors is the product of magnitude of one vector and the projection of another vector over the initial vector.

Note:

1. If α = 0° then both vectors $\vec{p}$ and $\vec{q}$ are collinear and their dot product will be equal to product of their magnitude, i.e.

    $\vec{p}.\vec{q} = pq\cos\alpha = pq\cos 0 = pq$

Thus *if $\hat{i}.\hat{j}$* and $\hat{k}$ are unit vectors along the rectangular coordinate axis X, Y and Z, then dot product between identical vector will be

$\hat{i}.\hat{i} = \hat{j}.\hat{j} = \hat{k}.\hat{k} = 1$ as $\alpha = 0°$

2. If α = 90° then both vectors $\vec{p}$ and $\vec{q}$ are perpendicular to each other and their dot product will be equal to zero i.e.

    $\vec{p}.\vec{q} = pq\cos\alpha = pq\cos 90° = 0$

Thus for unit vectors $\hat{i}.\hat{j}$ and $\hat{k}$, the dot product between different vectors will be

$\hat{i}.\hat{j} = \hat{j}.\hat{k} = \hat{k}.\hat{i} = 0$ as $\alpha = 90°$

3. The dot product is used to define **work** in dynamics Section.
4. If two vectors are given as

$$\vec{p} = p_x \check{i} + p_y \check{j} + p_z \check{k} \text{ and } \vec{q} = q_x \check{i} + q_y \check{j} + q_z \check{k}$$
$$\vec{p} \cdot \vec{q} = (p_x \check{i} + p_y \check{j} + p_z \check{k}) \cdot (q_x \check{i} + q_y \check{j} + q_z \check{k})$$
$$= (p_x q_x + p_y q_y + p_z q_z)$$

Cross or Vector Product

The cross product of two vectors is equal to the product of their magnitudes and the sine of the angle between them; however its direction is perpendicular to the plane containing vectors and can be determine with right hand screw rule. If $\vec{p}$ and $\vec{q}$ are two vectors then the cross product will be

$$\vec{p} \times \vec{q} = pq \sin \alpha \; \hat{n}$$

Where $\hat{n}$ is a unit vector and α ($\leq 180°$) is the angle between two vectors.
The magnitude of cross product is given by

$$|\vec{p} \times \vec{q}| = pq \sin \alpha$$

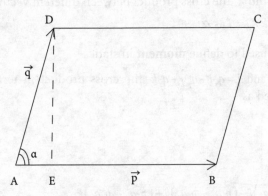

Fig. 1.25 Projection of Vectors in cross product

Consider Fig. 1.25 where the magnitude of cross product is shown graphically:

$$|\vec{p} \times \vec{q}| = pq \sin \alpha$$

where p and q are magnitude (of vectors $\vec{p}$ and $\vec{q}$) represented by AB and AD respectively and $DE = AD \sin \alpha$

$$|\vec{p} \times \vec{q}| = (AB)(q \sin \alpha)$$
$$|\vec{p} \times \vec{q}| = (AB)(AD \sin \alpha)$$
$$|\vec{p} \times \vec{q}| = (AB)(DE)$$
$$|\vec{p} \times \vec{q}| = \text{area of parallelogram ABCD}$$

Thus it can be concluded that the magnitude of cross product of two vectors is equal to the area of parallelogram whose adjacent sides are formed by two vectors.

Note:

1. If α = 0° then both vector $\vec{p}$ and $\vec{q}$ are collinear and their cross product will be equal to zero

   $\vec{p} \times \vec{q} = pq \sin\alpha \hat{n}$

   $\vec{p} \times \vec{q} = pq \sin 0 \hat{n} = 0$

   Thus *if $\hat{i}.\hat{j}$* and $\hat{k}$ are units vectors along the rectangular coordinate axis X, Y and Z then cross product between identical vector will be

   $\vec{i} \times \vec{i} = \vec{j} \times \vec{j} = \vec{k} \times \vec{k} = 0$ as $\alpha = 0°$

2. If α = 90° then both vectors $\vec{p}$ *and* $\vec{q}$ are perpendicular to each other and their cross product will be equal to product of their magnitude, i.e.

   $\vec{p} \times \vec{q} = pq \sin\alpha \hat{n}$

   $\vec{p} \times \vec{q} = pq \sin 90° \hat{n} = p.q\hat{n}$

   Thus for unit vectors $\hat{i}.\hat{j}$ and $\hat{k}$, the cross product between different vectors will be

   $\hat{i} \times \hat{j} = \hat{k}; \hat{j} \times \hat{k} = \hat{i}; \hat{k} \times \hat{i} = \hat{j}$ as $\alpha = 90°$

3. The cross product is used to define **moment** in statics.

4. If $\vec{p} = p_x\hat{i} + p_y\hat{j} + p_z\hat{k}$ and $\vec{q} = q_x\hat{i} + q_y\hat{j} + q_z\hat{k}$ the cross product in terms of rectangular components is defined as

   $$\vec{p} \times \vec{q} = \begin{vmatrix} \hat{i} & \hat{j} & \hat{k} \\ p_x & p_y & p_z \\ q_x & q_y & q_z \end{vmatrix}$$

   $\vec{p} \times \vec{q} = (p_y q_z - p_z q_y)\hat{i} - (p_x q_z - p_z q_x)\hat{j} + (p_x q_y - p_y q_x)\hat{k}$

1.8.4 Vectorial representation of component of force

Force is a vector quantity as it is defined by both magnitude and direction. Consider a force R is shown by vector OA. If vector OA is extended in three-dimensional space, than it can be expressed in term of its components as R_x, R_y and R_z as shown in the Fig. 1.26.

In term of unit vectors force R is expressed as

$\vec{R} = R_x\hat{i} + R_y\hat{j} + R_z\hat{k}$

and the magnitude is given by

$|\vec{R}| = \sqrt{R_x^2 + R_y^2 + R_z^2}$

Introduction

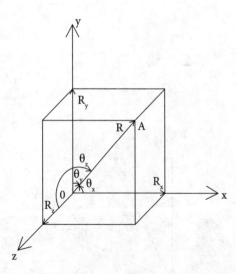

Fig. 1.26 Component of a force in three dimensions

If force R make angles θ_x, θ_y and θ_z with the rectangular coordinate axis X, Y and Z then it is expressed as

$$R = R\cos\theta_x \hat{i} + R\cos\theta_y \hat{j} + R\cos\theta_z \hat{k}$$

Force R can be expressed in terms of unit vector as

$$\vec{R} = R\hat{n}$$

Thus comparing two equations of R, unit vector is given by

$$\hat{n} = \cos\theta_x \hat{i} + \cos\theta_y \hat{j} + \cos\theta_z \hat{k}$$

As the magnitude of unit vector is unity thus

$$\cos^2\theta_x + \cos^2\theta_y + \cos^2\theta_z = 1$$

Where $\cos\theta_x$, $\cos\theta_y$ and $\cos\theta_z$ are known as direction of cosines of force R.

The direction of cosines can be further determined as

$$\cos\theta_x = \frac{R_x}{R}, \cos\theta_y = \frac{R_y}{R} \text{ and } \cos\theta_z = \frac{R_z}{R}$$

1.8.5 Vectorial representation of force passing through two points in space

The earlier sections has dealt with where force vector acting at origin in space but generally force passes through any two points in space instead through origin. Consider Fig. 1.27 where force R passes through two points P and Q which are having coordinates (x_1, y_1, z_1) and (x_2, y_2, z_2) respectively in space. Thus the position vectors of both points P and Q from origin will be

$\overrightarrow{OP} = x_1\hat{i} + y_1\hat{j} + z_1\hat{k}$

Similarly,

$\overrightarrow{OQ} = x_2\hat{i} + y_2\hat{j} + z_2\hat{k}$

Force $\vec{R}$ is given by $\overrightarrow{PQ}$

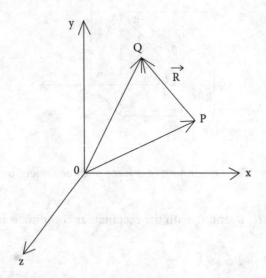

Fig. 1.27 Force passing through two points in space

The unit vector along PQ is given by $\hat{n}_{PQ} = \dfrac{\overrightarrow{PQ}}{|\overrightarrow{PQ}|}$

Thus Force $\vec{R} = R\hat{n}_{PQ} = (x_2 - x_1)\hat{i} + (y_2 - y_1)\hat{j} + (z_2 - z_1)\hat{k}$

Example: 1.1

Two vectors acting parallel are given by

$\vec{p} = 3\vec{i} + 5\vec{j} - 2\vec{k}$ and $\vec{q} = 2\vec{i} + \vec{j} + 4\vec{k}$

Determine:

(i) Magnitude of vector $\vec{p}$ and $\vec{q}$
(ii) Unit vectors along their directions

Solution:

(i) Magnitude of vector, $\vec{p} = |\vec{p}| = p = \sqrt{(3)^2 + (5)^2 + (-2)^2}$

$p = \sqrt{9 + 25 + 4}$

$p = \sqrt{38}$ units

Introduction

Similarly,

Magnitude of vector $\vec{q} = |\vec{q}| = q = \sqrt{(2)^2 + (1)^2 + (4)^2}$

$$q = \sqrt{4+1+16}$$
$$q = \sqrt{21} \text{ units}$$

(ii) Unit vector along the direction of vector $\vec{p}$ is given as

$$\hat{n} = \frac{\vec{p}}{|\vec{p}|} = \frac{3\vec{i} + 5\vec{j} - 2\vec{k}}{\sqrt{38}}$$

Similarly,
Unit vector along the direction of vector $\vec{q}$ is given by

$$\hat{n} = \frac{\vec{q}}{|\vec{q}|} = \frac{2\vec{i} + \vec{j} + 4\vec{k}}{\sqrt{21}}$$

Example: 1.2

In the previous question, if a vector $\vec{R} = 2\vec{p} - \vec{q}$ then determine magnitude and unit vector along its direction.

Solution:

$$\vec{R} = 2\vec{p} - \vec{q}$$
$$\vec{R} = 2(3\vec{i} + 5\vec{j} - 2\vec{k}) - (2\vec{i} + \vec{j} + 4\vec{k})$$
$$= 6\vec{i} + 10\vec{j} - 4\vec{k} - 2\vec{i} - \vec{j} - 4\vec{k}$$
$$\vec{R} = (4\vec{i} + 9\vec{j} - 8\vec{k})$$

The magnitude of $\vec{R} = |\vec{R}| = R = \sqrt{(4)^2 + (9)^2 + (-8)^2}$

$$= \sqrt{16 + 81 + 64}$$
$$= \sqrt{161} \text{ units}$$

Unit vector along its direction is given by

$$\hat{n} = \frac{\vec{R}}{|\vec{R}|}$$

$$\hat{n} = \frac{(4\vec{i} + 9\vec{j} - 8\vec{k})}{\sqrt{161}}$$

Example: 1.3

If $\vec{p} = \vec{i} + 2\vec{j} + 3\vec{k}$ and $\vec{q} = 2\vec{i} - 3\vec{j} + 4\vec{k}$ then find $\vec{p}.\vec{q}$ and the angle between them. Also determine the projection of $\vec{q}$ on $\vec{p}$.

Solution:

The dot product of two vectors is given by

$$\vec{p}.\vec{q} = (\vec{i} + 2\vec{j} + 3\vec{k}).(2\vec{i} - 3\vec{j} + 4\vec{k})$$
$$= 2 - 6 + 12$$
$$\vec{p}.\vec{q} = 8$$

However, magnitude of vector $\vec{p} = |\vec{p}| = \sqrt{(1)^2 + (2)^2 + (3)^2}$

$$p = \sqrt{14}$$

Simillarly, $q = \sqrt{(2)^2 + (-3)^2 + (4)^2}$
$$= \sqrt{29}$$

We know that the dot product is given by

$$\vec{p}.\vec{q} = |\vec{p}||\vec{q}|\cos\theta$$
$$= pq\cos\theta$$

or $\cos\theta = \dfrac{\vec{p}.\vec{q}}{pq} = \dfrac{8}{\sqrt{14}\sqrt{29}}$

$\cos\theta = 0.397$ or $\theta = 66.61°$

projection of $\vec{q}$ on $\vec{p} = \dfrac{\vec{p}.\vec{q}}{|\vec{p}|} = \dfrac{8}{\sqrt{14}} = 2.14$

Example: 1.4

Determine the angle between two vectors $(\vec{p} + \vec{q})$ and $(\vec{p} - \vec{q})$
if $\vec{p} = \vec{i} - 2\vec{j} + 3\vec{k}$ and $\vec{q} = 2\vec{i} + \vec{j} + \vec{k}$

Solution:

For given vectors $\vec{p}$ & $\vec{q}$,

$$\vec{p} + \vec{q} = (\vec{i} - 2\vec{j} + 3\vec{k}) + (2\vec{i} + \vec{j} + \vec{k})$$
$$= (3\vec{i} - \vec{j} + 4\vec{k})$$

and $\vec{p} - \vec{q} = (\vec{i} - 2\vec{j} + 3\vec{k}) - (2\vec{i} + \vec{j} + \vec{k})$
$$= (-\vec{i} - 3\vec{j} + 2\vec{k})$$

Introduction

We know that

$$(\vec{p}+\vec{q})\cdot(\vec{p}-\vec{q}) = |\vec{p}+\vec{q}||\vec{p}-\vec{q}|\cos\theta$$

$$\cos\theta = \frac{(\vec{p}+\vec{q})\cdot(\vec{p}-\vec{q})}{|\vec{p}+\vec{q}||\vec{p}-\vec{q}|} \qquad \ldots\ldots (1)$$

Thus, $|\vec{p}+\vec{q}| = \sqrt{(3)^2 + (-1)^2 + (4)^2} = \sqrt{26}$

and $|\vec{p}-\vec{q}| = \sqrt{(-1)^2 + (-3)^2 + (2)^2} = \sqrt{14}$

from equation (1),

$$\cos\theta = \frac{(-3+3+8)}{\sqrt{26}\sqrt{14}} = 0.42$$

$\theta = 65.21°$

Example: 1.5

Determine the cross product of two vectors and the angle between them if vectors are given by

$\vec{p} = \vec{i} + 2\vec{j} + 5\vec{k}$ and $\vec{q} = 2\vec{i} - 3\vec{j} + 2\vec{k}$

Solution:

The cross product is given by

$$\vec{p}\times\vec{q} = pq\sin\theta\,\hat{n}$$

or $|\vec{p}\times\vec{q}| = pq\sin\theta$ $\qquad \ldots\ldots (1)$

thus, $\vec{p}\times\vec{q} = \begin{vmatrix} \vec{i} & \vec{j} & \vec{k} \\ 1 & 2 & 5 \\ 2 & -3 & 2 \end{vmatrix}$

$$= \vec{i}(4+15) - \vec{j}(2-10) + \vec{k}(-3-4)$$

$$= (19\vec{i} + 8\vec{j} - 7\vec{k})$$

thus, $|\vec{p}\times\vec{q}| = \sqrt{(19)^2 + (8)^2 + (-7)^2}$

$$= 21.77$$

and $p = |\vec{p}| = \sqrt{1^2 + 2^2 + 5^2} = \sqrt{30}$

$q = |\vec{q}| = \sqrt{2^2 + (-3)^2 + 2^2} = \sqrt{17}$

From equation (1),

$$\sin\theta = \frac{|\vec{p}\times\vec{q}|}{pq}$$

$$= \frac{21.77}{(\sqrt{30})(\sqrt{17})}$$

$\sin\theta = 0.96$

$\theta = 74.58°$

Example: 1.6

Determine the area of a parallelogram whose adjacent sides are given by $\vec{i} - 4\vec{j}$ and $2\vec{i} - 3\vec{k}$.

Solution: Let

$\vec{p} = \vec{i} - 4\vec{j}$ and $\vec{q} = 2\vec{i} - 3\vec{k}$

Thus, area of parallelogram is given by the magnitude of cross product of two vectors i.e., $|\vec{p} \times \vec{q}|$,

$$\vec{p} \times \vec{q} = \begin{vmatrix} \vec{i} & \vec{j} & \vec{k} \\ 1 & -4 & 0 \\ 2 & 0 & -3 \end{vmatrix}$$

$$= \vec{i}(12-0) - \vec{j}(-3-0) + \vec{k}(0+8)$$
$$= (12\vec{i} + 3\vec{j} + 8\vec{k})$$

and $|\vec{p} \times \vec{q}| = \sqrt{(12)^2 + (3)^2 + (8)^2}$

$$= \sqrt{217}$$
$$= 14.73$$

Thus area of parallelogram = 14.73 units.

Example: 1.7

A force, R, is given as $3\vec{i} - 2\vec{j} + 5\vec{k}$ is acting at a point. Find out its magnitude and the angles it makes with x, y and z axis.

Solution:

The magnitude of the force, R is given by

$$|\vec{R}| = \sqrt{(3)^2 + (-2)^2 + (5)^2}$$
$$R = \sqrt{38}$$
$$= 6.16 \text{ units}$$

If θ_x, θ_y and θ_z are the angles made by force, R with x, y and z axis, respectively

then $R_x = R \cdot \cos\theta_x$

i.e. $\cos\theta_x = \dfrac{R_x}{R} = \dfrac{3}{6.16} = 0.487$

$\theta_x = 60.86°$

Simillarly $\cos\theta_y = \dfrac{R_y}{R} = \dfrac{(-2)}{6.16} = 0.325$

$\theta_y = 71.03°$

and $\cos\theta_z = \dfrac{R_z}{R} = \dfrac{5}{6.16} = 0.811$

$\theta_y = 35.81°$

Introduction

Example: 1.8

Three concurrent forces are given by $(2\vec{i} - 3\vec{j} + 2\vec{k})$, $(3\vec{i} + 2\vec{j} - 4\vec{k})$ and $(\vec{i} + 3\vec{j} + 3\vec{k})$. Determine their resultant force, magnitude and its direction cosines.

Solution:

The resultant force, $\vec{R}$ is given by

$$= (2\vec{i} - 3\vec{j} + 2\vec{k}) + (3\vec{i} + 2\vec{j} - 4\vec{k}) + (\vec{i} + 3\vec{j} + 3\vec{k})$$
$$= (6\vec{i} + 2\vec{j} + \vec{k})$$

Thus magnitude of resultant force, R

$$= |\vec{R}| = \sqrt{(6)^2 + (2)^2 + (1)^2}$$
$$= \sqrt{36 + 4 + 1}$$
$$= \sqrt{41}$$
$$= 6.40$$

Thus direction of cosines will be

$$\cos\theta_x = \frac{R_x}{|\vec{R}|} = \frac{R_x}{R} = \frac{6}{6.40} = 0.94$$

$$\theta_x = 19.95°$$

Similarly, $\cos\theta_y = \dfrac{R_y}{|\vec{R}|} = \dfrac{R_y}{R} = \dfrac{2}{6.4} = 0.31$

$$\theta_y = 71.94°$$

and $\cos\theta_y = \dfrac{R_z}{|\vec{R}|} = \dfrac{R_z}{R} = \dfrac{1}{6.4} = 0.16$

$$\theta_z = 80.79°$$

Example: 1.9

Prove that the two vectors are parallel to each other

$$\vec{p} = 2\vec{i} - 2\vec{j} - 4\vec{k} \text{ and } \vec{q} = -\vec{i} + \vec{j} + 2\vec{k}$$

Solution:

We know that

$$\vec{p} \times \vec{q} = pq \sin\theta \hat{n}$$
$$\text{or, } |\vec{p} \times \vec{q}| = pq \sin\theta$$
$$\text{or, } \sin\theta = \frac{|\vec{p} \times \vec{q}|}{pq} \quad \quad \ldots\ldots (1)$$

If $\theta = 0°$ them $\vec{p}$ and $\vec{q}$ will be parallel to each other

thus, $\vec{p} \times \vec{q} = \begin{vmatrix} \vec{i} & \vec{j} & \vec{k} \\ 2 & -2 & -4 \\ -1 & 1 & 2 \end{vmatrix}$

$= \vec{i}(-4+4) - \vec{j}(4-4) + \vec{k}(2-2)$
$= 0$

thus from equation (1),
$\sin\theta = 0°$
$\theta = 0°$

Hence, both vectors are parallel to each other.

Alternative Method:

we know that $\vec{p}.\vec{q} = p.q\cos\theta$

i.e. $\cos\theta = \dfrac{\vec{p}.\vec{q}}{pq}$ ..... (2)

$\vec{p}.\vec{q} = (2\vec{i} - 2\vec{j} - 4\vec{k}).(-\vec{i} + \vec{j} + 2\vec{k})$
$= (-2 - 2 - 8)$
$= -12$

$p = |\vec{p}| = \sqrt{2^2 + (-2)^2 + (-4)^2} = \sqrt{24}$
$q = |\vec{q}| = \sqrt{(-1)^2 + (1)^2 + (2)^2} = \sqrt{6}$

From equation (2),

$\cos\theta = \dfrac{-12}{\sqrt{24}\sqrt{6}} = \dfrac{-12}{\sqrt{4 \times 6}\sqrt{6}}$

$\cos\theta = -1$
$\theta = 180°$ or π

i.e., two vectors are parallel to each other but unlike in nature.

Example: 1.10

Prove that the given vectors are perpendicular to each other.
$\vec{p} = 8\vec{i} - 6\vec{j} + 5\vec{k}$ and $\vec{q} = 5\vec{i} + 10\vec{j} + 4\vec{k}$

Solution:

We know that $\vec{p}.\vec{q} = pq\cos\theta$

or, $\cos\theta = \dfrac{\vec{p}.\vec{q}}{pq}$ ..... (1)

Introduction

If θ = 90° then the vector will be perpendicular to each other.

$$\vec{p}.\vec{q} = (8\vec{i} - 6\vec{j} + 5\vec{k}).(5\vec{i} + 10\vec{j} + 4\vec{k})$$
$$= (8 \times 5) + (-6 \times 10) + (5 \times 4)$$
$$= 40 - 60 + 20$$
$$= 0$$

Thus from equation (1),
cos θ = 0   i.e θ = 90°

Thus, both vectors are perpendicular to each other.

Example: 1.11

A force of 48 kN passes through two points A (2, –4, 3) and B (–5, 2, 1) in space as shown in Fig. 1.28. Represent the force in terms of unit vectors $\vec{i}, \vec{j}$ and $\vec{k}$.

Solution:

Vector between two points is given by

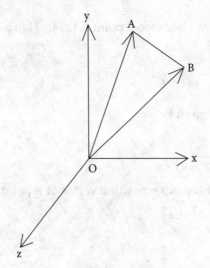

Fig. 1.28

$$\overrightarrow{AB} = \overrightarrow{OB} - \overrightarrow{OA}$$

where O is the origin in space

and $\overrightarrow{OA} = 2i - 4j + 3k$, $\overrightarrow{OB} = -5\vec{i} + 2\vec{j} + \vec{k}$

thus $\overrightarrow{AB} = (-5 - 2)\vec{i} + (2 + 4)\vec{j} + (1 - 3)\vec{k}$
$$= (-7\vec{i} + 6\vec{j} - 2\vec{k})$$

Unit vector along $\overrightarrow{AB} = \dfrac{(-7\vec{i} + 6\vec{j} - 2\vec{k})}{|\overrightarrow{AB}|}$

$$= \dfrac{(-7\vec{i} + 6\vec{j} - 2\vec{k})}{\sqrt{(-7)^2 + (6)^2 + (-2)^2}}$$

$$= \left(\dfrac{-7\vec{i} + 6\vec{j} - 2\vec{k}}{9.43}\right)$$

$$= \left(-0.742\vec{i} + 0.636\vec{j} - 0.212\vec{k}\right)$$

Thus force 48 kN in terms of unit vector will be

$\vec{R}$ = Magnitude of force × Unit vector

$= 48\left(-0.742\vec{i} + 0.636\vec{j} - 0.212\vec{k}\right)$

$\vec{R} = \left(-35.62\vec{i} + 30.53\vec{j} - 10.18\vec{k}\right)$

Example: 1.12

A force $\vec{R} = 4\vec{i} + 5\vec{j} - 2\vec{k}$ moves a body from point A (2, 4, -2) to point B (1, -2, 3). Determine work done by force on body.

Solution:

We know that work done is given by

W = $\vec{R}$. Displacement ..... (1)

The displacement will be given by

= position vector of A point − position vector of B point

$= (2\vec{i} + 4\vec{j} - 2\vec{k}) - (\vec{i} - 2\vec{j} + 3\vec{k})$

$= (\vec{i} + 6\vec{j} - 5\vec{k})$

Thus from equation (1),

$W = (4\vec{i} + 5\vec{j} - 2\vec{k}) . (\vec{i} + 6\vec{j} - 5\vec{k})$

$= (4 \times 1) + (5 \times 6) + (-2)(-5)$

$= 4 + 30 + 10$

$= 44$ units

Introduction

Theoretical Problems

T 1.1 Define rigid body.
T 1.2 Discuss briefly engineering mechanics, statics and dynamics.
T 1.3 Differentiate between fundamental unit and derived unit.
T 1.4 Define scalar and vector quantities with examples.
T 1.5 Define and classify the force system with examples.
T 1.6 What do you mean by Non concurrent coplanar force system?
T 1.7 Discuss collinear coplanar force system with example.
T 1.8 Define concurrent coplanar force system.
T 1.9 Illustrate the Gravitational law of attraction.
T 1.10 Define Parallelogram law, Triangle law and Polygon law of forces.
T 1.11 Define Sliding, Fixed and Free vectors with suitable examples.
T 1.12 Briefly explain unit vector and null vector
T 1.13 Discuss the dot product and cross product in multiplication of vectors.

Numerical Problems

N 1.1 Two parallel vectors are given as
$\vec{p} = \vec{i} - 2\vec{j} + 3\vec{k}$ and $\vec{q} = 3\vec{i} + 2\vec{j} - \vec{k}$
Determine: (i) Magnitude of both vectors
(ii) Unit vectors of each vector along their directions

N 1.2 Determine $\vec{p}.\vec{q}$ in the previous question. Also determine the angle between them.

N 1.3 Determine the angle between two vectors $(\vec{p}+\vec{q})$ and $(\vec{p}-\vec{q})$
If $\vec{p} = 2\vec{i} - 3\vec{j} + 2\vec{k}$ and $\vec{q} = 4\vec{i} - \vec{j} - 2\vec{k}$

N 1.4 Determine the cross product of vectors $\vec{p}$ and $\vec{q}$
$\vec{p} = \vec{i} + \vec{j} + 2\vec{k}$ and $\vec{q} = 3\vec{i} - 3\vec{j} + 2\vec{k}$
Also determine the angle between them.

N 1.5 The adjacent sides of a Parallelogram are given by
$2\vec{i} - 3\vec{j}$ and $\vec{i} + 2\vec{k}$
Determine area of such Parallelogram.

N 1.6 If three concurrent forces are given by
$(\vec{i} - \vec{j} - \vec{k}), (2\vec{i} + 2\vec{j} - 3\vec{k})$ and $(3\vec{i} - \vec{j} + \vec{k})$
Determine their resultant force, magnitude and its direction cosines.

N 1.7 Prove that following two vectors are perpendicular to each other,
$\vec{p} = 3\vec{i} + 4\vec{j} + 5\vec{k}$ and $\vec{q} = 7\vec{i} - 9\vec{j} + 3\vec{k}$

N 1.8 Prove that the following two vectors are parallel to each other.
$\vec{p} = 2\vec{i} - 6\vec{j} + 4\vec{k}$ and $\vec{q} = -\vec{i} + 3\vec{j} - 2\vec{i}$

N 1.9 Determine force 30kN in terms of unit vectors $\vec{i}, \vec{j}$ and $\vec{k}$ if passing through two points A (3, -2, 2) and B (-4, 3, 1) in space.

N 1.10 If a body is moved from point (1, 3, -2) to point B (2, -1, 4) by a force $\vec{p} = 3\vec{i} - 4\vec{j} + 5\vec{k}$ then determine work done by force.

Multiple Choice Questions

1. A body is called as rigid body if
   a. change in volume is significant
   b. change in volume is negligible
   c. distance between points remains variable
   d. all of these

2. Engineering mechanics deals with forces acting on
   a. elastic body
   b. plastic body
   c. rigid body
   d. all of these

3. Determine the vector quantity
   a. weight
   b. mass
   c. time
   d. none of these

4. A force is characterized by
   a. magnitude
   b. point of application
   c. line of action and sense
   d. all of these

5. If number of forces are acting in a common plane along a line, the force system will be called as
   a. collinear coplanar force system
   b. concurrent coplanar force system
   c. parallel coplanar force system
   d. none of these

6. If number of forces are acting at a point but in different planes, the force system will be called as
   a. collinear non coplanar force system
   b. concurrent coplanar force system
   c. concurrent non coplanar force system
   d. none of these

7. If number of forces are parallel in common plane, the force system will be called as
   a. collinear coplanar force system
   b. parallel non coplanar force system
   c. parallel coplanar force system
   d. none of these

8. Laws of Forces include following laws
   a. parallelogram law,
   b. triangle law
   c. polygon law
   d. all of these

9. Vectors are classified as
   a. Sliding vector
   b. Fixed vector
   c. Free vector
   d. all of these

Introduction

10. Which is a free vector quantity
    a. moment about a point
    b. couple
    c. force
    d. none of these

11. The moment value of a force about a point represents
    a. Sliding vector
    b. Fixed vector
    c. Free vector
    d. none of these

12. In engineering mechanics, a force acting on a rigid body represents
    a. Sliding vector
    b. Fixed vector
    c. Free vector
    d. none of these

Answers

1. b 2. c 3. a 4. d 5. a 6. c 7. c 8. d 9. d 10. b 11. b 12. a

Chapter 2
Two Dimensional Concurrent Force Systems

2.1 Resolution of Force and Force Systems

Resolution is a method of resolving the forces acting on a rigid body; in two perpendicular directions. The algebraic sum of all forces is determined in two perpendicular directions i.e., X and Y directions which are designated as ΣX and ΣY. Subsequently, the resultant force acting on the body can be determined as explained in section 2.2.

Points to be noted:

i. For ΣX take (+) sign if force is along X-direction and (–) sign if force is opposite to X-direction.
ii. For ΣY take (+) sign if force is along Y-direction and (–) sign if force is opposite to Y-direction.
iii. Angle must lie between line of action of force and any side of quadrant.
iv. If the force is resolved along angle side then use function "cosine" however use "sine" if it is not resolved along angle side.

Some of the examples are shown in Fig. 2.1 (a, b, c, d, e and f) to illustrate the resolution method.

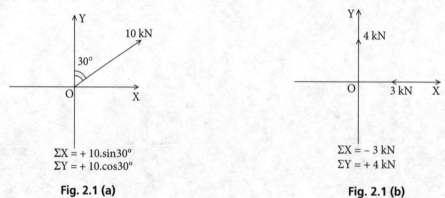

$\Sigma X = + 10.\sin 30°$
$\Sigma Y = + 10.\cos 30°$

Fig. 2.1 (a)

$\Sigma X = - 3$ kN
$\Sigma Y = + 4$ kN

Fig. 2.1 (b)

Two Dimensional Concurrent Force Systems

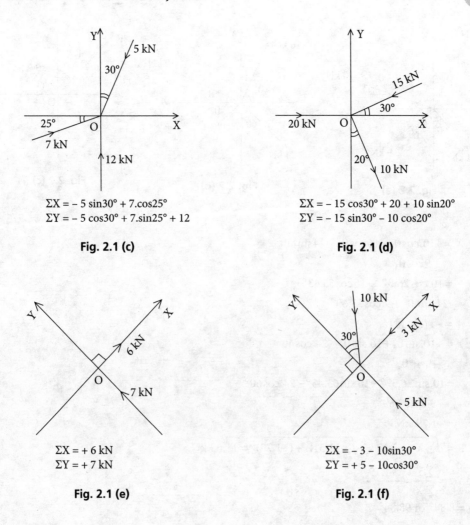

$\Sigma X = -5 \sin30° + 7.\cos25°$
$\Sigma Y = -5 \cos30° + 7.\sin25° + 12$

Fig. 2.1 (c)

$\Sigma X = -15 \cos30° + 20 + 10 \sin20°$
$\Sigma Y = -15 \sin30° - 10 \cos20°$

Fig. 2.1 (d)

$\Sigma X = +6$ kN
$\Sigma Y = +7$ kN

Fig. 2.1 (e)

$\Sigma X = -3 - 10\sin30°$
$\Sigma Y = +5 - 10\cos30°$

Fig. 2.1 (f)

Example 2.1

Determine the resultant of the concurrent coplanar force system acting at point 'O' as shown in Fig. 2.2.

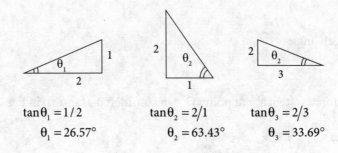

$\tan\theta_1 = 1/2$ $\tan\theta_2 = 2/1$ $\tan\theta_3 = 2/3$
$\theta_1 = 26.57°$ $\theta_2 = 63.43°$ $\theta_3 = 33.69°$

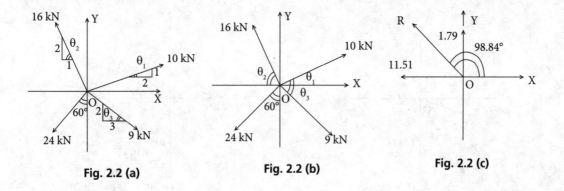

Fig. 2.2 (a)   Fig. 2.2 (b)   Fig. 2.2 (c)

$$\sum X = +10 \cdot \cos\theta_1 - 16\cos\theta_2 - 24\sin 60°$$
$$\quad + 9\cos\theta_3$$
$$= 10 \cdot \cos 26.57° - 16\cos 63.43° -$$
$$\quad 24\sin 60° + 9 \cdot \cos 33.69°$$
$$= -11.51 \text{ kN}$$
$$\sum Y = +10 \cdot \sin\theta_1 + 16 \cdot \sin\theta_2 - 24\cos 60°$$
$$\quad - 9\sin\theta_3$$
$$= 10 \cdot \sin 26.57° + 16\sin 63.43° - 24\cos 60°$$
$$\quad - 9 \cdot \sin 33.69°$$
$$= +1.79 \text{ kN}$$
$$R = \sqrt{\sum X^2 + \sum Y^2} = \sqrt{(-11.51)^2 + (+1.79)^2} = 11.65 \text{ kN}$$
$$\tan\alpha = \frac{\sum Y}{\sum X} = \frac{+1.79}{-11.51}$$
$$\alpha = 8.84° \text{ or } 98.84°$$

Example 2.2

A wheel has five equally spaced radial spokes. If the three consecutive spokes are in tension 800 N, 500 N and 300 N respectively, find the tensions in other two spokes.

Solution:

Each spoke is at an angle
$$= \frac{360°}{5} = 72°$$

The concurrent force system about point 'O' can considered as shown in Fig. 2.3. (a) and (b)

Two Dimensional Concurrent Force Systems

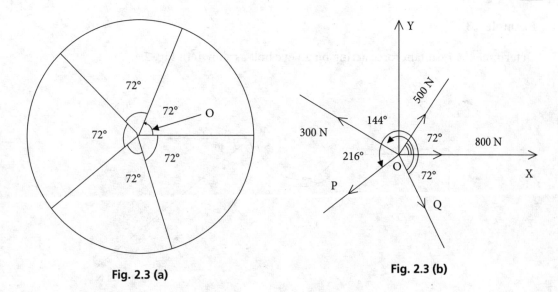

Fig. 2.3 (a)　　　　　　Fig. 2.3 (b)

$\Sigma X = 800 + 500.\cos 72° + 300.\cos 144° + P.\cos 216° + Q.\cos 72°$
$0 \quad = 711.8 - 0.81P + 0.31Q$ .....(1)
$\Sigma Y = 500.\sin 72° + 300\ \sin 144° + P.\sin 216° - Q.\sin 72°$
$0 \quad = 651.86 - 0.59P - Q \times 0.95$ .....(2)

Equation (1) and (2) can be further simplified as

$-P + 0.38Q = -878.77$ .....(3)
$P + 1.61Q = 1104.85$ .....(4)

Add equation (3) and (4)

$1.99\ Q = 226.08$
$Q = 113.61\ N$ substituting value of Q in equation (4)
$P + 1.61(113.61) = 1104.85$
$P = 921.94\ N$

2.2 Resultant of Two Dimensional Concurrent Forces

The resultant of concurrent forces acting in two dimensional plane can be determined by following ways:

a. Graphical methods- parallelogram law, triangle law and polygon law (stated in section 1.7.3 laws of forces)
b. Analytical methods- vector approach, resolution method, Lami's theorem, sine law and cosine law.

Example 2.3

Determine the resultant force acting on an eye bolt as shown in Fig. 2.4.

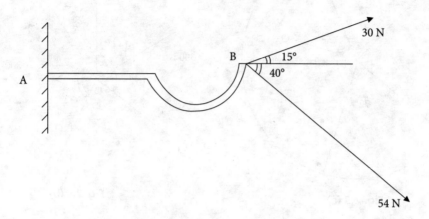

Fig. 2.4 Eye bolt

Solution:

(i) By Resolution Method:

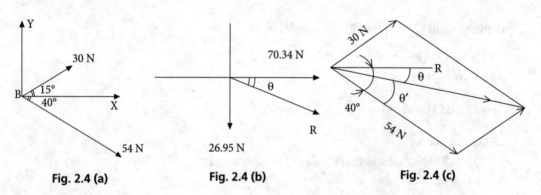

Fig. 2.4 (a) Fig. 2.4 (b) Fig. 2.4 (c)

$\Sigma X = 30 \cos 15° + 54 \cos 40°$
$= 70.34$ N
$\Sigma Y = +30 \sin 15° - 54 \sin 40°$
$= -26.95$ N
$R = \sqrt{\Sigma X^2 + \Sigma Y^2} = \sqrt{(70.34)^2 + (-26.95)^2}$
$= 75.32$ N

$\tan\theta = \dfrac{\Sigma Y}{\Sigma X} = \dfrac{-26.95}{70.34}$

$\theta = -20.96°$ (clockwise) or $339.04°$ (anticlockwise)

Two Dimensional Concurrent Force Systems

Note: As ΣX is +ve while ΣY is –ve. this shows that resultant is in fourth quadrant.

(ii) By using parallelogram law

$P = 54$ N, $Q = 30$ N, $\alpha = 55°$

$R = \sqrt{P^2 + Q^2 + 2PQ\cos\alpha}$

$= \sqrt{54^2 + 30^2 + 2 \times 54 \times 30 \cos 55°}$

$= 75.33$ N

$\tan\theta' = \dfrac{Q\sin\alpha}{P + Q\cos\alpha} = \dfrac{30\sin 55°}{54 + 30\cos 55°}$

$\theta' = 19.04°$

$\theta' = 19.04°$, from 54 N force or

$\theta = 40° - 19.04° = 20.96°$, from horizontal line BX.

Example 2.4

Two bars *AC and CB* are hinged together at *C* as shown in the Fig. 2.5. Find the forces induced in the bar. Assume weight of bars as negligible.

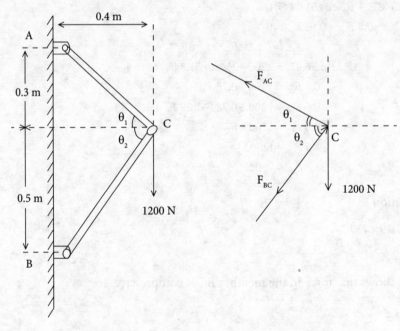

Fig. 2.5   Fig. 2.5 (a)

Solution:

Let both bar *AC* and *BC* are in tensions.

Then considering equilibrium of force system about C, (Fig. 2.5 (*a*)).

As this is a concurrent coplanar force system,

$\Sigma X = 0$,

$$F_{AC} \cos\theta_1 + F_{BC} \cos\theta_2 = 0 \quad \ldots\ldots(1)$$

$\Sigma Y = 0$,

$$F_{AC} \sin\theta_1 = 1200 + F_{BC} \sin\theta_2 \quad \ldots\ldots(2)$$

From Fig. 2.5,

$$\tan\theta_1 = \frac{0.3}{0.4} \text{ and } \tan\theta_2 = \frac{0.5}{0.4}$$
$$\theta = 36.87° \text{ and } \theta_2 = 51.34°$$

Substituting values of θ_1 and θ_2 in equations (1) and (2)

$$F_{AC} \cos 36.87° + F_{BC} \cos 51.34° = 0$$
$$0.8 \, F_{AC} + 0.62 \, F_{BC} = 0$$
$$F_{BC} = -1.29 \, F_{AC} \quad \ldots\ldots(3)$$
$$F_{AC} \sin 36.87° = 1200 + F_{BC} \sin 51.34°$$
$$0.6 \, F_{AC} = 1200 + 0.78 \, F_{BC}$$
$$0.6 \, F_{AC} = 1200 + 0.78 \, (-1.29 \, F_{AC})$$
$$0.6 \, F_{AC} = 1200 - F_{AC}$$
$$1.6 \, F_{AC} = 1200$$
$$F_{AC} = 750 \, N$$

From equation (3)

$$F_{BC} = -1.29 \times 750$$
$$F_{BC} = -967.5 \, N$$

–ve values shows that force in the member *BC* is compressive

Example 2.5

Determine the resultant of a force system consisting forces 3 N, $3\sqrt{3}$ N, 5 N, 6 N and 4 N from an angular point of a regular hexagon acting towards the other angular points in a regular order.

Two Dimensional Concurrent Force Systems

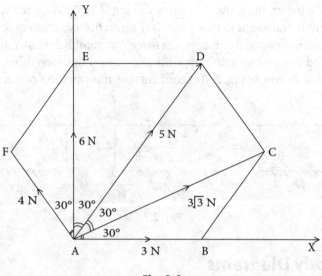

Fig. 2.6

Solution:

Consider Fig. 2.6 as shown. As the resultant is required to determine, this shows that force system about 'A' is not in equilibrium thus resolved all force in two parts.

$\Sigma X = 3 \cos 0° + 3\sqrt{3} \cos 30° + 5 \cos 60° + 6 \cos 90° + 4 \cos 120°$

$= 3 + 3\sqrt{3} \cdot \dfrac{\sqrt{3}}{2} + 5 \cdot \dfrac{1}{2} + 0 + 4\left(-\dfrac{1}{2}\right)$

$= 3 + 4.5 + 2.5 - 2 = + 8$

$\Sigma Y = 3 \sin 0° + 3\sqrt{3} \sin 30° + 5 \sin 60° + 6 \sin 90° + 4 \sin 120°$

$= 0 + 3\sqrt{3} \cdot \dfrac{1}{2} + \dfrac{5\sqrt{3}}{2} + 6.1 + \dfrac{4\sqrt{3}}{2}$

$= 6\sqrt{3} + 6 = 16.39 \, N$

$R = \sqrt{\Sigma x^2 + \Sigma y^2} = \sqrt{(8)^2 + (16.39)^2} = 18.24 \, N$

$\tan \theta = \dfrac{\Sigma Y}{\Sigma X} = \dfrac{16.39}{8}$

$\theta = 83.04°$

2.3 Principle of Transmissibility of Forces

This law states that force acting at a point on a body can be transferred to any new point but along its line of action only and its effect on the body remains unchanged. Using this law, forces are transferred to a suitable point and then force system is analyzed.

Consider a box lying on the ground as shown in Fig. 2.7. Where force R is acting at point A. Suppose we want to transfer it to new point B lying on line of action of force R as shown in Fig. 2.7(a). Let us introduce equal and opposite force R at point B as shown in Fig. 2.7(b). Now force at point A and B is collinear and opposite thus cancel out each other. Finally, the force remains at point B as shown in Fig. 2.7(c) confirms the transmission of force from A to B.

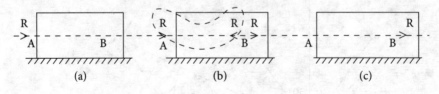

Fig. 2.7

2.4 Free Body Diagrams

Free Body Diagram is very crucial for all numerical problems as it represents the skeleton of the force system which can be analyzed easily by using law of forces. Following are the steps required to draw the free body diagram of a body:

i. Draw the figure of the body separately, i.e., draw the isolated body.
ii. Mention the weight of the body, if given.
iii. Mark the external forces as applied on the body, if given.
iv. Remove the contact surfaces one by one and mark their reactions on the body accordingly.

Some of the examples are shown in Fig. 2.8 (a, b and c) to illustrate the free body diagram:

i. Figure 2.8 (a) shows a cylinder of weight W rests in between two inclined planes which are inclined at angles α and β respectively from horizontal plane. The free body diagram of the cylinder is drawn by reactions R_A and R_B at both contact points and weight W as shown in Fig. 2.8 (b).
ii. Figure 2.8 (c) shows a cylinder of weight W is kept in equilibrium on inclined plane with the help of a string OA. The inclined plane is inclined at angle α from horizontal plane. The string OA remains under tension. The normal reaction of inclined plane on cylinder acts as R_B. The free body diagram of the cylinder is shown in Fig. 2.8 (d).
iii. Figure 2.8 (e) shows a beam supported on hinge and roller support bearing. The weight of beam is W and an external force P is acting at an angle α with the vertical as shown in the Fig. 2.8 (e). The hinge always restricts motion along both horizontal and vertical axis i.e., X and Y directions, thus offers reactions along both axis as R_{AH} and R_{AV}. However, the roller support bearing only restricts motion along vertical Y axis, thus offers reactions in Y axis as R_B. The free body diagram of the beam is shown in Fig. 2.8 (f).

Two Dimensional Concurrent Force Systems

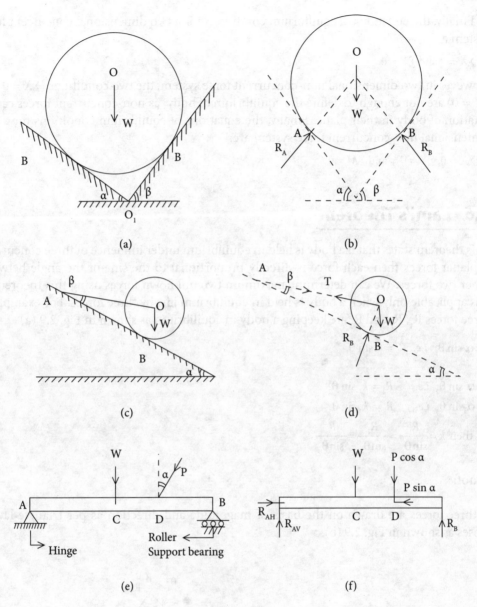

Fig. 2.8

2.5 Equations of Equilibrium Conditions

A body is said to be in equilibrium if the net resultant force (R) acting on the body is zero.

Thus, ΣX the algebraic sum of all forces in horizontal direction (horizontal component of resultant force) is equal to zero.

Similarly, ΣY the algebraic sum of all forces in vertical direction (vertical component of resultant force) is equal to zero.

Finally, the equations of equilibrium conditions for a two dimensional concurrent force system are:

$\Sigma X = 0$ and $\Sigma Y = 0$

However in two dimensional non-concurrent force system, the two conditions ($\Sigma X = 0$ and $\Sigma Y = 0$) are not enough to define the equilibrium of body; as non-concurrent forces causes rotation of body in their plane. Finally, the equations of equilibrium conditions for a two dimensional non-concurrent force system are:

$\Sigma X = 0, \Sigma Y = 0$ = and $\Sigma M = 0$

2.6 Lami's Theorem

This theorem states that if a body is held in equilibrium under influence of three concurrent coplanar forces then each force is directly proportional to the sine of the angle between other two forces. We can determine maximum two unknown forces using this theorem as it is applicable only when a body is held in equilibrium just by three forces. For example, if three forces R_1, R_2 and R_3 are keeping a body in equilibrium as shown in Fig. 2.9 (a).

$R_1 \alpha \sin \theta_1$ i.e., $R_1 = k.\sin \theta_1$

$R_2 \alpha \sin \theta_2$ i.e., $R_2 = k.\sin \theta_2$
$R_3 \alpha \sin \theta_3$ i.e., $R_3 = k.\sin \theta_3$

then $k = \dfrac{R_1}{\sin \theta_1} = \dfrac{R_2}{\sin \theta_2} = \dfrac{R_3}{\sin \theta_3}$

Proof:

If three forces are drawn on the basis of magnitude and direction as per triangle law, it closes as shown in Fig. 2.9 (b).

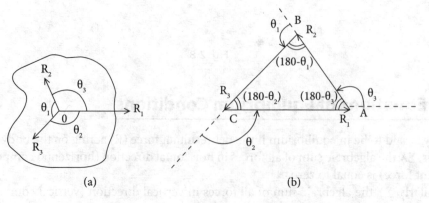

Fig. 2.9

Two Dimensional Concurrent Force Systems

Applying Law of Sines Fig. 2.9 (b).

$$\frac{AC}{\sin(180-\theta_1)} = \frac{AB}{\sin(180-\theta_2)} = \frac{BC}{\sin(180-\theta_3)}$$

$$\frac{AC}{\sin\theta_1} = \frac{AB}{\sin\theta_2} = \frac{BC}{\sin\theta_3}$$

thus, $\dfrac{R_1}{\sin\theta_1} = \dfrac{R_2}{\sin\theta_2} = \dfrac{R_3}{\sin\theta_3}$.

Example 2.6

A horizontal cylinder of weight 240 kN is held against a wall with the help of string as shown in Fig. 2.10. The string makes an angle 20° with the wall. Determine the reaction of the wall and tension induced in the string.

Solution:

Following steps are taken out:

(i) Draw free body diagram of cylinder as shown in Fig. 2.10 (*a*)
(ii) Apply Principle of Transmissibility so that Reaction R_B transmit to point 'O' as shown in Fig. 2.10 (*b*)
(iii) Draw skeleton of forces as shown in Fig. 2.10 (*c*)
(iv) Now Fig. 2.10 (*c*) represents concurrent coplanar force system which can be analyzed by two methods:

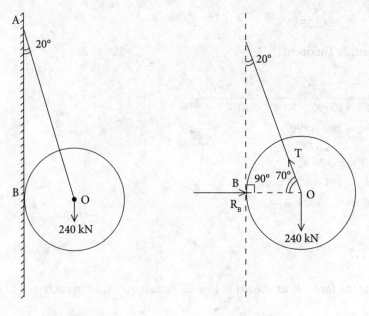

Fig. 2.10   Fig. 2.10 (a)

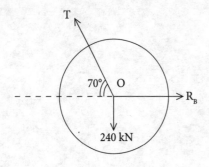
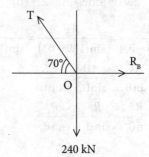

Fig. 2.10 (b) F.B.D. of Cylinder   **Fig. 2.10 (c)** Concurrent coplanar force system

(a) By Resolution Method

As cylinder is in equilibrium thus $\Sigma X = 0$,

$T \cos 70° = R_B$ .....(1)

and $\Sigma Y = 0$,

$T \sin 70° = 240$ .....(2)

$$T = \frac{240}{\sin 70°}$$

$T = 255.40$ kN

Substituting value of T in equation (1),

$255.40 \cos 70° = R_B$

$R_B = 87.35$ kN

(b) By Lami's Theorem

$$\frac{T}{\sin(90°)} = \frac{R_B}{\sin(90°+70°)} = \frac{240}{\sin(180°-70°)}$$

$$T = \frac{240 \cdot \sin 90°}{\sin 70°}$$

$T = 255.40$ kN

and $R_B = 240 \cdot \cos 70° / \sin 70°$

$R_B = 87.35$ kN

Example 2.7

Determine the force 'R' as shown in Fig. 2.11 causes tension in each part of the string 120 N.

Two Dimensional Concurrent Force Systems

Solution:

Given, $T_{OA} = T_{OB} = 120$ N.
Considering force system equilibrium about 'O'
Figure 2.11 (a) can further simplified about 'O' and can be easily resolved. As the force system is in equilibrium,

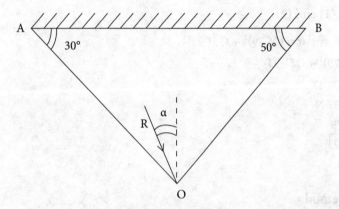

Fig. 2.11

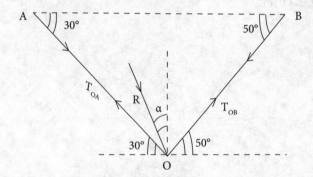

Fig. 2.11 (a)

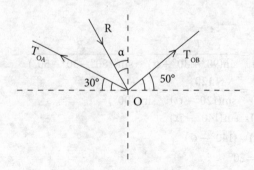

Fig. 2.11 (b)

$120 \cos 50° + R \sin \alpha - 120 \cos 30° = 0$

$R \sin \alpha = + 26.79$ .....(1)

$\sum Y = T_{OB} \sin 50° + T_{OA} \sin 30° - R \cos \alpha = 0$

$120 \sin 50° + 120 \sin 30° - R \cos \alpha = 0$

$R \cos \alpha = + 151.93 \text{ N}$ .....(2)

from equation (1) and (2)

$R^2 \sin^2 \alpha + R^2 \cos^2 \alpha = (26.79)^2 + (151.93)^2$

$R^2 = (26.79)^2 + (151.93)^2$

$R^2 = 23800.43$

$R = 154.27 \text{ N}$

$\tan \alpha = \dfrac{26.79}{151.93}$

$\alpha = 10°$

Alternative Method

The above problem can be solved by using Lami's theorem as shown below Fig. 2.11(*a*) is simplified as in Fig. 2.11 (*c*),

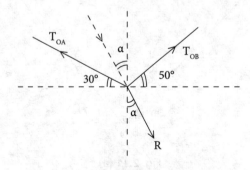

Fig. 2.11 (c)

$\dfrac{T_{OA}}{\sin(90° - \alpha + 50°)} = \dfrac{T_{OB}}{\sin(90° + 30° + \alpha)} = \dfrac{R}{\sin(180° - 30° - 50°)}$

$\dfrac{120}{\sin(140° - \alpha)} = \dfrac{120}{\sin(120° + \alpha)} = \dfrac{R}{\sin(100°)}$

i.e., $\sin(120° + \alpha) = \sin(140° - \alpha)$

or, $(120° + \alpha) = (140° - \alpha)$

$2\alpha = 20°$

$\alpha = 10°$

Two Dimensional Concurrent Force Systems

thus,
$$\frac{120}{\sin(120°+\alpha)} = \frac{R}{\sin(100°)}$$

$$R = \frac{120 \cdot \sin 100}{\sin(120°+10°)}$$

$$R = 154.269$$

$$R = 154.27\ N$$

Example 2.8

Determine the tension in the string *OA* and reaction of inclined plane when a roller of weight 600 N is supported as shown in Fig. 2.12.

Solution:

Considering Free body diagram of roller as shown in Fig. 2.12 (*a*).

The force system acting at point '*O*' can be further simplified as shown in Fig. 2.12 (*b*)

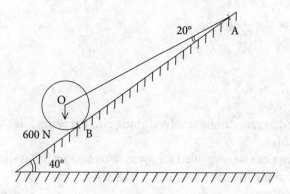

Fig. 2.12

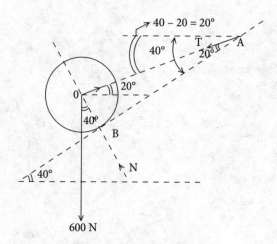

Fig. 2.12 (a) F.B.D. of Cylinder

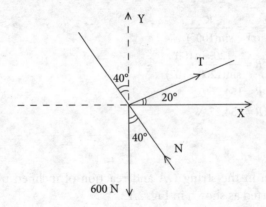

Fig. 2.12 (b)

As the roller is in equilibrium,

$\Sigma X = 0$ i.e. $T\cos 20° - N\sin 40° = 0$

$T = 0.684 \times N$ .....(1)

$\Sigma Y = 0$, i.e. $T\sin 20° + N\cos 40° - 600 = 0$

$T = 410.40 N$

$N = 600 N$

Alternative Method

The above free body diagram can be resolved along and perpendicular to the inclined plane as shown in Fig. 2.12 (c).

The Fig. 2.12 (c) further can be simplified for force system about 'O' as shown in Fig. 2.12 (d). Resolving the forces along the inclined plane i.e., $\Sigma X = 0$

$T\cos 20° - 600\sin 40° = 0$

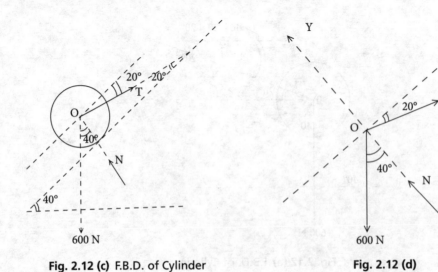

Fig. 2.12 (c) F.B.D. of Cylinder   **Fig. 2.12 (d)**

Two Dimensional Concurrent Force Systems

or, $T \cos 20° = 600 \sin 40°$

$T = 410.42\ N$

Resolving the forces perpendicular to the inclined plane,

i.e., $\sum Y = 0$,

$N = T \sin 20° + 600 \cos 40°$

$= 410.42 \sin 20° + 600 \cos 40°$

$N = 600\ N$

Example 2.9

A roller of weight 'W' rests over two inclined planes as shown in the Fig. 2.13. Determine support reactions at points A and B.

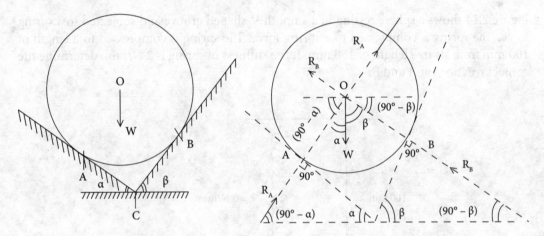

Fig. 2.13

Fig. 2.13 (a) F.B.D. of Cylinder

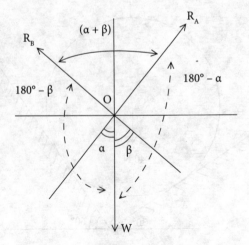

Fig. 2.13 (b)

Solution:

With application of Principle of Transmissibility the reactions R_A and R_B will be transferred through centre 'O'. The force system acting at centre 'O' can be simplified as shown in Fig. 2.13 (a) and (b) Applying Lami's theorem,

$$\frac{R_A}{\sin(180°-\beta)} = \frac{R_B}{\sin(180°-\alpha)} = \frac{W}{\sin(\alpha+\beta)}$$

$$R_A = \frac{W.\sin\beta}{\sin(\alpha+\beta)}$$

and $\quad R_B = \dfrac{W.\sin\alpha}{\sin(\alpha+\beta)}$

Example 2.10

Figure 2.14 shows a sphere resting in a smooth V-shaped groove and subjected to a spring force. The spring is compressed to a spring force. The spring is compressed to a length of 100 mm from its free length of 150 mm. It the stiffness of spring is 2 N/mm, determine the contact reactions at A and B.

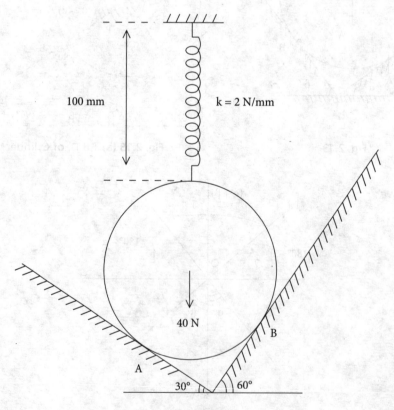

Fig. 2.14

Solution:

The spring is compressed by = 150 − 100 = 50 mm.

Thus compression force in the spring = 50 × stiffness of spring = 50 × 2 = 100 N

The free body diagram of both spring and sphere are as shown in Fig. 2.14 (a) and (b), respectively

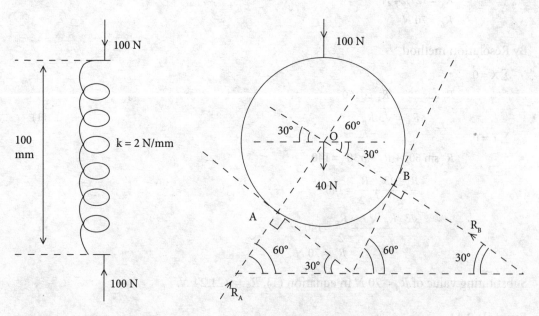

Fig. 2.14 (a) F.B.D. of Spring   **Fig. 2.14 (b)** F.B.D. of Cylinder

Figure 2.14 *(b)* further simplified as shown in Fig. 2.14 *(c)* where force 100 N will shift to center O by using principle of transmissibility.

Figure 2.14 *(c)* can be evaluated further by using Lami's theorem or resolution.

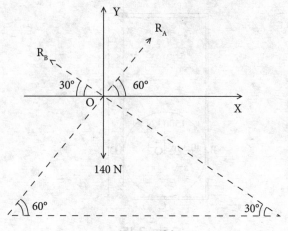

Fig. 2.14 (c)

By Lami's theorem,

$$\frac{R_A}{\sin(90°+30°)} = \frac{R_B}{\sin(90°+60°)} = \frac{140}{\sin(180°-30°-60°)}$$

$$\frac{R_A}{\cos 30°} = \frac{R_B}{\cos 60°} = \frac{140}{\sin 90°}$$

$$R_A = 121.24 \ N$$
$$R_B = 70 \ N$$

By Resolution method,

$\Sigma X = 0,$

$$R_B \cos 30° = R_A \cos 60°$$
$$R_A = \sqrt{3} \cdot R_B \qquad \qquad \text{.....(1)}$$

$\Sigma Y = 0,$

$$R_A \sin 60° + R_B \sin 30° = 140$$

$$\frac{R_A \cdot \sqrt{3}}{2} + \frac{R_B}{2} = 140$$

$$\frac{\sqrt{3} \cdot R_B \cdot \sqrt{3}}{2} + \frac{R_B}{2} = 140$$

$$R_B = 70 \ N$$

Substituting value of $R_B = 70 \ N$ in equation (1), $R_A = 121.24 \ N$.

Example 2.11

A 1500 kN cylinder is supported by the frame ABC as shown in Fig. 2.15. The frame is hinged to the wall at A. Determine the reactions at A, B, C and D. Take weight of the frame as negligible.

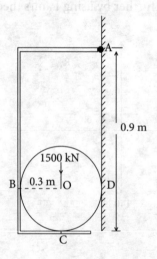

Fig. 2.15

Solution:

Considering free body diagram of cylinder as shown in Fig. 2.15 (a)

$\Sigma X = 0$,

$R_B = R_D$ .....(1)

$\Sigma Y = 0$,

$R_C = 1500 \text{ kN}$

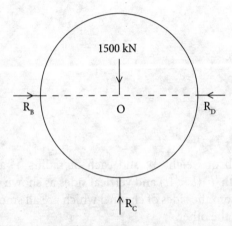

Fig. 2.15 (a) F.B.D. of Cylinder

Considering F.B.D. of channel, Fig. 2.15 (b) the force at B and C will be balanced by reaction of hinge A and this is possible only when three forces are concurrent.

$\tan \theta = \dfrac{0.6}{0.3}$

$\theta = 63.43°$

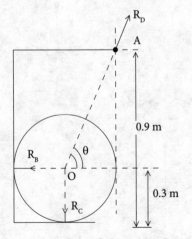

Fig. 2.15 (b) F.B.D. of Channel

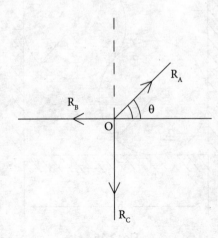

Fig. 2.15 (c)

$\sum X = 0,$
$$R_B = R_A \cos \theta$$
$\sum Y = 0,$
$$R_C = R_A \sin \theta$$
$$1500 = R_A \sin \theta$$
$$R_A = \frac{1500}{\sin 63.43°}$$
$$R_A = 1677.05 \text{ kN}$$
$$R_B = 1677.05 \times \cos 63.43°$$
$$R_B = 750.13 \text{ kN}$$
from equation (1),
$$R_D = 750.13 \text{ kN}.$$

Example 2.12

Two smooth spheres each of weight W and each of radius 'r' are in equilibrium in a horizontal channel of width 'b' ($b < 4r$) and vertical sides as shown in Fig. 2.16.
Find the three reactions from the sides of channel which are all smooth. Also find the force exerted by each sphere on the other.

(UPTU I sem, 2005–06)

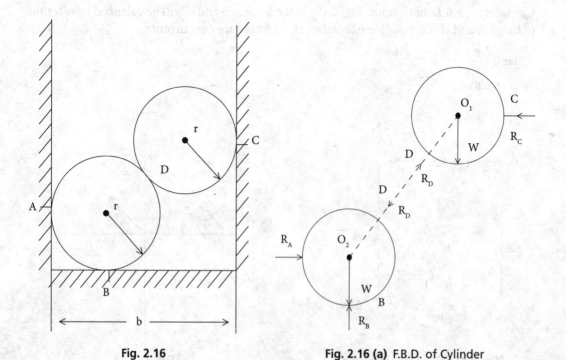

Fig. 2.16   Fig. 2.16 (a) F.B.D. of Cylinder

Two Dimensional Concurrent Force Systems

Solution:

The following are the steps:

(i) Draw the F.B.D. of both spheres as shown in Fig. 2.16 (a)
(ii) Apply principle of transmissibility and transfer R_A, R_B, R_C, and R_D to the respective centres of the sphere as shown in Fig. 2.16 (b).
(iii) Draw the skeleton of force system as shown in Fig. 2.16 (c)

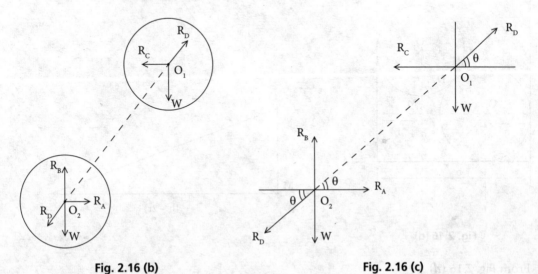

Fig. 2.16 (b) **Fig. 2.16 (c)**

Considering equilibrium of upper sphere.

$\Sigma X = 0$, $R_C = R_D \cdot \cos\theta$ .....(1)

$\Sigma Y = 0$,

$R_D \sin\theta = W$

i.e., $R_D = \dfrac{W}{\sin\theta}$ substituting in equation (1)

or, $R_C = \dfrac{W}{\sin\theta} \cdot \cos\theta$

$R_C = W \cot\theta$ and $R_D = W/\sin\theta$

Considering equilibrium of bottom sphere

$\Sigma X = 0$,

$R_D \cos\theta = R_A$ .....(2)

$\Sigma Y = 0$,

$R_B = W + R_D \sin\theta$ .....(3)

Substituting R_D value in equation (2),

$\dfrac{W}{\sin\theta} \cdot \cos\theta = R_A$

$R_A = W \cdot \cot\theta$

Substituting R_D value in equation (3),

$$R_B = W + \frac{W}{\sin\theta} \cdot \sin\theta$$
$$R_B = 2W$$

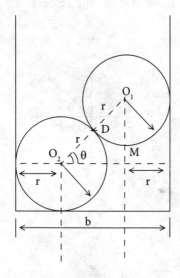

Fig. 2.16 (d)

From Fig. 2.16 (d)

$$O_2M = b - r - r$$
$$= (b - 2r)$$
$$O_1O_2 = r + r = 2r$$

Thus $\cos\theta = \dfrac{O_2M}{O_1O_2}$

$$\cos\theta = \dfrac{(b-2r)}{2r}$$

$$O_1M = \sqrt{(2r)^2 - (b-2r)^2}$$
$$= \sqrt{4r^2 - (b^2 + 4r^2 - 4br)}$$
$$O_1M = \sqrt{4br - b^2}$$

Thus, $\cot\theta = \left(\dfrac{(b-2r)}{\sqrt{4br-b^2}}\right)$ and

$$\sin\theta = \left(\dfrac{\sqrt{4br-b^2}}{2r}\right)$$

Finally, $R_A = R_C = W\cot\theta$

$$R_A = R_C = \left(\dfrac{W.(b-2r)}{O_1M}\right) \quad R_D = \dfrac{W}{\sin\theta}, \; R_D = \dfrac{2rW}{\sqrt{4br-b^2}}.$$

Two Dimensional Concurrent Force Systems

Example 2.13

Two cylinders *A* and *B* of weight 1000 *N* and 500 *N* rest on smooth inclined planes as shown in the figure. A bar of negligible weight is hinged to each cylinder at its geometric centre by smooth pins. Determine the force '*P*' as applied can hold the system in equilibrium for given position as shown in Fig. 2.17.

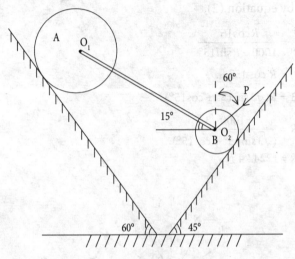

Fig. 2.17

Solution:

(i) Draw the F.B.D. of both cylinders and apply the Principle of Transmissibility.

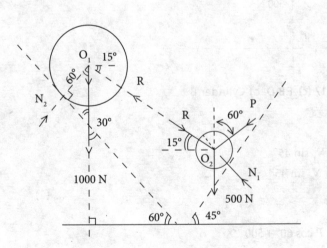

Fig. 2.17 (a) F.B.D. of Cylinder

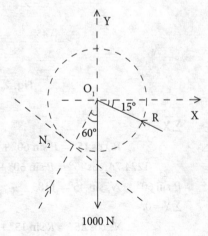

Fig. 2.17 (b) F.B.D. of Cylinder A

$\sum X = 0,$
$$N_2 \sin 60° = R \cos 15° \qquad .....(1)$$
$\sum Y = 0,$
$$N_2 \cos 60° + R \sin 15° = 1000 \; N$$
$$N_2 \cos 60° = 1000 - R \sin 15° \qquad .....(2)$$

Divide equation (1) by equation (2),

$$\frac{N_2 \sin 60°}{N_2 \cos 60°} = \frac{R \cos 15°}{1000 - R \sin 15°}$$

$$\sqrt{3}(1000 - R \sin 15°) = R \cos 15°$$

$$1000\sqrt{3} = R(\sqrt{3} \sin 15° + \cos 15°)$$

$$R = \frac{1000\sqrt{3}}{(\sqrt{3} \sin 15° + \cos 15°)}$$

$$R = 1224.74 \; N$$

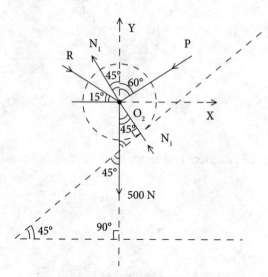

Fig. 2.17 (c) F.B.D. of Cylinder B

$\sum X = 0,$
$$R \cos 15° = P \sin 60° + N_1 \sin 45°$$
$$1224.74 \cos 45° = P \sin 60° + N_1 \sin 45°$$
$$P.\sin 60° + N_1 \sin 45° = 1183 \qquad .....(3)$$
$\sum Y = 0,$
$$N_1 \cos 45° = R \sin 15° + P \cos 60° + 500$$
$$N_1 \cos 45° = 1224.74 \sin 15° + P \cos 60° + 500$$
$$N_1 = (816.99 + P \cos 60°)\sqrt{2}$$

Substituting N_1 in equation (3),

$$P \sin 60° + (816.99 + P\cos 60°)\sqrt{2}.\sin 45° = 1183$$
$$P = 267.94 \ N.$$

Example 2.14

A roller of weight 2400 N, radius 180 cm is required to be pulled over a brick of height 90 cm as shown in the Fig. 2.18 by a horizontal pull applied at the end of a string wound round the circumference of the roller. Determine the value of P which will just tend to turn the roller over the brick.

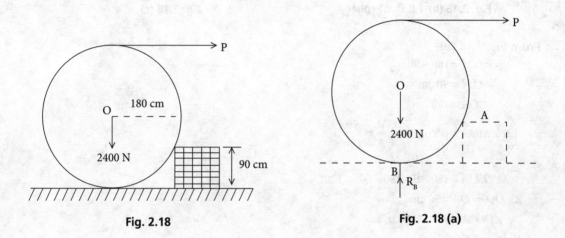

Fig. 2.18 Fig. 2.18 (a)

Solution:

The steps will be taken as follows:

(i) Draw F.B.D. of roller, when roller just about to turn over the brick at that instant, the contact between roller and ground will break-off. thus $R_B = 0$

(ii) The roller will be in equilibrium under three forces P, 2400 N and R_A.

Thus three forces can hold a body in equilibrium only when they pass through a common point i.e., represent a concurrent system. Which can be possible only when weight. of roller is transferred to O' by principle of transmissibility thus reaction R_A passes through O' and this way roller will be held in equilibrium.

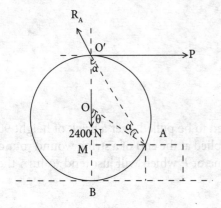

Fig. 2.18 (b) F.B.D. of roller

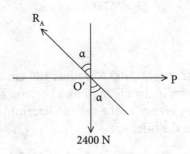

Fig. 2.18 (c)

From Fig. 2.18 (b),

$OM = 180 - 90$

$OM = 90 \text{ cm}$

$OA = 180 \text{ cm}$

Let $\angle MOA = \theta$, $\cos \theta = \dfrac{OM}{OA} = \dfrac{90}{180}$

$\theta = 60°$

$\angle O'OA = 180 - \theta = 180 - 60° = 120°$

As $OO' = OA = 90 \text{ cm}$

$\angle OO'A = \angle O'AO = \alpha$, say

thus for $\Delta OO'A$,

$\alpha + \alpha + 120° = 180°$

$\alpha = 30°$.

(iii) Draw the skeleton of forces as shown in Fig. 2.18 (c).
Consider equilibrium of skeleton of forces as shown in Fig. 2.18 (c)

$\Sigma X = 0$,

$R_A . \cos \alpha = 2400$ .....(1)

$\Sigma Y = 0$,

$R_A . \sin \alpha = P$ .....(2)

from equation (1),

$R_A = \dfrac{2400}{\cos 30°}$

$R_A = 2771.28 \ N$

Substituting value of R_A in equation (2),

$P = R_A \sin \alpha = 2771.28 \sin 30°$

$P = 1385.64 \ N$

Two Dimensional Concurrent Force Systems

Example 2.15

In the previous question, if pull, P is applied at centre of the roller then determine the minimum pull P, which will cause the roller to just turn over the brick.

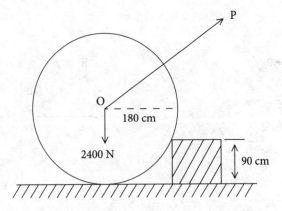

Fig. 2.19

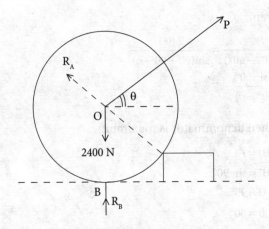

Fig. 2.19 (a)

Solution:

(i) Draw F.B.D. of the roller,

Let the minimum pull takes place at an angle θ from the horizontal line as shown in Fig. 2.19 (a)

$R_B = 0$ As stated in earlier question.

Roller will be in equilibrium under forces P, 2400 N and R_A. As two forces are acting at centre thus R_A will also go through O to formulate concurrent force system in equilibrium. Simplified F.B.D. as shown in Fig. 2.19 (b).

(ii) Draw the skeleton of force system as shown in Fig. 2.19 (c)

Let $\angle AOM = \alpha$,

$OM = 180 - 90 = 90$ cm

$\cos \alpha = \dfrac{90}{180} = \dfrac{1}{2}$

$\alpha = 60°$

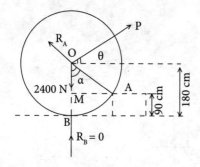

Fig. 2.19 (b)

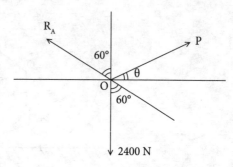

Fig. 2.19 (c)

(iii) By Lami's Theorem,

$$\dfrac{R_A}{\sin(90° + \theta)} = \dfrac{P}{\sin(180° - 60°)} = \dfrac{2400}{\sin(60 + 90 - \theta)}$$

$$P = \dfrac{2400 \cdot \sin 60°}{\sin(150° - \theta)} \quad \ldots(1)$$

P will be minimum when denominator is maximum.

thus, $\sin(150° - \theta) = 1$

$\sin(150° - \theta) = \sin 90°$

$150° - \theta = 90°$

$\theta = 60°$

Substituting 'θ' in equation (1),

$P = 2078.46$ N

By Resolution method,

$\Sigma X = 0$, $\quad R_A \sin 60° = P \cos \theta$ $\quad \ldots(2)$

$\Sigma Y = 0$, $\quad R_A \cos 60° + P \sin \theta = 2400$ $\quad \ldots(3)$

Substituting R_A from equation (2) to equation (3)

$\dfrac{P \cdot \cos \theta}{\sin 60°} \cdot \cos 60° + P \sin \theta = 2400$

$P[\cos \theta \cos 60° + \sin \theta \sin 60°] = 2400 \sin 60°$

$$P = \dfrac{2400\sqrt{3}}{(\cos \theta + \sqrt{3} \sin \theta)} \quad \ldots(4)$$

Two Dimensional Concurrent Force Systems

P will be minimum, if $(\cos\theta + \sqrt{3}\sin\theta)$ is maximum, for maxima and minima condition,

i.e., $\dfrac{d}{d\theta}(\cos\theta + \sqrt{3}\sin\theta) = 0$

$-\sin\theta + \sqrt{3}\cos\theta = 0$

$\tan\theta = \sqrt{3}$

i.e., $\theta = 60°$

if P is minimum at $\theta = 60°$ then second differentiation should be negative

i.e., $\dfrac{d^2}{d\theta^2}(\cos\theta + \sqrt{3}\sin\theta)$ should be $-ve$

$= \dfrac{d}{d\theta}(-\sin\theta + \sqrt{3}\cos\theta)$

$= (-\cos\theta - \sqrt{3}\sin\theta)$

$= (-\cos 60° - \sqrt{3}\sin 60°)$

$= -\dfrac{1}{2} - \dfrac{\sqrt{3} \cdot \sqrt{3}}{2} = -2$

thus P is minimum when $\theta = 60°$
from equation (4)

$P_{min} = \dfrac{2400 \cdot \sqrt{3}}{(\cos 60° + \sqrt{3}\sin 60°)}$

$P_{min} = 2078.46$ N

Example 2.16

Two identical spheres of weight 1000 N are held in equilibrium on inclined plane and against wall as shown in Fig. 2.20. Determine reactions at A, B, C and D.

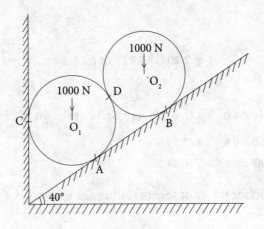

Fig. 2.20

Solution:

The steps are as follows:

(i) Draw F.B.D. of both spheres as shown in Fig. 2.20 (a). Both spheres push each other about D.
Hence R_D will be compressive force and parallel to the inclined plane since $O_1A = O_2B = r$.

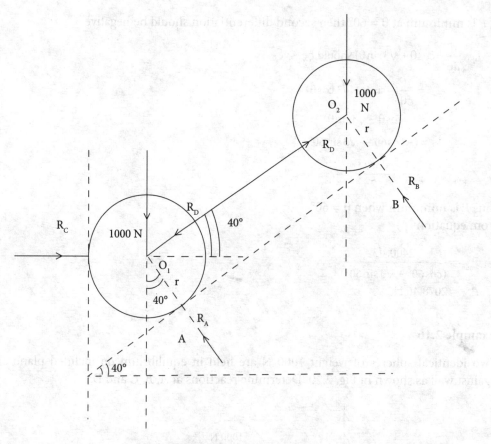

Fig. 2.20 (a) F.B.D. of rollers

(ii) Considering skeleton of F.B.D. of upper sphere as shown in Fig. 2.20 (b).

$\Sigma X = 0$, $R_D = 1000 \sin 40° = 642.79$ N
$\Sigma Y = 0$, $R_B = 1000 \cos 40° = 766.04$ N

Considering skeleton for forces of sphere resting against wall as shown in Fig. 2.20 (c)

Two Dimensional Concurrent Force Systems

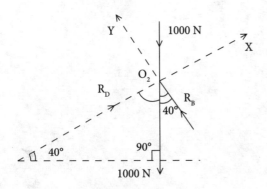

Fig. 2.20 (b)

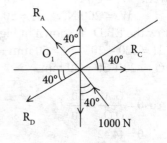

Fig. 2.20 (c)

$\sum X = 0$, $\quad R_A \sin 40° + R_D \cos 40° = R_C$ .....(1)
$\sum Y = 0$, $\quad R_A \cos 40° = R_D \sin 40° + 1000$ .....(2)

Substituting value of R_D in equation (2),

$R_A \cos 40° = 642.79 \sin 40° + 1000$

$R_A = 1844.77\ N$

Substituting R_A and R_D in equation (1),

$1844.77 \sin 40° + 642.79 \cos 40° = R_C$

$R_C = 1678.20\ N$

Example 2.17

For given Fig. 2.21, determine reactions at A, B, C and D.

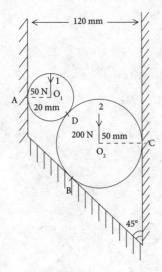

Fig. 2.21

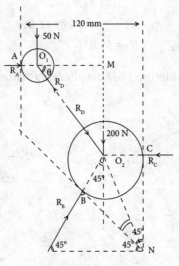

Fig. 2.21 (a)

Solution:

Given, $W_1 = 50$ N, $r_1 = 20$ mm
$W_2 = 200$ N, $r_2 = 50$ mm

Draw the F.B.D. as shown in Fig. 2.21 (a)

$O_1M = 120 - 20 - 50 = 50$ mm
$O_1O_2 = 20 + 50 = 70$ mm

$$\cos\theta = \frac{O_1M}{O_1O_2} = \frac{50}{70}$$

$\theta = 44.42°$

Considering FBD of sphere 1 and draw skeleton of forces as shown in fig. 2.21(b).

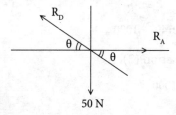

Fig. 2.21 (b)

$\sum X = 0$,
$R_D \cos\theta = R_A$
$\sum Y = 0$
$R_D \sin\theta = 50$
$R_D \sin 44.42° = 50$
$R_D = 71.44 N$
$R_A = 71.44 \times \cos 44.42°$
$R_A = 51.02 N$

Considering *FBD* of sphere 2 as shown in Fig. 2.21 (c)

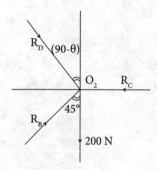

Fig. 2.21 (c)

$$\Sigma X = 0,$$
$$R_D \sin(90° - \theta) + R_B \sin 45° = R_C$$
$$R_D \cos\theta + R_B \times \sin 45° = R_C$$
$$71.44 \cos 44.42° + R_B \times \sin 45° = R_C$$
$$51.02 + R_B \times 1/\sqrt{2} = R_C \quad \quad \ldots(1)$$
$$\Sigma Y = 0,$$
$$R_D \cos(90° - \theta) + 200 = R_B \cos 45°$$
$$R_D \sin\theta + 200 = R_B \times 1/\sqrt{2}$$
$$71.44 \sin 44.42° + 200 = R_B \times 1/\sqrt{2}$$
$$R_B = 353.56 \ N, \text{Substituting in equation (1)}$$
$$R_C = 301.02 \ N$$

Example 2.18

A prismatic bar AB of weight 1000 N and length l *is* supported through hinge 'A' and cable BC as shown in the Fig. 2.22. Determine reaction of hinge and tension in the string.

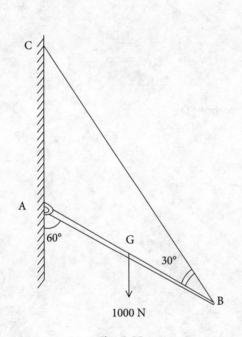

Fig. 2.22

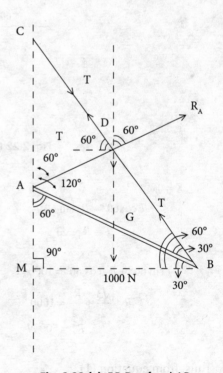

Fig. 2.22 (a) F.B.D. of rod AB

Solution:

Draw the F.B.D. of bar AB as shown in Fig. 2.22 (a)

Given, $\angle BAM = 60°$, $\angle ABC = 30°$

$\angle BAC = 180 - 60° = 120°$

In $\triangle ABC$, $\angle ACB = 180 - (120 + 30°)$

$= 30°$

as $\angle ACB = \angle ABC = 30°$

thus AD is bisector

hece $\angle CAD = \angle DAB = 60°$

As weight 1000 N and Tension T meets about point D thus hinge reaction will go through this point to maintain equilibrium of the force system.

Considering FBD about D, as shown in Fig. 2.22 (b).

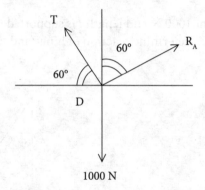

Fig. 2.22 (b) Force system about D

$\sum X = 0$,

$T \cos 60° = R_A \sin 60°$

$T = R_A \cdot \sqrt{3}$ .....(1)

$\sum Y = 0$,

$T \sin 60° + R_A \cos 60° = 1000$ .....(2)

$\sqrt{3} \cdot R_A \cdot \dfrac{\sqrt{3}}{2} + \dfrac{R_A}{2} = 1000$

$R_A = 500 \, N$

Thus, from equation (1),

$T = 500\sqrt{3}$

$T = 866.03 \, N$

Example 2.19

A cylinder of weight 'W' is held in equilibrium on smooth plane with the help of two strings passing over two smooth pulleys and supporting loads P and Q as shown in the Fig. 2.23. Determine reaction between cylinder and the horizontal surface, angle α and tensions in the both strings.

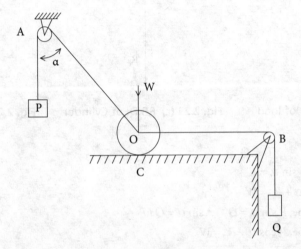

Fig. 2.23

Solution:

The given Fig. 2.23 can be considered as shown in Fig. 2.23 (a).

Considering F.B.D. of both loads P and Q, cylinder as shown below in Fig. 2.23 (b), (c) and (d)

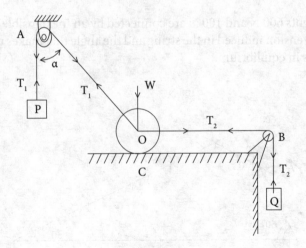

Fig. 2.23 (a)

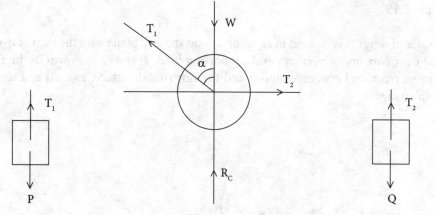

Fig. 2.23 (b) F.B.D. of load   Fig. 2.23 (c) F.B.D. of Cylinder   Fig. 2.23 (d) F.B.D. of load

$$T_1 = P \qquad T_2 = Q$$
$$\Sigma X = 0, \qquad T_1 \sin \alpha = T_2 \qquad \qquad \qquad \text{.....(1)}$$
$$\Sigma Y = 0, \qquad T_1 \cos \alpha + R_C = W \qquad \qquad \text{.....(2)}$$

equation (1) will be, $P \sin \alpha = Q \quad \therefore \sin \alpha = Q/P$

equation (2) will be, $P \cos \alpha + R_C = W$

$$R_C = W - P\sqrt{1 - \sin^2 \alpha}$$
$$R_C = W - P\sqrt{1 - \frac{Q^2}{P^2}}$$
$$\alpha = \sin^{-1}\left(\frac{Q}{P}\right)$$

Example 2.20

Two rollers of weights 600 N and 100 N are connected by an inextensible string as shown in Fig. 2.24. Find the tension induced in the string and the angle that makes with the horizontal when the system is in equilibrium.   *(SRTMU June, 1996)*

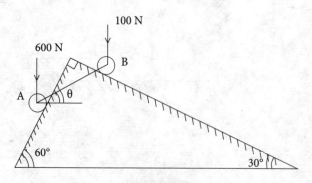

Fig. 2.24

Two Dimensional Concurrent Force Systems

Solution:

Considering *FBD* of body *A*, Resolving the forces along the inclined plane, as shown in Fig. 2.24 (*a*)

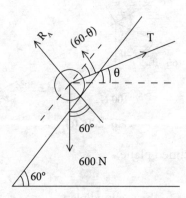

Fig. 2.24 (a) F.B.D. of roller A

$$T\cos(60° - \theta) = 600 \sin 60° \quad \ldots(1)$$
$$R_A = T\sin(60° - \theta) + 600 \cos 60° \quad \ldots(2)$$
$$\angle ABC = 180° - \theta - 90°$$
$$= (90° - \theta)$$
$$\angle BDE = 30°$$

and $\quad \angle EBD = 90°$

∴ $\quad \angle BED = 60°$

$$\angle EBN = 180° - 90° - 60° = 30°$$

thus $\quad \angle ABE = 90° - \theta - 30° = (60° - \theta)$

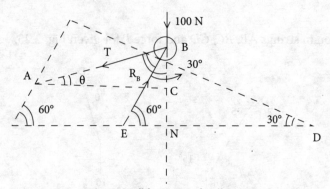

Fig. 2.24 (b) F.B.D. of roller B

Considering skeleton of force system about B, as shown in Fig. 2.24 (b) and (c)

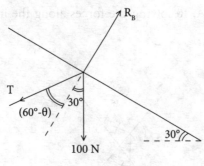

Fig. 2.24 (c)

Resolving forces along the inclined plane,
$T \sin(60° - \theta) = 100 \sin 30°$ .....(3)

Resolving forces perpendicular to the inclined plane,
$R_B = T \cos(60° - \theta) + 100 \cos 30°$ .....(4)

Considering equations (1) and (3),

$$\frac{\sin(60° - \theta)}{\cos(60° - \theta)} = \frac{100 \sin 30°}{600 \sin 60°}$$

$\tan(60° - \theta) = 0.962$

$(60° - \theta) = \tan^{-1}(0.962) = 43.89°$

$\theta = 1.61°$

Substituting value of θ in equation (3),

$T \sin(60° - 16.1) = 100 \sin 30°$

$T = 72.11 \, N$

Example 2.21

Determine tension in strings AB, BC, CD and force P for given Fig. 2.25.

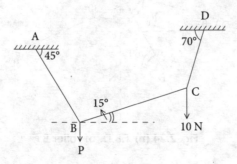

Fig. 2.25

Two Dimensional Concurrent Force Systems

Solution:

Considering *FBD* about points *B* and *C* as shown in Fig. 2.25 (*a*)
Applying Lami's theorem for force system about *B*,

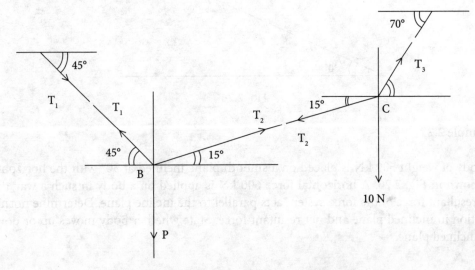

Fig. 2.25 (a) Force system about B and C

$$\frac{T_1}{\sin(90°+15°)} = \frac{T_2}{\sin(90°+45°)} = \frac{P}{\sin(180°-45°-15°)}$$

$$\frac{T_1}{0.96} = \frac{T_2}{0.71} = \frac{P}{0.87} \qquad \qquad \ldots\ldots(1)$$

Similarly about *C*,

$$\frac{T_2}{\sin(90°+70°)} = \frac{T_3}{\sin(90°-15°)} = \frac{10}{\sin(180°-70°+15°)}$$
$$= 12.2$$

$T_2 = 0.34 \times 12.2$ and $T_3 = 0.97 \times 12.2$
$T_2 = 4.15\ N \qquad T_3 = 0.97 \times 12.2$

From equation (1),

$$T_1 = T_2 \times \frac{0.96}{0.71} = \frac{4.15 \times 0.96}{0.71}$$
$$T_1 = 5.61\ N$$

and $\qquad P = \dfrac{0.87 \times T_2}{0.71}$

$\qquad\qquad P = 5.09\ N$

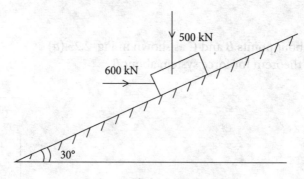

Fig. 2.26

Example 2.22

A body of weight 500 kN is placed on a smooth plane inclined at 30° with the horizontal as shown in Fig. 2.26. A horizontal force 600 kN is applied on a body in such a way that net resultant force of the force system acts parallel to the incline plane. Determine normal reaction in inclined plane and net resultant force. State whether body moves up or down the inclined plane.

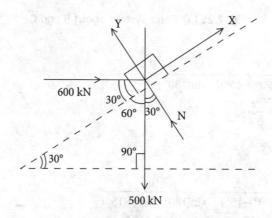

Fig. 2.26 (a) F.B.D. of body on inclined plane

Solution:

Consider FBD of the body, Fig. 2.26 (a) since net resultant force acts parallel to the incline plane this means that

$\Sigma X \neq 0$ but $\Sigma Y = 0$

thus resolving the forces perpendicular to the inclined plane,

$\Sigma Y = 0$,

i.e., $\qquad N = 600 \sin 30° + 500 \cos 30°$

Two Dimensional Concurrent Force Systems

$$N = 733.01 \ kN$$
$$N = 800 \ kN$$

and $\quad \Sigma X = +600 \cos 30° - 500 \sin 30°$
$\quad\quad \Sigma X = 269.62 \ kN$

Thus net resultant force,

$$R = \sqrt{\Sigma x^2 + \Sigma y^2} = \sqrt{\Sigma x^2 + 0}$$
$$R = \Sigma X$$
$$R = 269.62 \ kN$$

As ΣX is +ve this shows that the body will move up.

Example 2.23

A string supported at A and B carries a load of 20 kN at D and a load of W at C as shown in Fig. 2.27. Determine the value of W so that string CD remains horizontal.

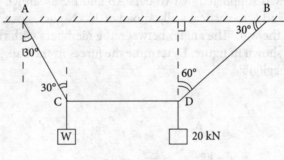

Fig. 2.27

Solution:

Considering Force system about point C and D as shown below in Fig. 2.27 (a) and (b).

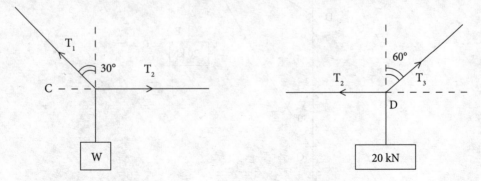

Fig. 2.27 (a) Force system about C **Fig. 2.27 (b)** Force system about D

Considering equilibrium Fig.2.27(a),, ΣX = 0,

$T_1 \sin 30° = T_2$

$T_1 = 2T_2$ .....(1)

ΣY = 0,

$T_1 \cos 30° = W$ .....(2)

equilibrium for Fig.2.27(b), ΣX = 0,

$T_2 = T_3 \sin 60°$ .....(3)

$T_3 \cos 60° = 20$

$T_3 = 40$ kN

$T_2 = 34.64$ kN

From equation (1) and (2)
Thus $T_1 = 2T_2 = 2 \times 34.64 = 69.28$ kN

Example 2.24

The frictionless pulley is supported by two bars *AB* and *BC* as shown in the Fig. 2.28. The bars are hinged at *A* and *C* to vertical wall. A load of 20 kN passes over pulley at *B* with the help of string tied to the wall. The angles between the members with the wall and between pulley and string are shown in figure. Determine the forces in the bars *AB* and *BC*. Take the size of the pulley as negligible.

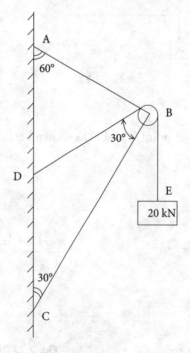

Fig. 2.28

Two Dimensional Concurrent Force Systems

Solution:

As the size of the pulley is negligible it can be idealized as a point B. The tension in both portion of string DB and BE will be equal as it is frictionless pulley. Thus the diagram can be simplified further as shown in Fig. 2.28 (a).

$T = 20$ kN

Let member AB is in tension then member BC will be in compression.
Considering equilibrium about point B,

$$\angle BDC = 180° - 30° - 30° = 120°$$
$$\therefore \quad \angle ADB = 180° - 120° = 60°$$

thus $\triangle ADB$ is an equilateral triangle. BM, a horizontal line will be a bisector
$\therefore \angle ABM = \angle DBM = 30°$, considering equilibrium of forces about B as shown in Fig. 2.28 (b)

$\Sigma X = 0$
$$F_{AB} \cos 30° + 20 \cos 30° = F_{BC} \sin 30° \qquad \text{.....(1)}$$
$\Sigma Y = 0,$
$$F_{AB} \sin 30° + F_{BC} \cos 30° = 20 + 20 \sin 30° = 30$$
$$\frac{F_{AB}}{2} + \frac{F_{BC} \sqrt{3}}{2} = 30$$
$$F_{AB} + \sqrt{3} F_{BC} = 60$$

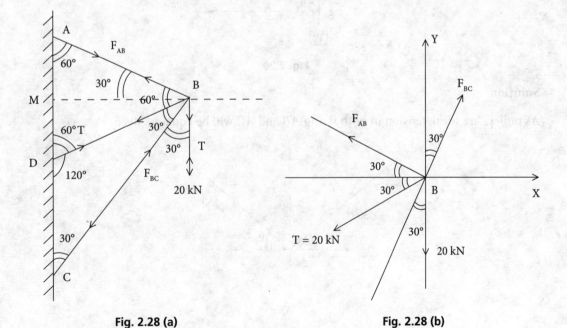

Fig. 2.28 (a)   Fig. 2.28 (b)

i.e., $\quad F_{AB} = (60 - \sqrt{3}F_{BC})$

Substituting F_{AB} in equation (1),

$$(60 - \sqrt{3}F_{BC}) \cdot \frac{\sqrt{3}}{2} + 20\frac{\sqrt{3}}{2} = F_{BC} \cdot \frac{1}{2}$$

$$60\sqrt{3} - 3F_{BC} + 20\sqrt{3} = F_{BC}$$

$$F_{BC} = 34.64 \text{ kN} \quad \text{and} \quad F_{AB} = 0$$

Example 2.25

Determine the value of weight 'W' that can be supported by 30 N over two ideal pulleys as shown in the Fig. 2.29.

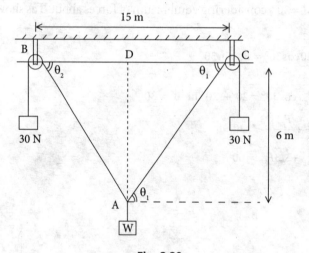

Fig. 2.29

Solution:

As pulleys are ideal, tension in both string AB and AC will be equal to 30 N.

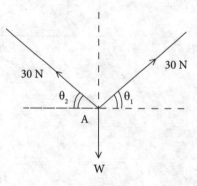

Fig. 2.29 (a)

Considering equilibrium about A,

Let $\angle BCA = \theta_1$, thus $\angle CBA = \theta_2$,

$\sum Y = 0$, $W = 30 \sin \theta_1 + 30 \sin \theta_2$

$\sum X = 0$, $30 \cos \theta_2 = 30 \cos \theta_1$

i.e., $\theta_1 = \theta_2$ thus $W = 60 \sin \theta_1$

Thus AD is a bisector,

$$\tan \theta_1 = \frac{AD}{DC} = \frac{6}{7.5}$$

$\theta_1 = 38.66°$ and $W = 37.48$ N

Example 2.26

A rope AB is tied up horizontally between two walls 24 m apart. When a mass of 60 kg is suspended at its midpoint C, then rope takes deflection of 6 m. Determine tensions in both AC and BC part.

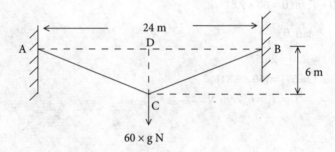

Fig. 2.30

Solution:

Draw the Fig. 2.30 (a) C is the mid-point of rope hence AC = BC thus CD will be bisector of AB and will make equal angle i.e., $\angle CAD = \angle CBD = \theta$, say

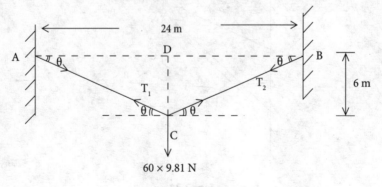

Fig. 2.30 (a)

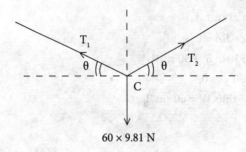

Fig. 2.30 (b)

Considering equilibrium about 'C'. Fig. 2.30 (b)

$\Sigma X = 0$,

$$T_1 \cos\theta = T_2 \cos\theta$$
$$T_1 = T_2$$

$\Sigma Y = 0$,

$$T_1 \sin\theta + T_2 \sin\theta = 60 \times 9.81$$

from fig (b), $\quad \tan\theta = \dfrac{6}{12}$

$$\therefore \theta = 26.57°$$
$$2T_1 \sin(26.57) = (60 \times 9.81)$$
$$T_1 = T_2 = 657.96 \ N$$

Example 2.27

A log of wood is lifted using two ropes AC and BC as shown in the Fig. 2.31. If the tensions in the ropes AC and BC are 1.8 kN and 2.1 kN, respectively. Determine the resultant force at C if AC and BC makes 35° and 30° with respect to vertical, respectively.

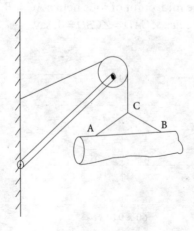

Fig. 2.31

Solution:

Consider force system about point 'C', Fig. 2.31 (a)

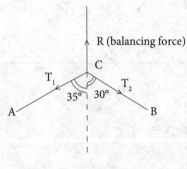

Fig. 2.31 (a)

Given, $T_1 = 1.8$ kN
$T_2 = 2.1$ kN
$\Sigma Y = 0$,
$R = T_1 \cos 35° + T_2 \cos 30°$
$= 1.8 \cos 35° + 2.1 \cos 30°$
$R = 3.29$ kN

Thus, the resultant force will be equal and opposite to balancing force.
Resultant force = 3.29 kN (vertically downward)

Example 2.28

A rigid prismatic bar AB hinged at wall and supports a load W at free end is held in equilibrium by string CD as shown in Fig. 2.32. Determine tension in the string CD. Take length of bar as L, angle of inclination θ form wall and weight of the bar as negligible.

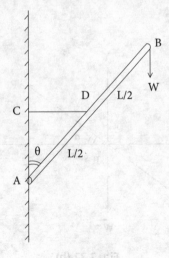

Fig. 2.32

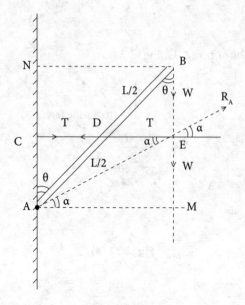

Fig. 2.32 (a)

Solution:

Consider Fig. 2.32 (a), the bar AB is in equilibrium under three forces T, R_A and W. The equilibrium will be maintained if three forces are concurrent.

Thus line of action of T and W cross each other at point E with the help of Principle of Transmissibility, where from the reaction of hinge A will pass through.

$EM = AC = L/2 \cos\theta$ and $CN = BE = L/2 \cos\theta$

$$\tan\alpha = \frac{EM}{AM} = \frac{L/2\cos\theta}{L\sin\theta}$$

$$\tan\alpha = \frac{\cot\theta}{2} \qquad \text{.....(1)}$$

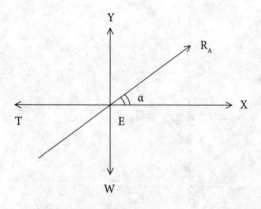

Fig. 2.32 (b)

Two Dimensional Concurrent Force Systems

Considering equilibrium of forces about point E as shown in Fig. 2.32 (b)

$\Sigma X = 0$, $T = R_A \cos\alpha$
and $W = R_A \sin\alpha$

$$T = \frac{W}{\sin\alpha} \cdot \cos\alpha$$

$$T = \frac{W}{\tan\alpha}$$

From equation (1),

$$T = \frac{W}{\frac{\cot\theta}{2}}$$

$T = 2W\tan\theta$

Example 2.29

A body of weight 500 kN is hung with the help of a member *AB* and string *BC*. The member is positioned at 45° with the help of another string *BD* makes 30° with the horizontal as shown in Fig. 2.33. Take weight of the member *AB* as negligible. Determine tension in the string and reaction at A.

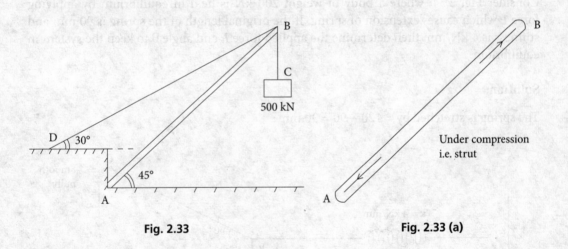

Fig. 2.33   Fig. 2.33 (a)

Solution:

Here when 500 kN is hanged at B, the member AB becomes under compression. The force system is shown in Fig. 2.33 (b)

The forces can be determined by using Lami's theorem

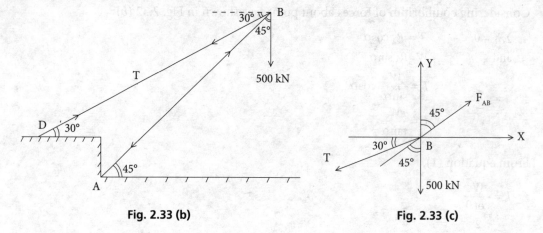

Fig. 2.33 (b)   Fig. 2.33 (c)

Using Lami's theorem,

$$\frac{T}{\sin(180°-45°)} = \frac{F_{AB}}{\sin(90°-30°)} = \frac{500}{\sin(30°+90°+45°)}$$

$T = 1366.03$ kN   and $F_{AB} = 1673.03$ kN.

Example 2.30

Consider Fig. 2.34, where a body of weight 200 kN is held in equilibrium by applying force P which causes extension of spring. If the original length of the spring is 90 mm and stiffness is 4 kN/mm then determine the applied force P and angle θ to keep the system in equilibrium.

Solution:

The spring is stretched by = 120 − 90 = 30 mm

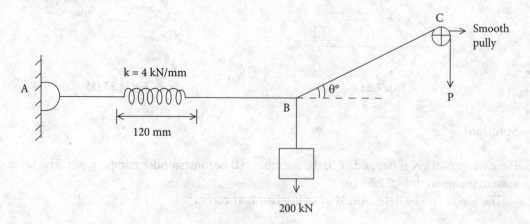

Fig. 2.34

thus, the tension in the spring will be = $k \times$ stretched amount

$$= 4 \times 30 \frac{kN}{mm} \times mm = 120 \text{ kN}.$$

As the pulley at C is smooth
thus tension in the string BC will be P.
Considering force system at B as shown in Fig. 2.34 (a),

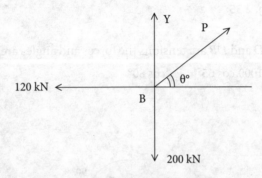

Fig. 2.34 (a)

$\Sigma X = 0$, $P\cos\theta = 120$ .....(1)
$\Sigma Y = 0$, $P\sin\theta = 200$ .....(2)

Dividing equation (2) by (1)

$$\tan\theta = \frac{200}{120}$$

$\therefore \theta = 59.04°$ and $P = 233.24$ kN

Example 2.31

Three bars AC, CD and BD are arranged to form a four linked mechanism with AB. The bars AC and BD are hinged at A and B and pinned at C and D respectively as shown in the Fig. 2.35. Determine the value of R that will keep the mechanism in equilibrium.

Solution:

Consider equilibrium of forces about point D. as shown in Fig. 2.35 (a).

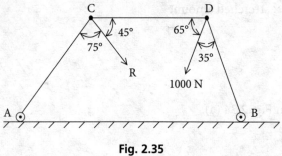

Fig. 2.35 Fig. 2.35 (a)

Let the forces in bars *CD* and *DB* are tensions the forces and angles are shown in Fig. 2.35 (*b*)

$\sum X = 0$, $\quad F_{CD} + 1000 \cdot \cos 65° = F_{DB} \cos 80°$ .....(1)

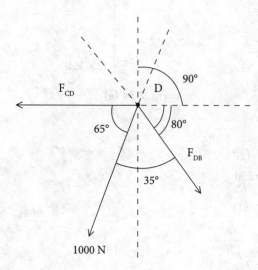

Fig. 2.35 (b)

$\sum Y = 0$, $\quad 1000 \sin 65° + F_{DB} \sin 80° = 0$ .....(2)

$\qquad F_{DB} = -920.29 \text{ N}$

– ve sign shows that force in bar *DB* is of compressive nature.
Substituting value of F_{DB} in equation (1),

$F_{CD} + 1000 \cdot \cos 65° = -920.29 \cos 80°$

$\qquad F_{CD} = -582.52 \text{ N}$

Considering equilibrium about '*C*'.
Let the forces in bars are tension as shown in Fig. 2.35 (*d*)

Two Dimensional Concurrent Force Systems

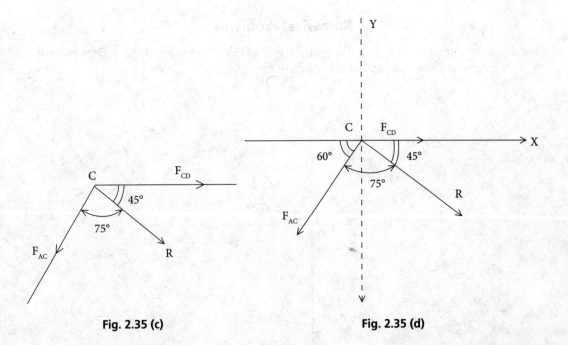

Fig. 2.35 (c)   Fig. 2.35 (d)

$\sum X = 0$,
$$F_{AC} \cos 60° = F_{CD} + R \cos 45° \qquad \ldots(3)$$

$\sum Y = 0$,
$$R \sin 45° + F_{AC} \sin 60° = 0 \qquad \ldots(4)$$

Substituting F_{CD} in equation (3),

$$F_{AC} \cos 60° = -582.42 + R \cos 45°$$

From equation (4),

$$F_{AC} = \frac{-R.\sin 45°}{\sin 60°}$$

$$\left(\frac{-R.\sin 45°}{\sin 60°}\right) \cos 60° = -582.42 + R \cos 45°$$

$$-R \times 0.41 = -582.42 + R \times 1.41$$
$$1.82 \times R = 582.42$$
$$R = 320 \, N$$

Numerical Problems

N 2.1 A concurrent force system is acting at a point O as shown in Fig. NP 2.1. Determine the magnitude and direction of the resultant force.

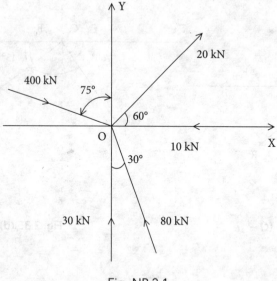

Fig. NP 2.1

N 2.2 A wire rope fixed at A and B supports 300 kN and 200 kN at D and C, respectively as shown in Fig. NP 2.2. Determine tension in AD, DC and CB. Also determine angle of inclination made by DC with horizontal.

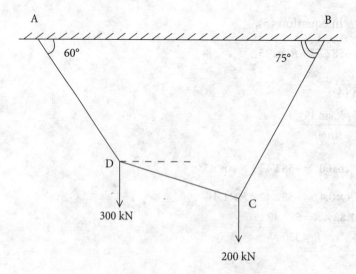

Fig. NP 2.2

N 2.3 Two wire ropes fixed at A and B supports a load 600 kN at C as shown in Fig. NP 2.3. Determine tension in each wire.

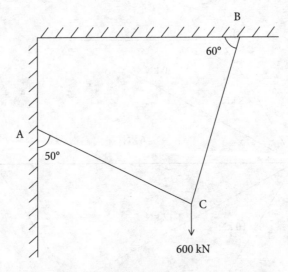

Fig. NP 2.3

N 2.4 Two cylinders of identical size but of different materials having weight 100 kN and 160 kN are lying on an inclined plane. If the angle of inclination is 30° as shown in Fig. NP 2.4, determine reactions at A, B and C.

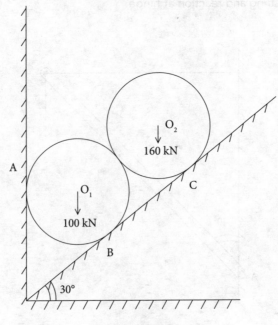

Fig. NP 2.4

N 2.5 A concurrent coplanar force system is acting at point O in a hexagon as shown in Fig. NP 2.5. Determine resultant force of the force system.

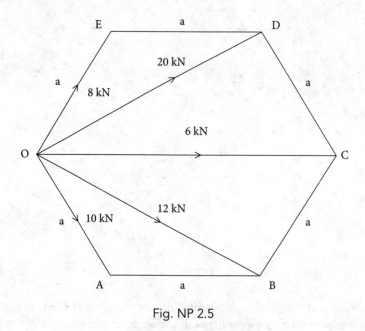

Fig. NP 2.5

N 2.6 A prismatic bar of negligible weight is hinged at wall and supports a load of 4 kN with the help of a rope wire BC fixed to wall horizontally as shown in Fig. NP 2.6. Determine tension in the string and reaction at hinge.

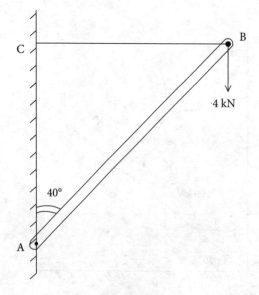

Fig. NP 2.6

Two Dimensional Concurrent Force Systems

N 2.7 Determine forces at A, B and C as shown in Fig. NP 2.7.
Take $W_1 = 50$ kN, $r_1 = 60$ mm
$W_2 = 100$ kN, $r_2 = 120$ mm

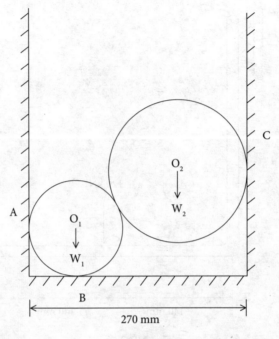

Fig. NP 2.7

N 2.8 A cylinder of weight 900 kN rests over two inclined planes as shown in Fig. NP 2.8. Determine reactions at points A and B.

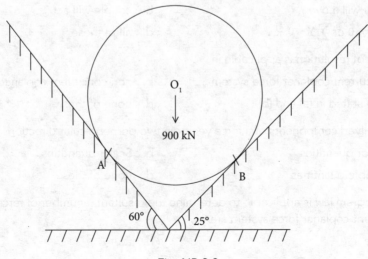

Fig. NP 2.8

N 2.9 An L-shaped plate of negligible weight is hinged at C, supports a load of 120 kN with the help of roller support bearing at A as shown in Fig. NP 2.9. Determine reactions at A and C.

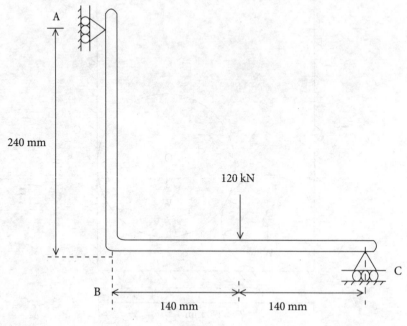

Fig. NP 2.9

Multiple Choice Questions

1. If two equal and opposite forces are concurrent on a body
   a. body will move
   b. couple will act
   c. $\sum X = 0$ or $\sum Y = 0$
   d. all of these

2. Method of resolution is applicable in
   a. concurrent coplanar force system
   b. non-concurrent coplanar force system
   c. both stated in (a) and (b)
   d. none of these

3. The resolved components of a force vector in two perpendicular directions are
   a. Vector quantities
   b. Scalar quantities
   c. Variable quantities
   d. none of these

4. Parallelogram law is applicable to determine the resultant if number of forces acting in a concurrent coplanar force system are
   a. three
   b. four
   c. two
   d. none of these

Two Dimensional Concurrent Force Systems

5. Principle of transmissibility of force is used to
   a. transfer force to a new point along the line of action of force
   b. transfer force to a new point perpendicular to the line of action of force
   c. remove the force
   d. none of these

6. Principle of transmissibility of force can be if forces are acting on
   a. deformable body
   b. particle
   c. rigid body
   d. none of these

7. Free Body Diagrams of a body includes
   a. weight of the body
   b. external forces acting on the body
   c. reactions of other bodies in contact
   d. all of these

8. Law of sine relates to
   a. parallelogram law
   b. triangle law
   c. polygon law
   d. all of these

9. Which law will be used to analyse a coplanar force system of four forces
   a. parallelogram law
   b. triangle law
   c. polygon law
   d. all of these

10. Lami's theorem is applicable when body is in under equilibrium of
    a. three concurrent coplanar forces
    b. three non-concurrent coplanar forces
    c. four concurrent coplanar forces
    d. all of these

Answers

1. c   2. c   3. a   4. c   5. a   6. c   7. d   8. b   9. c   10. a

Chapter 3
Two Dimensional Non-concurrent Force Systems

3.1 Introduction

In non-concurrent coplanar force system, the forces do not act at a common point in a plane of body. Such a situation causes rotation of body even if net forces in X and Y directions are zero i.e., if both ΣX and ΣY are zero, even then body will rotate and cannot be held in equilibrium.

Consider for an example a rigid body is under non-concurrent coplanar force system as shown in Fig. 3.1. Here the algebraic sum of force in X and Y directions is zero but it can be observed that body is under rotation and not in equilibrium. This situation took place due to different characteristic of force called moment stated in next heading.

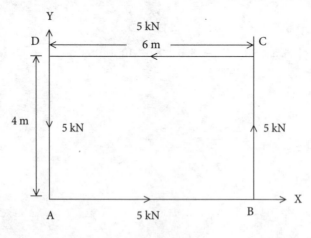

Fig. 3.1

3.2 Moment

The term moment is nothing but a special effect of force which causes rotation to a member or a link, or bending (if member or a link is restricted to rotate). The rotation or bending happens only, when line of action of force does not pass through point about which member is hinged or fixed at one end. Numerically, it is equal to the product of force and the minimum distance (perpendicular) between line of action of force and about which member is fixed or hinged.

Consider for an example where a cricket stump OA is fixed about point O. The stump height is 'h' as shown in Fig. 3.2. If a ball is hit in the direction east to west and perpendicular to the axis of stump at point A by force P then stump will turn along anti-clockwise direction and moment about O is expressed in magnitude by $M_o = P*h$.

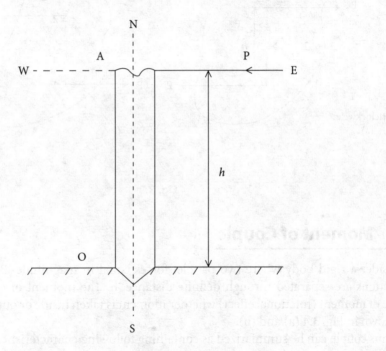

Fig. 3.2

Thus it is a vector quantity as the effect of moment (direction of turning stump) depends upon the direction of applying force. The stump height 'h' is also known as the arm length of moment. It can be realized that by increasing arm length, the effort P can be reduced for same amount of required moment. This is accomplished in various day to day life applications like loosening of nut can be easily done, breaking of lock with the help of long bar, operating hand pump, lofted shot played by batsman when ball is picked at the root of the bat and fixation of door lever at maximum distance from fixed end of the door, etc.

3.3 Couple

It is formed when two parallel forces equal in magnitude and opposite in directions are separated through definite distance on a body. The translatory effect of couple is zero as the algebraic sum of two forces causing couple is zero in any direction. Opening or closing of bottle cap, water tap and locking or unlocking the lock by using key are the example of couple. Couple causes rotation of body whether in clockwise or anti-clockwise depending upon the direction of parallel unlike forces. This is illustrated below in Fig. 3.3. (a) and (b).

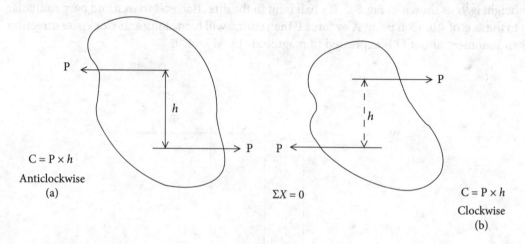

Fig. 3.3

3.4 Moment of Couple

Consider a rigid body where two parallel forces equal in magnitude 'P' and opposite in directions are separated through definite distance 'h'. The moment of couple gives same value of moment (rotational effect) whether moment is taken inside or outside of two forces as shown in Fig. 3.4 (a) and (b).

Thus couple can be summarized as containing following characteristics:

- The translatory effect of couple is zero in any direction.
- A couple is possible only when two parallel forces equal in magnitude and opposite in directions are separated by definite distance.
- The value of moment i.e., rotational effect remains same about any point in a plane of body and depends upon the product of magnitude and perpendicular distance of forces.

Two Dimensional Non-concurrent Force Systems

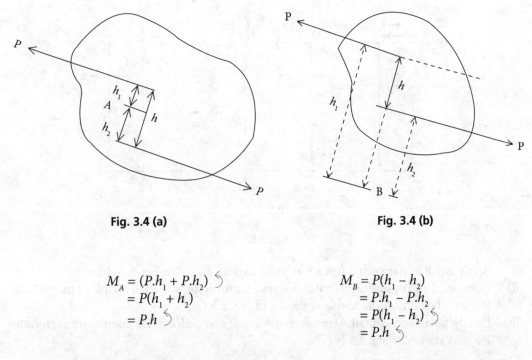

Fig. 3.4 (a) Fig. 3.4 (b)

$$M_A = (P.h_1 + P.h_2)$$
$$= P(h_1 + h_2)$$
$$= P.h$$

$$M_B = P(h_1 - h_2)$$
$$= P.h_1 - P.h_2$$
$$= P(h_1 - h_2)$$
$$= P.h$$

As the couple causes only rotational effect due to moment which is the same about all points, the effect of a couple remains same if:

- The couple is shifted to any other point in the same plane and due to this it is also known as Free vector quantity.
- The couple is rotated by any angle in same plane.
- The couple can be replaced by another set of forces which produces same rotational effect.

3.5 Transfer of a Force to Parallel Position

A force can be transferred to any new position parallel to its line of action and this results an introduction of couple along with transferred force. Consider Fig. 3.5 (a) where a force P is acting on a column at distance x from the polar axis. This force can be transferred to polar axis by following steps:

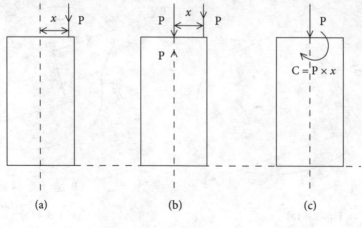

Fig. 3.5

(i) Apply equal and opposite force P at polar axis as shown in Fig. 3.5 (b).
(ii) Now equal and opposite force P at distance x causes formation of couple of magnitude P × x and direction clockwise as shown in Fig. 3.5 (c).
(iii) Finally polar axis contains transferred force P and a clockwise couple of magnitude P × x as shown in Fig. 3.5 (c).

This is also known as resolution of a force into a force and a couple.

3.6 Graphical Presentation of Moment

Consider Fig. 3.6 (a) where moment about O is

$M_O = P \times h$

The force P can be represented by length AB and h can be represented by OC as shown in Fig. 3.6 (b). Thus

$$M_O = P \times h$$
$$= AB \times OC$$
$$= 2\left(\frac{1}{2} \times AB \times OC\right)$$
$$M_O = 2 \cdot \text{Area of } \triangle OAB$$

M_O = Twice area of triangle formed by vertices of force (A and B) and centre of moment (O).

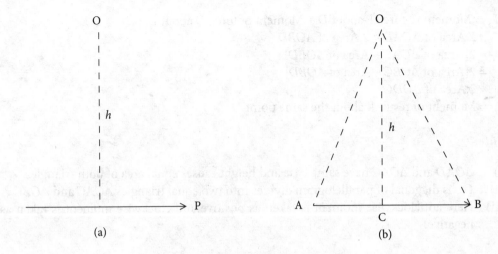

Fig. 3.6

3.7 Varignon's Theorem

This theorem states that, "if a body is subjected to number of forces, than algebraic sum of moment of all forces about a point will be equal to the moment of their resultant about the same point". However, it can also be stated as the moment of a force about a point will be equal to algebraic sum of moment of its component about the same point. This theorem is useful to determine the position of the resultant of a non-coplanar force system, centroid or center of gravity of a body.

Consider two forces P and Q are acting at point O in a body and R is their resultant as shown in Fig. 3.7 (a). Let the forces are represented by sides of a parallelogram as shown in Fig. 3.7 (b). Select a point D for moment of both forces

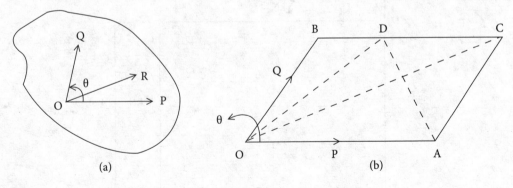

Fig. 3.7

M_D = Moment of force P about D + Moment of force Q about D
= 2.Area of $\triangle OAD$ − 2 Area of $\triangle OBD$
= 2[Area of $\triangle OAC$ − Area of $\triangle OBD$]
= 2[Area of $\triangle OBC$ − Area of $\triangle OBD$]
= 2.Area of $\triangle ODC$
= Moment of result R about the same point D.

Note:

(i) $\triangle OAD$ and $\triangle OAC$ have same base and height causes equal area of both triangles.
(ii) OC is diagonal of parallelogram divides into two equal triangles $\triangle OAC$ and $\triangle OBC$.
(iii) Here anticlockwise moment is taken as positive and clockwise moment is taken as negative.

3.8 Equations of Equilibrium Conditions

As stated in Section 3.1, if a body is subjected to two dimensional non-concurrent coplanar force system it cannot be held in equilibrium even if two conditions are satisfied i.e.,

$\Sigma X = 0$
$\Sigma Y = 0$

Thus equilibrium can be maintained for such cases if third condition also satisfies that $\Sigma M = 0$. Consider for an example a rigid body is under non-concurrent coplanar force system as shown in Fig. 3.8 (a). Here algebraic sum of force in X and Y directions is zero but one can observe that body is under rotation and not in equilibrium.

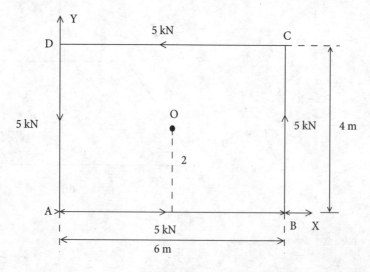

Fig. 3.8 (a)

The algebraic sum of moment of all the forces acting on rectangular lamina about any vertices or at center O as shown in the Fig. 3.8 (a) is $\Sigma M_o = (5 \times 2 + 5 \times 3 + 5 \times 2 + 5 \times 3)$

$\Sigma M = 50$ kNm anticlockwise which states that the rectangular lamina is rotating anticlockwise with moment of magnitude 50 kNm about O.

Consider Fig. 3.8 (b) where a clockwise couple of magnitude 50 kN-m is applied at the center O, now the algebraic sum of moment will be

$\Sigma M_o = (5 \times 2 + 5 \times 3 + 5 \times 2 + 5 \times 3 - 50)$

$\Sigma M_o = 0$

This shows that the lamina is under equilibrium.

Being free vector quantity clockwise couple 50kNm can be applied anywhere in lamina and lamina will remain in equilibrium

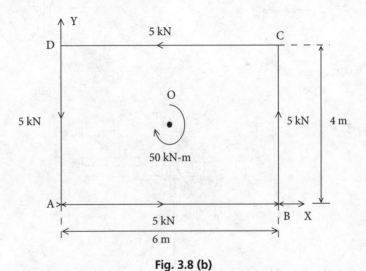

Fig. 3.8 (b)

3.9 Types of Supports and their Reactions on Beams

Beam is a structural element whose one dimension i.e., length is very large in comparison to other dimensions i.e., width and depth. Depending upon the type of supports they are categorized further into simply supported beam, cantilever beam, propped beam, continuous beam and overhanging beam etc. Beams are subjected to transverse loading. The beams are kept in equilibrium by supports like hinge, roller support bearing etc. which offer reactions against external loading. The different types of supports offer different type of reactions which are illustrated below:

Hinged Support: The hinge always restricts motion of beam in both horizontal-X and vertical-Y axis, thus offers reactions along both axis. However it does not restrict beam to rotate. Thus it does not offer resisting moment. The reaction of hinge is resolved into horizontal and vertical components during free body diagram of beam for the analysis purpose. Hinged support is represented by two ways as shown in Fig. 3.9 (a) and (b).

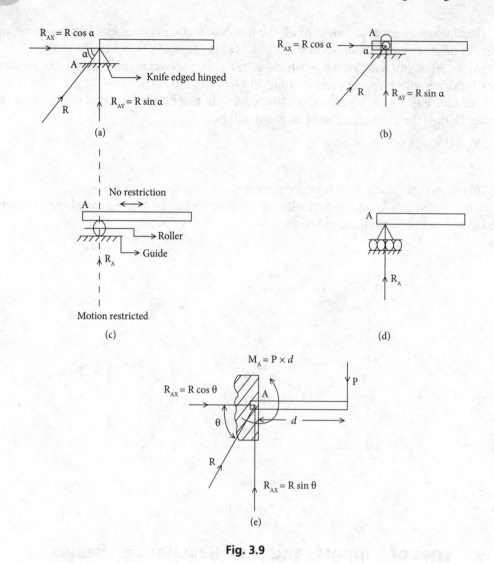

Fig. 3.9

Roller Support: In this support the beam is allowed to slide due to frictionless rolling of roller between guides (X-axis) of the support but restricted to move perpendicular to guides (Y-axis). Thus it offers reaction only perpendicular to guides (Y-axis). Like hinged support, it does not restrict beam to rotate and thus it does not offer resisting moment. Roller Support is represented by two ways as shown in Fig. 3.9 (c) and (d).

Fixed Support: This kind of support does not allow any motion to beam neither translator nor rotary. Thus like hinge it offers restriction in both horizontal-X and vertical-Y axis and offers reactions in both axes. Unlike to beam, it restricts the beam to rotate by offering a resisting moment. The reaction of fixed support is resolved into horizontal and vertical components and rotation is restricted by resisting moment as shown in Fig. 3.9 (e). The external load P on beam at distance d produces clockwise moment of magnitude P*d about fixed end A which offers resisting moment in anti-clockwise direction with same magnitude as shown in the Fig. 3.9 (e).

3.10 Types of Beams

Cantilever Beam: The beam which is fixed at one end and remains free at other end called cantilever beam as shown in Fig. 3.10 (a).

Propped Cantilever Beam: The beam which is fixed at one end and remains simply supported at other end called propped cantilever beam as shown in Fig. 3.10 (b).

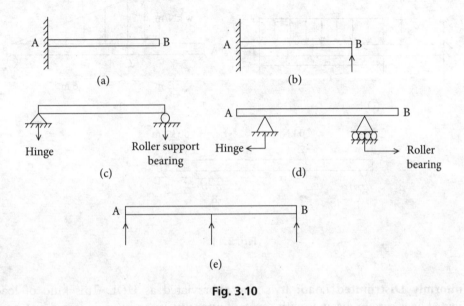

Fig. 3.10

  Simply Supported Beam: The beam which is simply supported at both ends called simply supported beam as shown in Fig. 3.10 (c). The supports used are combination of hinge and roller bearings.

  Overhanging Beam: The beam which extends beyond its one or two supports is called overhanging beam as shown in Fig. 3.10 (d). The supports used are combination of hinge and roller bearings.

  Continuous Beam: The beam which is supported throughout its length by more than two supports called continuous beam as shown in Fig. 3.10 (e).

3.11 Types of Loading on Beams

There are four types of loading acts on beams. The beam may be subjected to any type of single loading or in combination of two or more. The different types of loading are described below:

  Point load or Concentrated Load: This kind of load is concentrated at a point or acts at a point hence called as Point load or Concentrated load. It is represented by an arrow. Consider Fig. 3.11 (a) where W kN Point load is acting on a simply supported beam.

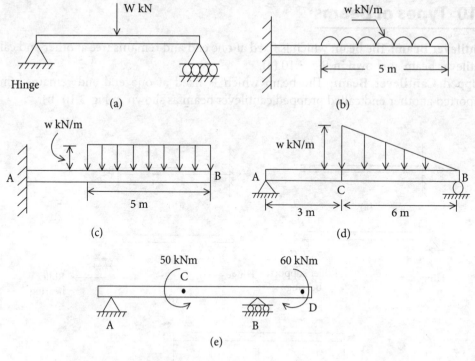

Fig. 3.11

Uniformly Distributed Load: It is also abbreviated as UDL. This kind of load is distributed over certain distance with constant intensity. It can be represented by two ways as shown in Fig. 3.11 (b) and (c) where load of constant intensity w kN/m is distributed over certain distance 5 m on a cantilever beam. For analysis purpose total load is computed i.e., area of rectangle (w*5) and assumed that acting on the middle of length of its length (Centroid of rectangle) i.e., 2.5 m from end B.

Uniformly Varying Load: It is also abbreviated as UVL. This kind of load is distributed over certain distance with uniformly varying intensity. Consider Fig. 3.11 (d) where load of uniformly varying intensity w kN/m is distributed over certain distance 6 m on a simply supported beam. For analysis purpose total load is computed i.e., area of triangle $\left(\frac{1}{2}*w*6\right)$ and assumed that acting on the Centroid of triangle i.e., 5 m from end A $\left(3+\frac{1}{3}*6=5\right)$, 4 m from end $\left(\frac{2}{3}*6=4\right)$ B and 2 m from C $\left(\frac{1}{3}*6=2\right)$.

Applied External Moment or Couple: This kind of loading tends to rotate the beam as shown in Fig. 3.11 (e) where two moments 50 kNm and 60 kNm are acting at points C and D anticlockwise and clockwise respectively an on an overhanging beam.

Example 3.1

A non-concurrent coplanar force system is acting on a rectangular lamina as shown in the Fig. 3.12. Determine the moments about A, B, C and D, respectively.

Note: Let clockwise moment is positive.

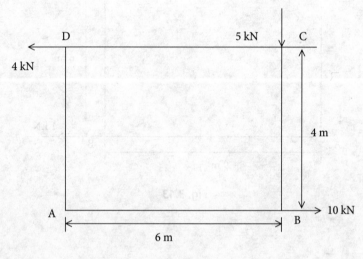

Fig. 3.12

Solution:

$M_A = 10 \times 0 + 5 \times 6 - 4 \times 4$
　　$= 30 - 16$
　　$= +14$ kNm clockwise.

$M_B = 10 \times 0 + 5 \times 0 - 4 \times 4$
　　$= -16$ kNm anticlockwise

$M_C = 5 \times 0 - 10 \times 4 + 4 \times 0$
　　$= -40$ kNm anticlockwise.

$M_D = 4 \times 0 + 5 \times 6 - 10 \times 4$
　　$= 30 - 40$
　　$= -10$ kNm anticlockwise.

Example 3.2

Similar to example 3.1, Determine values of moments about A, B, C and D.

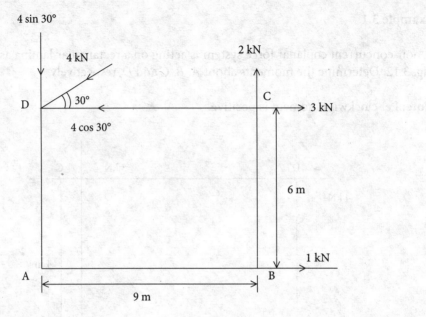

Fig. 3.13

Solution:

$M_A = 1 \times 0 + 3 \times 6 - 2 \times 9 - 4 \cos 30° \times 6 + 4 \sin 30° \times 0$
$ = 18 - 18 - 24 \cos 30°$
$ = -24 \cos 30°$
$ = -20.78 \text{ kNm, anticlockwise.}$

$M_B = 1 \times 0 + 3 \times 6 + 2 \times 0 - 4 \cos 30° \times 6 - 4 \sin 30° \times 9$
$ = 0 + 18 + 0 - 24 \cos 30° - 18$
$ = -24 \cos 30°$
$ = -20.78 \text{ kNm, anticlockwise.}$

$M_C = 2 \times 0 + 3 \times 0 - 1 \times 6 + 4 \cos 30° \times 0 - 4 \sin 30° \times 9$
$ = -6 - 18$
$ = -24 \text{ kNm, anticlockwise.}$

$M_D = 4 \sin 30° \times 0 + 4 \cos 30° \times 0 + 3 \times 0 - 2 \times 9 - 1 \times 6$
$ = -18 - 6$
$ = -24 \text{ kNm, anticlockwise.}$

Example 3.3

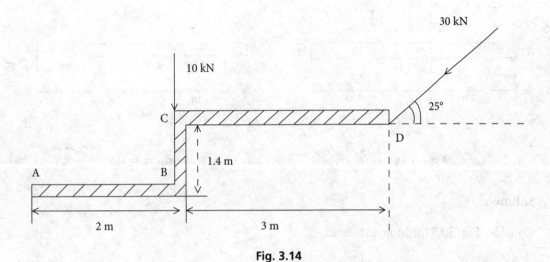

Fig. 3.14

Determine the values of moments about A, B, C and D for force system acting on lever ABCD as shown in Fig. 3.14.

Note: Resolve the force 30 kN acting at D and then take moment.

Solution:

Moment about A, $M_A = +10 \times 2 + 30 \sin 25° \times 5 - 30 \cos 25° \times 1.4$
$\qquad = 45.33$ kNm, clockwise

Here clockwise moment is considered as positive.

Moment about B, $M_B = 10 \times 0 + 30 \sin 25° \times 3 - 30 \cos 25° \times 1.4$
$\qquad = -0.03$ kNm, anticlockwise

Moment about C, $M_C = 10 \times 0 + 30 \sin 25° \times 3 - 30 \cos 25° \times 0$
$\qquad = 90 \sin 25°$
$\qquad = 38.04$ kNm, clockwise.

Moment about D, $M_D = 30 \times 0 - 10 \times 3$
$\qquad = -30$ kNm, anticlockwise

Example 3.4

A point load of 12 kN is acting on a rigid bar as shown in Fig. 3.15. Determine the moment about A and C.

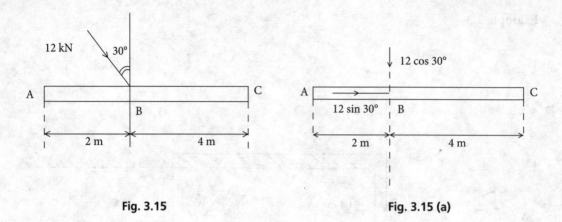

Fig. 3.15 Fig. 3.15 (a)

Solution:

Consider Fig. 3.15 (*a*) Moment about *A*,

$M_A = 12 \cos 30° \times 2 + 12 \sin 30° \times 0$
$\quad = 24 \cos 30°$
$\quad = 20.78 \text{ kNm, clockwise}$
$M_C = -12 \cos 30° \times 4 + 12 \sin 30° \times 0$
$\quad = -48 \cos 30°$
$\quad = -41.57 \text{ kNm, anticlockwise}$

Example 3.5

A 240 kN vertical force is applied at point *A* of a lever by a mechanic to tighten the nut at '*O*' as shown on the shaft Fig. 3.16. Determine.

(i) The smallest force to be applied at the end of lever which produces same moment about '*O*'

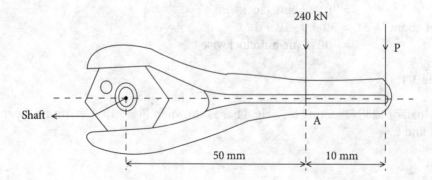

Fig. 3.16

(ii) If 600 kN vertical force is applied, then at what distance from O it will produce same moment about O.

Solution:

(i) Let the smallest force 'P' is applied at the end of a lever.
Moment applied by mechanic at 'O'
$$= 240 \times 50 \text{ kNmm}$$
The moment produced by smallest force P at 'O'
$$= P \times 60 \text{ kNmm}$$
for same moment to produce on 'O',

$$P \times 60 = 240 \times 50$$
$$P = 200 \text{ kN}$$

(ii) If 600 kN force is applied at distance x mm from 'O'

Then moment at O will be $= 600 \times x$ kNmm
Thus, $600 \times x = 240 \times 50$
$$x = 20 \text{ mm from '}O\text{'}.$$

Example 3.6

A parallel coplanar force system is acting on a rigid bar as shown in Fig. 3.17.

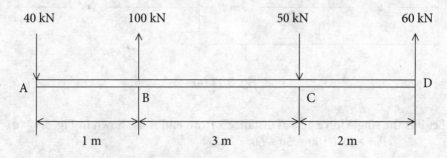

Fig. 3.17

Reduce this force system into:

(i) a single force and couple at B.
(ii) a single force and couple at D.
(iii) a single force.

Solution:

Let upward force is considered as positive and downward as negative.
the algebraic sum of force on bar
$$= -40 + 100 - 50 + 60$$
$$= +70 \text{ kN, upward}$$

(i) Thus single force at B = 70 kN, upward and couple will be
= algebraic sum of moment of all forces about B

Note: Anticlockwise moment is considered as positive while clockwise as negative.
$$= -50 \times 3 + 60 \times 5 + 40 \times 1$$
$$= -150 + 300 + 40$$
$$= +190 \text{ kNm, anticlockwise.}$$

(ii) Single force at 'D' will be = 70 kN upward.
and couple at D will be $= + 50 \times 2 - 100 \times 5 + 40 \times 6$
$$= 100 - 500 + 240$$
$$= -160 \text{ kNm, clockwise.}$$

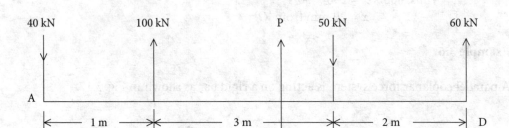

Fig. 3.17 (a)

(iii) Let P is the single force acts at distance x from end A as shown in Fig. 3.17 (a)
$$P = -40 + 100 - 50 + 60 = 70 \text{ kN}$$

Thus using Varignon's theorem, algebraic sum of moment of all forces about A
= moment of single force about A,
$$(40 \times 0 + 100 \times 1 - 50 \times 4 + 60 \times 6) = P \times x$$
$$100 - 200 + 360 = 70 \times x$$
$$x = 3.71 \text{ m}$$

Two Dimensional Non-concurrent Force Systems

Example 3.7

Find the equivalent force-couple system at A for force applied on L shaped plate. Assume width of plate as negligible.

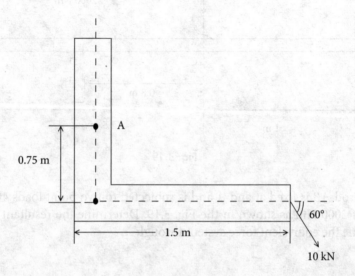

Fig. 3.18

Solution:

Consider Fig. 3.18 (a), the equivalent force-couple system at A will be algebraic sum
   = Moment about A of all forces acting on plate.
   = $+10 \sin 60° \times 1.5 - 10 \cos 60° \times 0.75$
   = 9.24 kNm, clockwise.

Consider Fig. 3.18 (b) shows the equivalent force couple system at A.

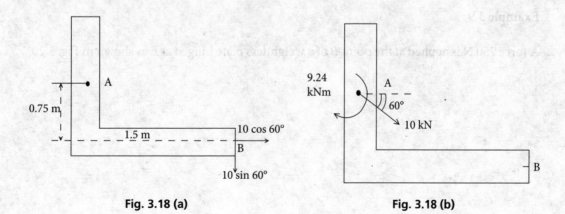

Fig. 3.18 (a)   Fig. 3.18 (b)

Example 3.8

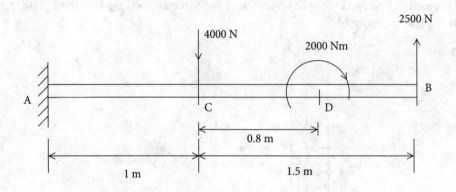

Fig. 3.19

A horizontal rod AB is fixed at and A and is subjected to two point loads 4000 N, 2500 N and couple of 2000 Nm as shown in the Fig. 3.19. Determine the resultant of the system. Also determine the equivalent force system through A.

(UPTU-2001)

Solution:

Total force on the rod = algebraic sum of load
= +4000 − 2500 = 1500 N downward
Total moment on the beam AB about A,
= algebraic sum of moment about A
= 4000 × 1 + 2000 − 2500 × 2.5 = −250 Nm *i.e.,* anticlockwise
The equivalent force system through A will be
 single force at A = 1500 N downward
 and couple at A = 250 Nm, anticlockwise.

Example 3.9

A force 750 N is applied at the point B of a weightless plate hinged at E as shown in Fig. 3.20.

Two Dimensional Non-concurrent Force Systems

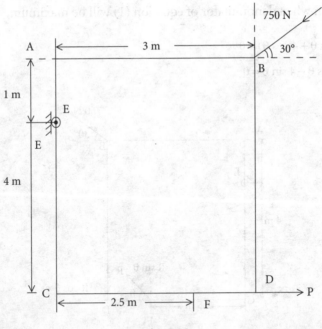

Fig. 3.20

Determine:

(i) the moment of the force about E.
(ii) the horizontal force applied at F by which plate can be held in equilibrium.
(iii) the smallest force required at F to keep the plate in equilibrium.

Solution:

First, Resolve the force 750 N at B.

(i) the moment about $E = 750 \sin 30° \times 3 - 750 \cos 30° \times 1$
$\qquad = +475.48$ Nm, clockwise

(ii) Let P horizontal force is applied at F. The plate will be in equilibrium, when P produces the same amount of moment about E in anticlockwise direction,
Thus moment of P about $E = P \times 4$ anticlockwise
i.e., $\qquad P \times 4 = 475.48$
$\qquad\qquad P = 118.87$ N
in the direction from F to D.

(iii) Let the smallest force required at F is R at an angle θ as shown below in Fig. 3.20 (a)

$$\sum M_E = 0,$$
$$R \sin\theta \times 2.5 + R\cos\theta \times 4 = 475.48$$

$$R = \left(\frac{475.48}{2.5 \sin\theta + 4\cos\theta}\right) \qquad\qquad \ldots\ldots (1)$$

R will be minimum when denominator of equation (1) will be maximum,

Thus $\dfrac{d}{d\theta}$ $(2.5 \sin \theta + 4 \cos \theta) = 0$

$2.5 \cos \theta - 4 \sin \theta = 0$

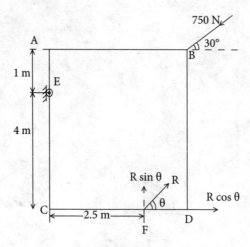

Fig. 3.20 (a)

$\tan \theta = \dfrac{2.5}{4}$

R will be minimum at 32° because denominator is maximum at 32° as second derivative of equation (1) is negative

$\theta = 32°$ substituting in equation (1),

$R = \dfrac{475.48}{2.5 \sin 32° + 4 \cos 32°}$

$R = 100.80$ N

Example 3.10

A square block of wood of mass M is hinged at A and rests on a roller at B. It is pulled by means of a string attached at D and inclined at an angle 30° with the horizontal. Determine the force P which should be applied to the string to just lift the block off the roller Fig. 3.21.

(U.P.T.U. 2004-05)

Solution:

Draw free body diagram of block, as shown below in Fig. 3.21 (a).

When block just lift off the roller, the contact between block and roller breakup. Thus, there is no reaction acting at B i.e., $R_B = 0$

Two Dimensional Non-concurrent Force Systems

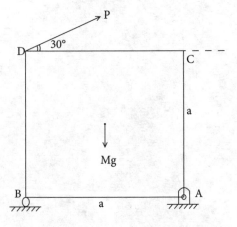

Fig. 3.21

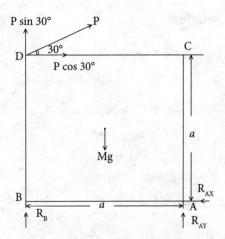

Fig. 3.21 (a) F.B.D. of Block

Since the block is in equilibrium,

$\Sigma M_A = 0$

$Mg \times \dfrac{a}{2} = P\cos 30° \times a + P\sin 30° \times a$

$P = \dfrac{Mg.a}{2a(\cos 30° + \sin 30°)} = 0.366\ Mg$

$P = 0.37\ Mg.$

Example 3.11

The cross section of a block is an equilateral triangle. It is hinged at *A* and rests on a roller at *B*. It is pulled by means of a string attached at *C*. If the weight of the block is *Mg* and the string is horizontal, determine the force *P* which should be applied through string to just lift the block off the roller.

(U.P.T.U. 2002–03)

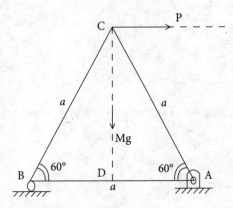

Fig. 3.22

Solution:

Similar to example 3.10, $R_B = 0$

Consider right angle triangle ADC

$$\sin 60° = \frac{CD}{AC}$$
$$\Rightarrow \quad CD = AC \sin 60°$$
$$= a \sin 60°,$$

Taking moment about A,

$$Mg \times \frac{a}{2} = P \times CD$$

$$Mg \times \frac{a}{2} = P \times a \sin 60°$$

$$P = 0.577 \ Mg$$

Example 3.12

Find the magnitude and direction of the resultant of the system of coplanar forces acting on a square lamina as shown in the Fig. 3.23.

(U.P.T.U. 2000–01)

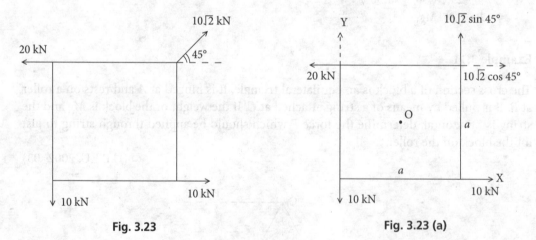

Fig. 3.23   Fig. 3.23 (a)

Solution:

First, determine the algebraic sum of all force in X and Y direction, consider Fig. 3.23 (a)

$$\sum X = +10 + 10\sqrt{2} \cos 45° - 20 = 0$$
$$\sum Y = +10\sqrt{2} \sin 45° - 10 = 0$$

As $\sum X$ and $\sum Y$ both are zero thus

$$R = \sqrt{\sum X^2 + \sum Y^2} = 0$$

Two Dimensional Non-concurrent Force Systems

As the force system is non concurrent and coplanar it may rotate the lamina i.e. $\Sigma M = 0$,

$$\Sigma M_O = +10 \cdot \frac{a}{2} + 10 \cdot \frac{a}{2} - 10\sqrt{2}\cos 45° \times \frac{a}{2} + 10\sqrt{2}\sin 45° \frac{a}{2} + 20 \cdot \frac{a}{2}$$
$$= +20.a \text{ (anticlockwise)}.$$

Example 3.13

Four forces having magnitudes of 20 N, 40 N 60 N and 80 N, respectively, are acting along the four sides (1 m each), of a square *ABCD*, taken in order as shown in the Fig. 3.24. Determine the magnitude, direction and position of the resultant force.

(U.P.T.U., 2005–06)

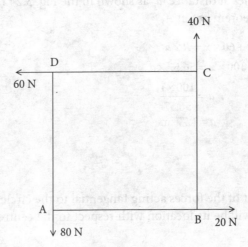

Fig. 3.24

Solution:

Resolve the forces in X and Y direction.

$$\Sigma X = +20 - 60 = -40 \text{ N}$$
$$\Sigma Y = +40 - 80 = -40 \text{ N}$$

Resultant, $R = \sqrt{\Sigma X^2 + \Sigma Y^2} = \sqrt{(-40)^2 + (-40)^2}$

$R = 56.57$ N

$$\tan\theta = \frac{\Sigma Y}{\Sigma X} = \frac{-40}{-40}$$

$\theta = 45°$ or $225°$

As both ΣX and ΣY are negative this means resultant is lying in third quadrant. But as the force system is non-concurrent, their resultant position is not known.

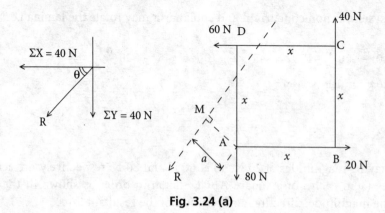

Fig. 3.24 (a)

Let the resultant force lies at distance 'a' as shown in the Fig. 3.24 (a). Applying Varignon's theorem about A,

$$80 \times 0 + 20 \times 0 + 40 \times x + 60 \times x = R \times a$$
$$100 \times x = 56.57 \times a$$

as $x = 1$ m

$$\therefore a = \frac{100 \times 1}{56.57}$$
$$a = 1.77 \text{ m}$$

Example 3.14

Determine the resultant of the forces acting tangential to the circle of radius 3 m as shown in the Fig. 3.25. What will be its location with respect to the centre of the circle? Consider Fig. 3.25 (a).

(U.P.T.U. 2003–04)

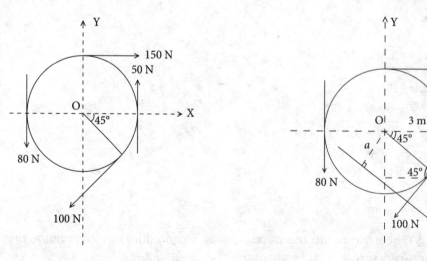

Fig. 3.25   Fig. 3.25 (a)

Solution:

Consider Fig. 3.25 (a),

$\Sigma X = +150 - 100 \cos 45°$
$\quad = 79.29 \text{ N}$

$\Sigma Y = +50 - 100 \sin 45° - 80$
$\quad = -100.71 \text{ N}$

Resultant, $R = \sqrt{\Sigma X^2 + \Sigma Y^2} = \sqrt{(79.29)^2 + (-100.71)^2}$
$\quad R = 128.18 \text{ N}$

$\tan \theta = \dfrac{\Sigma Y}{\Sigma X} = \dfrac{100.71}{79.29}, \quad \theta = 51.79°$

Let Resultant, R is at distance 'a' from center O as shown in Fig. 3.25 (a),

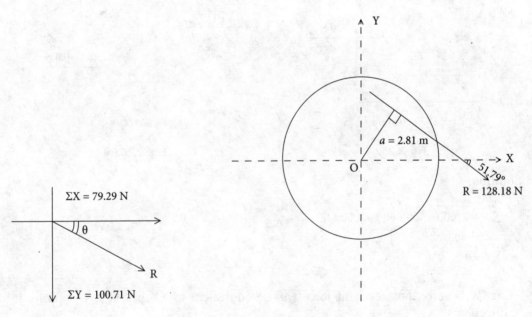

Fig. 3.25 (b)

Applying Varignon's theorem about O,
$+50 \times 3 + 80 \times 3 - 150 \times 3 - 100 \times 3 = R \times a$
$\quad 150 + 240 - 450 - 300 = 128.18 \times a$
$\quad \quad \quad \quad \quad \quad \quad a = -2.81 \text{ m}$

Note:

(i) Anticlockwise moments are considered as positive while clockwise moments as negative.

(ii) –ve sign shows that resultant is acting at distance 2.81 m from O as shown in Fig. 3.25 (b).

Example 3.15

Determine the resultant of the parallel coplanar force system shown in Fig. 3.26.

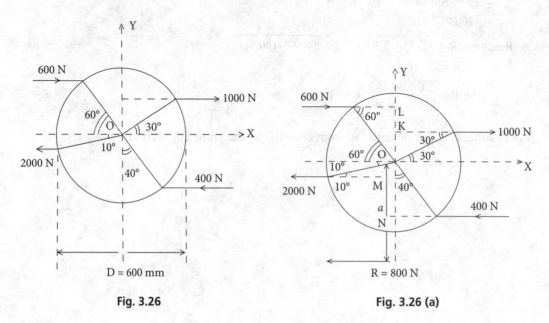

Fig. 3.26   Fig. 3.26 (a)

Solution:

$$\Sigma X = +1000 - 400 + 600 - 2000$$
$$= -800 \text{ N}$$
$$\Sigma Y = 0$$

As ΣY is zero, thus resultant force will be equivalent to ΣX i.e., 800 N opposite in X-direction.

Let R is acting at distance 'a' from the centre as shown in Fig. 3.26 (a).
Applying Varignon's theorem about 'O'.

$$1000 \times OK + 600 \times OL + 2000 \times OM + 400 \times ON = 800 \times a$$
$$1000 \times 300 \sin 30° + 600 \times 300 \sin 60° + 2000 \times 300 \sin 10°$$
$$+ 400 \times 300 \cos 40° = 800 \times a$$
$$a = 627.46 \text{ mm}$$

Example 3.16

Determine the resultant of a non-concurrent coplanar force system acting on grid as shown in the Fig. 3.27. Take each side of the grid as 6 mm.

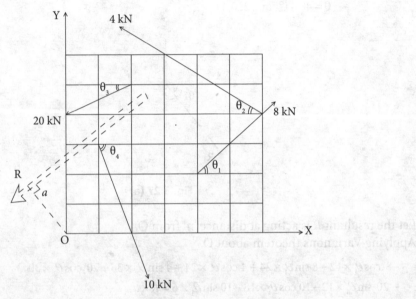

Fig. 3.27

Solution:

Let the forces 8 kN, 4 kN, 20 kN and 10 kN are making angles θ_1, θ_2, θ_3, and θ_4 respectively from horizontal line.

thus, $\tan \theta_1 = \dfrac{12}{12}$, $\tan \theta_2 = \dfrac{12}{18}$

$\tan \theta_1 = 1$, $\tan \theta_2 = \dfrac{2}{3}$

$\theta_1 = 45°$, $\theta_2 = 33.69°$

$\tan \theta_3 = \dfrac{6}{12}$, $\tan \theta_4 = \dfrac{18}{6}$

$\theta_3 = 26.57°$, $\theta_4 = 71.57°$

$\Sigma X = +8 \cos \theta_1 - 4 \cos \theta_2 - 20 \cos \theta_3 + 10 \cos \theta_4$

$= +8 \cos 45° - 4 \cos 33.69° - 20 \cos 26.57° + 10 \cos 71.57°$

$\Sigma X = -12.39 \text{ kN}.$

$\Sigma Y = +8 \sin \theta_1 + 4 \sin \theta_2 - 20 \sin \theta_3 - 10 \sin \theta_4$

$= 8 \sin 45° + 4 \sin 33.69° - 20 \sin 26.57° - 10 \sin 71.57°$

$= -10.56 \text{ kN}.$

$$R = \sqrt{\Sigma X^2 + \Sigma Y^2} = \sqrt{(-12.39)^2 + (-10.56)^2}$$
$$R = 16.28 \text{ kN.}$$
$$\tan\theta = \frac{\Sigma Y}{\Sigma X} = \frac{-10.56}{-12.39}$$
$$\theta = 40.44° \text{ or } 220.44°$$

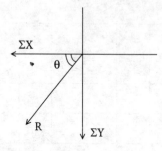

Fig. 3.27 (a)

Let the resultant R is acting at distance 'a' from O.
Applying Varignon's theorem about O.

$-8\ \cos\theta_1 \times 12 + 8\ \sin\theta_1 \times 24 + 4\ \cos\theta_2 \times 24 + 4\ \sin\theta_2 \times 36 + 20\ \cos\theta_3 \times 30$
$- 20\ \sin\theta_3 \times 12 - 10\ \cos\theta_4 \times 18 - 10\ \sin\theta_4 \times 6 = R \times a$
$-(8\ \cos 45° \times 12) + (8\ \sin 45° \times 24) + (4\ \cos 33.69° \times 24) + (4\ \sin 33.69° \times 36)$
$+ (20\ \cos 26.57° \times 30) - (20\ \sin 26.57° \times 12) - (10\ \cos 71.57° \times 18) - (10\ \sin 71.57°) \times 6$
$= 16.28 \times a$

$a = 33.36$ mm as shown in the Fig. 3.27

Example 3.17

A bar AB of weight 180 kN is hinged at A and held in equilibrium by force 'P' as shown is the Fig. 3.28. Determine the force P and the hinge reaction at A for equilibrium position of the bar AB. Take length of the bar as 10 m.

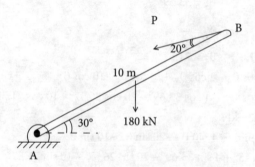

Fig. 3.28

Two Dimensional Non-concurrent Force Systems

Solution:

The *FBD* of bar *AB* Fig. 3.28 (*a*) will be,
$AM = BD$ and $BD = 10 \sin 30°$, $AC = 5 \cos 30°$ and $AD = 10 \cos 30°$
As the bar is in equilibrium, $\Sigma M_A = 0$

$$P \cos 10° \times AM = 180 \times AC + P \sin 10 \times AD$$
$$P \cos 10° \times 10 \sin 30° = 180 \times 5 \cos 30° + P \sin 10° \times 10 \cos 30°$$
$$P = 227.89 \text{ kN}$$

For hinge reaction,
$\Sigma X = 0$, $R_{AX} = P \cos 10° = 227.89 \cos 10° = 224.43$ kN

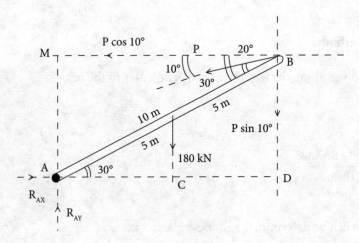

Fig. 3.28 (a)

$\Sigma Y = 0$, $\quad R_{AY} = 180 + P \sin 10°$
$\qquad\qquad\qquad = 180 + 227.89 \times \sin 10°$
$\qquad\qquad\qquad = 219.57$ kN

$\qquad\qquad R_A = \sqrt{R^2_{AH} + R^2_{AV}}$
$\qquad\qquad\quad = \sqrt{(224.43)^2 + (219.57)^2}$, $\quad R_A = 313.97$ kN

$\qquad\qquad \tan \theta = \dfrac{\Sigma Y}{\Sigma X}$ or $\dfrac{R_{Ay}}{R_{Ax}} = \dfrac{219.57}{224.43}$

$\qquad\qquad \theta = 44.37°$

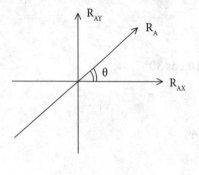
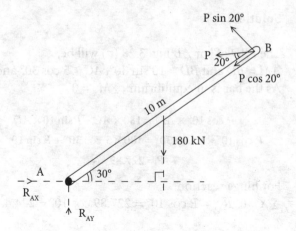

Fig. 3.28 (b)

Alternative method:

P could be resolved along the length and perpendicular to the bar,

$\Sigma M_A = 0$,
$P \cos 20° \times 0 + P \sin 20° \times 10 = 180 \times 5 \cos 30°$

$$P = \frac{180 \times 5 \cos 30°}{10 \sin 20°}$$
$P = 227.89 \ kN$

Hinge reaction can be determined as done earlier.

Example 3.18

A Crane is pivoted at 'O' and supported with the help of a guide at C via tie member as shown in Fig. 3.29. Determine tension in the member BC and reaction at 'O' when it supports a load 240 kN at D as shown. Take weight of the member OCD as negligible.

Solution:

The free body diagram of the member OCD will be:
 Here reaction of the pivoted end O will pass through E because, a body can be held in equilibrium by three forces if all forces are concurrent.

Two Dimensional Non-concurrent Force Systems

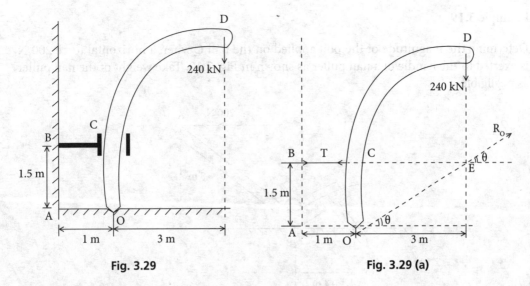

Fig. 3.29   Fig. 3.29 (a)

Thus using Principle of transmissibility, forces at C and D meet at E and hence R_o passes through there and let it makes angle 'θ' from horizontal.

Thus, $\tan\theta = \dfrac{1.5}{3}$, $\theta = 26.57°$

The figure can be further focused about point E as shown in Fig. 3.29 (b) which can be further solved by using Lami's theorem or by resolution method.

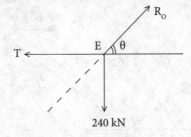

Fig. 3.29 (b)

Lami's theorem

$$\frac{T}{\sin(90°+\theta)} = \frac{R_o}{\sin(90°)} = \frac{240}{\sin(180°-\theta)}$$

$$T = \frac{240\cos\theta}{\sin\theta} = \frac{240}{\tan\theta} = \frac{240}{0.5}$$

$$T = 480 \text{ kN}$$

$$R_o = \frac{240}{\sin\theta} = \frac{240}{\sin(26.57°)}$$

$$R_o = 536.56 \text{ kN}.$$

Example 3.19

Determine the magnitude of the pull applied on the nail C when a horizontal force 200 N is exerted to the handle of a nail puller as shown in Fig. 3.30. Take weight of the nail puller as negligible.

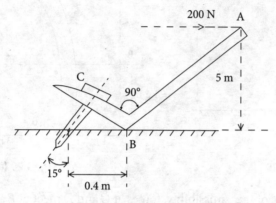

Fig. 3.30

Solution:

Consider F.B.D. of nail puller as shown in Fig. 3.30 (a)
  Let nail is applying force R_C on the nail puller.
  Taking moment about B,

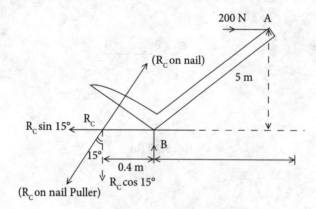

Fig. 3.30 (a) F.B.D. of Nail puller

$$M_B = 0$$
$$R_C \cos 15° \times 0.4 = 200 \times 5$$
$$R_C = 2588.19 \text{ N}$$

Thus pull applied on nail will be

$$R_C = 2588.19 \text{ N}.$$

Example 3.20

Determine the forces exerted on the cylinder at A and B by the spanner wrench due to vertical forces applied at C. Take friction as negligible at B. Consider weight of spanner wrench also as negligible.

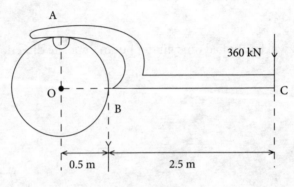

Fig. 3.31

Solution:

As the applied force is acting on spanner wrench i.e., 360 kN, thus considering free body diagram of the same as shown in Fig. 3.31 (*a*)

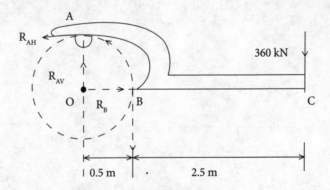

Fig. 3.31 (a) F.B.D. of Spanner Wrench

Using equations,

$\Sigma X = 0$,
$R_{AH} = R_B$ ..... (1)
$\Sigma Y = 0$,
$R_{AV} = 360$ kN

$\Sigma M_A = 0$,
$R_B \times 0.5 = 360 \times 3$
$\quad R_B = 2160$ kN $\qquad$ Thus from equation (1),
$\quad R_{AH} = 2160$ kN
$\therefore \quad R_A = \sqrt{R^2_{AH} + R^2_{AV}} = \sqrt{(2160)^2 + (360)^2}$
$\quad R_A = 2189.79$ kN

Thus these reactions will be applied on cylinder but in opposite directions as shown below

$R_B = 2160$ kN
$R_A = 2189.79$ kN

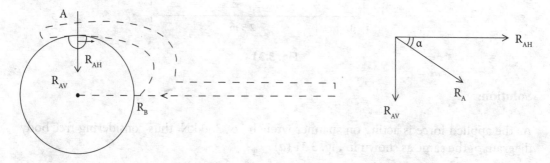

Fig. 3.31 (b) F.B.D. of Cylinder

$\tan \alpha = \dfrac{R_{AV}}{R_{AH}}$

$\quad = \dfrac{360}{2160}$

$\alpha = 9.46°$ $\qquad$ in fourth quadrant as shown above.

Example 3.21

A roller shown in Fig. 3.32 is of mass 150 kg. What force P is necessary to start the roller over the block A?

(UPTU, 2008–09)

Two Dimensional Non-concurrent Force Systems

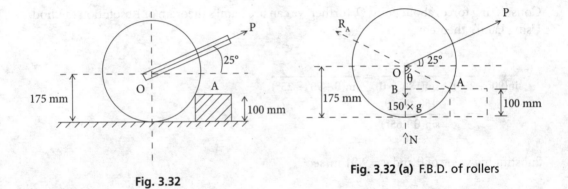

Fig. 3.32

Fig. 3.32 (a) F.B.D. of rollers

Solution:

Consider the free body diagram of roller, shown in Fig. 3.32 (a). The forces acting on roller will be P, N, weight of roller and reaction of block (R_A) when roller tends to move over the block, its contact from the floor terminates and thus reaction of the floor also disappears

i.e., $N = 0$

Thus only three forces kept the roller in equilibrium thus all forces have to be concurrent, that's why reaction of block R_A passes through the center of the roller.

Let $\angle AOB = \theta$, In $\triangle AOB$,

$$\cos\theta = \frac{OB}{OA} = \frac{175-100}{175}$$

$$\theta = 64.62°$$

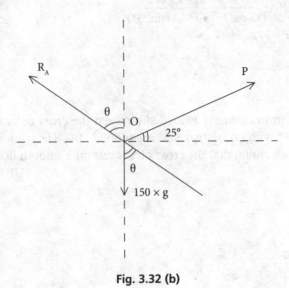

Fig. 3.32 (b)

Considering forces about point O, further we can use Lami's theorem or Resolution method. Using Lami's theorem,

$$\frac{R_A}{\sin(90°+25°)} = \frac{P}{\sin(180°-\theta)} = \frac{150 \times g}{\sin(\theta+90°-25°)}$$

$$P = \frac{150g \sin\theta}{\sin(\theta+65°)}$$

Substituting value of θ and g = 9.81 m/sec²

$$P = \frac{150g \sin 64.62°}{\sin(64.62°+65°)}$$
$$P = 1725.94 \text{ N}$$

Using Moment method,

Taking moment about A, as we need only P,

$P.\cos 25° \times OB + P.\sin 25° \times AB = 150 \times 9.81 \times AB$
$P[75.\cos 25° + 158.11 \sin 25°] = 150 \times 9.81 \times 158.11$
$OB = 175 - 100$
$\quad = 75$ mm
$AB = \sqrt{175^2 - 75^2}$
$\quad = 158.11$ mm

$$P = \frac{150 \times 9.81 \times 158.11}{(75.\cos 25° + 158.11 \sin 25°)}$$
$$P = 1726.04 \text{ N}$$

Example 3.22

A 500 N cylinder, 1 m in diameter is loaded between the cross pieces AE and BD which makes an angle of 60° with each other and are pinned at C. Determine the tension in the horizontal rope DE, assuming that the cross pieces rest on a smooth floor.

(UPTU I Sem, 2001–02)

Two Dimensional Non-concurrent Force Systems

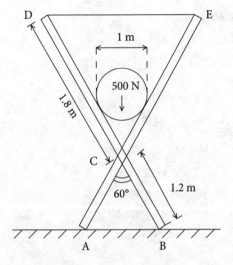

Fig. 3.33

Solution:

Let the reaction at A and B is N_A and N_B, respectively
Thus, if total setup is considered for equilibrium
Then, $\Sigma Y = 0$,
    $N_A + N_B = 500$ N
As the setup is symmetrical thus $N_A = N_B$.
$\therefore$ $N_A = N_B = 250$ N
Considering FBD of cylinder as shown in Fig. 3.33 (a)

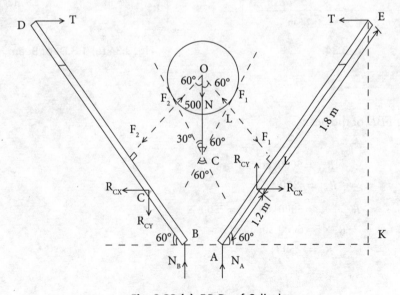

Fig. 3.33 (a) F.B.D. of Cylinder

$\Sigma Y = 0$, $F_1 \cos 60° + F_2 \cos 60° = 500$
$\Sigma X = 0$, $F_1 \sin 60° = F_2 \sin 60°$
i.e., $F_1 = F_2$
$$2F_1 \cos 60° = 500$$
$$F_1 = F_2 = 500 \text{ N}$$
$\Sigma M_C = 0$, consider FBD of AE of Fig. 3.33 (a),

ΔOLC, $\tan 60° = \dfrac{CL}{0.5}$, $CL = 0.5 \tan 60°$

$$N_A \times 1.2 \cos 60° + F_1 \times CL = T \times CE \sin 60°$$
$$N_A \times 1.2 \cos 60° + F_1 \times 0.5 \tan 60° = T \times 1.8 \sin 60°$$
$$250 \times 1.2 \cos 60° + 500 \times 0.5 + 60° = T \times 1.8 \sin 60°$$
$$T = 374 \text{ N}$$

Example 3.23

A force $P = 5000$ N is applied at the center C of the beam AB of length 5 m as shown in the figure. Find the reactions of the hinge and roller support bearing.

(UPTU 2002)

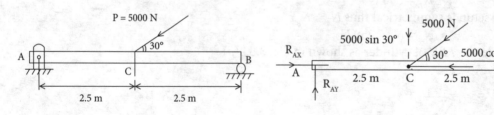

Fig. 3.34 **Fig. 3.34 (a)** F.B.D. of Beam

Solution:

Considering FBD of the beam

$\Sigma X = 0$, $R_{AX} = 5000 \cos 30°$
$\qquad R_{AX} = 4330.12$ N ..... (1)

$\Sigma Y = 0$, $R_{AY} + R_B = 5000 \sin 30° = 2500$ N
$\Sigma M_A = 0$, $R_B \times 5 = 5000 \sin 30° \times 2.5$
$\qquad R_B = 1250$ N

Two Dimensional Non-concurrent Force Systems

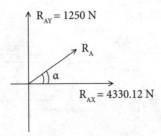

Fig. 3.34 (c)

from equation (1), $R_{AY} = 2500 - 1250 = 1250$ N

$$R_A = \sqrt{R^2_{AH} + R^2_{AV}}$$
$$= \sqrt{(4330.12)^2 + (1250)^2}$$
$$R_A = 4506.93 \text{ N}$$

$$\tan \alpha = \frac{R_{AY}}{R_{AX}} = \frac{1250}{4330.12}$$
$$\alpha = 16.10°$$

Example 3.24

A beam 8 m long is hinged at A and supported on rollers over a smooth surface inclined at 30° to the horizontal at B as shown in Fig. 3.35. Determine support reactions.

(UPTU 2003)

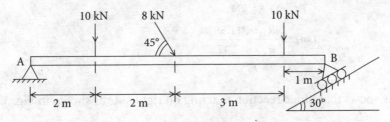

Fig. 3.35

Solution:

Consider *FBD* of beam as showing Fig. 3.35 (*a*)

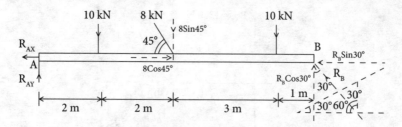

Fig. 3.35 (a) F.B.D. of Beam

$\Sigma Y = 0$, $R_{AY} + R_B \cdot \cos 30° = 10 + 8 \sin 45° + 10$

$\qquad\qquad\qquad = 20 + 4\sqrt{2}$ ..... (1)

$\Sigma M_A = 0$, $R_B \cos 30° \times 8 = 10 \times 7 + 8 \sin 45° \times 4 + 10 \times 2$

$\qquad\qquad R_B = 16.25$ kN

Substituting value of R_B in equation (1),

$\qquad\qquad R_{AY} = 11.58$ kN

$\Sigma X = 0$, $R_{AX} + R_B \cdot \sin 30° = 8 \cos 45°$

$\qquad R_{AX} + 16.25 \sin 30° = 8 \cos 45°$

$\qquad\qquad R_{AX} = -2.47$ kN

$\qquad\qquad R_A = \sqrt{R_{AX}^2 + R_{AY}^2}$

$\qquad\qquad\quad = \sqrt{(-2.47)^2 + (11.58)^2}$

$\qquad\qquad R_A = 11.84$ kN

$\qquad\qquad \tan \alpha = \dfrac{R_{AY}}{R_{AX}} = \dfrac{11.58}{2.47}$

$\qquad\qquad \alpha = 77.96°$

negative value shows that hinge reaction is acting on right side as shown in Fig. 3.35(b)

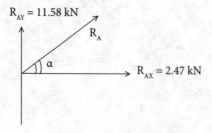

Fig. 3.35 (b)

Example 3.25

Two weights $C = 2000$ N and $D = 1000$ N are located on a horizontal beam AB as shown in Fig. 3.36. Find the distance of weight C from support A, i.e., x so that the support reaction at A is twice that at B.

(UPTU-2001–01 IInd Sem.)

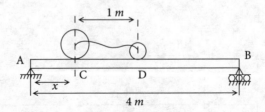

Fig. 3.36

Solution:

First, draw the F.B.D. of the beam AB as shown below in Fig. 3.36 (a)

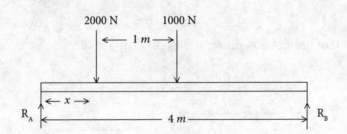

Fig. 3.36 (a) F.B.D. of Beam

Given that, $\quad R_A = 2.\ R_B$

$\Sigma Y = 0,\quad R_A + R_B = 2000 + 1000$

$\qquad 2R_B + R_B = 3000$

$\qquad R_B = 1000$ N

$\qquad R_A = 2000$ N

$\Sigma M_A = 0,\quad R_B \times 4 = 2000 \times x + 1000\ (x+1)$

$\qquad 1000 \times 4 = 2000x + 1000x + 1000$

$\qquad x = 1\text{m}$

Example 3.26

Two rollers of mass 20 kg and 10 kg rest on a horizontal beam as shown in the Fig. 3.37 with a massless wire fixing the two centers. Determine the distance x of the load 20 kg from the support A, if the reaction at A is twice of that at B. The length of the beam is 2 m and the length of the connecting wire is 0.5 m. Neglect weight of beam. Assume the rollers to be point masses neglecting its dimensions.

(UPTU-2008)

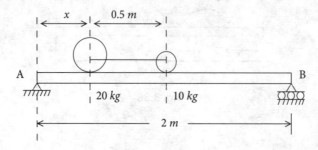

Fig. 3.37

Solution:

Considering FBD of beam AB as shown in Fig. 3.37 (a)

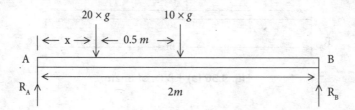

Fig. 3.37 (a) F.B.D. of Beam

..... (1)

$\Sigma Y = 0$,
$$R_A + R_B = 20 \times g + 10 \times g$$
$$= 30 \times g$$
$$R_A + R_B = 30 \times 9.81$$

Given that $\quad R_A = 2R_B$

$\therefore \quad 2R_B + R_B = 30 \times 9.81$

$\quad R_B = 98.1$ N and $R_A = 196.2$ N

$\Sigma M_A = 0$,

$$R_B \times 2 = 20 \times g \times x + 10 \times g(x + 0.5)$$
$$98.1 \times 2 = 20 \times g \times x + 10g \times x + 10g \times 0.5$$

$$20 = 20x + 10x + 5$$
$$15 = 30x$$
$$x = 0.5 \ m$$

Example 3.27

Two rollers C and D produce vertical forces P and Q on the horizontal beam AB as shown in the Fig. 3.38. Determine the distance 'x' of the load P from support A if the reaction at A is twice as great as the reaction at B. The weight of beam is to be neglected.

(UPTU 2008–09 II Sem)

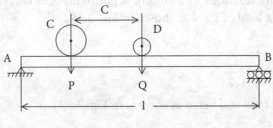

Fig. 3.38

Given:

$P = 18$ kN, $Q = 9$ kN, $l = 3.6$ m and $C = 0.9$ m

Solution:

Consider FBD of beam AB, as shown in Fig. 3.38 (a)

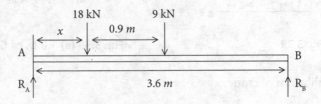

Fig. 3.38 (a) F.B.D. of Beam

$\Sigma Y = 0$, $\quad R_A + R_B = 18 + 9$
$\quad\quad\quad\quad\quad = 27$ kN ...(1) and given that $R_A = 2R_B$

thus $\quad 2R_B + R_B = 27$, $R_B = 9$ kN, $R_A = 18$ kN

$\Sigma M_A = 0$, $\quad R_B \times 3.6 = 18 \times x + 9(x + 0.9)$

$\quad\quad\quad\quad 9 \times 3.6 = 18x + 9x + 8.1$

$\quad\quad\quad\quad 32.4 = 18x + 9x + 8.1$

$\quad\quad\quad\quad 27x = 24.3$

$\quad\quad\quad\quad x = 0.9\ m$

Example 3.28

Determine the support reactions of a simply supported beam carrying a crane of weight 5 kN and supporting a load of 1 kN as shown in Fig. 3.39.

(UPTU-2000-01)

Solution:

Considering FBD of crane and beam as shown, in Fig. 3.39 (a)

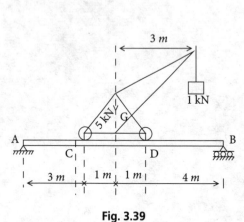

Fig. 3.39

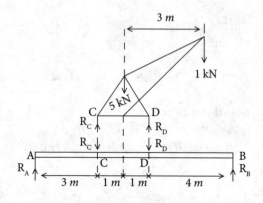

Fig. 3.39 (a) F.B.D. of Crane and Beam

Considering equilibrium of crane,

$\Sigma Y = 0$, $\quad R_C + R_D = 5 + 1$

$\quad\quad\quad\quad R_C + R_D = 6$ kN $\quad\quad\quad\quad\quad\quad\quad$ ..... (1)

$\Sigma M_C = 0$, $\quad R_D \times 2 = 5 \times 1 + 1 \times 4 = 9$

$\quad\quad\quad\quad R_D = 4.5$ kN

Substituting the value of R_D in equation (1),

$$R_C = 6 - R_D = 6 - 4.5$$
$$R_C = 1.5 \text{ kN}$$

Considering equilibrium of beam AB

$\Sigma Y = 0$, $\qquad R_A + R_B = R_C + R_D$
$\qquad\qquad\qquad\qquad = 1.5 + 4.5 = 6 \text{ kN}$ ..... (2)

$\Sigma M_A = 0$, $\qquad R_B \times 9 = 1.5 \times 3 + 4.5 \times 5$
$\qquad\qquad\qquad\qquad = 27$
$\qquad\qquad\qquad R_B = 3 \text{ kN}$
$\qquad\qquad\qquad R_A = 3 \text{ kN}.$

Example 3.29

Determine support reactions at A and B for beam given below in Fig. 3.40.

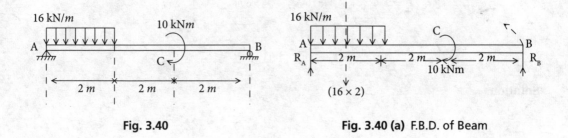

Fig. 3.40 　　　　　　　　　　　　Fig. 3.40 (a) F.B.D. of Beam

Solution:

$\Sigma Y = 0$, $\qquad R_A + R_B = 16 \times 2 \text{ (kN/m} \times \text{m)}$
$\qquad\qquad\qquad\qquad = 32 \text{ kN}$ ..... (1)

Note:

(i) U.D.L. of 16 kN/m acts at mid of its length and load is given by multiplying its intensity to the length it is spread or covered.
(ii) In ΣX or ΣY equations, couple never affects.

$\Sigma M_A = 0$, $\qquad R_B \times 6 = 10 + (16 \times 2) \times 1 = 42$
$\qquad\qquad\qquad R_B = 7 \text{ kN}$

Substituting in equation (1),

$R_A = 32 - R_B = 32 - 7$
$R_A = 25$ kN

Example 3.30

Determine reactions of hinge and roller bearing for the given Fig. 3.41.

(UPTU-2007)

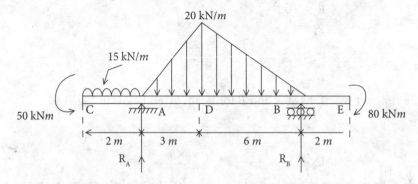

Fig. 3.41

Solution:

$\Sigma Y = 0$,

$R_A + R_B$ = U.D.L. on AC + U.V.L. on AD + U.V.L. on DB portion

$$= (15 \times 2) + \left(\frac{1}{2} \times 3 \times 20\right) + \left(\frac{1}{2} \times 6 \times 20\right)$$

$$= 30 + 30 + 60$$

$R_A + R_B = 120$ ..... (1)

$\Sigma M_A = 0$,

$R_B \times 9 + (15 \times 2) \times 1 + 50 = 80 + \left(\frac{1}{2} \times 6 \times 20\right) \times \left(3 + \frac{1}{3} \times 6\right) + \left(\frac{1}{2} \times 3 \times 20\right)\left(\frac{2}{3} \times 3\right)$

$9R_B + 80 = 80 + 60 \times 5 + 30 \times 2$

$9 R_B = 300 + 60 = 360$

$R_B = 40$ kN

Substituting value of R_B in equation (1), $R_A = 80$ kN

Example 3.31

Determine the reactions at A, B and D of the system as shown in Fig. 3.42.

(UPTU 2001–02 Ist sem)

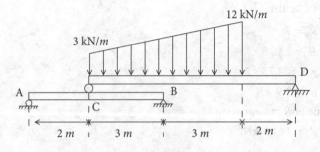

Fig. 3.42

Solution:

First draw the F.B.D. of both beams and apply equilibrium equations to beam CD and AB respectively as shown in Fig. 3.42 (a).

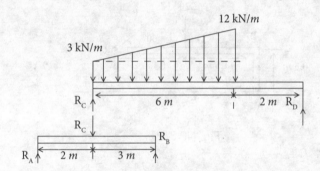

Fig. 3.42 (a) F.B.D. of Beams

Load acting on beam CD is a combination of U.D.L and U.V.L.

thus, $\Sigma Y = 0$,

$$R_C + R_D = (3 \times 6) + \left(\frac{1}{2} \times 6 \times (12-3)\right)$$

$$= 18 + \frac{1}{2} \times 6 \times 9$$

$$= 45$$

$$\Sigma M_C = 0, \quad R_D \times 8 = (3 \times 6) \times \left(\frac{6}{2}\right) + \left(\frac{1}{2} \times 6 \times 9\right) \times \left(\frac{2}{3} \times 6\right)$$

$$R_D = 20.25 \text{ kN}$$

Substituting value of R_D in equation (1),

$$R_C = 45 - 20.25$$
$$R_C = 24.75 \text{ kN}$$

Considering FBD of beam AB,

$\Sigma Y = 0$, $\quad R_A + R_B = R_C$
$\qquad R_A + R_B = 24.75$ ..... (2)
$\Sigma M_A = 0$, $\quad R_B \times 5 = 24.75 \times 2$
$\qquad R_B = 9.9 \text{ kN}$

Substituting value of R_B in equation (2),

$$R_A = 14.85 \text{ kN}.$$

Numerical Problems

N 3.1 A lever of negligible weight is hinged at A and supports a load of 220 kN as shown in Fig. NP 3.1 Determine value of force P by which lever can be held in equilibrium.

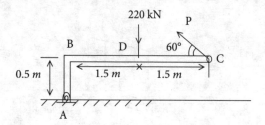

Fig. NP 3.1

N 3.2 A rigid bar of negligible weight experiences a parallel coplanar force system as shown in Fig. NP 3.2. Determine the equivalent reduced force system:
A. a single force
B. a sing force and couple at A
C. a single force and couple at C

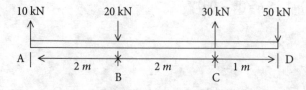

Fig. NP 3.2

N 3.3 A rectangular gate of weight 600 kN is supported over a hinge and roller support bearing as shown in Fig. NP 3.3. Determine minimum pull, P by which gate just lift off the roller bearing.

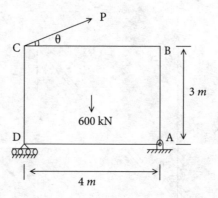

Fig. NP 3.3

N 3.4 Four forces are acting along sides of a square as shown in Fig. NP 3.4. Determine their resultant's magnitude, direction and position.

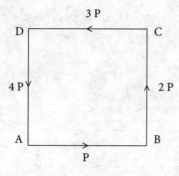

Fig. NP 3.4

N 3.5 A coplanar force system is acting on an equilateral triangular lamina as in Fig. NP 3.5. Determine resultant's magnitude, direction and position.

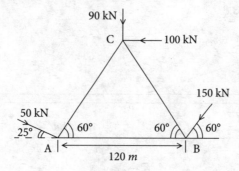

Fig. NP 3.5

N 3.6 A rod AB of weight 50 kN is held in equilibrium with horizontal string BC and wall. If 10 kN is supported at end B as shown in Fig. NP 3.6., determine tension in the string and reaction of wall.

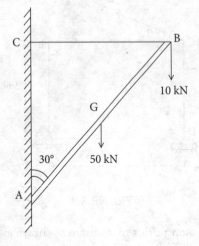

Fig. NP 3.6

N 3.7 Determine reaction of hinge and roller support bearing of simple supported beam as shown in Fig. NP 3.7.

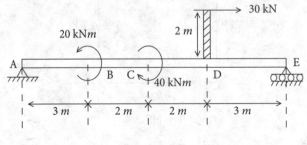

Fig. NP 3.7

N 3.8 Determine reaction of hinge and roller support bearing of simple supported beam as shown in Fig. NP 3.8.

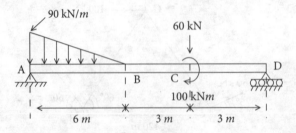

Fig. NP 3.8

Two Dimensional Non-concurrent Force Systems

N 3.9 Determine reaction of hinge and roller support bearing of over hanging beam as shown in Fig. NP 3.9.

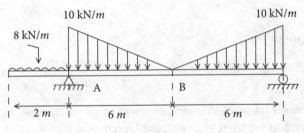

Fig. NP 3.9

N 3.10 Determine reaction of hinge and roller support bearing of simple supported beam as shown in Fig. NP 3.10.

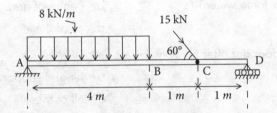

Fig. NP 3.10

N 3.11 Two beams AD and EG are supported as shown in Fig. NP 3.11. Determine reactions at A, C, E and G.

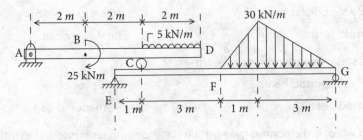

Fig. NP 3.11

Multiple Choice Questions

1. In a non-concurrent coplanar force system, the equilibrium equations are given by
   a. $\sum X = 0, \sum Y = 0$ and $\sum M = 0$
   b. $\sum X = 0$ and $\sum Y = 0$
   c. $\sum X = 0$ or $\sum Y = 0$
   d. none of these

2. The moment of a force passing through the centre of moment will be
   a. maximum
   b. minimum
   c. zero
   d. none of these

3. Select the vector quantity
   a. force
   b. moment
   c. couple
   d. all of these

4. In graphical presentation of moment, moment is given by how much area of triangle formed by centre of moment and vertices of force
   a. half area
   b. equal to area
   c. twice area
   d. none of these

5. Varignon's Theorem is used for
   a. non-coplanar force system
   b. centre of gravity of a body
   c. centroid
   d. all of these

6. Which support does not offer horizontal reaction
   a. roller support
   b. hinged support
   c. fixed support
   d. all of these

7. The beam whose one end remains fixed and other end remains free, is called as
   a. simply supported beam
   b. cantilever beam
   c. propped beam
   d. overhanging beam

8. The beam which is supported by more than two supports, is called as
   a. simply supported beam
   b. continuous beam
   c. propped beam
   d. cantilever beam

9. Which is statically indeterminate beam
   a. simply supported beam
   b. cantilever beam
   c. propped beam
   d. overhanging beam

10. The load which is distributed over certain distance with uniform varying intensity is called as
    a. UVL
    b. UDL
    c. concentrated load
    d. all of these

Answers

1. a 2. c 3. d 4. c 5. d 6. a 7. b 8. b 9. c 10. a

Chapter 4

Friction

4.1 Introduction

So far we have studied the forces (action and reaction) acting between contact surfaces which were considered (idealized) as smooth surfaces. However, in actual practice tangential forces also act along rough surfaces. These tangential forces are called as friction forces between rough surfaces. The friction forces always act against the driving force or opposite to the applied force. Thus it can be treated as resistive forces always act opposite to the direction of applied force.

Generally there are three causes for the existence of friction forces. However out of these three causes, first one plays major role for friction. These causes are as follows:

a. The locking of irregularities of meeting surfaces. The microscopic examinations of all surfaces show irregularities i.e., continuous combination of crests and valleys. When a body is moved over other, these irregularities get interlocked and produce a resistive force called as friction force.
b. Adhesion between surfaces. This might takes place due to molecular attraction. Two smooth pieces of glass can stick together if wrung together.
c. Indentation of harder body over soft body. A resistance is experienced by the harder body when it comes out of very small area like crater formed on soft body.

Friction plays a vital role in various engineering components and day to day life activities. In some situations like motion of car (air friction blows over car body tends to reduce mileage), rotation of all type of bearing, gears, power screws and propulsion of rockets and missiles; friction tends to consume the power or reduce the applied efforts which shows friction is undesirable. However in some other areas, the maximum friction is desirable as it helps to perform functions like belt drive, brakes and clutches, rotation of wheel and simply walking of a person on road or ground.

4.2 Coulomb's Laws of Dry Friction

The rules of dry or coulomb friction were laid down from extensive experimental work of Charles A. Coulomb in 1781 and from work extended by Morin from 1831 to 1834:

a. The maximum frictional force produced is proportional to the normal reaction between two meeting surfaces i.e., mathematically,

   $F \alpha N$

   $F = \mu N$ ..... (4.1)

   where F is maximum frictional force, N is the normal reaction between two meeting surfaces and μ is the coefficient of static friction.

b. The maximum frictional force produced does not depend on the magnitude of area of contact.
c. The maximum frictional force produced depends upon the roughness value of the surfaces.
d. The frictional force produced between sliding bodies does not depend on speed of sliding bodies.

4.3 Static Friction, Limiting Friction, Kinetic Friction

Consider a solid body of weight W lying on a rough horizontal surface subjected to a force P as shown in Fig. 4.1. If applied force P is increased gradually starting from zero then it is observed that resistive force (friction) also increased proportionally up to a certain stage as shown in Fig. 4.2. The value of friction force up to this stage is called as **Static friction**. However, the maximum value of static friction where body tends to move or impend is called as **Limiting friction**. If force P is increased further, the body comes under motion and the value of friction reduces to some extent and then becomes almost constant. This value of friction force is called as **Kinetic friction**.

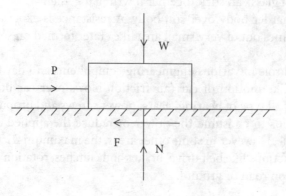

Fig. 4.1

Friction

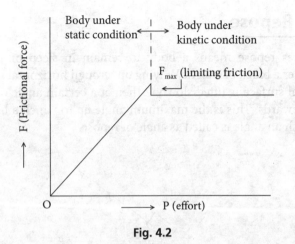

Fig. 4.2

4.4 Angle of Friction

Consider a block of weight W lying on a rough horizontal surface subjected to a force P as shown in Fig. 4.3. Where F is the maximum friction force and N is the normal reaction of ground on the block. If applied force P is increased such that body tends to move or impends than maximum frictional force is induced. If normal reaction N of ground on block and maximum friction force F is replaced by resultant reaction R than the angle made by resultant reaction R with the normal reaction N is called as **angle of friction**. Generally it is denoted by ϕ.

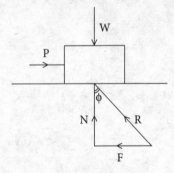

Fig. 4.3

From Fig. 4.3,

$$\tan \phi = \frac{F}{N}$$

We know that $F = \mu N$

Thus $\tan \phi = \dfrac{\mu N}{N}$

i.e., $\tan \phi = \mu$ ..... (4.2)

4.5 Angle of Repose

As dictionary defines repose means a body to remain in sleep or rest i.e., a state of equilibrium. Consider a block of weight W lying on a rough horizontal surface as shown in Fig. 4.4. If horizontal surface is tilted slowly-2 then at a certain angle of inclination, body tends to move downwards. This is the maximum angle up to which a body can rest on an inclined surface. Such an angle is called as angle of repose.

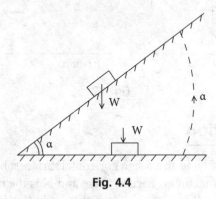

Fig. 4.4

Please note that no external force is acting on the body. Let the horizontal surface is tilted by angle α at which body tends to move downwards. The free body diagram shows weight W, normal reaction N of inclined surface and upward limiting friction F are acting on body as shown in Fig. 4.5.

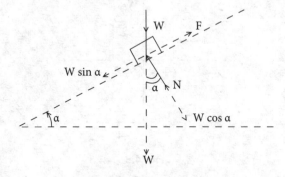

Fig. 4.5 F.B.D. of Block

As the body is in equilibrium,
 $\Sigma X = 0$ i.e., algebraic sum of all forces along the inclined surface must be zero
 i.e., $W \sin \alpha = F$
 $\Sigma Y = 0$ i.e., algebraic sum of all forces normal to the inclined surface must be zero
 i.e., $W \cos \alpha = N$

or, $\dfrac{W \sin \alpha}{W \cos \alpha} = \dfrac{F}{N} = \mu$

Friction

i.e., tan α = μ = tan φ from equation (4.2)
i.e., α = φ
Thus a body can repose on an inclined surface up to angle of friction.

4.6 Cone of Friction

Consider a block of weight W lying on a rough horizontal surface subjected to a force P as shown in Fig. 4.6.

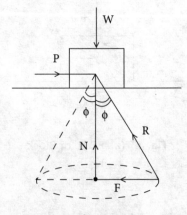

Fig. 4.6

If applied force P is increased such that body tends to move or impends than maximum frictional force is induced. The resultant reaction makes angle of friction φ from normal reaction N as studied in section 4.3. If force P is rotated from 0 to 360° along horizontal surface then resultant reaction also rotates about normal reaction by angle φ and makes a cone. Such a cone whose semi vertical angle is equal to angle of friction φ is called as **Cone of friction**.

Example 4.1

A wooden block rests on a horizontal plane as shown in the Fig. 4.7. If the coefficients of limiting and dynamic friction between the block and the horizontal plane are 0.3 and 0.24, respectively, determine:

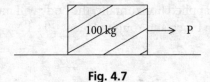

Fig. 4.7

(i) The force required to impend the block right side.
(ii) Status of motion if 200 N force is applied horizontally.
(iii) Status of motion, if 300 N force is applied horizontally.

Solution:

Consider, the F.B.D. of the block as showing in Fig. 4.7 (*a*)

(i) The body will impend right side when *P* will be equal to the limiting friction.

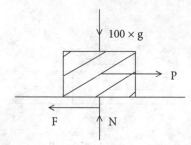

Fig. 4.7 (a) F.B.D. of Block

i.e., $P = F_{limiting}$ or F_{max}
as, $F = \mu \times N$
$F = 0.3 \times N$ ..... (i)

$\mu = 0.3$ for limiting friction

$\Sigma Y = 0$,
$N = 100 \times g$

Substituting value of N in equation (i),
$F = 0.3 \times 100 \times g$
$F = 294.3\ N$
$\therefore \quad P = 294.3\ N$

(ii) When 200 N, force is applied horizontally, the body will not move, it'll remain static because
$200\ N < F_{max}\ (294.3\ N)$
(iii) When 300 N, force is applied horizontally, the body will move because the maximum friction value between block and rough plane is 294.3 N.

Friction

Example 4.2

The force required to pull the body of weight 50 N on a rough horizontal surface in 20 N when it is applied at an angle of 25° with the horizontal. Determine the co-efficient of friction and the magnitude of reaction R between the body and the horizontal surface. Does the reaction pass through the C.G. of the body?

(U.P.T.U Ist Sem 2002–2003)

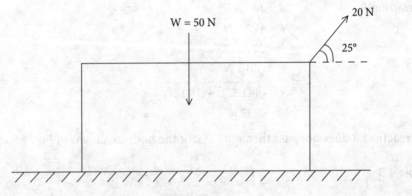

Fig. 4.8

Solution:

Consider F.B.D. of the body as shown in Fig. 4.8 (a). As the body impends right side, frictional force reaches to maximum (limiting) in opposite direction,

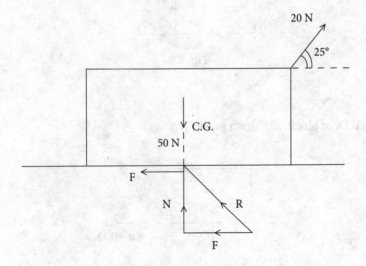

Fig. 4.8 (a) F.B.D. of Block

$\Sigma X = 0$, $20 \cos 25° = F$ ..... (1)
$\Sigma Y = 0$, $N + 20 \sin 25° = 50$ ..... (2)
$$N = 50 - 20 \sin 25°$$
$$N = 41.55 \text{ N Substituting in equation (1)}$$
$$20 \cos 25° = \mu \times N$$
$$20 \cos 25° = \mu \times 41.25$$
$$\mu = 0.44$$

for reaction R,
$$R = \sqrt{N^2 + F^2}$$
$$N = 41.55 \text{ N}$$
$$F = 20 \cos 25°$$
$$F = 18.12 \text{ N}$$
$$R = \sqrt{(41.55)^2 + (18.12)^2}$$
$$R = 45.33 \text{ N}$$

The reaction R does not pass through C.G. of the body as shown in Fig. 4.8 (a).

Example 4.3

A block of weight 5 kN is pulled by a force P as shown in Fig. 4.9. The coefficient of friction between the contact surface is 0.35. Find the magnitude θ for which P is minimum and find the corresponding value of P.

(U.P.T.U. IInd Sem, 2003–2004)

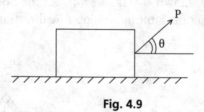

Fig. 4.9

Solution:

Considering F.B.D. of block, as shown in Fig. 4.9 (a)

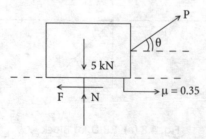

Fig. 4.9 (a) F.B.D. of Block

Friction

$\Sigma X = 0$,

$$P\cos\theta = F$$
$$P\cos\theta = \mu N \quad\quad \ldots\ldots(1)$$

$\Sigma Y = 0$,

$$P\sin\theta + N = 5$$

i.e., $\quad N = (5 - P\sin\theta)$

Substituting the value of N in equation (1)

$$P\cos\theta = 0.35(5 - P\sin\theta)$$
$$P = \frac{1.75}{(\cos\theta + 0.35\sin\theta)} \quad\quad \ldots\ldots(2)$$

P will be minimum it denominator of equation (2) becomes maximum.

$$\frac{d}{d\theta}(\cos\theta + 0.35\sin\theta) = 0$$
$$-\sin\theta + 0.35\cos\theta = 0$$
$$\tan\theta = 0.35$$
$$\theta = 19.29°$$

If second differential value comes negative, denominator will be maximum.

$$= \frac{d^2}{d\theta^2}(\cos\theta + 0.35\sin\theta)$$
$$= \frac{d}{d\theta}(-\sin\theta + 0.35\cos\theta)$$
$$= -\cos\theta - 0.35\sin\theta$$
$$= -\cos 19.29° - 0.35\sin 19.29°$$
$$= -1.06$$

Thus denominator is maximum when $\theta = 19.29°$

from equation (2), $P = \dfrac{1.75}{(\cos 19.29 + 0.35\sin 19.29°)} = 1.65$ kN.

Example 4.4

A block of 150 kg mass placed on a rough horizontal plane. The coefficient of static friction between the block and plane is 0.33. Determine the force required to impend the block for each of the following case:

a. if block is pushed by a horizontal force.
b. if block is pulled by a horizontal force.
c. if block is pulled by an inclined force, inclined at 20° to the horizontal.

Solution:

$m = 150$ kg, $W = 150 \times g$, $\mu = 0.33$

a. when the block is pushed by a horizontal force P,

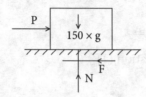

Fig. 4.10 (a) F.B.D. of Block

$\Sigma X = 0$, $P = F$
$\Sigma Y = 0$, $N = 150 \times g$
as $F = \mu \times N$
            $P = F = \mu \times N$
            $P = 0.33 \times 150 \times 9.81$
            $P = 485.60 \ N$

b. When the block is pulled by a horizontal force, the direction of friction of frictional forces changes but magnitude remains same.

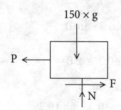

Fig. 4.10 (b) F.B.D. of Block

then, $P = 485.60 \ N$

c. When body is pulled at 20° by force,

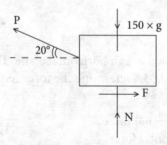

Fig. 4.10 (c) F.B.D. of Block

Friction

$\Sigma X = 0$,
  $P.\cos 20° = F$ .... (1)

$\Sigma Y = 0$,
  $P.\sin 20° + N = 150 \times g$
    $N = (150 \times g - P \sin 20°)$ Substituting in equation (1),
  $P.\cos 20° = F = \mu \times N$
  $P.\cos 20° = 0.33 (150 \times g - P \sin 20°)$
    $P = 461.34\ N$

Example 4.5

A cuboid block of weight 240 N rests on a rough floor as shown in Fig. 4.11. Determine the minimum force which can impend the block? Take co-efficient of friction between the contact surfaces 0.28.

Determine the maximum height at which this minimum force can be applied to just impend the block without toppling.

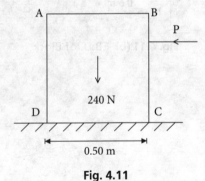

Fig. 4.11

Solution:

Consider FBD of the block, as showing Fig. 4.11 (a)

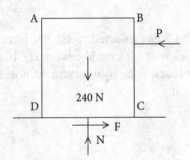

Fig. 4.11 (a) F.B.D. of Block

$\Sigma X = 0$, $P = F$ .... (1)
$\Sigma Y = 0$, $N = 240$
           $P = F = \mu N$
           $P = 0.28 \times 240$
           $P = 67.2$ N

when minimum force P is applied at maximum height, the block tends to topple and slide about D, thus block remain in contact with surface about D as shown in Fig. 4.11 (b).

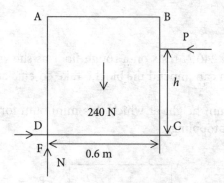

Fig. 4.11 (b) F.B.D. of Block

Thus $\Sigma M_D = 0$,
$$240 \times \frac{0.6}{2} = P \times h$$
$$h = \frac{240 \times 0.3}{P}$$
$$h = \frac{240 \times 0.3}{67.2}$$
$$h = 1.07 \text{ m}$$

Example 4.6

A wooden block of weight 10 kN is subjected to 1.5 kN force tangential to inclined plane as shown in Fig. 4.12. If coefficients of friction between the block and the plane are $\mu_s = 0.3$ and $\mu_k = 0.25$, determine the block is in equilibrium or not. Find out the magnitude and direction of frictional force acting.

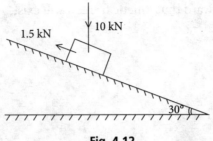

Fig. 4.12

Solution:

Consider FBD of block, Fig. 4.12 (a). Resolving forces perpendicular to inclined plane, $\Sigma Y = 0$,

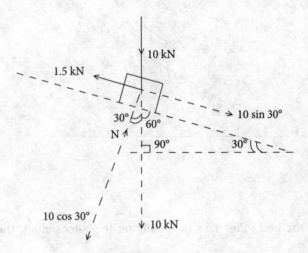

Fig. 4.12 (a) F.B.D. of Block

$\quad N = 10 \cdot \cos 30°$

$\quad N = 8.66$ kN

Thus maximum frictional force between contact surfaces

$\quad F_{limiting} = \mu_s \times N = 0.3 \times 8.66$

$\quad F_{max} = 2.59$ kN

Tangential forces along the inclined plane,

$\quad \Sigma X = +10 \sin 30° - 1.5$

$\quad\quad = 5 - 1.5 = 3.5$ kN downwards

As $F_{max} < \Sigma X$ thus frictional force value

cannot hold the block in equilibrium hence block moves or slides downward.

As the block slides downward thus kinetic friction will exist,

$$F = F_{kinetic} = \mu_k \times N$$
$$= 0.25 \times 8.66$$
$$F_{kinetic} = 2.17 \text{ kN}$$

Example 4.7

A body of weight 500 N is pulled up along an inclined plane having an inclination of 30° with the horizontal. If the coefficient of friction between the body and the plane is 0.3 and the force is applied parallel to the inclined plane, determine the force required.

(U.P.T.U. IInd sem, 2005–2006)

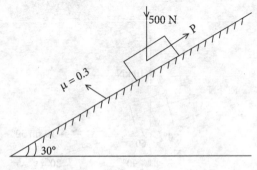

Fig. 4.13

Solution:

Draw the F.B.D. of the body Fig. 4.13 (a) resolving the forces along the inclined plane, i.e., $\Sigma X = 0$,

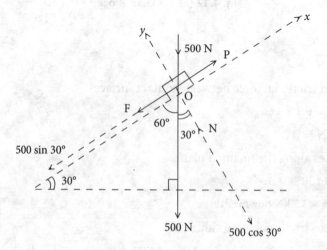

Fig. 4.13 (a) F.B.D. of Block

$P = F + 500 \sin 30°$

$P = \mu.N + 250$ .... (1)

Resolving the forces perpendicular to the inclined plane,
i.e., $\Sigma Y = 0$

$N = 500.\cos 30°$

From equation (1),

$P = 0.3 (500 \cos 30°) + 250$

$P = 379.90 \, N$

Example 4.8

Tangential force P is applied on a 16 kN wooden block held on inclined plane as shown in Fig. 4.14. Find out the force P when (i) body impends upward (ii) body impends downward. Take $\mu_s = 0.28$.

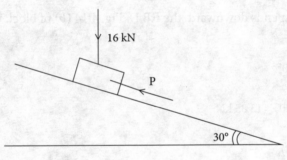

Fig. 4.14

Solution:

(i) When body impends upward, the F.B.D. of block Fig. 4.14 (a) will be,

$\Sigma X = 0$,

$P = F + 16 \sin 30°$ .... (1)

$\Sigma Y = 0$,

$N = 16.\cos 30° = 13.85 \, kN$

From equation (1).

$P = \mu N + 16 \sin 30°$

$= 0.28 \times 13.85 + 8$

$P = 11.87 \, kN$

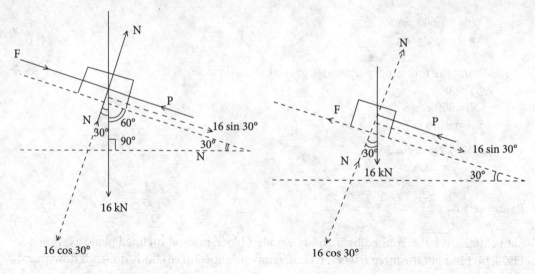

Fig. 4.14 (a) F.B.D. of Block   **Fig. 4.14 (b)** F.B.D. of Block

(ii) When body inpends downward, the F.B.D. Fig. 4.14 (b) of block will be,

$\Sigma X = 0$,

$P + F = 16 \sin 30°$ .... (2)

$\Sigma Y = 0$,

$N = 16 . \cos 30° = 13.85$ kN

From equation (2),

$P + \mu . N = 16 \sin 30°$

$P + 0.28 \times 13.85 = 8$

$P = 4.12$ kN

Example 4.9

For the system of blocks as shown in Fig. 4.15 find force P required to move the lower block A and tension in the cable C. Take coefficient of friction for all contact surfaces to be 0.3.

(U.P.T.U. Ist Sem, 2002-003)

Friction

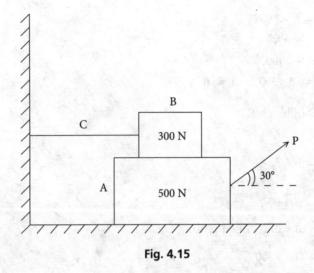

Fig. 4.15

Solution:

Draw the F.B.D. of both blocks together as shown below in Fig. 4.15 (a).

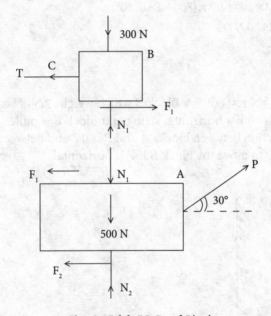

Fig. 4.15 (a) F.B.D. of Blocks

When force P pull the block A, the block B and floor opposes its motion causing frictional force F_1 and F_2 towards left. Because block B applies F_1 towards left thus block A applies F_1 in right side as shown above in F.B.D. of block B.

First consider F.B.D. of block B,

$\Sigma X = 0$, $T = F_1$ .... (1)
$\Sigma Y = 0$, $N_1 = 300\ N$ .... (2)
$T = F_1 = \mu_1 N_1$
$T = 0.3 \times 300 \Rightarrow T = 90\ N$

Considering *FBD* of block *B*,

$\Sigma X = 0$,
$P.\cos 30° = F_1 + F_2$ .... (3)
$\Sigma Y = 0$,
$P.\sin 30° + N_2 = 500 + N_1$ .... (4)

Substituting the value of N_1 in equation (4),

$P \sin 30° + N_2 = 500 + 300$
$P \sin 30° + N_2 = 800$

Substituting value of N_1 and N_2 is equation (3)

$P \cos 30° = \mu_1 N_1 + \mu_2 N_2$
$P \cos 30° = 0.3 \times 300 + 0.3 \times (800 - P \sin 30°)$
$P = 324.80\ N$

Example 4.10

Block *A* weighing 1.25 kN rests over a Block *B* which weighs 2.5 kN as shown in Fig. 4.16. Block *A* is tied to a wall with a horizontal string and Block *B* is pulled by a horizontal pull *P*. The coefficient of friction between blocks *A* and *B* is 0.3 and between *B* and floor is 0.35. Calculate the value of *P* to move the block *B* if *P* is horizontal.

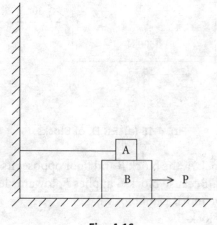

Fig. 4.16

Friction

Solution:

Draw the F.B.Ds of both blocks as shown in Fig. 4.16 (a)
Considering equilibrium of block A,

$\Sigma X = 0$, $\quad T = F_1$
$\qquad\qquad T = \mu_1 N_1$ .... (1)
$\Sigma Y = 0$, $\quad N_1 = 1.25$ kN

from equation (1),
$\qquad T = 0.3 \times 1.25$
$\qquad T = 0.375$ kN

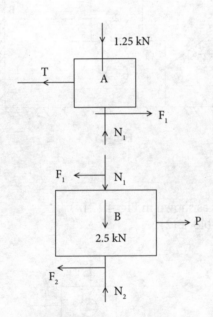

Fig. 4.16 (a) F.B.D. of Blocks

Considering equilibrium of block B,

$\Sigma X = 0$, $\qquad P = F_1 + F_2$ or $\mu_1 N_1 + \mu_2 N_2$ .... (2)
$\Sigma Y = 0$, $\qquad N_2 = N_1 + 2.5 = 1.25 + 2.5$
$\qquad\qquad N_2 = 3.75$ kN

Substituting μ_1, μ_2, N_1 and N_2 in equation (2),
$\qquad P = 0.3 \times 1.25 + 0.35 \times 3.75$
$\qquad P = 1.69$ kN

Example 4.11

Block 2 rests on block 1 and is attached by horizontal rope *AB* on the wall as shown in Fig. 4.17. What force *P* is necessary to cause motion of the block 1 to impend? The coefficient of friction between the blocks is 1/4 and between the floor and the block (1) is 1/3. Mass of the blocks 1 and 2 are 14 kg and 9 kg, respectively.

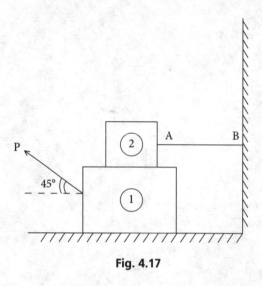

Fig. 4.17

Solution:

Draw F.B.Ds of both blocks as shown in Fig. 4.17 (*a*).
considering equilibrium of block (2),

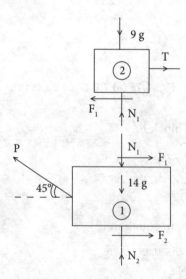

Fig. 4.17 (a) F.B.D. of Blocks

Friction

$\Sigma X = 0$,
$\quad T = F_1$
$\quad T = \mu_1 N_1$ .... (1)

$\Sigma Y = 0$,
$\quad N_1 = 9 \times g = 88.29$ N
$\quad T = \mu_1 N_1$

Thus, $\quad T = \dfrac{1}{4} \times 88.29$
$\quad T = 22.07$ N

Considering equilibrium of block (1),

$\Sigma X = 0$,
$\quad P\cos 45° = F_1 + F_2$
$\quad P\cos 45° = \mu_1 N_1 + \mu_2 N_2$
$\quad P\cos 45° = \left(\dfrac{1}{4}\right) \times 88.29 + \left(\dfrac{1}{3}\right) \times N_2$ .... (2)

$\Sigma Y = 0$,
$\quad P\sin 45° + N_2 = N_1 + 14g = 88.29 + 14 \times 9.81$
$\quad P\sin 45° + N_2 = 225.63$ N .... (3)

From equations (2) and (3),
$\quad N_2 = 152.67$ N
$\quad P = 103.18$ N

Example 4.12

Two wooden blocks A and B are connected by a string passing over a smooth pulley as shown in Fig. 4.18. Determine the range of force P by which both blocks can be held in equilibrium. Take weight of blocks A and B as 240 kN and 600 kN respectively and coefficient of friction of all contact surfaces to be 0.30.

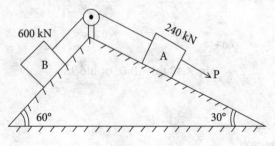

Fig. 4.18

Solution:

As pulley is smooth, the tension in the string will be equal on both blocks.

The range of force P shows the maximum and minimum value of force due to which block tends to impend.

(i) When force P is kept maximum, both blocks tend to impend in the direction of P i.e., 240 kN downward side and 600 kN upward side as shown in Fig. 4.18 (a).

Consider *FBD* of block *B*,

$\Sigma X = 0$,

$$T = 600 \sin 60° + F_B \qquad \qquad \dots (1)$$

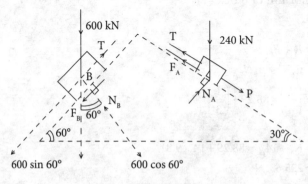

Fig. 4.18 (a)

$\Sigma Y = 0$,

$$N_B = 600 \cos 60° = 300 \text{ kN}$$

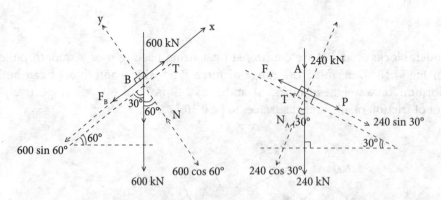

Fig. 4.18 (b) F.B.D. of Blocks

From equation (1),

$T = 600 \sin 60° + \mu \times N$

$ = 600 \sin 60° + 0.3 \times 300$

$T = 609.62$ kN

Consider F.B.D. of block A,

$\Sigma X = 0$,
$\quad T + F_A = P + 240 \sin 30°$ .... (2)

$\Sigma Y = 0$,
$\quad N_A = 240 \cos 30°$

From equation (2),

$\quad T + \mu \cdot N_A = P + 240 \sin 30°$
$\quad 609.62 + 0.3 \times 240 \cos 30° = P + 240 \sin 30°$
$\quad P_{max} = 551.97$ kN

(ii) When force P is applied minimum, it just holds both blocks i.e., block B tends to impend downward and pulls block A to impend upward side as shown in Fig. 4.18 (c). Consider FBD of both blocks,

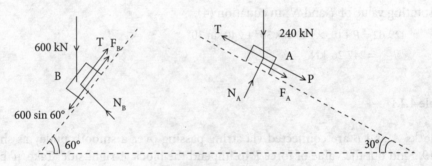

Fig. 4.18 (c) F.B.D. of Blocks

Consider block B for equilibrium,

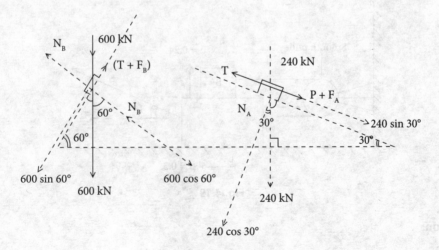

Fig. 4.18 (d) F.B.D. of Blocks

$\Sigma X = 0$,
$$T + F_B = 600 \sin 60° \qquad \dots (3)$$
$\Sigma Y = 0$,
$$N_B = 600 \cos 60°$$
$$N_B = 300 \text{ kN}$$

From equation (3),
$$T + \mu.N_B = 600 \sin 60°$$
$$T + 0.3 \times 300 = 600 \sin 60°$$
$$T = 429.62 \text{ kN}$$

Consider block A for equilibrium,

$\Sigma X = 0$, $\quad T = P + F_A + 240 \sin 30° \qquad \dots (4)$
$\Sigma Y = 0$, $\quad N_A = 240 \cos 30°$

Substituting value of T and N_A in equation (4),
$$429.62 = P + 0.3 \times 240 \cos 30° + 240 \sin 30°$$
$$P_{min} = 247.26 \text{ kN}$$

Example 4.13

Two blocks A and B are connected via string passing over a smooth pulley as shown in Fig. 4.19. Find out the value of force P to impend the block A right side. Take 'μ' between blocks as 0.24 and between block A and floor as 0.3.

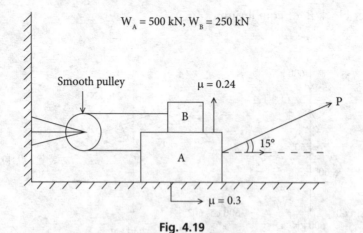

Fig. 4.19

Solution:

Due to smooth pulley, tension between block A and B will remain same.

Friction

Draw *FBD* of both blocks as shown in Fig. 4.19 (*a*).
When block *A* impends right side, block *B* applied friction force on left side on block A. Thus block *A* applies same friction force on block *B* right side.
Considering equilibrium of block *B*,

$\Sigma X = 0$, $T = F_1$ .... (1)

$\Sigma Y = 0$, $N_1 = 250$ kN

$T = \mu_1 N_1 = 0.24 \times 250$

$T = 60$ kN

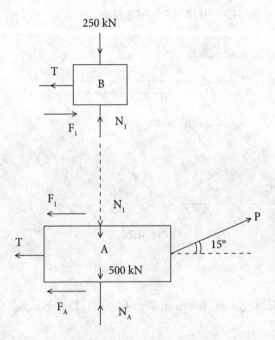

Fig. 4.19 (a) F.B.D. of Blocks

Considering equilibrium of block *A*,

$\Sigma X = 0$,
   $P \cos 15° = T + F_A + F_1$ .... (2)

$\Sigma Y = 0$,
   $P \sin 15° + N_A = N_1 + 500$
               $= 250 + 500 = 750$ .... (3)

Substituting values of T and F_1 in equation (2),

   $P \cos 15° = 60 + \mu_A . N_A + 60$
   $P \cos 15° = 120 + 0.3 . N_A$ .... (4)

From equation (3),

$N_A = (750 - P \sin 15°)$ Substituting in equation (4)

$P \sin 15° = 120 + 0.3 (750 - P \sin 15°)$

$P = 330.59$ kN

Example 4.14

Determine value of force P to impend the block A right side. Take weight of blocks A and B as 200 kN and 500 kN, respectively. Assume coefficient of friction between all contact surfaces as 0.30. Determine tension in the string also.

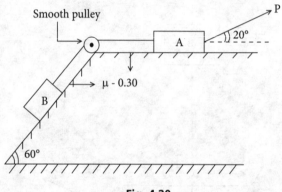

Fig. 4.20

Solution:

Consider FBDs of both blocks as shown in Fig. 4.20 (a). Take block B for equilibrium,

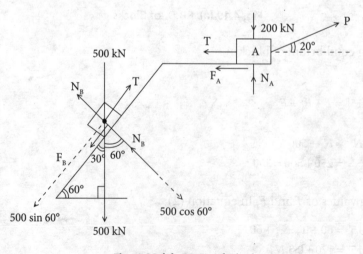

Fig. 4.20 (a) F.B.D. of Blocks

Friction

$\Sigma X = 0,$
$$500 \sin 60° + F_B = T \quad \ldots (1)$$
$\Sigma Y = 0,$
$$N_B = 500 \cos 60°$$
$$N_B = 250 \text{ kN}$$

From equation (1),
$$500 \sin 60° + \mu \times N_B = T$$
$$500 \sin 60° + 0.3 \times 250 = T$$
$$T = 508 \text{ kN}$$

Considering equilibrium of block A,

$\Sigma X = 0,$
$$P \cos 20° = T + F_A \quad \ldots (2)$$
$\Sigma Y = 0,$
$$P \sin 20° + N_A = 200 \quad \ldots (3)$$

from equation (2),
$$P \cos 20° = 508 + 0.3 \, N_A$$
$$N_A = \left(\frac{P \cos 20° - 508}{0.3} \right)$$

from equation (3),
$$P \sin 20° + \frac{P \cos 20° - 508}{0.3} = 200$$
$$0.3 \, P \sin 20° + P \cos 20° - 508 = 200 \times 0.3$$
$$P = 544.95 \text{ kN}$$

Example 4.15

Two blocks A and B of equal weighing 2000 N are lying on inclined planes as shown in Fig. 4.21. Both blocks are connected by a uniform horizontal rod of weight 800 N. If the coefficient of friction for all surfaces is 0.29, determine the value of force P to impend the block A upward.

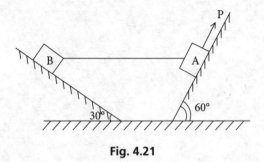

Fig. 4.21

Solution:

Given, $W_A = W_B = 2000$ N
$\mu_A = \mu_B = 0.29$

The rod connecting both blocks is of weight 800 N. Considering F.B.Ds of all bodies, as shown in Fig. 4.21 (a)

Let reaction between rod and block A and B are R_1, R_2 respectively due to weight of rod and S_1, S_2 respectively due to compression of rod. Considering F.B.D. of rod for equilibrium,

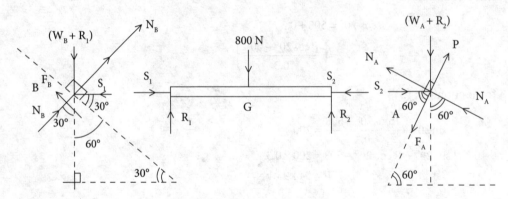

Fig. 4.21 (a) F.B.D. of Blocks and Rod

$R_1 + R_2 = 800$ N .... (1)

As rod is uniform, weight will act on middle of it.
Thus $R_1 = R_2$ i.e., $R_1 = R_2 = 400$ N
and $\qquad S_1 = S_2$
Considering skeleton of forces for both blocks A and B.

Friction

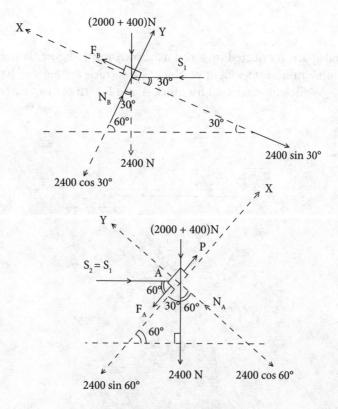

Fig. 4.21 (b) F.B.D. of Blocks

Resolving forces along the inclined plane i.e., $\Sigma X = 0$,
$$F_B + S_1 \cos 30° = 2400 \sin 30°$$
$$\mu_B . N_B + S_1 \cos 30° = 1200 \quad \quad ....(1)$$

Resolving forces perpendicular to inclined plane i.e., $\Sigma Y = 0$,
$$N_B = 2400 \cos 30° + S_1 \sin 30° \quad \quad ....(2)$$

Substituting N_B, μ_B values in equation (1)
$$0.29 \times (2400 \cos 30° + S_1 \sin 30°) + S_1 \cos 30° = 1200$$
$$S_1 = 590.73 \text{ N}$$

Considering equilibrium of block A,

$\Sigma X = 0$,
$$S_1 \cos 60° + P = F_A + 2400 \sin 60°.$$
$$S_1 \cos 60° + P = \mu_A N_A + 2400 \sin 60° \quad \quad ....(3)$$

$\Sigma Y = 0, \quad N_A = S_1 \sin 60° + 2400 \cos 60° \quad \quad ....(4)$

Substituting in equation (5)
$$S_1 \cos 60° + P = 0.29 (S_1 \sin 60° + 2400 \cos 60°) + 2400 \sin 60°$$
$$590.73 \cos 60° + P = 0.29 (590.73 \sin 60° + 2400 \cos 60°) + 2400 \sin 60°$$
$$P = 2279.45 \text{ N}$$

Example 4.16

Two blocks A and B are connected by a rod as shown in the figure. If weight of block A is 240 kN, determine minimum weight of block B by which system can be maintained in equilibrium. Take coefficient of friction for block A and B with contact surfaces as 0.34 and 0.29 respectively.

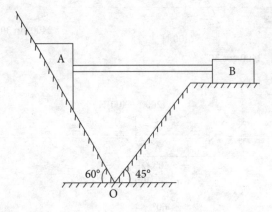

Fig. 4.22

Solution:

Since $\mu_A = 0.34$
We know that $\quad \mu = \tan \phi$
$\quad\quad\quad\quad\quad\quad 0.34 = \tan \phi$
angle of friction, $\quad \phi = 18.77°$
since angle of inclination of inclined plane, α is 60°
thus $\alpha > \phi$, block A tends to move downwards and pushes block B to impend right side.
Considering FBDs of both blocks as given in Fig. 4.22 (a)

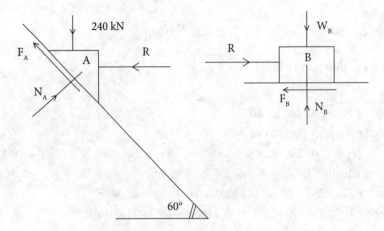

Fig. 4.22 (a) F.B.D. of Blocks

Friction

Considering skeleton of force for block A,

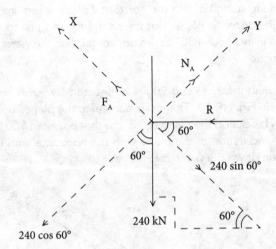

Fig. 4.22 (b)

$\Sigma X = 0$,
$$F_A + R \cos 60° = 240 \sin 60° \quad ....(1)$$
$\Sigma Y = 0$,
$$N_A = R \sin 60° + 240 \cos 60°$$

Substituting value of N_A in equation (1)

$$\mu_A . N_A + R \cos 60° = 240 \sin 60°$$
$$0.34 \, (R \sin 60° + 240 \cos 60°) + R \cos 60° = 240 \sin 60°$$
$$R = 210.26 \text{ kN}$$

For block B,

$\Sigma X = 0$,
$$R = F_B \quad ....(2)$$
$\Sigma Y = 0$,
$$N_B = W_B \quad ....(3)$$

from equation (2),

$$210.26 = \mu_B \times N_B$$
$$210.26 = 0.29 \times N_B$$
$$N_B = 725.03 \text{ kN}$$
$$W_B = 725.03 \text{ kN}$$

Numerical Problems

N 4.1 A pull of 250 N inclined at 30° to the horizontal plane is required just to move a body kept on a rough horizontal plane. But the push required just to move the body is 300 N. If the push is in inclined at 30° to the horizontal, find the weight of the body and the coefficient of friction.

(A.M.U. May/June 2009)

N 4.2 Two blocks of weight 500 N and 900 N connected by a rod are kept on an inclined plane as shown in Fig. NP 4.1. The rod is parallel to the plane. The coefficient of friction between 500 N block and the plane is 0.3 and that between 900 N block and the plane is 0.4. Find the inclination of the plane with the horizontal and the tension in the rod when the motion down the plane is just about to start.

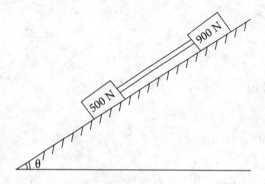

Fig. NP 4.1

N 4.3 What should be the value of θ in Fig. NP 4.2 which will make the motion of 900 N block down the plane to impend? The coefficient of friction for all contact surfaces is 0.3.

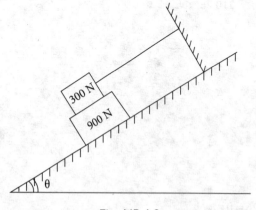

Fig. NP 4.2

Friction

N 4.4 Two blocks A and B, of weight 150 N and 200 N, respectively rest on an inclined plane as shown in Fig. NP 4.3. The coefficient of friction between the two blocks is 0.3 and between blocks A and inclined plane is 0.4. Find the value of θ for which either one or both the blocks start slipping. At what instant, what is the friction force between B and A, and the same between A and inclined plane?

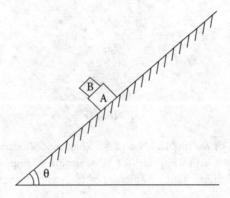

Fig. NP 4.3

N 4.5 A rectangular block of mass 30 kg rests on a rough plane as shown in Fig. NP 4.4. The coefficient of friction between the block and the plane is 0.32. Determine the highest position to which a horizontal force of 100 N can cause to just move the block without tipping.

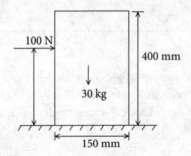

Fig. NP 4.4

N 4.6 Two blocks A and B are kept in equilibrium with the help of rod of negligible weight as shown in Fig. NP 4.5. The coefficient of friction between block A and horizontal plane is 0.36 and between block B and inclined plane is 0.30. Determine the minimum weight of block A which can hold the system in equilibrium.

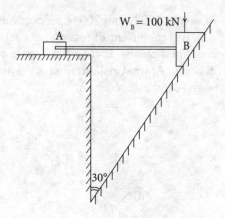

Fig. NP 4.5

N 4.7 Two blocks A and B of weight 12 kN and 4.5 kN are connecting by a string passing over an ideal pulley as shown in Fig. NP 4.6. Determine minimum force P which can hold the system in equilibrium. Take coefficient of friction between contact surface as 0.28.

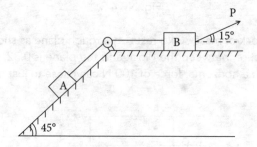

Fig. NP 4.6

N 4.8 Two blocks A and A of weight 50 kN each are held in equilibrium with the help of a rod of negligible weight as shown in Fig. NP 4.7. If coefficient of friction between all contact surfaces is same and block B impends downward then determine the value of coefficient of friction.

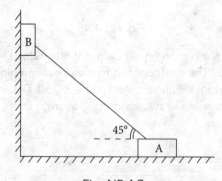

Fig. NP 4.7

Friction

N 4.9 A homogeneous cylinder of weight 120 kN rests on a horizontal plane in contact with a wall as shown in Fig. NP 4.8. Determine the couple M required on the cylinder to cause anticlockwise rotation. Take coefficient of friction for all contact surfaces as 0.30.

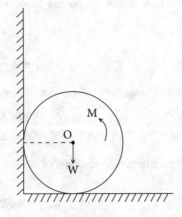

Fig. NP 4.8

Multiple Choice Questions

1. If a person is walking from point A to point B then direction of frictional force will be
    a. from point A to point B
    b. from point B to point A
    c. cannot be predict
    d. none of these

2. The direction of frictional force remains
    a. Normal to the meeting surfaces
    b. inclined to the meeting surfaces
    c. tangential to the meeting surfaces
    d. none of these

3. A box of weight 40 N is lying on a rough surface. If the coefficient of friction between meeting surfaces is 0.25 then the frictional force will be
    a. zero
    b. 10 N
    c. Less than 10 N
    d. none of these

4. The frictional force does not depend on
    a. area of meeting surfaces
    b. roughness of meeting surfaces
    c. normal reaction between meeting surfaces
    d. all of these

5. The maximum static frictional force is known as
    a. kinetic friction
    b. coefficient of friction
    c. limiting friction
    d. none of these

6. A body under impending motion consists frictional force
   a. zero
   b. limiting friction
   c. kinetic friction
   d. none of these

7. The maximum angle of friction is the angle made by resultant reaction from normal reaction when
   a. frictional force is static
   b. frictional force is limiting
   c. frictional force is kinetic
   d. none of these

8. The angle of inclination of inclined plane when a body impends downward is called as
   a. angle of static friction
   b. angle of repose
   c. angle of cone of friction
   d. none of these

9. If a body is placed on inclined plane whose angle of inclination is equal to angle of friction the body will
   a. stay on inclined plane
   b. move down
   c. impend downward
   d. none of these

10. The frictional force is desirable in
    a. automobile engine
    b. brakes and clutches
    c. ball bearings
    d. all of these

Answers

1. a 2. c 3. a 4. a 5. c 6. b 7. b 8. b 9. c 10. b

Chapter 5

Application of Friction

5.1 Ladder Friction

A ladder is a device which is widely used in small factories, homes and shops to perform day-to-day functions. Ladder bears normal reactions and frictional forces offered by both wall and horizontal floor when it is placed between them. However the frictional force becomes zero if it is placed against smooth wall or floor. Ladder can be considered as a body subjected to non-concurrent coplanar force system because normal reactions, frictional forces, weight and external force act on different points on the ladder in a common plane. Thus unknown quantities can be determined by using equilibrium conditions of a non-concurrent coplanar force system

i.e., $\sum X = 0$, $\sum Y = 0$ and $\sum M = 0$

Example: 5.1

A ladder 6 m long, weighting 300 N, is resting against a wall at an angle of 60° with the horizontal plane as shown in Fig. 5.1. A man weighting 750 N climbs the ladder from position B towards A. At what position along the ladder from the bottom of the ladder does he induce slipping? The coefficient of friction for both the wall and the ground with the ladder is 0.2.

(M.T.U. Ist Sem 2012–13)

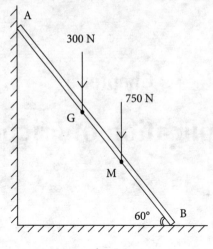

Fig. 5.1

Solution:

Let the ladder is uniform of weight W, thus it acts at midpoint of ladder. Now draw the F.B.D. of ladder, when man climbs on ladder by distance x, the ladder begins to slip. The point A tends to move downward and point B tends to slide about right side, thus frictional forces act opposite to this on ladder as shown in the figure.

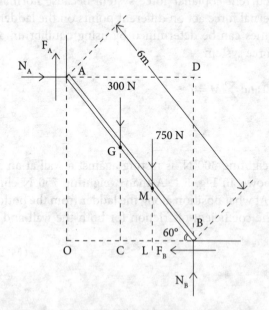

Fig. 5.1 (a) Free Body Diagram of Ladder AB

As all the forces, weights are acting at different points, this represents a coplanar, non-concurrent force system thus three equilibrium conditions will be used.

Application of Friction

$\sum X = 0,$

$N_A = F_B$

$N_A = \mu_B \cdot N_B$ ..... (1)

$\sum Y = 0,$

$F_A + N_B = 300 + 750$

$\mu_A N_A + N_B = 1050$ ..... (2)

Let the man climbs by x distance on the ladder,

$\sum M_B = 0,$

$750 \times BL + 300 \times CB = N_A \times BD + F_A \times OB$

$750 \times x \cos 60° + 300 \times GB \cos 60° = N_A \times OA + F_A \times 6 \cos 60°$

as $GB = \dfrac{AB}{2} = \dfrac{6}{2} = 3$ mt

$OA = 6 \sin 60°, \ \mu_A = \mu_B = 0.2$

$750 \times x \cos 60° + 300 \times 3 \cos 60° = N_A \times 6 \sin 60° + \mu_A \cdot N_A \times 6 \cos 60°$ ..... (3)

In equation (3), if N_A value is placed, we will get the value of 'x'.

The N_A will be determined by solving equations (1) and (2).

from equation (2), $N_B = (1050 - \mu_A N_A)$

substitute value of N_B in equation (1),

$N_A = 0.2(1050 - 0.2 N_A)$

$N_A = 201.92 N$

From equaiton (3),

$750 \times x \cos 60° + 300 \times 3 \cos 60°$

$\qquad = 201.92 \times 6 \sin 60° + 0.2 \times 201.92 \times 6 \cos 60°$

$x = 1.92 m$

Example: 5.2

A uniform ladder of length 10 m and weighting 20 N is placed against a smooth vertical wall with its lower end 8m from the wall as shown in Fig. 5.2. In this position the ladder is just to slip. Determine:

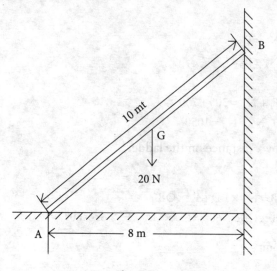

Fig. 5.2

(i) Co-efficient of friction between the ladder and the floor.
(ii) Frictional force acting on the ladder at the point of contact between ladder and floor.

(U.P.T.U. Ist Sem 2004–2005)

Solution:

Note: Wall is given as smooth thus friction at point B will be zero.
Let ladder AB is making an angle of θ with horizontal as shown in Fig. 5.2(a). Considering F.B.D. of ladder AB for equilibrium,

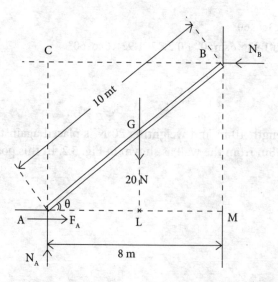

Fig. 5.2 (a) Free Body Diagram of Ladder AB

Application of Friction

$\sum X = 0$,
   $F_A = N_B$
   $\mu_A \cdot N_A = N_B$ ..... (1)

$\sum Y = 0$, $N_A = 20N$

$\sum M_A = 0$, $20 \times AL = N_B \times AC$ as $AC = BM = AB \sin\theta$
   $20 \times AG \cos\theta = N_B \times AB \sin\theta$ In $\triangle AMB$,
   $20 \times 5 \cos\theta = N_B \times 10 \sin\theta$ $\cos\theta = \dfrac{AM}{AB} = \dfrac{8}{10}$
   $20 \times 5 \cos 36.86° = N_B \times 10 \sin 36.86°$ $\theta = 36.86°$
   $N_B = 13.33N$

Substituting value of N_A and N_B in equation (1),
   $\mu_A \times 20 = 13.33$
   $\mu_A = 0.66$

(ii) If frictional force between ladder and the floor is F_A,

from equation (1),
$F_A = N_B$
$F_A = 13.33N$

Example: 5.3

A ladder of length 'L' rests against a wall, the angle of inclination being 45°. If the coefficient of friction between the ladder and the ground, and that between the ladder and the wall is 0.5 each, what will be the maximum distance along ladder to which a man whose weight is 1.5 times the weight of the ladder may ascend before the ladder being to slip?

(U.P.T.U. Ist Sem 2005–2006)

Solution:

Let the ladder is uniform of weight W, acting middle of ladder and the man ascend on ladder by distance 'x' as shown in Fig. 5.3.

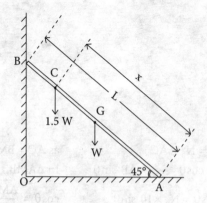

Fig. 5.3

Consider F.B.D. of the ladder as shown in Fig 5.3(a),

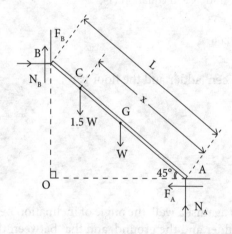

Fig. 5.3 (a) Free body diagram of ladder AB

$\sum X = 0$,
$\qquad N_B = F_A$ ..... (1)
$\qquad N_B = \mu_A . N_A$

$\sum Y = 0$,
$\qquad F_B + N_A = 1.5W + W$ ..... (2)
$\qquad \mu_B N_B + N_A = 2.5W$

$\sum M_A = 0$,
$W \times AG\cos45° + 1.5W \times AC\cos45° = N_B \times L\sin45° + F_B \times L\cos45°$

$W \times \dfrac{L}{2} \times \dfrac{1}{\sqrt{2}} + 1.5W \times x \times \dfrac{1}{\sqrt{2}} = N_B \times L \times \dfrac{1}{\sqrt{2}} + F_B \times L \times \dfrac{1}{\sqrt{2}}$

$W \times \dfrac{L}{2} + 1.5W \times x = N_B \times L + F_B \times L$ ..... (3)

Application of Friction

Given, $\mu_A = \mu_B = 0.5$

Substituting N_B in equation (1),

$\mu_B \cdot \mu_A \cdot N_A + N_A = 2.5W$

$(0.5 \times 0.5 + 1)N_A = 2.5W$

$N_A = 2W$

From equation (1), $N_B = \mu_A \cdot N_A$

$= 0.5 \times 2W$

$N_B = W$

Substituting values of N_B in equation (3),

$W \times \dfrac{L}{2} + 1.5W \times x = W \times L + \mu_B \times N_B \times L$

$W \times \dfrac{L}{2} + 1.5W \times x = W \times L + 0.5 \times W \times L$

$1.5W \times x = W \times L$

$x = \dfrac{2}{3} \cdot L$

Example: 5.4

A uniform ladder 8 m long rests on a horizontal floor and placed against a vertical wall at an angle of 30° with the vertical wall. The weight of the ladder is 100 kN. A man weighting 80 kN stands at a distance of 2 m from the foot of the ladder when the ladder is on the point of sliding. Determine the co-efficient of friction between the wall and the ladder, when that between the floor and ladder is 0.21.

(M.T.U. IInd Sem 2012–13)

Solution:

As figure is not given, first draw Fig. 5.4.

$\mu_B = ?, \mu_A = 0.21$

Consider F.B.D. of ladder as shown in Fig. 5.4(a).

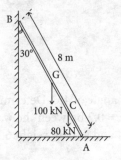

Fig. 5.4

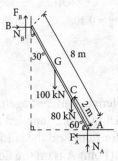

Fig. 5.4 (a) Free body diagram of ladder AB

$\sum X = 0$,
$$N_B = F_A$$
$$N_B = \mu_A . N_A \qquad \ldots (1)$$

$\sum Y = 0$,
$$F_B + N_A = 100 + 80$$
$$\mu_B . N_B + N_A = 180 \qquad \ldots (2)$$

$\sum M_A = 0$,
$$80 \times 2\cos 60° + 100 \times 4\cos 60°$$
$$= N_B \times 8\sin 60° + F_B \times 8\cos 60°$$
$$80 \times 2 \times \frac{1}{2} + 100 \times 4 \times \frac{1}{2} = N_B \times 8\sin 60° + \mu_B . N_B \times 8\cos 60°$$
$$280 = N_B \times 8\sin 60° + \mu_B \times N_B \times 8\cos 60° \qquad \ldots (3)$$

From equation (2),
$$N_A = (180 - \mu_B N_B)$$
From equation (1),
$$N_B = \mu_A . N_A$$
$$N_B = 0.21(180 - \mu_B . N_B)$$
$$N_B = 37.8 - 0.21 \mu_B . N_B$$
$$N_B = \left(\frac{37.8}{1 + 0.21 \mu_B} \right)$$

Substituting in equation (3),
$$280 = N_B . 8\sin 60° + \mu_B . N_B . 8\cos 60°$$
$$280 = \left(\frac{37.8}{1 + 0.21 \mu_B} \right) . 8\sin 60° + \mu_B \left(\frac{37.8}{1 + 0.21 \mu_B} \right) . 8\cos 60°$$
$$280(1 + 0.21 \mu_B) = 37.8 \times 8\sin 60° + 37.8 \mu_B \times 8\cos 60°$$
$$18.11 = 92.4 \mu_B$$
$$\mu_B = 0.20$$

Example: 5.5

For a ladder of length 4 m, rest against a vertical wall making an angle of 45°. Determine the minimum horizontal force applied at A to prevent slipping. If $\mu = 0.2$ between the wall and ladder, $\mu = 0.3$ for floor and the ladder. The ladder weight 200 N and a man weight 600 N is at 3 m from A. (Point A is on the floor)

(M.T.U. Sem I 2013–14)

Application of Friction

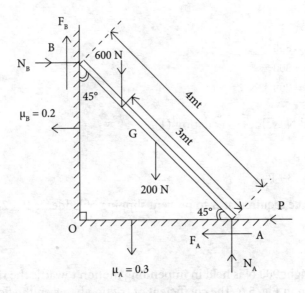

Fig. 5.5 Free body diagram of ladder AB

Solution:

Given $\angle OBA = 45°$ thus $\angle OAB = 45°$

Consider F.B.D. of ladder as shown in the Fig. 5.5. Let minimum force required at A to prevent slipping of ladder is P. At this instant, ladder still tends to slip i.e. impends to rightward about A and impends downward about B. Thus frictional forces at A and B act in opposite direction to impend.

Considering equilibrium of ladder,

$\sum X = 0$,

$\qquad N_B = F_A + P$

$\qquad N_B = \mu_A \cdot N_A + P$ ..... (1)

$\sum Y = 0$,

$\qquad F_B + N_A = 600 + 200$

$\qquad \mu_B \cdot N_B + N_A = 800$ ..... (2)

$\sum M_A = 0$,

$\angle OAB = 45°$

$200 \times 2\cos 45° + 600 \times 3\cos 45° = N_B \times 4\sin 45° + F_B \times 4\cos 45°$

$\qquad 400 + 1800 = 4N_B + 4\mu_B \cdot N_B$

$\qquad 2200 = 4N_B + 4 \times 0.2 \times N_B$

$\qquad N_B = 458.33 N$

From equation (2)

$$\mu_B . N_B + N_A = 800$$
$$0.2 \times 458.33 + N_A = 800$$
$$N_A = 708.33 N$$

Substituting N_A and N_B values in equation (1),

$$N_B = \mu_A . N_A + P$$
$$458.33 = 0.3 \times 708.33 + P$$

Thus, minimum force required at A to prevent slipping of ladder is P = 245.83 N

Example: 5.6

A ladder 'AB' of weight 900 N is held in impending motion towards the right by a rope tied to the wall as shown in Fig. 5.6. The coefficient of friction between the floor and the ladder is 0.25 and that between the wall and ladder is 0.40. Calculate the tension in rope.
Take AB = 10 m, AC = 2.5 m, AD = 5 m.

(U.P.T.U Sem II. C.O. 2008–09)

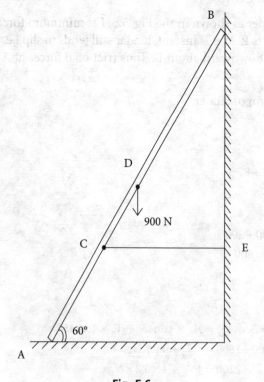

Fig. 5.6

Application of Friction

Solution:

Given,

$\mu_A = 0.25$
$\mu_B = 0.40$

Draw the F.B.D. of ladder as shown in Fig. 5.6(a) and let the tension in the rope CE is T. As the rope causes impending motion of ladder right side, the frictional force at A works left side and at B downward side.

Considering equilibrium of the ladder,

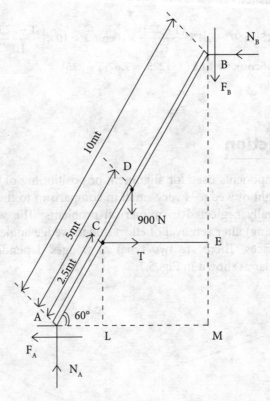

Fig. 5.6 (a) Free body diagram of ladder AB

$\sum X = 0$,

$T = F_A + N_B$
$T = \mu_A \cdot N_A + N_B$ ..... (1)

$\sum Y = 0$,

$N_A = 900 + F_B$
$N_A = 900 + \mu_B \cdot N_B$ ..... (2)

$\sum M_A = 0,$

$T \times CL + 900 \times 5\cos60° + F_B \times AM = N_B \times BM$

$T \times 2.5\sin60° + 900 \times 5\cos60° + \mu_B.N_B \times 10\cos60° = N_B \times 10\sin60°$

$T \times 2.5\sin60° + 900 \times 5\cos60° + 0.40 \times N_B \times 10\cos60° = 10.N_B.\sin60°$ ..... (3)

Substituting value of N_A in equation (1),

$T = \mu_A N_A + N_B$

$T = 0.25(900 + 0.4N_B) + N_B$

$T = 225 + 1.1N_B$

$N_B = \left(\dfrac{T-225}{1.1}\right)$ in equation (3),

$T \times 2.5\sin60° + 900 \times 5\cos60° + 0.4 \times \dfrac{(T-225)}{1.1} \times 10\cos60° = 10 \times \left(\dfrac{T-225}{1.1}\right)\sin60°$

$T \times 2.5\sin60° + 900 \times 5\cos60° + 1.82\,(T-225) = 7.87(T-225)$

$3611.25 = 3.885 \times T$

$T = 929.59$ N

5.2 Wedge Friction

Wedges are small components used for alignment or positioning of the big size members of machines. The weight of wedge is very small in comparison to the weight to be moved or aligned and generally neglected in numerical problems. The wedge remains under equilibrium (self-locking) after removal of effort due to wedge angle and friction existing between contact surfaces. There are two types of wedges depending upon shape, i.e., trapezoidal or triangular as shown in Fig. 5.7.

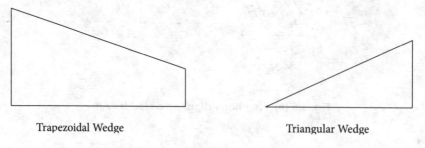

Trapezoidal Wedge   Triangular Wedge

Fig. 5.7

Example: 5.7

A block of weight 150 kN is to be lifted up with the help of a wedge as shown in Fig. 5.8. Determine the value of P for impending motion of weight upwards. Take coefficient of friction for all contact surfaces as 0.32 and weight of wedge as negligible.

Application of Friction

Solution:

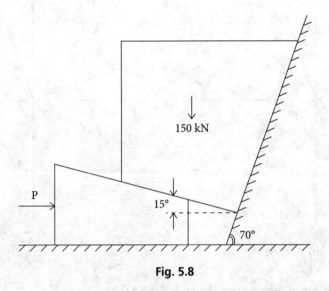

Fig. 5.8

Draw F.B.Ds of both block and wedge. As the wedge impends toward right side, the block and ground offers frictional force towards left side on it. Thus wedge offers frictional force on block right side as shown in Fig. 5.8(a). Further the Figs. 5.8(b) and (c) show the skeleton of forces acting on block and wedge, respectively.

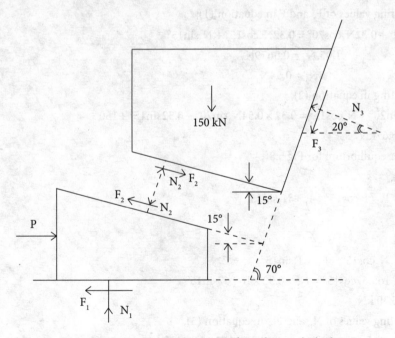

Fig. 5.8 (a) Free body diagram of Wedge and Block

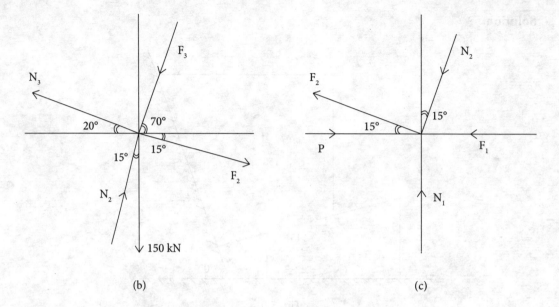

(b)　　　　　　　　　　　　　　(c)

Consider equilibrium of force system shown in Fig 5.8 (b),

$\sum X = 0$,

$N_3\cos20° + F_3\cos70° = F_2\cos15° + N_2\sin15°$ ..... (1)

$\sum Y = 0$,

$N_3\sin20° + N_2\cos15° = F_3\sin70° + F_2\sin15° + 150$ ..... (2)

Substituting values of F_2 and F_3 in equation (1),

$N_3\cos20° + 0.32N_3\cos70° = 0.32N_2\cos15° + N_2\sin15°$

$1.05.N_3 = 0.5679N_2$

$N_3 = 0.54N_2$

Substituting in equation (2),

$0.54N_2\sin20° + N_2\cos15° = 0.32 \times 0.54N_2\sin70° + 0.32\sin15° + 150$

$N_2 = 165.67 \text{kN}$

Consider equilibrium for Fig 5.8(c)

$\sum X = 0$,

$P = F_2\cos15° + F_1 + N_2\sin15°$ ..... (3)

$\sum Y = 0$,

$N_2\cos15° = N_1 + F_2\sin15°$ ..... (4)

$165.67\cos15° = N_1 + 0.32 \times 165.67\sin15°$

$N_1 = 146.30$ kN

Substituting values of N_1 and N_2 in equation (3),

$P = 0.32 \times 165.67\cos15° + 0.32 \times 146.3 + 165.67\sin15°$

$P = 140.90$ kN

Application of Friction

Alternative method

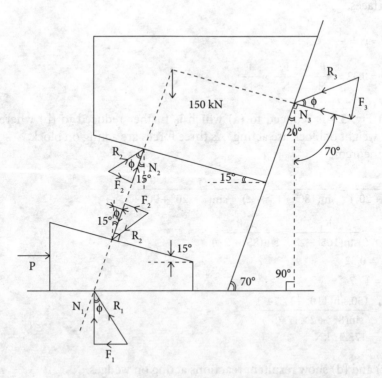

Fig. 5.9 Free body diagram of Wedge and Block

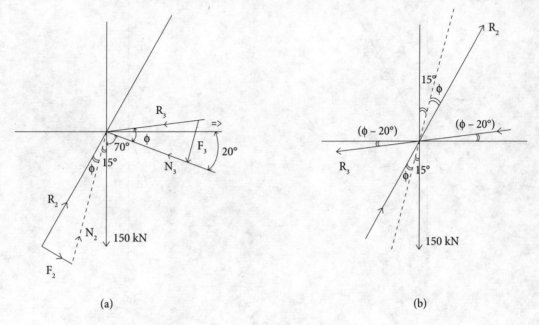

(a)   (b)

Fig. 5.9

We can combine the normal reaction and frictional force in terms of resultant reaction at all contact surfaces,

where $\tan\phi = \mu = \dfrac{F}{N}$

or $\tan\phi = 0.32 = \dfrac{F_3}{N_3} = \dfrac{F_2}{N_2} = \dfrac{F_1}{N_1}$

Consider that Fig. 5.9 is reduced to (a) which is further reduced to (b) where resultant reactions and weight of block are acting. As three forces are acting on block,
Using Lami's theorem,

$$\dfrac{R_2}{\sin(90°-\phi+20°)} = \dfrac{R_3}{\sin(180°-15°-\phi)} = \dfrac{150}{\sin(\phi-20°+90°+15°+\phi)}$$

$$\dfrac{R_2}{\sin(110°-\phi)} = \dfrac{R_3}{\sin(165°-\phi)} = \dfrac{150}{\sin(85°+2\phi)}$$

as $\tan\phi = \mu = 0.32$

$\phi = 17.74°$

$R_2 = \dfrac{150.\sin(110°-17.74°)}{\sin(85°+2\times17.75°)}$

$R_2 = 173.92$ kN

Figures 5.9 (c) and (d) show resultant reactions acting on wedges.

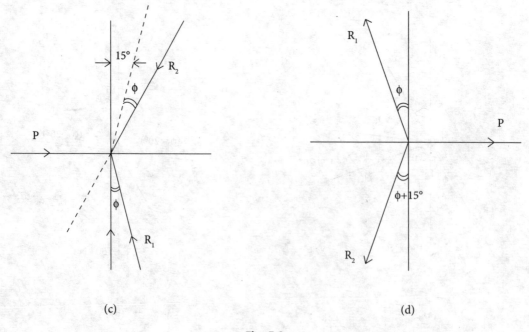

(c)   (d)

Fig. 5.9

Application of Friction

Using Lami's theorem,

$$\frac{P}{\sin(180°-2\phi-15°)} = \frac{R_2}{\sin(90°+\phi)} = \frac{R_1}{\sin(90°+\phi+15°)}$$

$$P = \frac{R_2 \sin(165°-2\phi)}{\sin(90°+\phi)}$$

Substituting $\phi = 17.74°$ and R_2 value for P,

$$P = \frac{173.92 \sin(165° - 2 \times 17.74°)}{\sin(90° + 17.74°)}$$

P = 140.86 kN

Example: 5.8

Two wedges of negligible weights are used to raise a block of weight 240 kN as shown in Fig. 5.10. If angle of friction is 18° for all contact surfaces then determine the value of force P which can impend the block upward.

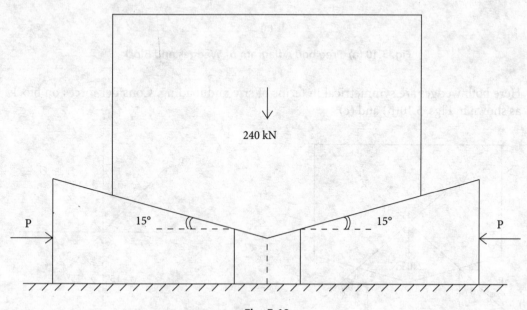

Fig. 5.10

Solution:

Consider F.B.Ds of block and two wedges as shown in Fig. 5.10 (a)

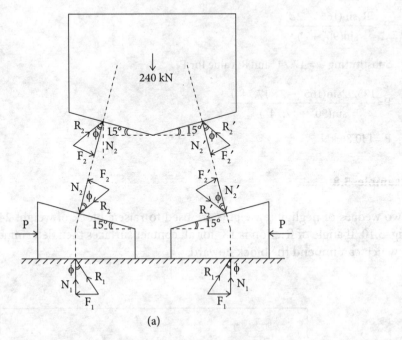

Fig. 5.10 (a) Free body diagram of Wedges and Block

Here both wedges are symmetrical in terms of size and loading. Consider forces on block as shown in Figs. 5.10(b) and (c),

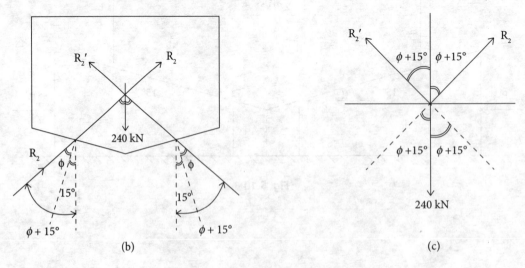

Fig. 5.10 Free body diagram of Block

Application of Friction

$$\sum X = 0, \ R_2' \sin(\phi+15°) = R_2\sin(\phi+15°)$$
$$\text{as } \sin(\phi+15°) \neq 0, \text{ thus } R_2' = R_2$$
$$\sum Y = 0, \ 2R_2\cos(\phi+15°) = 240$$
$$2R_2\cos(18°+15°) = 240$$
$$R_2 = 143.08 \text{ kN}$$

Consider forces acting on a single wedge as shown in Figs. 5.10 (d) and (e),

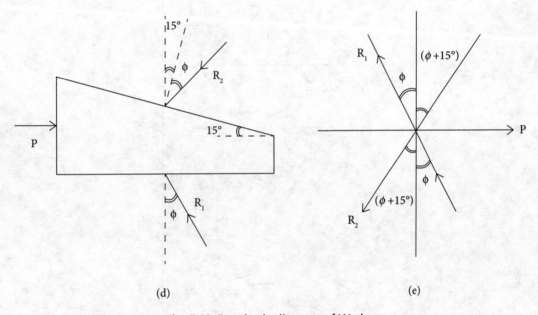

Fig. 5.10 Free body diagram of Wedge

Using Lami's theorem,

$$\frac{P}{\sin(180° - 2\phi - 15°)} = \frac{R_2}{\sin(90° + \phi)} = \frac{R_1}{\sin(90° + \phi + 15°)}$$

$$P = \frac{R_2 \cdot \sin(165° - 2\phi)}{\cos\phi}$$

Substituting values of R_2 and ϕ for P,

$$P = \frac{143.08 \sin(165° - 36°)}{\cos 18°}$$

$$P = 116.92 \text{ kN}$$

Example: 5.9

A body is supporting a load of 160 kN. Just to tighten body A, the triangular wedge B is pushed down as shown in fig 5.11. Determine the minimum value of force P to perform the task. Take angle of triangular wedge as 15° and coefficient of friction for all contact surfaces is 0.28.

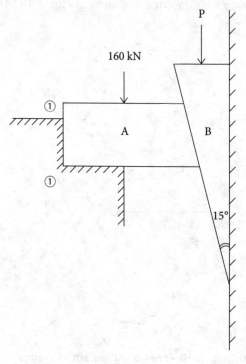

Fig. 5.11

Solution:

The minimum force P applied on wedge B just tightens the body A, which means the body A just impends to get tightened against surface (1)-(1). Thus no reaction is offered by surface (1)-(1) on body A. Consider F.B.Ds of both bodies as shown in Fig. 5.11(a).

$\tan\phi = 0.28$

$\phi = 15.64°$

Application of Friction

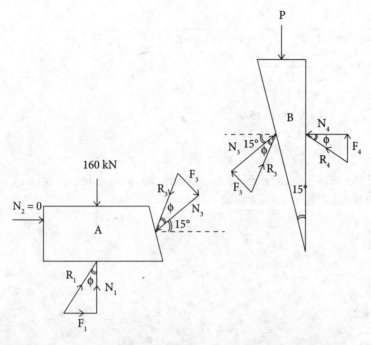

Fig. 5.11 (a) Free body diagram of Block and Wedge

Consider the skeleton of forces from F.B.D. of body A as shown in Fig. 5.11(b). Using Lami's theorem,

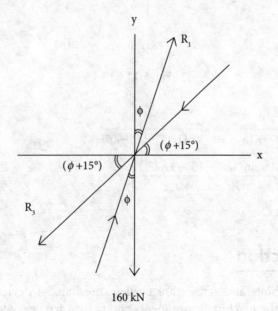

Fig. 5.11 (b)

$$\frac{R_1}{\sin(90°-\phi-15°)} = \frac{R_3}{\sin(180°-\phi)} = \frac{160}{\sin(90°+2\phi+15°)}$$

$$R_3 = \frac{160\sin(180°-\phi)}{\sin(105°+2\phi)}$$

since $\phi = 15.64°$

$$R_3 = \frac{160\sin(180°-15.64°)}{\sin(105°+31.28°)}$$

$$R_3 = 62.41 \text{ kN}$$

Consider the skeleton of forces from F.B.D. of body B as shown in Fig. 5.11(c). Using Lami's theorem,

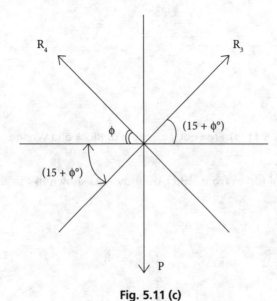

Fig. 5.11 (c)

$$\frac{R_3}{\sin(90°+\phi)} = \frac{R_4}{\sin(90°+15°+\phi)} = \frac{P}{\sin(180°-15°-2\phi)}$$

$$P = \left[\frac{62.41 \times \sin(180°-15°-31.28°)}{\sin(90°+15.64°)}\right]$$

$$P_{min} = 46.84 \text{ kN}$$

5.3 Screw Friction

Fasteners like nuts, bolts and screws are used in mechanical devices generally for two purposes, i.e. to join the machine components and to lift vary heavy loads. These fasteners have either V-threads or square threads of helical pattern. Square threads are used to lift

Application of Friction

very heavy loads, while V-threads are used for fastening work. Screw friction is a kind of resistive property which is utilized in Screw jack to lift the heavy loads.

Relation between lead and pitch
The axial distance travelled by screw during one rotation about its axis is called as *Lead (L)*.

The distance between two consecutive threads of a screw is called as *Pitch (p)*.

If a screw is threaded by 'n' times then the Lead is given by

Lead = n × Pitch

Thus for a single threaded screw, *Lead=Pitch*
However for double threaded screw, *Lead = 2 × Pitch*

Relation between lead angle and friction angle
The raising or lowering of the load on the screw jack can be treated as the raising or lowering of the load on an inclined plane whose slope is equal to the slope of the threads. This slope of threads or inclined plane is called as the lead angle or helix angle. The lead angle can be determine by the ratio of (axial upward movement) the lead (L) and (distance travelled in horizontal direction) the circumference of the mean circle ($2\pi R$) during one rotation of the screw as shown in Fig. 5.12.

Thus the lead angle or helix angle is given by,

$\tan \alpha = L/2\pi R$

where 'α' and 'R' are the lead angle and mean radius, respectively.

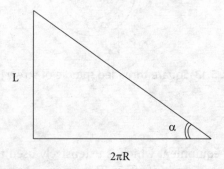

Fig. 5.12 Inclined plane of sloped lead angle

Consider Fig. 5.13; if R_o and R_i are the outer and inner radius of a screw then the mean radius is given by $R = \dfrac{R_o + R_i}{2}$ and pitch, p is given by $p/2 = R_o - R_i$

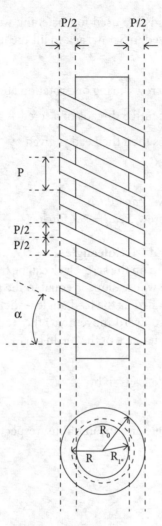

Fig. 5.13 Square threaded spindle of screw jack

Screw jack

Screw jack is mechanical equipment which is extensively used to raise, hold or lower the heavy loads by using low magnitude of effort. This equipment contains a fixed nut through which a larger head spindle passes. The spindle contains square threads of helical pattern over its surface. A long bar is inserted in to the spindle head which contains load 'W'. The spindle is rotated in the nut, by applying little effort to the long bar to raise or lower the load 'W' as shown in the Fig. 5.14.

Since coulomb's law states that friction does not depend on the area of contact, thus the load 'W' can be considered as acting at the mean radius of the square thread in screw jack. If a little effort 'T' is applied on the bar that acts as an equivalent force 'S' acting at the mean radius will satisfy $T \times l = S \times R$.

Application of Friction　　　　　　　　　　　　　　　　　　　　　　　　　209

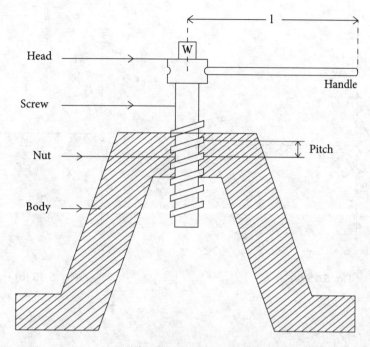

Fig. 5.14 Screw jack

Expression of effort to lift the load

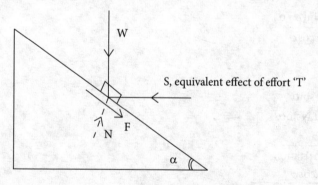

Fig. 5.15

Let the load 'W' is to be raised by using effort 'T' applied on the bar. The equivalent effort S causes to impend the block up at mean radius R. Thus frictional force F acts downwards as shown in the Fig. 5.15.

This concurrent coplanar force system can be considered as shown in Figs. 5.15(a) and (b),

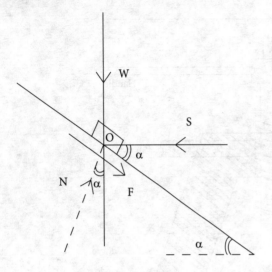
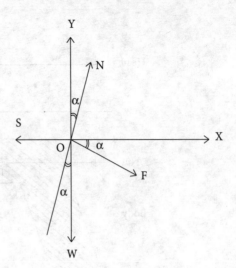

Fig. 5.15 (a)   Fig. 5.15 (b)

$\sum X = 0$,

$S = N\sin\alpha + F\cos\alpha$

$S = N\sin\alpha + \mu.N\cos\alpha$ as, $\mu = \tan\phi$

$S = N\sin\alpha + \tan\phi.N\cos\alpha$

$S = \dfrac{N\sin\alpha\cos\phi + N.\sin\phi.\cos\alpha}{\cos\phi}$

$S = \dfrac{N\sin(\phi+\alpha)}{\cos\phi}$

$\sum Y = 0$,

$N.\cos\alpha = W + F\sin\alpha$

$N.\cos\alpha = W + \mu.N\sin\alpha$

$N = \dfrac{W}{(\cos\alpha - \tan\phi.\tan\alpha)}$

$N = \dfrac{W.\cos\phi}{(\cos\phi.\cos\alpha - \sin\phi.\sin\alpha)}$

$N = \dfrac{W.\cos\phi}{\cos(\phi+\alpha)}$

Substituting equation (2) in equation (1),

$S = \dfrac{W.\cos\phi}{\cos(\phi+\alpha)} \cdot \dfrac{\sin(\phi+\alpha)}{\cos\phi}$

$S = W.\tan(\phi+\alpha)$

Application of Friction

as $T \times l = S \times R$

$$T \times l = W \tan(\phi + \alpha) \times R$$

$$T = \frac{W.R}{l} \tan(\phi + \alpha)$$

Similarly, the expression of effort to lower the load 'W' can be derived and is given by

$$T = \frac{W.R}{l} \tan(\phi - \alpha)$$

It is to be noted that if the angle of helix (α) is less than angle of friction (ϕ), the load W will remain hold and screw jack is considered in state of self-locking. However, if the angle of helix (α) exceeds the angle of friction (ϕ) then the load will come down in the absence of effort.

Expression of efficiency of screw jack

We know that the effort required to raise the load is given by $T = \frac{WR}{l} \tan(\phi + \alpha)$

The minimum effort will be required if the threads are frictionless, i.e. $\phi = 0$

i.e., $T_{mim} = \frac{WR}{l}.\tan\alpha$

Thus the efficiency of screw jack can be determined by using,

$$\eta = \frac{T_{min}}{T_{actual}} \times 100$$

$$\eta = \left(\frac{\tan\alpha}{\tan(\phi + \alpha)} \times 100\right)$$

The maximum efficiency of a screw jack is given by,

$$\frac{d\eta}{d\alpha} = 0$$

where angle of friction, ϕ is constant, further by solving, we get, $\alpha = 45° - \phi/2$

Substituting value of α in $\eta = \left(\frac{\tan\alpha}{\tan(\phi + \alpha)} \times 100\right)$

$$\eta_{max} = \left(\frac{1 - \sin\phi}{1 + \sin\phi}\right) \times 100$$

Example: 5.10

A screw spindle of single square threaded has a mean radius 60 mm and a pitch of 20 mm. If a user wants to raise a load of 150 kN then determine the force required to be applied at the end of a 75 cm long lever. Determine the efficiency of the jack also.

State screw spindle is under self-locking or not? If yes, then determine the force required to be applied at the end of lever to lower the same weight. Consider the co-efficient of static friction is 0.23.

Solution:

R = 60 mm, p = 20 mm, W = 150 kN, l = 75 cm = 750 mm
η = ?, T_{raise} = ?, T_{lower} = ?, μ = 0.23

The force required to raise the load is given by,

$$T_{raise} = \frac{W.R}{l}\tan(\phi + \alpha) \quad \ldots\ldots (1)$$

where $\phi = \tan^{-1}(\mu) = \tan^{-1}(0.23)$
$\phi = 12.95°$

As screw jack is single threaded,
$\therefore$ L = p = 20 mm

$$\tan\alpha = \frac{L}{2\pi R}$$

$$\tan\alpha = \frac{20}{25 \times 60} \quad \therefore \alpha = 3.04°$$

Using equaiton (1),

$$T_{raise} = \frac{150 \times 10^3 \times 60}{750}\tan(12.95° + 3.04°)$$

$$T_{raise} = 3.44 \times 10^3 \, N$$

The efficiency of a screw jack is given by

$$\eta = \left(\frac{1-\sin\phi}{1+\sin\phi}\right) \times 100 = \left(\frac{1-\sin 12.95°}{1+\sin 12.95°}\right) \times 100$$

$\eta = 63.39\%$

As, $\alpha < \phi$, the screw is self locked.

Force required to lower the load is given by

$$T_{lower} = \frac{WR}{l}\tan(\phi - \alpha)$$

$$= \frac{150 \times 10^3 \times 60}{750}\tan(12.95° - 3.04°)$$

$$T_{lower} = 2.09 \times 10^3 \, N$$

Application of Friction

Example: 5.11

A single square threaded screw jack has a pitch of 18 mm and a mean radius of 45 mm. Determine the force required to raise a load of 10 kN. Take length of the rod as 80cm and coefficient of static friction as 0.18.
Determine the efficiency of the screw jack also.

Solution:

For single square threaded screw, lead = pitch = p =18 mm,
R= 45 mm, T_{raise} =?, W=10 kN, l = 80 cm = 800 mm, μ = 0.18

$$T_{raise} = \frac{W.R}{l}\tan(\phi+\alpha)$$

$$\tan\alpha = \frac{lead}{2\pi R}, \quad and \quad \phi = \tan^{-1}(\mu)$$

$$\tan\alpha = \frac{18}{2\pi \times 45}, \quad \phi = \tan^{-1}(0.18)$$

$$\alpha = 3.64 \quad and \quad \phi = 10.2°$$

$$T_{raise} = \frac{10 \times 10^3 \times 45}{800}\tan(10.2+3.64)$$

$$T_{raise} = 138.58 N$$

The efficiency of a screw jack is given by

$$\eta = \left(\frac{1-\sin\phi}{1+\sin\phi}\right) \times 100 = \left(\frac{1-\sin 10.2°}{1+\sin 10.2°}\right) \times 100$$

$$\eta = 69.91\%$$

Example: 5.12

A screw spindle of screw jack has an inner diameter of 50mm and a pitch of 10 mm. If coefficient of friction between the screw and nut is 0.20, then determine:

(i) Moment required to raise a load of 12 kN
(ii) The efficiency of the screw jack
(iii) Moment required to lower the same load of 12 kN
(iv) pitch required for maximum efficiency of the screw

$$R_i = \frac{D_i}{2} = \frac{50}{2} = 25 \text{ mm}, \, p = 10 \text{ mm}, \, \mu = 0.20$$

Solution:

(i) W = 12 kN
Let the screw spindle is single-square threaded,
thus, lead = pitch = p = 10 mm

To calculate lead angle, first we have to determine mean radius, as $\tan\alpha = \left(\dfrac{p}{2\pi R}\right)$

Outer radius, $R_o = R_i + p/2$
$= 25 + 10/2$
$= 30$ mm

mean radius, $R = \dfrac{R_o + R_i}{2}$
$= \dfrac{30 + 25}{2}$
$= 27.5$ mm

$\tan\alpha = \dfrac{10}{2\pi \times 27.5}$ ∴ $\alpha = 3.31°$

and $\phi = \tan^{-1}(\mu)$
$= \tan^{-1}(0.20)$
$\phi = 11.31°$

(i) As force required to raise the load is given by,

$$T_{raise} = \dfrac{W.R}{l}\tan(\phi + \alpha)$$

Thus moment required to raise the load is given by,

$M_{raise} = T_{raise} \times l = WR\tan(\phi + \alpha)$
$M_{raise} = 12 \times 10^3 \times 27.5 \ \tan(11.31° + 3.31°)$
$= 86.08 \times 10^3$ Nmm
$M_{raise} = 86.08$ Nm

(ii) The efficiency of screw jack is given by

$\eta = \left(\dfrac{1 - \sin\phi}{1 + \sin\phi}\right) \times 100$

$= \left(\dfrac{1 - \sin 11.31°}{1 + \sin 11.31°}\right) \times 100$

$\eta = 78.59\%$

(iii) Moment required to lower the load is given by,

$M_{lower} = T_{lower} \times l$
$= WR\tan(\phi - \alpha)$
$= 12 \times 10^3 \times 27.5 \ \tan(11.31° - 3.31°)$
$= 46.38 \times 10^3$ Nmm
$= 46.38$ Nm

Application of Friction

(iv) Pitch required for maximum efficiency of the screw will be given by

$$\tan\alpha = \frac{p}{2\pi R}$$

where $\alpha = (45° - \phi/2)$

$$= \left(45° - \frac{11.31°}{2}\right)$$

$\alpha = 39.35°$

$$\tan 39.35° = \frac{p}{2\pi \times 27.5}$$

$p = 141.68$ mm

Example: 5.13

A double threaded screw jack of pitch 20 mm, mean radius 60 mm is used to raise and lower the load. Determine the load that can be raised and lowered by exerting moment of 750 Nm. Take coefficient of static friction as 0.16.

Solution:

Lead = 2 × pitch = 2 × 20 = 40 mm

R = 60 mm, $M_{raise} = M_{lower} = 750$ Nm $= 750 \times 10^3$ Nmm

$$\tan\alpha = \frac{\text{Lead}}{2\pi R}$$

$$\tan\alpha = \frac{40}{2\pi \times 60}$$

$\alpha = 6.06°$

$\phi = \tan^{-1}(0.16)$

$\phi = 9.09°$

$(Moment)_{raise} = T_{raise} \times l$

$750 \times 10^3 = WR \tan(\phi + \alpha)$

$750 \times 10^3 = W \times 60 \tan(9.09° + 6.06°)$

$W_{raise} = 46.17 \times 10^3$ N

$W_{raise} = 46.17$ kN

Similarly

$(M)_{lower} = T_{lower} \times l$

$= W_{lower} R \tan(\phi - \alpha)$

$750 \times 10^3 = W_{lower} \times 60 \tan(9.09° - 6.06°)$

$W_{lower} = 236.15 \times 10^3$ N

$W_{lower} = 236.15$ kN

Example: 5.14

A screw jack carries a load of 400 N. It has a square thread single start screw of 20 mm pitch and 50 mm mean diameter. The coefficient of friction between the screw and its nut is 0.27. Calculate torque required to raise the load and efficiency of the screw. What is the torque required to lower the load?

(UPTU, IInd Sem, 2003–04)

Solution:

W = 400N, lead = pitch = 20 mm
d_{mean} = 50mm, R_{mean} = 25mm, μ = 0.27

(i) Torque required to raise the load is given by

$(Torque)_{raise} = T_{raise} \times l$
$= W.R.\tan(\phi + \alpha)$
$\phi = \tan^{-1}(\mu) = \tan^{-1}(0.27)$
$\phi = 15.11°$

$\tan\alpha = \dfrac{P}{2\pi R}$
$\tan\alpha = \dfrac{20}{2\pi \times 25}$
$\alpha = 7.26°$
$(Torque)_{raise} = 400 \times 25 \tan(15.11° + 7.26°)$
$= 4.12 \times 10^3 \text{Nmm}$
$(Torque)_{raise} = 4.12 \text{Nm}$

$\eta = \left(\dfrac{1-\sin\phi}{1+\sin\phi}\right) \times 100$

$= \left(\dfrac{1-\sin 15.11°}{1+\sin 15.11°}\right) \times 100$

$\eta = 58.65\%$

$(Torque)_{lower} = WR \tan(\phi - \alpha)$
$= 400 \times 25 \tan(15.11° - 7.26°)$
$= 1.38 \times 10^3 \text{ Nmm}$

$(Torque)_{lower} = 1.38 \text{ Nm}$

5.4 Belt Friction

Whenever friction exists between belt and pulley, the tension does not remain the same throughout the belt, as at one side the belt becomes tight which is called as tight side; however at another side it becomes loose, which is called as the slack side.

Let the tension on tight side and slack sides are T_t and T_s respectively.

Application of Friction

Consider the figure where driven pulley is pulled by driver pulley, the tension on this side of

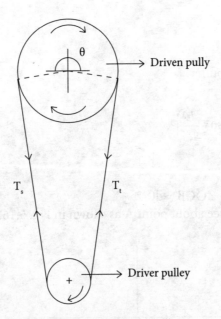

Fig. 5.16 Belt and Pulley

belt is more than the tension on another side due to frictional force between belt and pulley. Let the angle of contact between belt and pulley is θ. Consider an elemental portion of belt for free body diagram as shows below in Figs. 5.16(a) and (b). Let the elemental portion makes angle of contact $d\theta$ and if tension on slack side is T then tension on tight side will greater i.e., $T + dT$. As the belt impends clockwise (right-side), frictional force, F, acts left side. The pulley offers normal reaction N on belt as shown Fig. 5.16 (b).

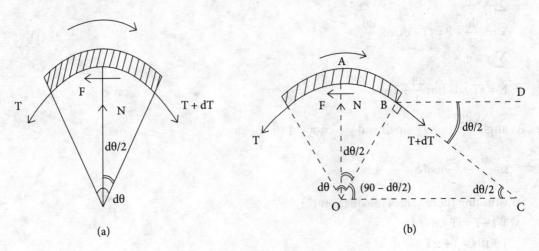

Free body diagram of elemental belt

as $\angle AOB = \dfrac{d\theta}{2}$

thus $\angle BOC = \left(90° - \dfrac{d\theta}{2}\right)$

Since BC is tangent to the pulley,
Thus $\angle OBC = 90°$

$\therefore \angle OCB = 180° - 90° - \left(90 - \dfrac{d\theta}{2}\right)$

$= \dfrac{d\theta}{2}$

As BD || OC thus $\angle DBC = \angle OCB = d\theta/2$
Considering skeleton of force about point A as shown in Fig. 5.16(c)

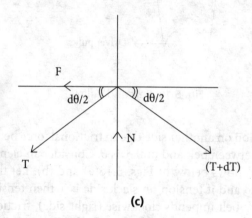

(c)

$\sum X = 0$,

$T.\cos\dfrac{d\theta}{2} + F = (T + dt)\cos\dfrac{d\theta}{2}$ ..... (1)

$\sum Y = 0$,

$N = (T + dT)\sin\dfrac{d\theta}{2} + T.\sin\dfrac{d\theta}{2}$ ..... (2)

As angle $\dfrac{d\theta}{2}$ is very small and we know in such a case,

$\sin\dfrac{d\theta}{2} \to \dfrac{d\theta}{2}$ and $\cos\dfrac{d\theta}{2} \to 1$

Substituting these values in equation (1),

$T.1 + F = (T + dT).1$

$T + \mu N = T + \alpha T$

$\mu N = dT$ ..... (3)

Application of Friction

where, μ is the coefficient of friction between belt and pulley.
From equation (2),

$$N = (T + dT) \cdot \frac{d\theta}{2} + T \cdot \frac{d\theta}{2}$$

$$= T \cdot \frac{d\theta}{2} + dT \cdot \frac{d\theta}{2} + T \cdot \frac{d\theta}{2}$$

$$N = T \cdot d\theta \qquad \qquad \ldots\ldots (4)$$

As dT and $\frac{d\theta}{2}$ are very small individually and their product will further become very small, it can be neglect.

Substituting the value of N in equation (3),

$$\mu \times T \times d\theta = dT$$

$$\frac{dT}{T} = \mu \cdot d\theta$$

Integrating both sides with limits where angles increases from zero to θ and tension varies from T_s to T_t during that

$$\int_{T_s}^{T_t} \frac{dT}{T} = \int_0^{\theta} \mu \cdot d\theta$$

$$\left[\log_e T\right]_{T_s}^{T_t} = \mu [\theta]_0^{\theta}$$

$$\log_e T_t - \log_e T_s = \mu \cdot \theta$$

$$\log_e \frac{T_t}{T_s} = \mu \cdot \theta$$

$$\frac{T_t}{T_s} = e^{\mu \cdot \theta}$$

Note:

(i) θ is in radians only
(ii) If $\mu = 0$ i.e., for smooth pulley

$$\frac{T_t}{T_s} = e^0 = 1$$

thus $\frac{T_t}{T_s} = 1$

$T_t = T_s$

(iii) The difference of tension on both side = $(T_t - T_s)$ causes rotation of pulley.

Example: 5.15

A person wants to support a load of 120 N by using a rope making $2\frac{1}{4}$ turns around a horizontal drum as shown in the Fig. 5.17. Determine the range of pull 'P' required by person to keep the load in equilibrium, if $\mu = 0.28$.

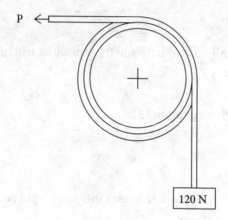

Fig. 5.17

Solution:

Given,

$$\theta = 2\frac{1}{4} \text{ turns}$$

$$= \frac{9}{4} \text{ turns}$$

$$\theta = \frac{9}{4} \times 2\pi \quad (1 \text{ turn} = 2\pi)$$

$$\theta = 4.5\pi \text{ and } \mu = 0.28$$

To determine range of pull P, there will be two cases.

Case-1: When P is applied maximum then 120 N impends upwards.
i.e., $T_t = P$ and $T_s = 120 N$
Using formula,

$$\frac{T_t}{T_s} = e^{\mu.\theta}$$

$$\frac{P_{max}}{120} = e^{(4.5 \times \pi \times 0.28)}$$

$$P_{max} = 6284.86 N$$

Application of Friction

Case-II: Where P is just applied to hold the load i.e., P_{min}

$T_t = 120N$, $T_s = P_{min}$

$$\frac{120}{P_{min}} = e^{(4.5 \times \pi \times 0.28)}$$

$P_{min} = 2.29 N$

Thus the range of pull is 2.29N to 6284.66N.

Example: 5.16

Determine the maximum weight that can he hold by a man applying 280 N pull on the rope. The rope is wrapped by $1\frac{1}{2}$ turns around the horizontal drums. Take coefficient of friction between rope and drum as 0.26.

Solution:

Given,

$$\theta = 1\frac{1}{2} \text{ turns}, \ \mu = 0.26$$

$$= \frac{3}{2} \text{ turns}$$

$$= \frac{3}{2} \times 2\pi$$

$$\theta = 3\pi$$

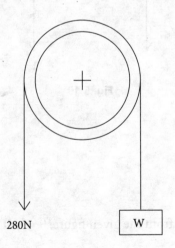

Fig. 5.18

As supported load is given maximum
thus, $T_t = W$ and $T_s = 280 N$

Using formula, $\dfrac{T_t}{T_s} = e^{\mu\theta}$

$$\dfrac{W}{280} = e^{(0.26 \times 3\pi)}$$

$$W = 3246.17 \text{ N}$$

Example: 5.17

A horizontal drum of a belt drive carries the belt over a semicircle around it. It is rotated anti-clockwise to transmit a torque of 300 Nm. If the coefficient of friction between the belt and drum is 0.3, calculate the tension in the limbs 1 and 2 of the belt as shown in figure and the reaction on the bearings. The drum has a mass of 20 kg and the belt is assumed to be massless.

(U.P.T.U. IInd Sem, 2001–2002)

Solution:

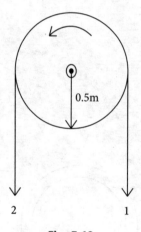

Fig. 5.19

Given,
Torque, M = 300 Nm
$\mu = 0.3$
$T_2 = T_t = ?$
$T_1 = T_s = ?$
Mass of drum, m_d = 20 kg and from the given figure, θ = half turn, i.e., 180°

or, $\theta = \pi$

$$\dfrac{T_t}{T_s} = e^{\mu\theta}$$

Application of Friction

$$\frac{T_2}{T_1} = e^{(0.3 \times \pi)}$$

$T_2 = 2.56 \times T_1$ ..... (1)

Torque on pulley = Axial force on pulley × radius

$300 = (T_2 - T_1) \times 0.5$ ..... (2)

Substituting value of T_2 in equation (2),

$300 = (2.56 T_1 - T_1) 0.5$

$T_1 = 384.62$ N and $T_2 = 984.62$ N

Reaction on the bearing will be as shown in Fig. 5.19 (a).

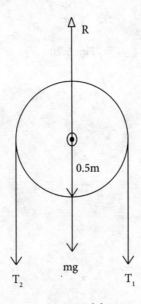

Fig. 5.19 (a)

$\sum Y = 0,$

$R = T_2 + T_1 + m_d \times g$

$= 984.62 + 384.62 + 20 \times 9.81$

$= 1565.44$ N

Example: 5.18

A horizontal drum of diameter 700 mm requires torque of 360 Nm to start rotating. If the coefficient of friction between belt and the drum is 0.32 then determine the pull required on drum to produce sufficient tension in the belt to start rotation of the drum. Take masses of the drum and belt as negligible.

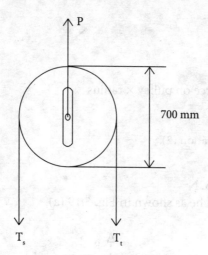

Fig. 5.20

Solution:

Given
$d = 700$ mm $= 0.7$ m
Torque, $M = 360$ Nm
$\mu = 0.32$
$P = ?$
We know that,

$$\text{Torque} = (T_t - T_s) \times \text{radius}$$
$$360 = (T_t - T_s) \times \frac{0.7}{2} \quad \quad \ldots(1)$$

$\dfrac{T_t}{T_s} = e^{\mu\theta}$ where $\mu = 0.32$ and $\theta = 180° = \pi$

$\dfrac{T_t}{T_s} = e^{(0.32 \times \pi)}$ i.e. $T_t = 2.73 \cdot T_s$ $\ldots(2)$

from equation (1), $360 = (2.73 T_s - T_s) \times \dfrac{0.7}{2}$

$T_s = 594.55$ N

from equation (2), $T_t = 2.73 \times 594.55 = 1623.12$

$P = T_t + T_s = 594.55 + 1623.12$

$P = 2217.67$ N

Application of Friction

Example: 5.19

A belt is stretched over two identical pulleys of diameter D meter. The initial tension in the belt throughout is 2.4 kN when the pulleys are at rest. In using these pulleys and belt to transmit torque, it is found that the increase in tension on one side is equal to the decrease on the other side. Find the maximum torque that can be transmitted by the belt drive, given that the coefficient of friction between belt and pulley is 0.30.

(U.P.T.U. Ist Sem 2002–03)

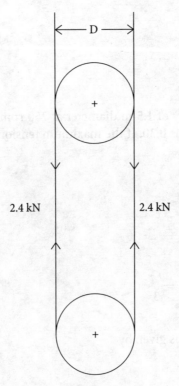

Fig. 5.21

Solution:

When pulleys are at rest, the tension on both sides is 2.4 kN. During transmission of torque, let tension increase by dT on the tight side; then the tension on the slack side will decrease by the same amount dT. Thus,

$$T_t = (2.4 + dT)$$
$$\text{and } T_s = (2.4 - dT)$$
$$\frac{T_t}{T_s} = e^{\mu\theta}, \text{ and } \theta = 180° = \pi$$

$$\frac{(2.4+d\mathrm{T})}{(2.4-d\mathrm{T})} = e^{0.3\times\pi}$$

$$d\mathrm{T} = 1.052\ kN$$

Maximum torque that can be transmitted, M
$$= (T_t - T_s) \times \text{radius}$$
$$= [(2.4+d\mathrm{T}) - (2.4-d\mathrm{T})] \times \text{radius}$$
$$= 2.d\mathrm{T} \times \frac{D}{2}$$
$$M = 1.052.D\ kNm$$

Example: 5.20

A belt is running over a pulley of 1.5 m diameter at 250 rpm. The angle of contact is 120° and the coefficient of friction is 0.30. If the maximum tension in the belt is 400 N, find the power transmitted by the belt.

(U.P.T.U. Ist Sem, C.O. 2003)

Solution:

Given,
$$d = 1.5\ m,\ N = 250\ rpm,$$
$$\mu = 0.3,\ \theta = 120° = 120° \times \frac{\pi}{180} = \frac{2\pi}{3}$$
$$T_t = 400\ N$$

Power transmitted by the belt is given by
$$P = \frac{2\pi NT}{60}\ \text{watt}$$

where, T = Average Torque in Nm
N = number of revolution per min

as $T = M = (T_t - T_s) \times \text{radius}$

$$\frac{T_t}{T_s} = e^{\mu\theta}$$

$$\frac{400}{T_s} = e^{\left(0.3 \times \frac{2\pi}{3}\right)}$$

$$T_s = 213.90\ N$$

Application of Friction

$$\therefore P = \frac{2\pi N.(T_t - T_s) \times \text{radius}}{60} \text{ watt}$$

$$= \frac{2\pi \times 250(400 - 213.9)}{60} \times \frac{1.5}{2}$$

P = 3654.06 watt

P = 3.65 kw

Example: 5.21

Consider the Fig. 5.22 where coefficient of friction is 0.2 between the rope and fixed pulley, and between other surfaces of contact, $\mu = 0.3$. Determine minimum weight W to prevent the downward motion of the 100 N body.

(U.P.T.U Ist Sem, 2001–2002)

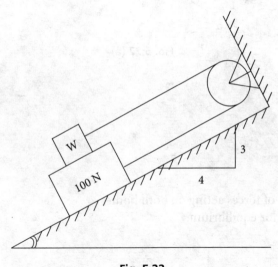

Fig. 5.22

Solution:

If W is minimum then 100N impends downward, thus T_t and T_s will operate on 100 N and W side respectively.

The F.B.Ds of all bodies will be as given below:

Let the angle of inclination is α.
tan α = 3/4

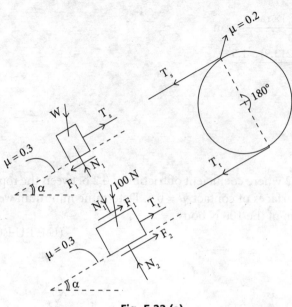

Fig. 5.22 (a)

from pulley,

$\theta = 180° = \pi$

$\dfrac{T_t}{T_s} = e^{\mu\theta} = e^{(0.2 \times \pi)}$

$T_t = 1.87 \times T_s$ ..... (1)

Considering skeleton of forces acting on both bodies,
Consider lower body for equilibrium,

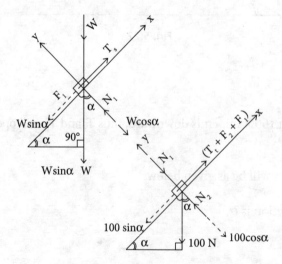

Fig. 5.22 (b) Free body diagrams of both weights

Application of Friction

$$\sum X = 0$$
$$100\sin\alpha = T_t + F_2 + F_1$$
$$100\sin\alpha = T_t + \mu_2 N_2 + \mu_1 N_1 \quad \ldots (2)$$
$$\sum Y = 0,$$
$$N_2 = N_1 + 100\cos\alpha$$

Substituting value of N_2 in equation (2),
$$100\sin\alpha = T_t + 0.3(N_1 + 100\cos\alpha) + 0.3N_1$$
$$T_t = (100\sin\alpha - 0.3N_1 - 0.3N_1 - 30\cos\alpha) \quad \ldots (3)$$

Consider upper block for equilibrium,
$$\sum X = 0,$$
$$W\sin\alpha + F_1 = T_s$$
$$W\sin\alpha + \mu_1 N_1 = T_s \quad \ldots (4)$$

as $\tan\alpha = \dfrac{3}{4}, \sin\alpha = \dfrac{3}{5}, \cos\alpha = \dfrac{4}{5}$

$$\sum Y = 0,$$

from equation (4),
$$W\sin\alpha + 0.3W\cos\alpha = T_s \quad \ldots (5)$$

Substitute value of N_1, T_t and T_s in equation (1),
$$100\sin\alpha - 0.3W\cos\alpha - 0.3W\cos\alpha - 30\cos\alpha = 1.87(W\sin\alpha + 0.3W\cos\alpha)$$

Substituting values of $\sin\alpha$ and $\cos\alpha$,
$$100 \times \frac{3}{5} - 0.6W \times \frac{4}{5} - 30 \times \frac{4}{5} = 1.87\left(W \times \frac{3}{5} + 0.3W \times \frac{4}{5}\right)$$
$$W = 17.55 \text{ N}$$

5.5 Band Brakes

Band brakes are known as power absorption device. It involves a simple mechanism of flexible band, rotating drum and lever as shown in Fig. 5.23. In a band brake the flexible band passes over the drum and is tied to the lever which is kept hinged. The thickness of band and weight of lever are assumed as negligible. It is used to maintain the constant torque on drum or to stop the drum by an effort applied at the free end of the lever. During effort, the band grips the rotating drum and reduces the speed continuously due to frictional force produced between the band and drum.

Consider an example where a drum is rotating anticlockwise as shown in Fig. 5.23. The free body diagram of drum, band and lever is detailed below:

In this case the band applies friction in clockwise direction on the drum as shown in the Fig. 5.24 (a), which means the drum applies the same friction on the band in anticlockwise direction as shown in the Fig. 5.24 (b). This causes more tension (T_t) at right side of the band than at the left side (T_s).

The ratio of tensions in the band brake is same as of belt drive i.e. $\dfrac{T_t}{T_s} = e^{\mu\theta}$

where, 'θ' is the angle of contact between band and drum in radian;
and 'μ' is the coefficient of kinetic friction due to slippage between band and drum unlike the belt drive.

The torque on the rotating drum is given by $Torque = (T_t - T_s) \times r$

where, 'r' is the radius of the drum.

The free body diagram of lever is as shown in Fig. 5.25 which can be analysed further to evaluate any unknown force T_t or T_s or P by taking moment through suitable point.

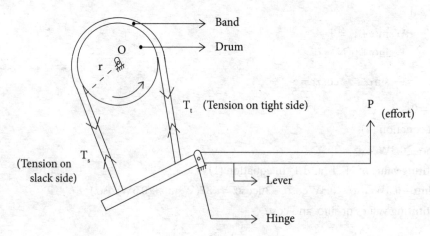

Fig. 5.23 Band brake

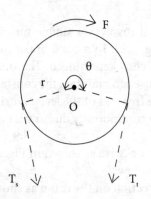

Fig. 5.24 (a) F.B.D. of drum

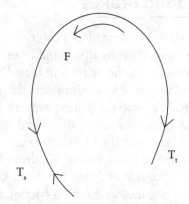

Fig. 5.24 (b) F.B.D. of Band

Application of Friction

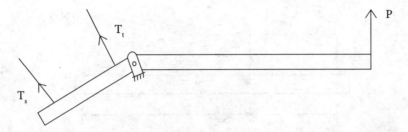

Fig. 5.25 F.B.D. of lever

Example: 5.22

Consider Fig. 5.26; determine the braking moment acting on drum if:

(i) The drum is rotating anticlockwise
(ii) The drum is rotating clockwise

Take coefficient of friction between band and drum as 0.28.

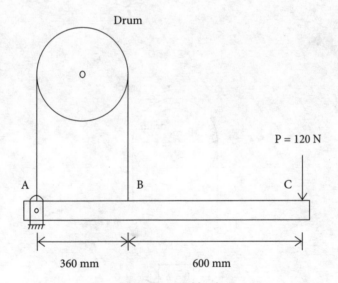

Fig. 5.26 Band brake

Solution:

(i) When drum is rotating anticlockwise:
In this situation, the band will apply friction clockwise on drum. Similarly, the drum applied friction on band anticlockwise which causes tension of larger value on B. Thus F.B.D of lever ABC will be as shown in Fig. 5.26 (a)

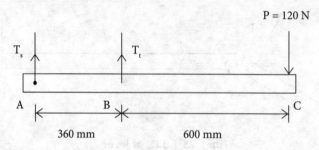

Fig. 5.26 (a) F.B.D. of lever

$\sum M_A = 0,$
$\quad T_t \times 360 = 120 \times 960$
$\quad T_t = 320 \, N$

Using ratio of tensions of belt friction,

$\dfrac{T_t}{T_s} = e^{\mu\theta}$

$\dfrac{320}{T_s} = e^{(0.28 \times \pi)}$

$T_s = 132.78 \, N$

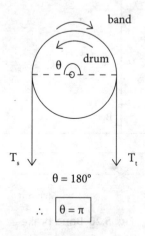

$\theta = 180°$

$\therefore \boxed{\theta = \pi}$

Fig. 5.26 (b) F.B.D. of drum

Braking moment or resisting torque on the drum will be given by

$= (T_t - T_s) \times r$
$= (320 - 132.78) \times \dfrac{360}{2}$
$= 33681.6 \, Nmm$
or $33.68 \, Nm$

Application of Friction

(ii) When drum is rotating clockwise:

In this situation, friction will act clockwise on the band, thus tension of larger value will be on A. Thus F.B.D of lever will be as shown in Fig. 5.26 (c)

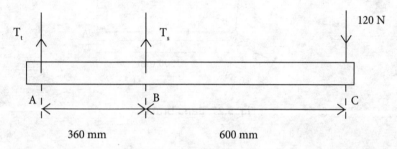

Fig. 5.26 (c) F.B.D. of lever

$\sum M_A = 0,$

$T_s \times 360 = 120 \times 960$

$T_s = 320 \, N$

$\dfrac{T_t}{T_s} = e^{\mu\theta}$

$\dfrac{T_t}{320} = e^{(0.28 \times \pi)}$

$T_t = 771.21 \, N$

The breaking moment or resisting torque is given by

$= (T_t - T_s) \times r$

$= (771.21 - 320) \times \dfrac{360}{2}$

$= 81218.66 \, Nmm$

$= 81.22 \, Nm$

Example: 5.23

In a differential band brake as shown in Fig. 5.27, determine force P to be applied at lever to maintain torque of 240 Nm on the drum if:

(i) the drum rotates anticlockwise
(ii) the drum rotates clockwise

Take angle of contact and coefficient of friction between drum and band brake as 250° and 0.24, respectively. The radius of drum is 210 mm.

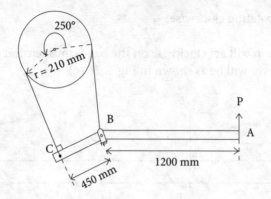

Fig. 5.27 Band brake

Solution:

(i) When drum is rotating anticlockwise, the friction on band brake will also act anticlockwise due to which tension at B will be higher than at C.

Thus F.B.D. of lever will be as shown in Fig. 5.27 (a)

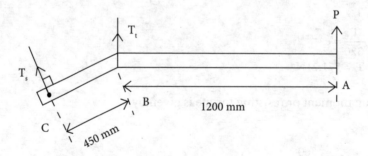

Fig. 5.27 (a) F.B.D. of lever

$\sum M_B = 0$,

$T_s \times 450 = P \times 1200$

$T_s = 2.67P$ ..... (1)

We know that,

$\dfrac{T_t}{T_s} = e^{\mu\theta}$

$\dfrac{T_t}{2.67P} = e^{\left(\dfrac{0.24 \times 250 \times \pi}{180}\right)}$

$T_t = 7.61P$ ..... (2)

Application of Friction

Torque maintained on the drum = $(T_t - T_s) \times r$
Substituting values from equation (1) and (2),

$$= (7.61P - 2.67P) \times \frac{210}{2}$$
$$= 518.7P \, Nmm$$
$$240 = 0.519P \, Nm$$
$$P = 462.43 \, N$$

(ii) When drum is rotating clockwise, the tension will be higher at C than at B.

The F.B.D. of lever will be as shown in Fig. 5.27 (b)

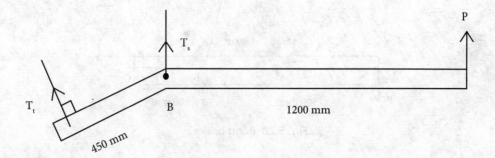

Fig. 5.27 (b) F.B.D. of lever

$\sum M_B = 0,$

$$T_t \times 450 = P \times 1200$$
$$T_t = 2.67 P \quad\quad\quad\quad\quad ..... (3)$$
$$\frac{T_t}{T_s} = e^{\mu\theta}$$
$$\frac{2.67P}{T_s} = e^{\left(\frac{0.24 \times 250 \times \pi}{180°}\right)}$$
$$T_s = 0.94P \quad\quad\quad\quad\quad ..... (4)$$

Toque on the drum = $(T_t - T_s) \times r$

$$= (2.67P - 0.94P) \times \frac{210}{2}$$
$$= 181.65P \, Nmm$$
$$240 = 0.18P \, Nm$$
$$P = 1333.33 \, N$$

Example: 5.24

Consider a band brake as shown in Fig. 5.28. The radium of drum is 150 mm and coefficient of friction between drum and brake as 0.25.

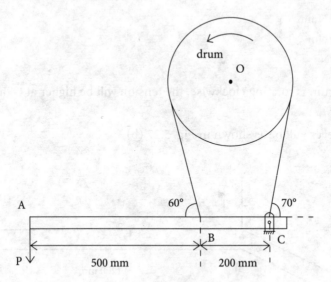

Fig. 5.28 Band brake

If the torque to be maintained at drum is 320 Nm then determine the required value of P to be applied on lever at A.

Solution:

As the drum is rotating anticlockwise, the tension will be larger at C than at B. The F.B.D. of lever is as shown in Fig. 5.28(a)

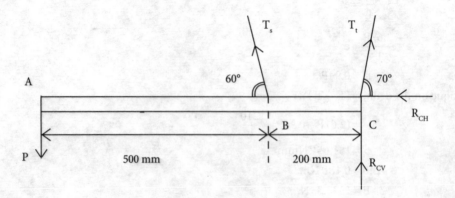

Fig. 5.28 (a) F.B.D. of lever

Application of Friction

$$\sum M_c = 0,$$
$$T_s \sin 60° \times 200 = P \times 700$$
$$T_s = 4.04P \quad\quad\quad\quad\quad\quad\quad\quad\quad\quad\quad\quad\quad\quad\quad\quad ..... (1)$$

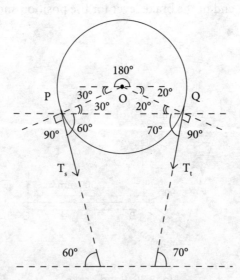

Fig. 5.28 (b) F.B.D. of drum

We know that,

$$\frac{T_t}{T_s} = e^{\mu\theta}$$

$$\frac{T_t}{T_s} = e^{0.25 \times \theta}$$

$$\theta = \angle POQ = 30° + 180° + 20°$$

$$\theta = 230° \text{ as shown in Fig. 5.28(b)}$$

$$\frac{T_t}{T_s} = e^{\left(0.25 \times 230 \times \frac{\pi}{180}\right)}$$

$$T_t = 2.73 T_s$$

Using equation (1),

$$T_t = 2.73 \times 4.04P$$
$$T_t = 11.03P$$

Torque required on brake drum

$$= (T_t - T_s).r$$

$$320 = (11.03P - 4.04P) \times \frac{150}{1000}$$

$$P = 305.19 N$$

Example: 5.25

A torque of 300 Nm acts on the brake drum is shown in Fig. 5.29. If the brake band is in contact with the brake drum through 250° and the coefficient of friction is 0.3, determine the force P applied at the end of the brake lever for the position shown in Fig. 5.29.

(MTU IInd Sem 2011–12)

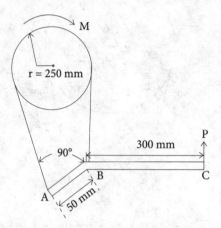

Fig. 5.29 Band brake

Solution:

$$\theta = 250° \times \frac{\pi}{180°} = 4.36 \text{ radian}$$

As drum is rotating clockwise, tension will be more on side A than B.

$\sum M_B = 0,$

$$T_t \times 50 = P \times 300$$
$$T_t = 6P \qquad \qquad \text{..... (1)}$$

We know that,

$$\frac{T_t}{T_s} = e^{\mu\theta}$$

$$\frac{6P}{T_s} = e^{(0.3 \times 4.36)}$$

$$T_s = 1.62P \qquad \qquad \text{..... (2)}$$

Torque on brake drum = $(T_t - T_s) \times r$

$$300 = (6P - 1.62P) \times \frac{250}{1000}$$

$$P = 273.97 \text{ N}$$

Application of Friction

Theoretical Problems

T 5.1 Briefly discuss the principle on which a screw jack works and state its applications.

T 5.2 Explain the following terms:
(i) Lead
(ii) Pitch
(iii) Lead angle

T 5.3 What is belt friction? State its applications.

T 5.4 Determine an expression for the ratio of belt tensions in a flat belt drive.

T 5.5 Describe the wedge briefly and state its significance.

T 5.6 State the applications of a band brake in engineering.

Numerical Problems

N 5.1 A uniform ladder of length 10 m rests against a vertical wall, with which it makes an angle of 30°, and on a floor. If a person, whose weight is two times of the ladder, climbs it. If the coefficients of friction between ladder and the wall is 0.35 and between ladder and floor is 0.32, determine the distance along the ladder will be travelled by person, when the ladder is about to slip.

N 5.2 A man climbs on a 5 m long ladder. The ladder makes an angle of 60° from the horizontal. The other end of the ladder is supported on a vertical wall. The coefficient of friction between the ladder and the wall is 0.2 and between the ladder and floor is 0.3. The weight of ladder and man are 150 N and 800 N, respectively. How far can the man climb along the ladder?

N 5.3 A non-homogeneous ladder of mass 50 kg and length 10 m is placed against a smooth wall and rough horizontal floor. If the mass is concentrated at 4 m from the floor then determine whether the ladder may be held in equilibrium with the floor at 60°. Take coefficient of friction between ladder and floor as 0.30.

N 5.4 Determine horizontal force P applied on 80 kN wedge (a) to raise block 300 kN upward (b) to lower block 300 kN downward. Consider coefficient of friction for all contact surfaces as 0.32.

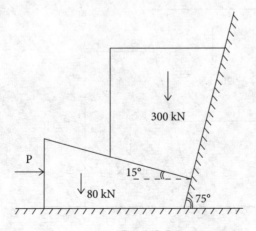

Fig. NP 5.1

N 5.5 A wedge and block arrangement is shown in Fig. NP 5.2. Determine minimum horizontal force P by which block can impend upward.

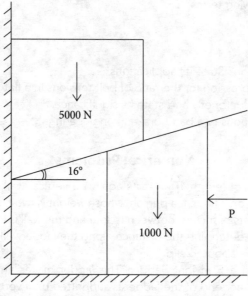

Fig. NP 5.2

N 5.6 Two blocks are to the held in equilibrium by applying minimum force P as shown in Fig. NP 5.3. Take coefficients of friction at floor as 0.3, 0.33 at wall and 0.24 between the blocks, determine minimum force P.

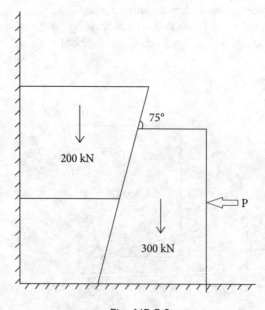

Fig. NP 5.3

Application of Friction

N 5.7 A screw jack has a screw, consisting square thread of two threads to 24 mm. The outer diameter of the screw is 90 mm. Determine the force required at the end of 750 mm long handle to raise the load. Take coefficient of friction between nut and the screw as 0.20.

N 5.8 A screw spindle of mean radius 50 mm has pitch of the thread as 18 mm. If the length of the handle is 800 mm, determine the force required to raise a load of 10 kN and the efficiency of the screw jack. Take coefficient of friction between the screw and nut as 0.22.

N 5.9 A screw jack has square threads of mean diameter 60 mm and pitch 12 mm. If the coefficient of friction between screw and nut is 0.08, determine:
  (i) Force to be applied at 300mm radius of jack to lift a load of 900 N.
  (ii) Check the self-locking of jack.

N 5.10 Determine the rage of weight can be supported by a person by applying 750 N force as shown in Fig. NP 5.4. The rope is wrapped by $2\frac{1}{4}$ turns around the horizontal drum. Take coefficient of friction between rope and drum as 0.24.

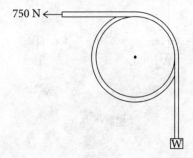

Fig. NP 5.4

N 5.11 Rod AB of weight 500 N supported by a cable wrapped around a semi cylinder having coefficient of friction 0.2. A weight C weighing 100 N can slide without friction on rod AB. What maximum range x from centreline, the mass C can be place without causing slipping.
(MTU 2010–11)

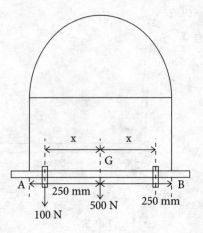

Fig. NP 5.5

N 5.12 A rope is wrapped on a horizontal drum by $1\frac{1}{2}$ turns. If the coefficient of friction between drum and rope is 0.27, determine the range of weight can be support by a man applying 800N at one end as shown in Fig. NP 5.6.

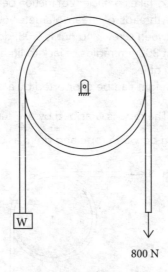

Fig. NP 5.6

N 5.13 A torque of 600Nm is acting on the brake drum as shown in Fig. NP 5.7. If coefficient of friction between brake band and brake drum is 0.26, determine the force P to be applied at the end of the brake lever if drum rotates in the clockwise direction.

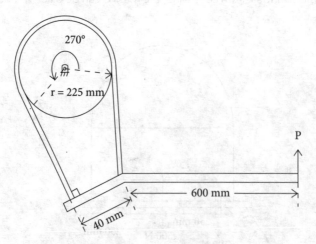

Fig. NP 5.7

Application of Friction

Multiple Choice Questions

1. A ladder can be analyse by using equilibrium equations
   a. $\sum X = 0, \sum M = 0$
   b. $\sum X = 0, \sum Y = 0$
   c. $\sum Y = 0, \sum M = 0$
   d. $\sum X = 0, \sum Y = 0, \sum M = 0$

2. In a Screw jack, the lead is defined as
   a. number of threads
   b. distance between two consecutive threads
   c. axial distance travelled by screw in a single rotation
   d. none of these

3. In a Screw jack, the pitch is defined as
   a. number of threads
   b. distance between two consecutive threads
   c. axial distance travelled by screw in a single rotation
   d. none of these

4. In a single threaded screw jack
   a. Lead = Pitch
   b. Lead > Pitch
   c. Lead < Pitch
   d. none of these

5. If a screw is threaded by n times, then relation between lead (L), and pitch (P) is given by
   a. $Lead = \dfrac{Pitch}{n}$
   b. $Lead = n \times Pitch$
   c. $Lead = \dfrac{n}{Pitch}$
   d. none of these

6. In a Screw jack, the lead angle is analogous to
   a. angle of repose
   b. angle of friction
   c. helix angle
   d. none of these

7. A screw jack has lead (L), mean radius (R) and pitch (p), its lead angle (α) will be given by
   a. $\tan\alpha = \dfrac{L}{2\pi R}$
   b. $\tan\alpha = \dfrac{R}{2\pi L}$
   c. $\tan\alpha = \dfrac{P}{2\pi R}$
   d. none of these

8. In a Screw jack, the relation between effort 'T' applied on the bar of length 'l' and its equivalent force 'S' acting at the mean radius 'R' is given by
   a. $T \times l = S \times R$
   b. $T \times l < S \times R$
   c. $T \times l > S \times R$
   d. none of these

9. In a belt drive, the ratio of tensions in tight side and slack side is given by relation
   a. $\dfrac{T_t}{T_s} = e^{\mu\theta}$
   b. $\dfrac{T_s}{T_t} = e^{\mu\theta}$
   c. $\dfrac{T_t}{T_s} = e^{\mu/\theta}$
   d. none of these

10. In a band brake, the torque on rotating drum of radius is given by
    a. $(T_t + T_s) \times r$
    b. $(T_t - T_s) \times r$
    c. $(T_s - T_t) \times r$
    d. none of these

Answers

1. d 2. c 3. b 4. a 5. b 6. c 7. a 8. a 9. a 10. b

Chapter 6
Analysis of Trusses

6.1 Introduction

Engineering structures are built by joining large number of members to support heavy loads. These are broadly classified as truss, frame and machine, depending upon loading conditions and applications. Truss is built up by joining various prismatic bars at different joints. The members may be of different cross-sections like angle section, channel section or circular section, etc. Trusses are called economic structure as these can support heavy load over the large span. The examples of trusses are mobile or electrical transmissions towers, bridges and roof truss, supporting roof of buildings and workshops, etc.

6.2 Classification of Trusses

Trusses are broadly classified in two categories:

a. **Depending upon joining of members in planes:**

Two-Dimensional Truss: This type of truss is also called plane truss as its all members lie in one plane. Thus, it can be considered as a structure of coplanar members. Bridges and roof trusses are examples of this type of truss.

Three-Dimensional Truss: This type of truss is also called space truss and its all members lie in different planes. It is a structure which consists of collection of non-coplanar members. Mobile or electrical transmission towers are examples of this type of truss.

b. **Depending upon quantity of members and joints:**

Perfect Truss: This type of truss does not change its shape under external loading. It utilizes the optimum numbers of members and joints. It is considered as just rigid truss because removal of any member can cause failure of the truss.

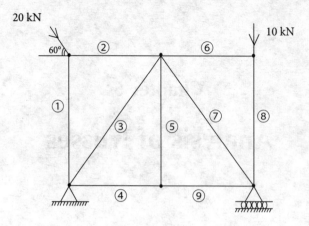

Fig. 6.1 Perfect truss

Mathematically the condition for a perfect truss is

$m = (2j - 3)$

Consider Fig. 6.1 where $m = 9$ and $j = 6$

$(2j - 3) = (2 \times 6 - 3) = 9$

As equality exist between 'm' and '$(2j - 3)$' it called perfect truss.

Imperfect Truss: This type of truss does not retain its shape under external loading due to joining of fewer members as compare to perfect truss. It is also called deficient truss.

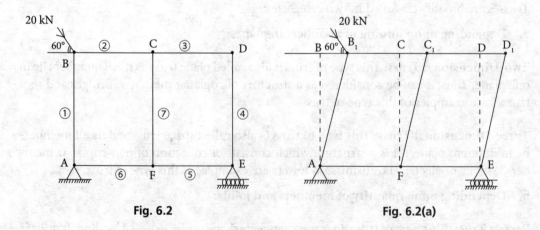

Fig. 6.2 **Fig. 6.2(a)**

Imperfect truss

Analysis of Trusses

Mathematically, the condition for an imperfect truss is

$m < (2j - 3)$

Consider Fig. 6.2 where $m = 7$ and $j = 6$

$(2j - 3) = (2 \times 6 - 3) = 9$

As $m < (2j - 3)$ exist for this truss, it is called imperfect truss.
Consider Fig. 6.2 (a) where joints B, C, and D have shifted to new position B_1, C_1 and D_1 under the application of load 20 kN which shows that this truss does not retain its shape under external loading.

Redundant Truss: This type of truss also does not change its shape under external loading. It is considered as over rigid truss because it contains more members than a perfect truss and becomes statically indeterminate truss (forces in the members cannot be determined by using the equilibrium equations).
Mathematically the condition for a redundant truss is
$m > (2j - 3)$
Consider Fig. 6.3 where $m = 8$ and $j = 5$
$(2j - 3) = (2 \times 5 - 3) = 7$
As $m > (2j - 3)$ exist for this truss, it is called redundant truss.

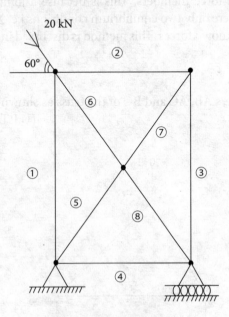

Fig. 6.3 Redundant truss

6.3 Assumptions for the Analysis of Perfect Truss

i. All members of the truss are connected by frictionless pins, i.e., the shape of truss cannot resist any moment.
ii. All members of the truss are connected at the joint.
iii. External forces act at the joints only and during analysis of joints, force is not allowed to shift to any other joint by using principle of transmissibility.
iv. All members of the truss are taken as weightless.
v. All members of the truss are straight and called two force members. The members may be either in tension (T) or compression (C).

6.4 Analysis of Forces in the Members of the Truss

There are two methods to analyze the forces in the members of truss:

I. Analytical method
II. Graphical method

Analytical method includes Method of Joint and Method of Section.

6.4.1 Method of joint

In this method, first a suitable joint is selected. The suitable joint is one which contains maximum two unknown force members. This is because a joint represents a concurrent coplanar force system where only two equilibrium conditions i.e. $\Sigma X = 0$ and $\Sigma Y = 0$ can be used to determine the unknown forces. This method is discussed in detail in solved examples.

Example: 6.1

Find the forces in members AB, AC and BC of the truss as shown in Fig. 6.4.

(U.P.T.U IInd Sem, 2002–2003)

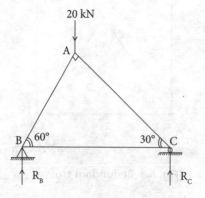

Fig. 6.4

Analysis of Trusses

Solution:

The given truss is a perfect truss and can be analysed by method of joint or method of section. Here we are using method of joint.

In this truss, total three joints A, B and C are available. Consider their F.B.D.s,

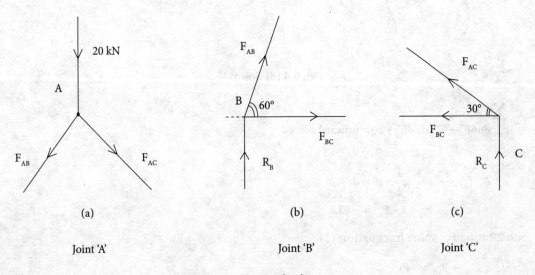

(a) Joint 'A'   (b) Joint 'B'   (c) Joint 'C'

Fig. 6.4 (a-c)

Let all the members AB, BC and CA are in tension. Thus force of each member will be marked away from the joint. As all the forces are concurrent in all joints, hence each joint represent a Concurrent coplanar force system. We know that in such a force system only two equilibrium conditions can be used i.e. $\sum X = 0$ and $\sum Y = 0$. Thus out of three joints, joint A can be considered as suitable joint because there are two unknown forces, i.e. F_{AB} and F_{AC}.

Considering joint A for equilibrium,

$$\sum X = 0,$$
$$F_{AB} \cdot \cos 60° = F_{AC} \cdot \cos 30°$$
$$F_{AB} \cdot \frac{1}{2} = F_{AC} \cdot \frac{\sqrt{3}}{2}$$
$$F_{AB} = \sqrt{3} \cdot F_{AC} \qquad \ldots (1)$$

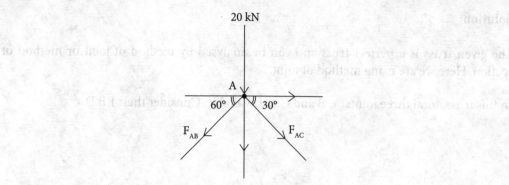

Fig. 6.4 (d) Joint A

$\sum Y = 0$,
$F_{AB} \sin 60° + F_{AC} \sin 30° + 20 =$ upward forces
$= 0$

$$F_{AB} \cdot \frac{\sqrt{3}}{2} + F_{AC} \cdot \frac{1}{2} + 20 = 0$$

$$F_{AB} \sqrt{3} + F_{AC} = -40 \quad \ldots\ldots (2)$$

Substituting F_{AB} value in equation (2),

$\sqrt{3} \cdot \sqrt{3} F_{AC} + F_{AC} = -40$

$F_{AC} = -10 \text{ kN (Compression)}$

Negative sign shows that force in the member AC is opposite to our assumption, i.e. it is of Compressive in nature.
From equation (1),

$F_{AB} = \sqrt{3}(-10)$

$F_{AB} = -17.32 \text{ kN (Compression)}$

Considering joint B for equilibrium,

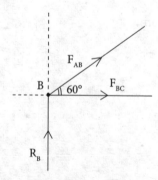

Fig. 6.4 (e) Joint B

Analysis of Trusses

$$\sum X = 0,$$
$$F_{AB} \cos 60° + F_{BC} = 0$$

Substituting value of F_{AB}

$$(-17.32).\cos 60° + F_{BC} = 0$$
$$F_{BC} = +8.66 \text{ kN (Tension)}$$

The positive sign shows that assumed direction is correct thus member is having tensile force or in tension.

Note:

1. In F_{AB}, the subscript is representing the name of member not a vector quantity, thus $F_{AB} = F_{BA}$ which is valid for all members of truss.
2. This question could be solved by determining support reactions first and then any joint could be selected for equilibrium.

Alternative Method:
As stated above, this question can be solved by determining support reactions first. Consider F.B.D. of whole truss,

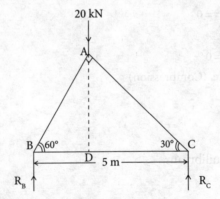

Fig. 6.4 (f)

As $\triangle$ ADB is right angled triangle,
BD = AB cos60°
Similarly $\triangle$ BAC is right angled triangle
AB = BC cos60° = 5 cos60°
Thus, BD = 5 cos60°. cos60°

$$\sum Y = 0,$$
$$R_B + R_C = 20 \text{ kN} \quad \quad \quad \text{..... (3)}$$

$\sum M_B = 0,$
$R_C \times 5 = 20 \times BD$
$R_C \times 5 = 20 \times 5.\cos 60°.\cos 60°$
$R_C = 5 \text{ kN}, R_B = 15 \text{ kN}$

Considering joint 'C' for equilibrium,

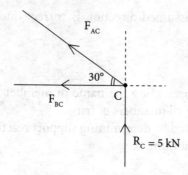

Fig. 6.4 (g) Joint C

$\sum X = 0,$
$F_{BC} + F_{AC} \cos 30° = 0$
$\sum Y = 0,$
$F_{AC} \sin 30° + 5 = 0$
$F_{AC} = -10 \text{ kN i.e. (Compression)}$

$F_{BC} + (-10).\cos 30° = 0$
$F_{BC} = +8.66 \text{ kN (Tension)}$

Considering joint 'B' for equilibrium,

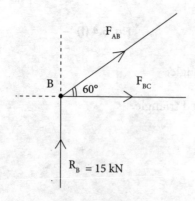

Fig. 6.4 (h) Joint B

Analysis of Trusses

$\sum X = 0,$
$F_{AB} \cos 60° + F_{BC} = 0$
$F_{AB} \cos 60° + 8.66 = 0$
$F_{AB} = -17.32 \text{ kN (Compression)}$

Example: 6.2

A truss is shown in the Fig. 6.5. Find the forces in all the members of the truss and indicate whether it is in tension or compression.

(U.P.T.U. Ist Sem 2000–2001)

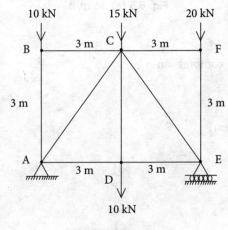

Fig. 6.5

Solution:

It is observed from the figure that analysis can be done by determining support reactions first and then joints can be analysed. However two joints B and F can be analysed earlier as there are two unknown forces. Thus considering joint B for equilibrium,

Note: All members are assumed to be in tension.

Considering joint B for equilibrium,

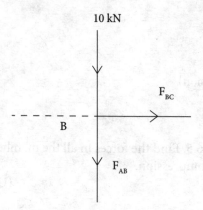

Fig. 6.5 (a) Joint B

$\sum X = 0$,
$F_{BC} = 0$ i.e. No tension or compression
$\sum Y = 0$,
$F_{AB} + 10 = 0$
$F_{AB} = -10 \text{ kN (Compression)}$

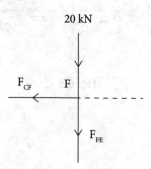

Fig. 6.5 (b) Joint F

Considering joint F for equilibrium,

$\sum X = 0$,
$F_{CF} = 0$
$\sum Y = 0$,
$F_{FE} + 20 = 0$
$F_{FE} = -20 \text{ kN (Compression)}$

Now no joint can be selected as all are having there unknown forces. Thus we've to determine support reactions.

Analysis of Trusses

Considering F.B.D. of whole truss,

Here hinge at A will offer only vertical reaction because all external forces are vertical.

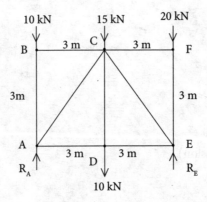

Fig. 6.5 (c)

$\sum Y = 0$,

$R_A + R_E = 10 + 15 + 20 + 10$

$\qquad = 55$ kN ..... (1)

$\sum M_A = 0$,

$R_E \times 6 = 10 \times (0) + 15 \times (3) + 20 \times (6) + 10 \times (3)$

$\qquad = 195$

$\qquad R_E = 32.5$ kN

From equation (1),
$\qquad R_A = 22.5$ kN

Now joint A or E can be considered.
Considering joint A for equilibrium,

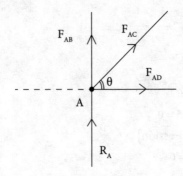

Fig. 6.5 (d) Joint A

Let $\angle DAC = \theta$

$$\tan\theta = \frac{CD}{AD} = \frac{3}{3}$$
$$\theta = 45°$$

$\sum X = 0,$
$F_{AD} + F_{AC}\cos\theta = 0$ ..... (2)

$\sum Y = 0,$
$F_{AB} + F_{AC}\sin\theta + R_A = 0$
$(-10) + F_{AC}\sin 45° + 22.5 = 0$
$F_{AC} = -17.68 \text{ kN}(\text{Compression})$

From equation (2),

$F_{AD} + (-17.68).\cos 45° = 0$
$F_{AD} = +12.5 \text{ kN}(\text{Tension})$

Considering joint E for equilibrium,

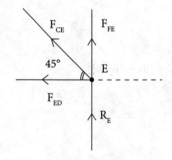

Fig. 6.5 (e) Joint E

$\sum X = 0,$

$F_{ED} + F_{CE}\cos 45° = 0$

$\sum Y = 0,$

$F_{CE}\sin 45° + F_{FE} + R_E = 0$
$F_{CE}\sin 45° + (-20) + 32.5 = 0$
$F_{CE} = -17.68 \text{ kN}(\text{Compression})$
$F_{ED} + (-17.68).\cos 45° = 0$
$F_{ED} = +12.5 \text{ kN}(\text{Tension})$

Analysis of Trusses

Considering joint D for equilibrium,

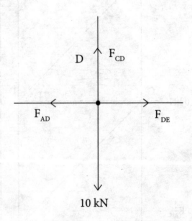

Fig. 6.5 (f) Joint D

$\sum X = 0,$

$F_{AD} = F_{DE}$
$\therefore F_{AD} = +12.5 \text{ kN} (\text{Tension})$

$\sum Y = 0,$

$F_{CD} = +10 \text{ kN} (\text{Tension})$

Now there is no need to analyse Joint 'C' as forces of all members have been determined.

Example: 6.3

Find out the axial forces in all the members of a truss with loading as shown in Fig. 6.6.
(U.P.T.U. IInd Sem 2001–2002)

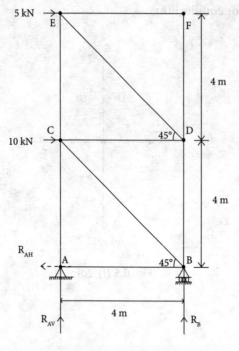

Fig. 6.6

Solution:

Considering joint F for equilibrium,

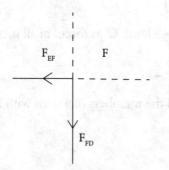

Fig. 6.6 (a) Joint F

$\sum X = 0,$
$F_{EF} = 0$
$\sum Y = 0,$
$F_{FD} = 0$

Analysis of Trusses

Considering joint E for equilibrium,

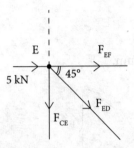

Fig. 6.6 (b) Joint E

$\sum X = 0,$
$\quad 5 + F_{EF} + F_{ED} \cos 45° = 0$
$\quad F_{ED} = -7.07 \text{ kN (Compression)}$
$\sum Y = 0,$
$\quad F_{ED} \sin 45° + F_{CE} = 0$
$\quad (-7.07).\sin 45° + F_{CE} = 0$
$\quad F_{CE} = +5 \text{ kN (Tension)}$

Considering joint D for equilibrium,

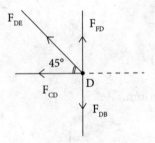

Fig. 6.6 (c) Joint D

$\sum X = 0,$
$\quad F_{DE} \cos 45° + F_{CD} = 0$
$\quad (-7.07).\cos 45° + F_{CD} = 0$
$\quad F_{CD} = +5 \text{ kN} \quad \text{(Tension)}$

$\sum Y = 0$
$F_{DE} \sin 45° + F_{FD} = F_{DB}$
$(-7.07).\sin 45° + 0 = F_{DB}$
$F_{DB} = -5$ kN   (Compression)

Considering Joint C for equilibrium,

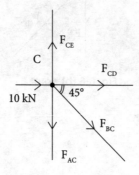

Fig. 6.6 (d) Joint C

$\sum X = 0,$
$10 + F_{CD} + F_{BC} \cos 45° = 0$
$10 + 5 + F_{BC} \cos 45° = 0$
$F_{BC} = -21.21$ kN   (Compression)

$\sum Y = 0,$
$F_{CE} = F_{BC} \sin 45° + F_{AC}$
$5 = (-21.21).\sin 45° + F_{AC}$
$F_{AC} = +20$ kN   (Tension)

Considering Joint B for equilibrium,

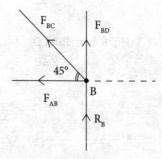

Fig. 6.6 (e) Joint B

Analysis of Trusses

$\sum X = 0$,
$F_{AB} + F_{BC} \cos 45° = 0$
$F_{AB} + (-21.21)\cos 45° = 0$
$F_{AB} = +15$ kN (Tension)

$\sum Y = 0$,
$F_{BC} \sin 45° + F_{BD} + R_B = 0$
$(-21.21)\sin 45° + (-5) + R_B = 0$
$R_B = 20$ kN (Tension)

If full truss is considered for equilibrium,

$\sum Y = 0$,
$R_{AV} + R_B = 0$
$R_{AV} + 20 = 0$
$R_{AV} = -20$ kN

$\sum X = 0$,
$R_{AH} = 10 + 5$
$R_{AH} = 15$ kN

Note:
This question could be solved by determining support reactions first by using equilibrium conditions,

$\sum X = 0, \sum Y = 0,$ and $\sum M = 0$

and then suitable joints can be analysed one by one.

Example: 6.4

Determine forces in all the members of the truss as shown in Fig. 6.7 and indicate the nature of force also.

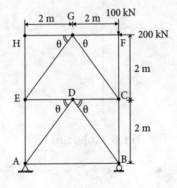

Fig. 6.7

Solution:

Consider joint H for equilibrium,

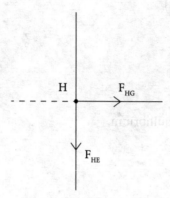

Fig. 6.7 (a) Joint H

$\sum X = 0$,
   $F_{HG} = 0$
$\sum Y = 0$,
   $F_{HE} = 0$

Consider joint F for equilibrium,

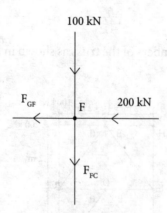

Fig. 6.7 (b) Joint F

Analysis of Trusses

$\sum X = 0,$
$F_{GF} + 200 = 0$
$F_{GF} = -200\,\text{kN}$ (Compression)

$\sum Y = 0,$
$100 + F_{FC} = 0$
$F_{FC} = -100\,\text{kN}$ (Compression)

Consider Joint G for equilibrium,

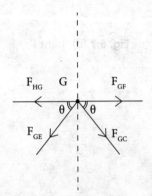

Fig. 6.7 (c) Joint G

$\tan\theta = 2/2$
$\theta = 45°$

$\sum X = 0$
$F_{HG} + F_{GE}\cos\theta = F_{GF} + F_{GC}\cos\theta$
$0 + F_{GE}\cos\theta = (-200) + F_{GC}\cos\theta$
$F_{GE}\cos 45° = -200 + F_{GC}\cos 45°$ ..... (1)

$\sum Y = 0$
$F_{GC}\sin\theta + F_{GE}\sin\theta = 0$
or $F_{GC} = -F_{GE}$ in equation (1),
$F_{GE}\cos 45° = -200 - F_{GE}\cos 45°$
$2F_{GE}\cos 45° = -200$
$F_{GE} = -141.42\,\text{kN}$ (C)
and $F_{GC} = +141.42\,\text{kN}$ (T)

Consider joint E for equilibrium,

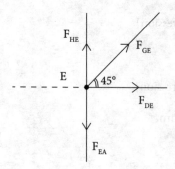

Fig. 6.7 (d) Joint E

$\sum X = 0,$
$F_{GE} \cos 45° + F_{DE} = 0$
$(-141.42).\cos 45° + F_{DE} = 0$
$F_{DE} = +100 \, kN \quad \text{(Tension)}$

$\sum Y = 0,$
$F_{HE} + F_{GE} \sin 45° = F_{EA}$
$0 + (-141.42).\sin 45° = F_{EA}$
$F_{EA} = -100 \, kN \, \text{(Compression)}$

Consider joint C for equilibrium,

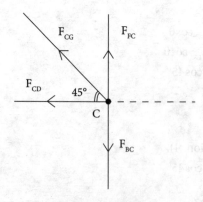

Fig. 6.7 (e) Joint C

Analysis of Trusses

$\sum X = 0,$
$F_{CD} + F_{CG} \cos 45° = 0$
$F_{CD} + 141.42 \cos 45° = 0$
$F_{CD} = -100 \text{ kN}$ (Compression)

$\sum Y = 0,$
$F_{CG} \sin 45° + F_{FC} = F_{BC}$
$141.42 \sin 45° + (-100) = F_{BC}$
$100 - 100 = F_{BC}$
$F_{BC} = 0$

Consider joint D for equilibrium,

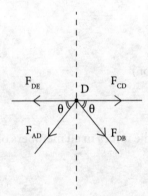

Fig. 6.7 (f) Joint D

$\sum X = 0,$
$F_{DE} + F_{AD} \cos\theta = F_{CD} + F_{DB} \cos\theta$
$100 + F_{AD} \cos 45° = -100 + F_{DB} \cos 45°$
$F_{DB} = F_{AD} + 282.84$ ..... (2)

$\sum Y = 0,$
$F_{AD} \sin\theta + F_{DB} \sin\theta = 0$
as $\sin\theta \neq 0,$ $F_{AD} = -F_{DB}$
From equation (2), $F_{DB} = -F_{DB} + 282.84$
$F_{DB} = +141.42 \text{ kN}$ (Tension)
and $F_{AD} = -141.42 \text{ kN}$ (Compression)

Consider joint B for equilibrium,

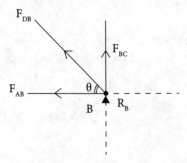

Fig. 6.7 (g) Joint B

$\sum X = 0,$
$F_{AB} + F_{DB} \cos\theta = 0$
$F_{AB} + 141.42 \cos 45°$
$F_{AB} = -100 \text{ kN}$ (Compression)

Example: 6.5

A truss is shown in Fig. 6.8. Find the forces in all members of the truss and indicate, whether it is in tension or compression.

(U.P.T.U. Ist Sem, c.o. 2003)

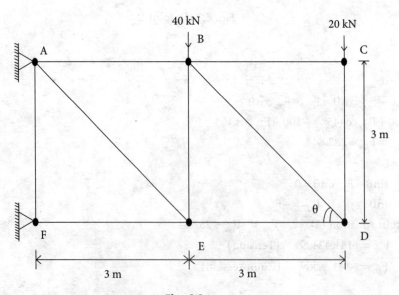

Fig. 6.8

Analysis of Trusses

Solution:

Considering Joint C for equilibrium,

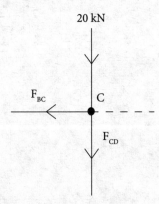

Fig. 6.8 (a) Joint C

$\sum X = 0,$
  $F_{BC} = 0$
$\sum Y = 0,$
  $F_{CD} + 20 = 0$
  $F_{CD} = -20$ kN (Compression)

Considering Joint D for equilibrium,

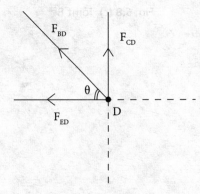

Fig. 6.8 (b) Joint D

Consider △ EDB,

$\tan\theta = \dfrac{3}{3}$

$\theta = 45°$

$\sum X = 0,$
$\quad F_{ED} + F_{BD}\cos\theta = 0$
$\sum Y = 0,$
$\quad F_{BD}\sin\theta + F_{CD} = 0$
$\quad F_{BD}\sin 45° + (-20) = 0$
$\quad F_{BD} = +28.28\,\text{kN}\,(\text{Tension})$
$\quad F_{ED} + (28.28)\cos 45° = 0$
$F_{ED} = -20\,\text{kN} \quad (\text{Compression})$

Considering Joint B for equilibrium,

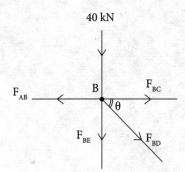

Fig. 6.8 (c) Joint B

$\angle CBD = \angle EDB = 45°$

$\sum X = 0,$
$F_{AB} = F_{BC} + F_{BD}\cos\theta,$
$F_{AB} = 0 + 28.28\cos 45°$
$F_{AB} = +20\,\text{kN} \quad (\text{Tension})$
$\sum Y = 0,$
$F_{BE} + F_{BD}\sin\theta + 40 = 0$
$F_{BE} + 28.28\sin 45° + 40 = 0$
$F_{BE} = -60\,\text{kN} \quad (\text{Compression})$

Analysis of Trusses

Considering Joint E for equilibrium,

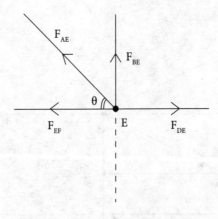

Fig. 6.8 (d) Joint E

$\sum X = 0,$
$F_{EF} + F_{AE} \cos\theta = F_{DE} = -20$

$\sum Y = 0,$
$F_{AE} \sin\theta + F_{BE} = 0$
$F_{AE} \times \sin 45° + (-60) = 0$
$F_{AE} = +84.85 \text{ kN} \quad \text{(Tension)}$
$F_{EF} + (84.85).\cos 45° = -20$
$F_{EF} = -80 \text{ kN} \quad \text{(Compression)}$

Example: 6.6

Determine the magnitude and nature of forces in the members of the truss as shown in Fig. 6.9.
(U.P.T.U Ist Sem, 2001–2002)

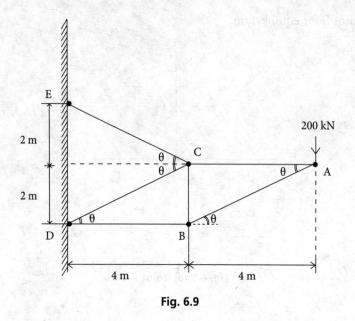

Fig. 6.9

Solution:

Let $\angle CAB = \theta$
Considering joint 'A' for equilibrium,

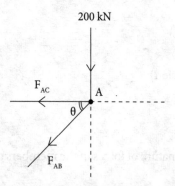

Fig. 6.9 (a) Joint A

$\sum X = 0$,
$F_{AC} + F_{AB} \cos\theta = 0$
$\sum Y = 0$,
$F_{AB} \sin\theta + 200 = 0$

Analysis of Trusses

Consider right angled triangle BCA,

$\tan\theta = 2/4, \theta = 26.57°$

$F_{AB} = \dfrac{-200}{\sin(26.57)}$

$F_{AB} = -447.14$ kN (Compression)

$F_{AC} + (-447.14).\cos(26.57°) = 0$

$F_{AC} = +399.92$ kN (Tension)

Considering joint 'B' for equilibrium,

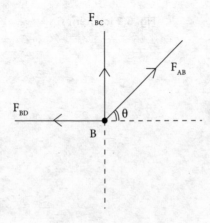

Fig. 6.9 (b) Joint B

$\sum X = 0,$

$F_{BD} = F_{AB}.\cos\theta$

$F_{BD} = (-447.14)\cos(26.57°)$

$F_{BD} = -399.92$ kN (Compression)

$\sum Y = 0,$

$F_{BC} + F_{AB}\sin\theta = 0$

$F_{BC} + (-447.14).\sin 26.57° = 0,$

$F_{BC} = +200$ kN (Tension)

Considering joint C for equilibrium,

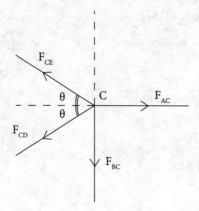

Fig. 6.9 (c) Joint C

$\sum X = 0$,
$F_{CE} \cos\theta + F_{CD} \cos\theta = F_{AC}$ ..... (1)

$\sum Y = 0$,
$F_{CE} \sin\theta = F_{CD} \sin\theta + F_{BC}$ ..... (2)

$F_{CE} = \left(F_{CD} + \dfrac{200}{\sin\theta} \right)$ ..... (3)

From equation (1),

$\left(F_{CD} + \dfrac{200}{\sin\theta} \right) \cos\theta + F_{CD} \cdot \cos\theta = F_{AC}$

$\left(F_{CD} + \dfrac{200}{\sin 26.57} + F_{CD} \right) \cos\theta = F_{AC}$

$2F_{CD} + 447.14 = \dfrac{399.32}{\cos 26.57°}$

$F_{CD} = 0$

From equation (3),

$F_{CE} = 0 + \dfrac{200}{\sin 26.57°}$

$F_{CE} = +\,447.14 \text{ kN} \quad \text{(Tension)}$

Example: 6.7

Find the axial forces in all members of a truss as shown in Fig. 6.10.

(U.P.T.U. Sem-II, 2008–2009)

Analysis of Trusses

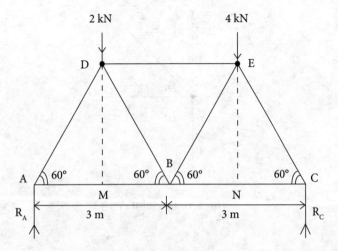

Fig. 6.10

Solution:

Here we cannot consider any joint without determining the support reactions as all the joints are accompanied with three unknown forces. Thus considering equilibrium of the entire truss,

$\sum Y = 0$,
$R_A + R_C = 2 + 4$
$\qquad = 6 \text{ kN}$ ..... (1)

$\sum M_A = 0$,
$\quad R_C \times 6 = 2 \times AM + 4 \times AN$
$\qquad = 2 \times AD\cos 60° + 4 \times (AB + BN)$
$\qquad = 2 \times AB\cos 60° + 4 \times (AB + BE\cos 60°)$
$\qquad = 2 \times 3\cos 60° + 4(3 + 3\cos 60°)$
$\qquad = 21$
$\quad R_C = 3.5 \text{ kN}$

From equation (1),
$\quad R_A = 2.5 \text{ kN}$

Note: $\triangle$ ABD and $\triangle$ BCE both are equilateral triangles.

Thus all sides of triangle i.e. length of all members are equal.

Now only joint A or C can be considered.

Considering joint A for equilibrium,

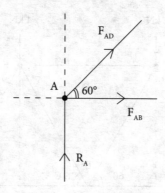

Fig. 6.10 (a) Joint A

$\sum X = 0,$
$F_{AB} + F_{AD} \cos 60° = 0$
$\sum Y = 0,$
$R_A + F_{AD} \sin 60° = 0$
$2.5 + F_{AD} \sin 60° = 0$
$F_{AD} = -2.88 \text{ kN} \quad \text{(Compression)}$
$F_{AB} + (-2.88) \cos 60° = 0$
$F_{AB} = +1.44 \text{ kN} \quad \text{(Tension)}$

Considering joint C for equilibrium,

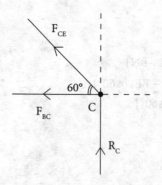

Fig. 6.10 (b) Joint C

Analysis of Trusses

$\sum X = 0$,
$F_{BC} + F_{CE} \cos 60° = 0$
$\sum Y = 0$,
$F_{CE} \sin 60° + R_C = 0$
$F_{CE} \sin 60° + 3.5 = 0$
$F_{CE} = -4.04$ kN   (Compression)
$F_{BC} + (-4.04) \cos 60° = 0$
$F_{BC} = +2.02$ kN   (Tension)

Considering joint B for equilibrium,

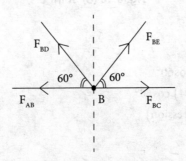

Fig. 6.10 (c) Joint B

$\sum X = 0$,
$F_{AB} + F_{BD} \cos 60° = F_{BC} + F_{BE} \cos 60°$
$1.44 + F_{BD} \cos 60° = 2.02 + F_{BE} \cos 60°$
$F_{BD} = F_{BE} + 1.16$
$\sum Y = 0$,
$F_{BD} \sin 60° + F_{BE} \sin 60° = 0$
$F_{BD} = -F_{BE}$
$-F_{BE} = F_{BE} + 1.16$
$F_{BE} = -0.58$ kN   (Compression)
$F_{BD} = +0.58$ kN   (Tension)

Considering joint E for equilibrium,

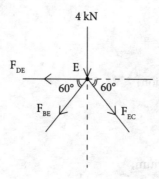

Fig. 6.10 (d) Joint E

$\sum X = 0,$
$F_{DE} + F_{BE} \cos 60° = F_{EC} \cos 60°$
$F_{DE} + (-0.58)\cos 60° = (-4.04).\cos 60°$
$F_{DE} = -1.73 \text{ kN} \quad \text{(Compression)}$

Example: 6.8

Find the forces in all members of a truss as shown in Fig. 6.11, which carries a horizontal load of 12 kN at joint D and a vertical downward load of 18 kN at joint C.

(U.P.T.U Ist Sem 2009–2010)

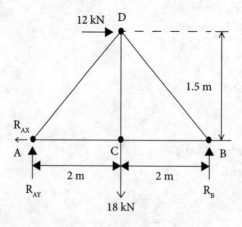

Fig. 6.11

Analysis of Trusses

Solution:

First, support reactions have to be determined. Here, CD is bisector of AB, thus $\triangle$ ACD and $\triangle$ BCD are equal triangles.
Considering equilibrium of entire truss,

$\sum X = 0$, $R_{AX} = 12$ kN
$\sum Y = 0$, $R_{AY} + R_B = 18$ kN ..... (1)
$\sum M_A = 0$, $R_B \times 4 = 18 \times 2 + 12 \times 1.5$

$R_B = 13.5$ kN
$R_{AY} = 4.5$ kN

Considering joint A for equilibrium,

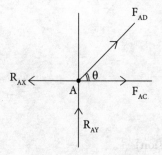

Fig. 6.11 (a) Joint A

Consider $\triangle ACD$, let $\angle DAC = \theta$

$$\tan\theta = \frac{1.5}{2}$$
$$\theta = 36.86°$$

$\sum X = 0$,
$R_{AX} = F_{AD} \cos\theta + F_{AC}$
$12 = F_{AD} \cos 36.86 + F_{AC}$

$\sum Y = 0$,
$R_{AY} + F_{AD} \sin\theta = 0$
$4.5 + F_{AD} \sin 36.86 = 0$
$F_{AD} = -7.5$ kN   (Compression)

$12 = F_{AD} \cos\theta + F_{AC}$
$12 = (-7.5).\cos 36.86 + F_{AC}$
$F_{AC} = +18$ kN   (Tension)

Considering joint B for equilibrium,

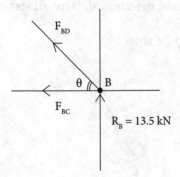

Fig. 6.11 (b) Joint B

$\sum X = 0,$
$F_{BC} + F_{BD} \cos\theta = 0$

$\sum Y = 0,$
$F_{BD} \sin\theta + 13.5 = 0$
$F_{BD} \sin 36.86° + 13.5 = 0$
$F_{BD} = -22.51 \text{ kN} \quad \text{(Compression)}$
$F_{BC} + (-22.51).\cos 36.86° = 0$
$F_{BC} = +18 \text{ kN} \quad \text{(Tension)}$

Considering joint C for equilibrium,

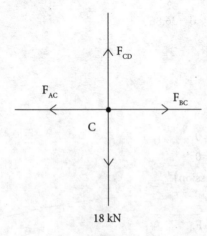

Fig. 6.11 (c) Joint C

Analysis of Trusses

$\sum Y = 0$,
$F_{CD} = +18$ kN   (Tension)

Example: 6.9

Determine the forces in all the members of the truss shown in Fig. 6.12.

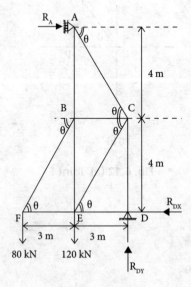

Fig. 6.12

Solution:

As, BC = FE = ED = 3m and AB = BE = CD = 4m, all triangles in truss are equal triangles.

Let $\angle CED = \theta$

then $\angle BFE = \theta$ and $\angle ECB = \angle BCA = \theta$

$\tan\theta = 4/3$

$\theta = 53.13°$

Considering equilibrium of complete truss,

$\sum X = 0$, $R_A = R_{DX}$
$\sum Y = 0$, $R_{DY} = 80 + 120$
$R_{DY} = 200$ kN
$\sum M_D = 0$,
$R_A \times 8 = 120 \times 3 + 80 \times 6$
$R_A = 105$ kN

i.e., R_{DX} = 105 kN, which is acting towards left side as shown in Fig. 6.12.

Considering joint F for equilibrium,

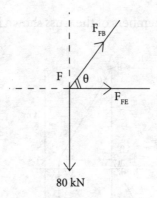

Fig. 6.12 (a) Joint F

$\sum X = 0$, $F_{FB} \cos\theta + F_{FE} = 0$
$\sum Y = 0$, $F_{FB} \sin\theta = 80$
$\qquad\qquad F_{FB} \cdot \sin 53.13° = 80$
i.e. $F_{FB} = +100$ kN   (Tension)
$100 \cos\theta + F_{FE} = 0$
$100 \cos 53.13° + F_{FE} = 0$
$F_{FE} = -60$ kN   (Compression)

Considering joint D for equilibrium,

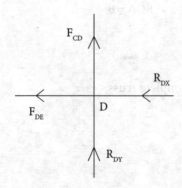

Fig. 6.12 (b) Joint D

Analysis of Trusses

$\sum X = 0,$
$R_{DX} + F_{DE} = 0$
$105 + F_{DE} = 0$
$F_{DE} = -105 \text{ kN} \quad \text{(Compression)}$
$\sum Y = 0,$
$R_{DY} + F_{CD} = 0$
$200 + F_{CD} = 0$
$F_{CD} = -200 \text{ kN} \quad \text{(Compression)}$

Considering joint E for equilibrium,

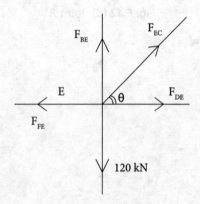

Fig. 6.12 (c) Joint E

$\sum X = 0,$
$F_{FE} = F_{DE} + F_{EC} \cos\theta$
$-60 = -105 + F_{EC} \cos 53.13°$
$F_{EC} = +75 \text{ kN} \quad \text{(Tension)}$
$\sum Y = 0,$
$F_{BE} + F_{EC} \sin\theta = 120$
$F_{BE} + 75 \sin 53.13° = 120$
$F_{BE} = +60 \text{ kN} \quad \text{(Tension)}$

Considering joint B for equilibrium,

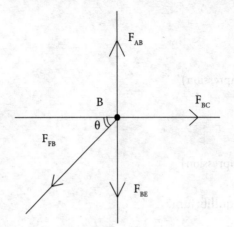

Fig. 6.12 (d) Joint B

$\sum X = 0,$
$F_{FB} \cdot \cos\theta = F_{BC}$
$100 \cdot \cos 53.13° = F_{BC}$
$F_{BC} = +60$ kN   (Tension)
$\sum Y = 0,$
$F_{AB} = F_{BE} + F_{FB} \sin\theta$
$F_{AB} = 60 + F_{FB} \sin 53.13°$ ..... (1)

Consider joint A for equilibrium,

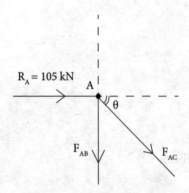

Fig. 6.12 (e) Joint A

Analysis of Trusses

$\sum X = 0,$
$105 + F_{AC} \cdot \cos\theta = 0$
$105 + F_{AC} \cdot \cos 53.13° = 0$
$F_{AC} = -175$ kN   (Compression)
$\sum Y = 0,$
$F_{AB} + F_{AC} \sin\theta = 0$
$F_{AB} + (-175) \cdot \sin 53.13° = 0$
$F_{AB} = +140$ kN   (Tension)

From equation (1),
$F_{AB} = 60 + F_{FB} \cdot \sin 53.13°$
$140 = 60 + F_{FB} \cdot \sin 53.13°$
$F_{FB} = +100$ kN   (Tension)

Example: 6.10

Determine force in all the members of the truss as shown in Fig. 6.13.

(M.T.U. Ist Sem, 2012–2013)

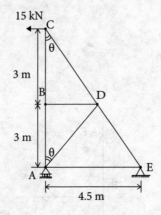

Fig. 6.13

Solution:

Consider right angled $\triangle$ CAE,

$\tan\theta = \dfrac{AE}{AC}$

$\tan\theta = \dfrac{4.5}{6}$

$\theta = 36.86°$

Considering joint C for equilibrium,

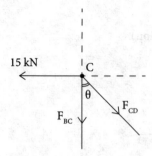

Fig. 6.13 (a) Joint C

$\sum X = 0$,
$15 = F_{CD} \sin\theta$
$15 = F_{CD} . \sin 36.86°$
$F_{CD} = +25$ kN   (Tension)
$\sum Y = 0$,
$F_{BC} + F_{CD} \cos\theta = 0$
$F_{BC} + 25.\cos 36.86° = 0$
$F_{BC} = -20$ kN   (Compression)

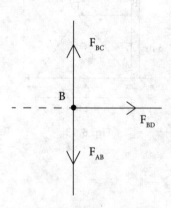

Fig. 6.13 (b) Joint B

Analysis of Trusses

Considering joint B for equilibrium,

$\sum X = 0, \quad F_{BD} = 0$
$\sum Y = 0, \quad F_{AB} = F_{BC}$
$\quad\quad\quad F_{AB} = -20 \text{ kN} \quad \text{(Compression)}$

Considering F.B.D. of the complete truss as shown in Fig. 6.13 (c),

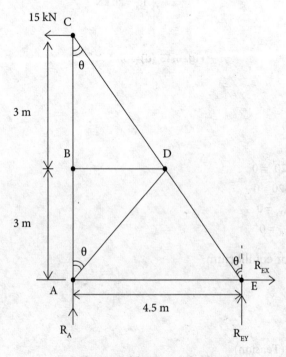

Fig. 6.13 (c) F.B.D. of truss

$\sum X = 0,$
$R_{EX} = 15 \text{ kN}$
$\sum Y = 0,$
$R_A + R_{EY} = 0$ ..... (1)
$\sum M_A = 0,$
$R_{EY} \times 4.5 + 15 \times 6 = 0$
$R_{EY} = -20 \text{ kN}$ i.e. downward

From equation (1), $R_A = +20 \text{ kN}$

Considering joint A for equilibrium,

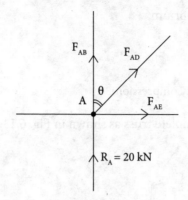

Fig. 6.13 (d) Joint A

$\sum X = 0,$
$F_{AD} \sin\theta + F_{AE} = 0$
$\sum Y = 0,$
$F_{AB} + F_{AD} \cos\theta + 20 = 0$
$-20 + F_{AD} \cos\theta + 20 = 0$
$$F_{AD} = 0$$
$$F_{AE} = 0$$

Considering joint E for equilibrium,

$\sum Y = 0,$
$F_{ED} \cos\theta - 20 = 0$
$F_{ED} \cdot \cos 36.86° = 20$
$F_{ED} = +25 \text{ kN} \quad \text{(Tension)}$

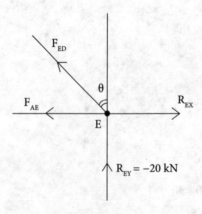

Fig. 6.13 (e) Joint E

Analysis of Trusses

Example: 6.11

Evaluate the forces in all the members of the truss as shown in Fig. 6.14.

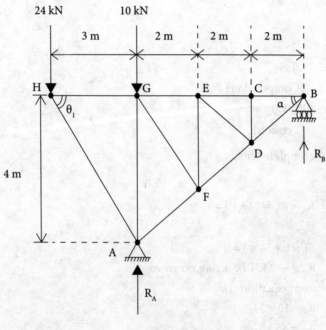

Fig. 6.14

Solution:

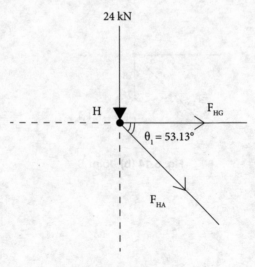

Fig. 6.14 (a) Joint H

Let $\angle GHA = \theta_1$, $\tan\theta_1 = \dfrac{4}{3} = 53.13°$

Consider joint H for equilibrium,

$\sum X = 0$,

$F_{HG} + F_{HA} \cos\theta_1 = 0$

$\sum Y = 0$,

$24 + F_{HA} \sin\theta_1 = 0$

$24 + F_{HA} \sin 53.13° = 0$

$F_{HA} = -30$ kN (Compression)

$F_{HG} + (-30).\cos 53.13° = 0$

$F_{HG} = +18$ kN (Tension)

Consider F.B.D. of complete truss,

$\sum Y = 0$,

$R_A + R_B = 24 + 10 = 34$ ..... (1)

$\sum M_A = 0$,

$R_B \times 6 + 24 \times 3 = 0$

$R_B = -12$ kN i.e acting downword

from equation (1),

$R_A - 12 = 34$

$R_A = 46$ kN

Considering joint B for equilibrium,

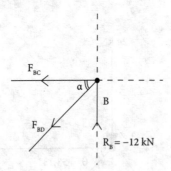

Fig. 6.14 (b) Joint B

Let $\angle CBD = \alpha$

Consider $\triangle GBA$,

$\tan\alpha = \dfrac{4}{6}$

$\alpha = 33.69°$

Analysis of Trusses

$\sum X = 0,$
$F_{BC} + F_{BD} \cos\alpha = 0$
$\sum Y = 0,$
$F_{BD} \sin\alpha = R_B$
$F_{BD} \sin 33.69° = -12$
$F_{BD} = -21.63$ kN   (Compression)
$F_{BC} + (-21.63)\cos 33.69° = 0$
$F_{BC} = +18$ kN   (Tension)

Considering joint C for equilibrium,

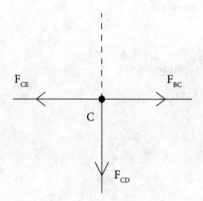

Fig. 6.14 (c) Joint C

$\sum X = 0,$
$F_{CE} = F_{BC}$
$F_{CE} = +18$ kN   (Tension)
$\sum Y = 0,$
$F_{CD} = 0$

Note: It can be concluded that if a joint is having three members and out of which two members are collinear and if no external load is acting on such joint then non collinear member has zero force and both collinear members contain like and equal force.

Consider joint D for equilibrium,

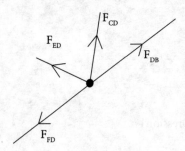

Fig. 6.14 (d) Joint D

as, $F_{CD} = 0$, it can be ignored. Now F_{ED} is a non collinear member; however F_{FD} and F_{DB} are collinear members,

thus $F_{ED} = 0$

and $F_{FD} = F_{DB}$

$F_{FD} = -21.63$ kN   (Compression)

Consider joint E for equilibrium,

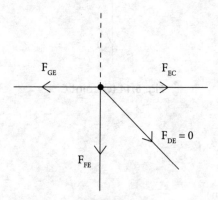

Fig. 6.14 (e) Joint E

as $F_{DE} = 0$,

$F_{FE} = 0$

and $F_{GE} = F_{EC}$

$F_{GE} = +18$ kN   (Tension)

Consider joint F for equilibrium,

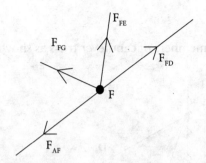

Fig. 6.14 (f) Joint F

as $F_{FE} = 0$
$F_{FG} = 0$
and $F_{AF} = F_{FD}$
i.e. $F_{AF} = -21.63$ kN   (Compression)

Consider joint G for equilibrium,

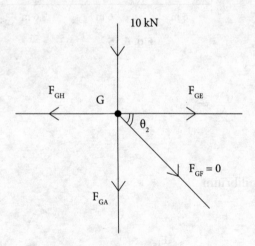

Fig. 6.14 (g) Joint G

$\Sigma Y = 0$,
$10 + F_{GA} + F_{GF} \sin\theta_2 = 0$
$F_{GA} = -10$ kN (Compression)

Example: 6.12

Determine forces in all the members of cantilever truss as shown in Fig. 6.15.

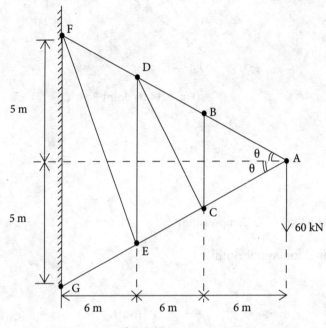

Fig. 6.15

Solution:

Let $\angle FAG = 2\theta$

$$\tan\theta = \frac{5}{18}$$

$$\theta = 15.52°$$

Consider joint A for equilibrium,

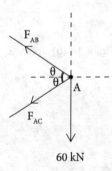

Fig. 6.15 (a) Joint A

$\sum X = 0$,
$F_{AB} \cos\theta + F_{AC} \cos\theta = 0$
as $\cos\theta \neq 0$,
$F_{AB} + F_{AC} = 0$
or,
$F_{AB} = -F_{AC}$ ..... (1)
$\sum Y = 0$,
$F_{AB} \sin\theta = F_{AC} \sin\theta + 60$
$-F_{AC} \sin 15.52° = F_{AC} \sin 15.52° + 60$
$-2.F_{AC} \sin 15.52° = 60$
$F_{AC} = -112.12$ kN   (Compression)

From equation (1),
i.e. $F_{AB} = +112.12$ kN   (Tension)

Consider joint B for equilibrium,

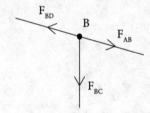

Fig. 6.15 (b) Joint B

As no external load is acting on joint B, thus

$F_{BC} = 0$
$F_{BD} = F_{AB} = +112.12$ kN   (Tension)

Consider joint C for equilibrium,

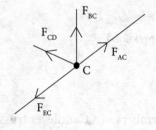

Fig. 6.15 (c) Joint C

as, $F_{BC} = 0$

$F_{CD} = 0$

and $F_{EC} = F_{AC} = -112.12$ kN   (Compression)

Similarly considering joint D for equilibrium,

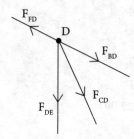

Fig. 6.15 (d) Joint D

$F_{CD} = 0$
$F_{DE} = 0$
$F_{FD} = F_{BD} = +112.12$ kN   (Tension)

Consider joint E for equilibrium,

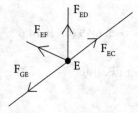

Fig. 6.15 (e) Joint E

$F_{ED} = 0$
$F_{EF} = 0$
$F_{GE} = F_{EC} = -112.12$ kN   (Compression)

Example: 6.13

Find the axial forces in all the members of a Cantilever truss as shown in Fig. 6.16.

(U.P.T.U. Ist Sem, 2002–2003)

Analysis of Trusses

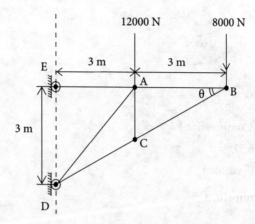

Fig. 6.16

Solution:

Let $\angle ABC = \theta$,
Consider $\triangle EBD$,

$$\tan\theta = \frac{3}{6}$$
$$\theta = 26.57°$$

Considering joint B for equilibrium,

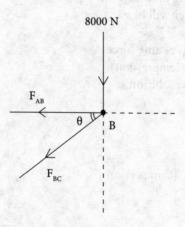

Fig. 6.16 (a) Joint B

$\sum X = 0,$
$F_{AB} + F_{BC} \cos\theta = 0$
$\sum Y = 0,$
$F_{BC} \sin\theta + 8000 = 0$
$F_{BC} \sin 26.57° = -8000$
$F_{BC} = -17885.45 \text{ N} \quad \text{(Compression)}$
$F_{AB} + (-17885.45)\cos 26.57° = 0$
$F_{AB} = +15996.54 \text{ N} \quad \text{(Tension)}$

Consider joint C for equilibrium,

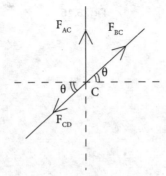

Fig. 6.16 (b) Joint C

Since no external load is acting on this joint
Force in non collinear member will be zero
i.e. $F_{AC} = 0$
and collinear member will have same force
i.e. $F_{CD} = F_{BC} = -17885.54\text{N} \quad \text{(Compression)}$
it can also be determined by resolution,

$\sum X = 0,$
$F_{CD} \cos\theta = F_{BC} \cos\theta$
as $\cos\theta \neq 0,$
or $F_{CD} = F_{BC} = -17885.54\text{N} \quad \text{(Compression)}$

$\sum Y = 0,$
$F_{AC} + F_{BC} \sin\theta = F_{CD} \sin\theta$
$F_{AC} + F_{CD} \sin\theta = F_{CD} \sin\theta$
$F_{AC} = 0$

Considering joint A for equilibrium,

Analysis of Trusses

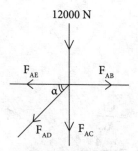

Fig. 6.16 (c) Joint A

Let EAD = α

$\tan\alpha = \dfrac{3}{3}$

$\alpha = 45°$

$\sum X = 0,$
$F_{AE} + F_{AD}.\cos\alpha = F_{AB}$
$\sum Y = 0,$
$F_{AD}.\sin\alpha + F_{AC} + 12000 = 0$
$F_{AD}\sin 45° + 0 + 12000 = 0$
$F_{AD} = -16970.56 \text{ N}$ (Compression)
$F_{AE} + (-16970.56)\cos 45° = 15996.54$
$F_{AE} = +27996.54 \text{ N}$ (Tension)

Example: 6.14

Determine forces in all the members of the simple supported truss as shown in Fig. 6.17.

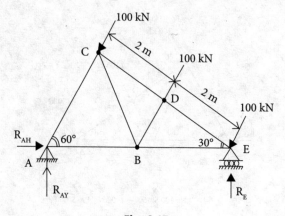

Fig. 6.17

Solution:

Consider △ ACE,
∠ACE = 180 − 60° − 30°
    = 90°
i.e. AE is hypotenuse.

$$\sin 60° = \frac{CE}{AE} \quad \text{and} \quad \tan 60° = \frac{CE}{AC}$$

$$AE = \frac{4}{\sin 60°} \quad\quad AC = \frac{4}{\tan 60°}$$

$$AE = 4.62 \text{m} \quad\quad AC = 2.31 \text{m}$$

as AC ∥ BD, thus ∠DBE = 60°
In △ CDB and △ BDE, BD is a bisector
thus ∠BCD = 30°
    ∠ACB = 90° − 30°
    ∠ACB = 60°
and ∠CBA = 60°
This shows that △ ABC is an equilateral triangle
i.e. AB = AC = 2.31m
or AB = BE = 2.31m
Consider F.B.D. of complete truss as shown in figure. 6.17(a),

$$\sum X = 0,$$
$$R_{AH} = 100\cos 60° + 100\cos 60° + 100\cos 60°$$
$$R_{AH} = 150 \text{ kN}$$

$$\sum Y = 0,$$
$$R_{AY} + R_E = 3 \times 100\sin 60° \quad\quad\quad\quad\quad\quad \text{..... (1)}$$

$$\sum M_A = 0,$$
$$R_E \times AE = 100\cos 60° \times 0 + 100\sin 60° \times AB + 100\cos 60° \times 0$$

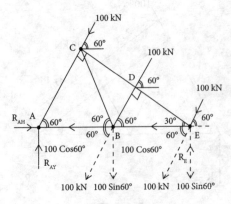

Fig. 6.17 (a)

Analysis of Trusses

Note:

To determine support reactions, we have used principle of transmissibility to shift force 100 kN acting at joint D to joint B. However, during analysis of joints we cannot use this principle as stated in section 6.3.

$R_E \times 4.62 = 100 \sin 60° \times 2.31 + 100 \sin 60° \times 4.62$
$R_E = 129.90$ kN
From equation (1),
$R_{AY} + 129.90 = 300 \sin 60°$
Substituting value of R_E in equation (1),
$R_{AV} = 129.90$ kN
Consider joint A for equilibrium,

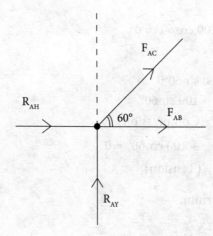

Fig. 6.17 (b) Joint A

$\sum X = 0,$

$\quad R_{AH} + F_{AC} \cos 60° + F_{AB} = 0$

$\sum Y = 0,$

$\quad F_{AC} \sin 60° + R_{AV} = 0$
$\quad F_{AC} \sin 60° + 129.90 = 0$

$\quad F_{AC} = -150$ kN $\quad$ (Compression)
$\quad 150 + F_{AC} \cos 60° + F_{AB} = 0$
$\quad 150 + (-150).\cos 60° + F_{AB} = 0$
$\quad 150 - 75 + F_{AB} = 0$
$\quad F_{AB} = -75$ kN $\quad$ (Compression)

Consider joint E for equilibrium,

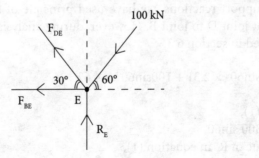

Fig. 6.17 (c) Joint E

$\sum X = 0$,
$$F_{DE} \cos 30° + F_{BE} + 100 \cos 60° = 0$$

$\sum Y = 0$,
$$F_{DE} \sin 30° + R_E = 100 \sin 60°$$
$$F_{DE} \sin 30° + 129.90 = 100 \sin 60°$$
$$F_{DE} = -86.59 \text{ kN} \quad \text{(Compression)}$$
$$(-86.59) \cos 30° + F_{BE} + 100 \cos 60° = 0$$
$$F_{BE} = +25 \text{ kN} \quad \text{(Tension)}$$

Consider joint D for equilibrium,

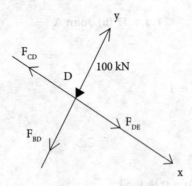

Fig. 6.17 (d) Joint D

$\sum X = 0$,
$$F_{CD} = F_{DE}$$
$$F_{CD} = -86.59 \text{ kN} \quad \text{(Compression)}$$

Analysis of Trusses

$\Sigma Y = 0$,

$F_{BD} + 100 = 0$

i.e. $F_{BD} = -100$ kN (Compression)

Consider joint B for equilibrium,

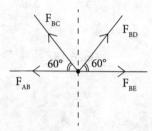

Fig. 6.17 (e) Joint B

$\Sigma Y = 0$,

$F_{BC} \sin 60° + F_{BD} \sin 60° = 0$

$F_{BC} = -F_{BD}$

$F_{BC} = +100$ kN (Tension)

6.4.2 Method of section

This method is preferred when forces are required to be determined in some particular members of a large truss. In this method the truss is cut in two portions by a suitable section. The section cuts the number of members whose forces are required to be determined. Such members of unknown forces should not be more than three as the sectioned truss represents a non-concurrent coplanar force system. To determine the maximum three unknown forces, free body diagram of sectioned truss is isolated and analyzed by using three equilibrium equations, i.e., $\Sigma X = 0$, $\Sigma Y = 0$ and $\Sigma M = 0$. However more sections can be taken further to determine unknown forces in other members. Further, this method is discussed in detail in solved examples.

Example: 6.15

Find the forces in members BC, BE and FE of the truss as shown in Fig. 6.18.

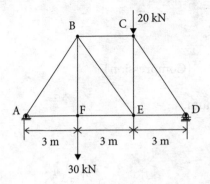

Fig. 6.18

(U.P.T.U. IInd Sem, 2003–2004)

Solution:

In this type of problem, method of section is more preferable as forces are required to be determined in three members only. However, this problem can also be solved by using method of joint.

In method of section truss is cut or sectioned in two parts. The cut may be applied horizontally, vertically or slanted but it should not cut any joint. Maximum three unknown members may be cut by one section due to availability of maximum three equations i.e. $\Sigma X = 0$, $\Sigma Y = 0$ and $\Sigma M = 0$

Consider a section 1-1 which cuts the truss vertically and produces in two parts as shown below:

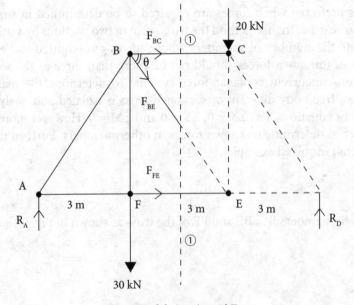

Fig. 6.18 (a) Sectioned Truss

Analysis of Trusses

Now any part whether left or side sectioned part can be considered for equilibrium.
If we consider left side sectioned truss, it can be observed that there are total four forces which are unknown i.e., R_A, F_{BC}, F_{BE} and F_{FE}. The same condition exists for right side sectioned truss.

Thus we've to determine support reactions first by considering F.B.D. of complete truss.

$\sum Y = 0,$
$R_A + R_D = 30 + 20 = 50 \text{ kN}$ ..... (1)

$\sum M_A = 0,$
$R_D \times 9 = 30 \times 3 + 20 \times 6$
$R_D = 23.33 \text{ kN}$
$R_A = 26.67 \text{ kN}$

Consider equilibrium of left side sectioned truss. Now forces acting on left side of the sectioned of truss will be considered.

$\sum Y = 0,$
$R_A = F_{BE} \sin\theta + 30$ ..... (1)

$\sum X = 0,$
$F_{BC} + F_{FE} + F_{BE} \cos\theta = 0$ ..... (2)

Assuming that CE = 3 m (Not Given)
$\sum M_B = 0,$
$R_A \times 3 = F_{FE} \times FB$
$26.67 \times 3 = F_{FE} \times 3$
$F_{FE} = +26.67 \text{ kN (Tension)}$
$\tan\theta = \dfrac{3}{3}$
$\theta = 45°$

From equation (1),

$26.67 = F_{BE} \sin 45° + 30$
$F_{BE} = -4.71 \text{ kN (Compression)}$

Substitute values of F_{FE} and F_{BE} in equation (2),

$F_{BC} + 26.67 + (-4.71).\cos 45° = 0$
$F_{BC} = -23.34 \text{ kN (Compression)}$

Note: We can take moment from any joint of the truss but forces are never considered acting outside of sectioned truss i.e. moment can be taken about joint E but 20 kN force acting on right side of sectioned truss will not be considered.

This is verified below

$$\sum M_E = 0,$$
$$R_A \times AE + F_{BC} \times EC = 30 \times FE$$
$$26.67 \times 6 + F_{BC} \times 3 = 30 \times 3$$
$$F_{BC} = -23.34 \text{ kN (Compression)}$$

which is same as determined earlier

Example: 6.16

Determine the support reactions and nature and magnitude of forces in members BC and EF of the diagonal truss as shown in Fig. 6.19.

(U. P. T. U, 2000–2001)

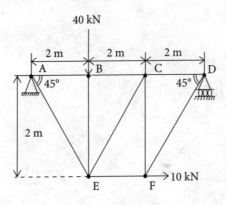

Fig. 6.19

Solution:

Consider F. B. D. of complete truss as shown in Fig. 6.19(a)

Analysis of Trusses

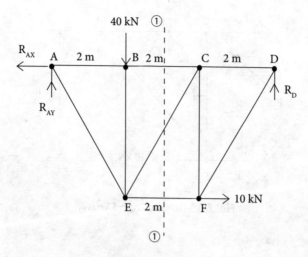

Fig. 6.19 (a)

$\sum X = 0,$

$R_{AX} = 10$ kN

$\sum Y = 0,$

$R_{AY} + R_D = 40$ ..... (1)

$\sum M_A = 0,$

$R_D \times 6 + 10 \times 2 = 40 \times 2$

$6R_D = 60$

$R_D = 10$ kN

From equation (1)

$R_{AY} = 30$ kN

As forces are to be determined in the members BC and EF, a section 1-1 will cut the truss in such a way that:

i. Maximum three unknown members will be cut.
ii. Truss is divided into two parts.
iii. Now left or right side any sectioned truss can be analysed.

Considering right side sectioned truss for equilibrium,

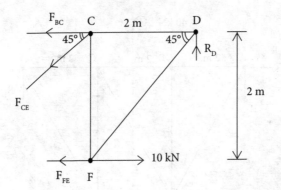

Fig. 6.19 (b) Sectioned Truss

$\sum X = 0,$
$F_{BC} + F_{CE} \cos 45° + F_{FE} = 10 \text{ kN}$ ..... (2)

$\sum Y = 0,$
$F_{CE} \sin 45° = R_D = 10 \text{ kN}$
$F_{CE} = +14.14 \text{ kN}$ (Tension)

$\sum M_D = 0,$
$F_{CE} \sin 45° \times 2 + 10 \times 2 = F_{FE} \times 2$
$\dfrac{14.14}{\sqrt{2}} \times 2 + 20 = F_{FE} \times 2$
$F_{FE} = +20 \text{ kN}$ (Tension)

Substituting value of F_{CE} and F_{FE} in equation (2),

$F_{BC} + 14.4 \cos 45° + 20 = 10$
$F_{BC} = -20 \text{ kN}$ (Compression)

Example: 6.17

A railway bridge truss is loaded as given in Fig. 6.20. Using method of section, determine force in the members FH, FG and EG.

Analysis of Trusses

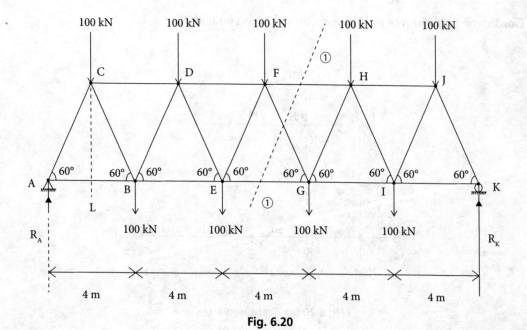

Fig. 6.20

Solution:

All triangles are equilateral triangles hence each side is 4 m. First support reactions have to be determined,

$\sum Y = 0$,
$R_A + R_K = 9 \times 100 = 900$ kN ..... (1)

$\sum M_A = 0$,
$R_K \times 20 = 100 \times AL + 100 \times AB + 100 \times (AL + CD)$
$\qquad + 100 \times AE + 100(AL + CF) + 100(AG) + 100 \times (AL + CH)$
$\qquad + 100 \times AI + 100(AL + CJ)$

$AL = AC \cos 60° = 4 \cos 60° = 2$ m.

$20 R_K = 100 \times 2 + 100 \times 4 + 100(2 + 4) + 100 \times 8 + 100(2 + 8)$
$\qquad + 100 \times 12 + 100 \times (2 + 12) + 100 \times 16 + 100(2 + 16)$

$R_K = 450$ kN and $R_A = 450$ kN

Note: In this problem support reactions can be determined directly without taking moment as it is a symmetrical loaded truss, thus support reactions will be equal i.e., $R_A = R_K$

Using equation (1),
$R_A = 450$ kN and $R_K = 450$ kN

Consider equilibrium of right side sectioned truss about section 1-1,

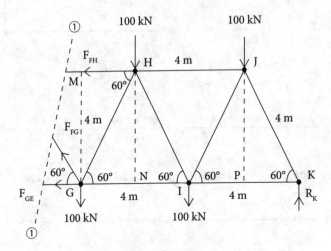

Fig.6.20 (a) Sectioned Truss

$\sum Y = 0$,
$F_{FG} \sin 60° + R_K = 400$ kN
$F_{FG} \sin 60° + 450 = 400$
$F_{FG} = -57.74$ kN (Compression)

$\sum X = 0$,
$F_{GE} + F_{FG} \cos 60° + F_{FH} = 0$ ..... (2)

$\sum M_G = 0$,
$F_{FH} \times GM + R_K \times 8 = 100 \times GN + 100 \times GI + 100(GP)$
$F_{FH} \times 4 \sin 60° + 450 \times 8 = 100 \times 4 \cos 60° + 100 \times 4 + 100(4 + 4\cos 60°)$
$F_{FH} = -692.82$ kN (Compression)

Substituting values of F_{FG} and F_{FH} in equation (2),

$F_{GE} + (-57.74)\cos 60° + (-692.82) = 0$
$F_{GE} = +721.69$ kN (Tension)

Example: 6.18

For the simply supported truss shown in Fig. 6.21, find the force in the members BD, DE, EG, BE and CE using method of sections.

(U.P.T.U. Ist Sem, 2003–2004)

Analysis of Trusses

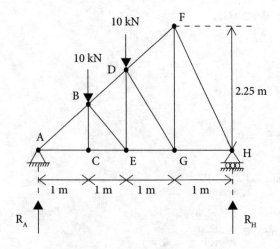

Fig. 6.21

Solution:

As all external loads are vertical, thus hinge will offer only vertical reaction. Consider complete truss for equilibrium,

$\sum Y = 0,$
$R_A + R_H = 20$ ..... (1)

$\sum M_A = 0,$
$R_H \times 4 = 10 \times 1 + 10 \times 2$
$R_H = 7.5 \text{ kN}$
$R_A = 12.5 \text{ kN}$

Consider section 1-1 cut the members BD, BE and CE as shown in Fig. 6.21(a)

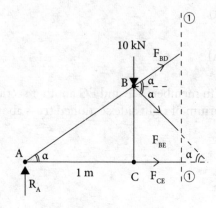

Fig. 6.21 (a) Sectioned Truss

Let $\angle BAC = \alpha$, Consider $\triangle AGF$,

$$\tan \alpha = \frac{2.25}{3}$$

i.e. $\alpha = 36.87°$

$\sum X = 0$,

$F_{BD} \cos \alpha + F_{BE} \cos \alpha + F_{CE} = 0$ ..... (2)

$\sum Y = 0$,

$F_{BD} \sin \alpha + 12.5 = F_{BE} \sin 36.87° + 10$ ..... (3)

Consider $\triangle ACB$, $\tan 36.87° = \dfrac{BC}{AC}$

$\tan 36.87° = \dfrac{BC}{1}$

$BC = 0.75 \, m$

$\sum M_B = 0$,

$R_A \times 1 = F_{CE} \times BC$

$R_A \quad = F_{CE} \times 0.75$

$12.50 = F_{CE} \times 0.75$

$F_{CE} \quad = +16.66 \, kN \, (\text{Tension})$

From equation (2),

$F_{BD} \cos 36.87° + F_{BE} \cos 36.87° + 16.66 = 0$

$F_{BD} = (-20.83 - F_{BE})$

substituting in equation (3),

$(-20.83 - F_{BE}) . \sin 36.87° + 12.5 = F_{BE} \sin 36.87° + 10$

$F_{BE} = -8.33 \, kN \, (\text{Compression})$

$F_{BD} = -20.83 - (-8.33)$

$F_{BD} = -12.5 \, kN \, (\text{Compression})$

Further to determine force in members DE and EG another section 2-2 will cut the truss in two parts. Consider equilibrium of right side sectioned truss about section 2-2 as shown in figure 6.21(b),

Analysis of Trusses

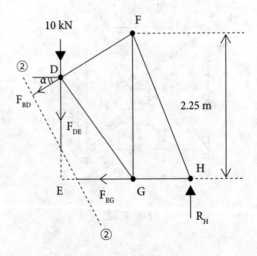

Fig. 6.21 (b) Sectioned Truss

$\sum X = 0$,
$F_{BD} \cos\alpha + F_{EG} = 0$ ..... (4)

$\sum Y = 0$,
$F_{BD} \sin\alpha + F_{DE} + 10 = 7.5$ ..... (5)

Substituting value of F_{BD} in equation (4),

$(-12.5).\cos 36.87° + F_{EG} = 0$

$F_{EG} = +10$ kN (Tension)

Substituting value of F_{BD} in equation (5),

$(-12.5).\sin 36.87° + F_{DE} + 10 = 7.5$

$F_{DE} = +5$ kN (Tension)

Example: 6.19

Determine forces in members AE, CD and FB of a square truss as shown in the Fig. 6.22 by using method of section.

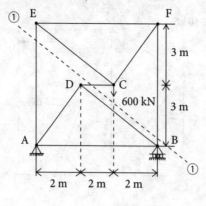

Fig. 6.22

Solution:

Consider section 1-1 cuts AE, CD and FB as shown in Fig. 6.22(a). Now truss is divided into two parts as upper side and lower side sectioned truss. If we consider equilibrium of lower side sectioned truss then we have to determine support reactions first. However if we consider upper side sectioned truss it can be directly analysed. Thus, consider the upper side sectioned truss for equilibrium,

$\sum X = 0,$
$F_{CD} = 0$
$\sum Y = 0,$
$F_{AE} + F_{FB} + 600 = 0$ ..... (1)
$\sum M_C = 0,$
$F_{AE} \times 4 = F_{FB} \times 2$
$F_{FB} = 2.F_{AE}$

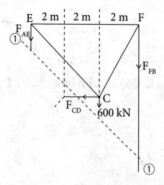

Fig. 6.22 (a) Sectioned Truss

Analysis of Trusses

From equation (1),

$F_{AE} + 2F_{AE} + 600 = 0$

$F_{AE} = -200$ kN (Compression)

$F_{FB} = -400$ kN (Compression)

Example: 6.20

The roof truss shown in Fig. 6.23 is supported at A and B and carries vertical loads at each of the upper joints. Using the method of sections determine the forces in the members CE and FG of truss.

(M.T.U, IInd Sem, 2011–12)

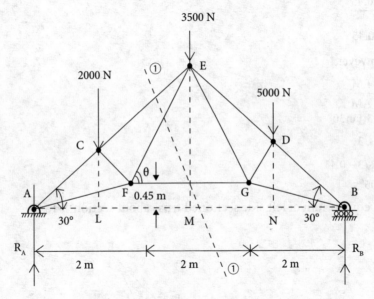

Fig. 6.23

Solution:

Consider F.B.D. of given truss for equilibrium,

$\Sigma Y = 0$,

$R_A + R_B = 10500$ N ..... (1)

Consider $\triangle$AME,

$$\cos 30° = \frac{AM}{AE}$$

$$\cos 30° = \frac{3}{AE}$$

$AE = 2\sqrt{3}$ m thus $AC = \sqrt{3}$ m,

Similarly, $BE = 2\sqrt{3}$ m thus $BD = \sqrt{3}$ m,
and $BN = BD\cos 30°$

$\sum M_A = 0$,
$R_B \times 6 = 2000 \times AL + 3500 \times AM + 5000 \times AN$
$R_B \times 6 = 2000 \times AC\cos 30° + 3500 \times 3 + 5000(6 - BN)$
$R_B \times 6 = 2000 \times \sqrt{3}.\cos 30° + 3500 \times 3 + 500(6 - \sqrt{3}\cos 30°)$
$R_B = 6000$ N and $R_A = 4500$ N

Let, $\angle EFG = \theta$

Consider $\triangle EFG$,

$\tan\theta = \dfrac{EM - 0.45}{1}$

$\tan\theta = EM - 0.45$

In $\triangle AME$ however,

$\tan 30° = \dfrac{EM}{AM}$

$EM = 3.\tan 30°$

$\quad = \sqrt{3}$

thus, $\tan\theta = \sqrt{3} - 0.45$

$\theta = 52.05°$

Considering equilibrium of left side sectioned truss about section 1-1,

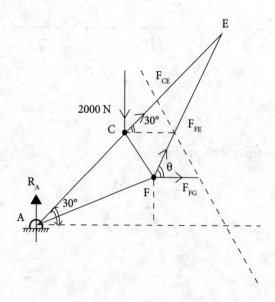

Fig. 6.23 (a) Sectioned Truss

$\sum X = 0$,
$F_{CE} \cos 30° + F_{FE} \cos \theta + F_{FG} = 0$ ...... (2)

$\sum Y = 0$,
$R_A + F_{CE} \sin 30° + F_{FE} \sin \theta = 2000$ ...... (3)

$\sum M_E = 0$,
$F_{FG} \times (EM - 0.45) + 2000 \times LM = R_A \times AM$
$F_{FG} (\sqrt{3} - 0.45) + 2000 \times (AM - AL) = 4500 \times 3$
$F_{FG} (1.732 - 0.45) + 2000 \times (3 - \sqrt{3}.\cos 30°) = 4500 \times 3$
$F_{FG} = +8190.33 \text{ N} \quad \text{(Tension)}$

Consider equation (2),
$F_{CE} \cos 30° + F_{FE} \cos 52.05° + 8190.33 = 0$
$F_{FE} = (-F_{CE} \times 1.41 - 13318.1)$

Substitute in equation (3),
$R_A + F_{CE} \sin 30° + (-F_{CE} \times 1.41 - 13318.18) \sin 52.05° = 2000$
$4500 + F_{CE} \sin 30° + (-1.41 F_{CE} - 13318.18) \sin 52.05° = 2000$
$4500 + F_{CE} . \sin 30° - 1.11 F_{CE} - 10502 = 2000$
$F_{CE} = -13118.0 \text{ N (Compression)}$

Theoretical Problems

T 6.1 Define a truss. What are its uses? How trusses are classified?
T 6.2 Differentiate between perfect, deficient and redundant trusses with suitable examples.
T 6.3 Name the different methods of finding out the forces in members of perfect frame.
T 6.4 Outline the steps involved while making an analysis of a simple truss by the
  (i) Method of Joints
  (ii) Method of Sections
T 6.5 Give the methods used to analyze plane truss. [MTU, 2012–13]
T 6.6 Write assumptions made in the analysis of Truss? [MTU, 2012–13]
T 6.7 Explain the following:
  (i) Types of trusses.
  (ii) Steps followed in analysis of truss by method of section. [MTU, 2012–13]
T 6.8 What is perfect truss? How it differ from an imperfect truss? [MTU, 2013–14]

Numerical Problems

N 6.1 Determine the forces in all the members of the truss shown in Fig. NP 6.1

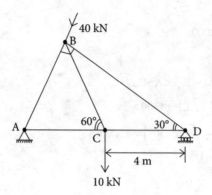

Fig. NP 6.1

N 6.2 Determine forces in all the members of the truss as shown in Fig. NP 6.2.

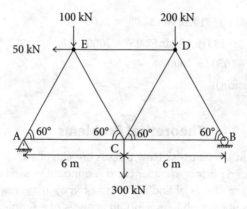

Fig. NP 6.2

N 6.3 Determine the forces in all the members of Cantilever truss as shown in Fig. NP 6.3.

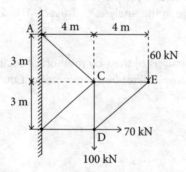

Fig. NP 6.3

Analysis of Trusses

N 6.4 Determine the forces in all the members of the truss as shown in Fig. NP 6.4.

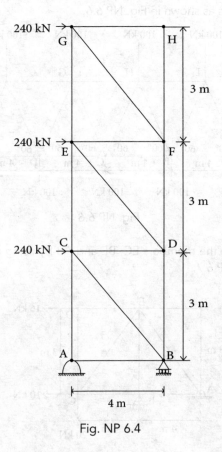

Fig. NP 6.4

N 6.5 Determine the forces in all the members of the truss as shown in Fig. NP 6.5.

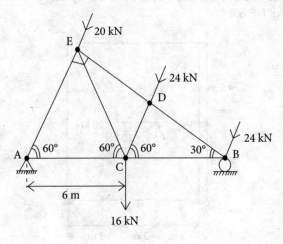

Fig. NP 6.5

N 6.6 Determine forces in the members HG, HC and BC by using method of section for railway bridge truss as shown in Fig. NP 6.6.

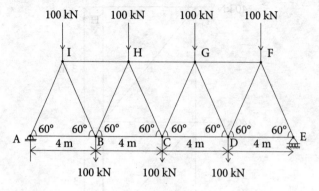

Fig. NP 6.6

N 6.7 Determine force in the members EC, BE and AB by using method of section for truss as shown in Fig. NP 6.7.

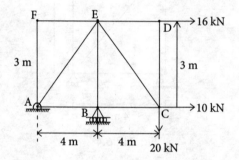

Fig. NP 6.7

N 6.8 Determine force in the members AB, EF and CD by using method of section for truss as shown in Fig. NP 6.8

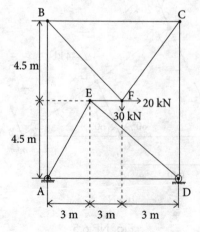

Fig. NP 6.8

Analysis of Trusses

N 6.9 Determine force in the members EC, CF and FB for truss as shown in Fig. NP 6.9

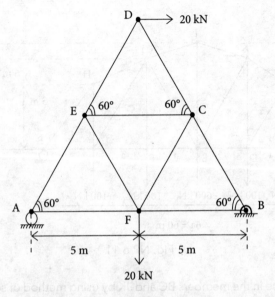

Fig. NP 6.9

N 6.10 Determine force in the members CD, BD and AB by using method of section as shown in Fig. NP 6.10.

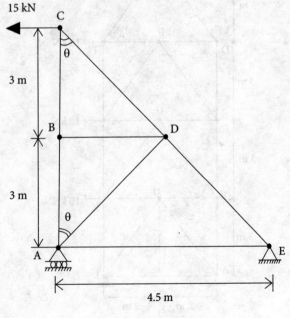

Fig. NP 6.10

N 6.11 Determine force in the member JK, KE and DE by using method of section for truss as shown in Fig. NP 6.11

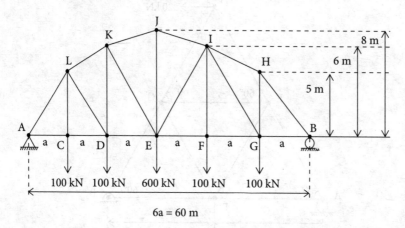

Fig. NP 6.11

N 6.12 Determine force in the members BC and JK by using method of section for truss shown in Fig. NP 6.12.

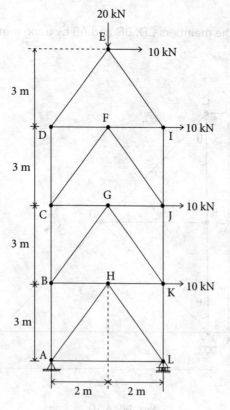

Fig. NP 6.12

Multiple Choice Questions

1. The equilibrium equations required to analyse the joint will be
   a. $\sum X = 0, \sum Y = 0, \sum M = 0$
   b. $\sum X = 0, \sum Y = 0$
   c. $\sum Y = 0, \sum M = 0$
   d. none of these

2. The equilibrium equations required to analyse the sectioned truss will be
   a. $\sum X = 0, \sum Y = 0, \sum M = 0$
   b. $\sum X = 0, \sum Y = 0$
   c. $\sum Y = 0, \sum M = 0$
   d. none of these

3. The members of a truss should be
   a. prismatic
   b. any cross-sections
   c. straight
   d. all of these

4. Select the plane truss
   a. electrical transmission tower
   b. tower used for signals
   c. roof truss
   d. none of these

5. Select the space truss
   a. electrical transmission tower
   b. bridge truss
   c. roof truss
   d. none of these

6. The mathematical expression for a redundant truss is
   a. $m = (2j - 3)$
   b. $m < (2j - 3)$
   c. $m > (2j - 3)$
   d. none of these

7. The mathematical expression for a deficient truss is
   a. $m = (2j - 3)$
   b. $m < (2j - 3)$
   c. $m > (2j - 3)$
   d. none of these

8. The mathematical expression $m = (2j - 3)$ is valid for
   a. deficient truss
   b. perfect truss
   c. redundant truss
   d. none of these

9. Which is valid assumption for the analysis of truss
   a. Weight of the members act at the joints
   b. Weight of the members act at midpoint of the members
   c. Weight of the members are considered as negligible
   d. none of these

10. In method of joint, forces acting at a joint represents
    a. concurrent coplanar force system
    b. concurrent non-coplanar force system
    c. concurrent collinear force system
    d. none of these

11. In method of section, how much maximum members of unknown forces can be cut by a section
    a. two members
    b. three members
    c. four members
    d. none of these

Answers

1. b 2. a 3. d 4. c 5. a 6. c 7. b 8. b 9. c 10. a 11. b

Chapter 7

Centroid and Centre of Gravity

7.1 Introduction

So far we have dealt with the point loads or concentrated loads which do not exist practically. Generally loads are distributed over a line, surface area or total volume of a body. Some examples of load distributions are weight of a body, uniformly distributed load, uniformly varying load, etc. To apply the conditions of equilibrium, such distributive forces are replaced by equivalent resultant force R which acts at a certain coordinates and produces the same effect as produces by distributive forces. These certain coordinates are the point of application and termed as centroid or centre of gravity, G. It is characterized as centroid for line, surface area or volume and centre of gravity, G remains unaltered if body is rotated.

The centroid or centre of gravity is very useful to determine the stability of fast moving vehicles like truck, ship, etc. during turning. These vehicles are supposed to topple if the height of centre of gravity is not close to their bottom portion. Apart this, it also plays a vital role to conduct sports events, stability of dams and safe lifting of heavy loads by cranes, etc.

7.2 Centre of Gravity, Centroid of Line, Plane Area and Volume

A body can be treated as a collection of very large numbers of particles of infinitesimal size. Each particle has its own weight which acts at certain coordinates. The weight of each particle acts vertically downwards towards the centre of the earth. Thus the weight of all particles represents a non-coplanar parallel force system. This force system can be replaced by their single resultant force which is nothing but weight, whose coordinates can be determined by using Varignon's theorem.

Consider a body of weight W consisting of n Particles of individual weights w_1, w_2, w_3... w_n whose coordinates are (x_1, y_1), (x_2, y_2), and (x_3, y_3), ... (x_n, y_n), respectively, from reference axis as shown in Fig. 7.1. Let the coordinate of weight W is $(\overline{X}, \overline{Y})$.

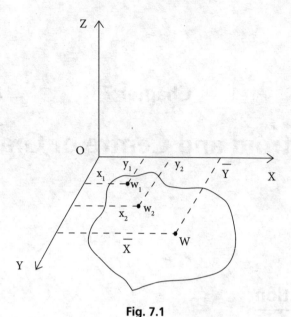

Fig. 7.1

Applying Varignon's theorem about axis OY we get,

$$W \times \overline{X} = (w_1 \times x_1) + (w_2 \times x_2) + (w_3 \times x_3) + ... (w_n \times x_n)$$

$$\overline{X} = \frac{(w_1 \times x_1) + (w_2 \times x_2) + (w_3 \times x_3) + ... (w_n \times x_n)}{W}$$

as, $W = (w_1 + w_2 + w_3 + ... w_n)$

$$\overline{X} = \frac{(w_1 \times x_1) + (w_2 \times x_2) + (w_3 \times x_3) + ... (w_n \times x_n)}{(w_1 + w_2 + w_3 + .. w_n)} = \sum_{i=1}^{n} \frac{w_i x_i}{w_i}$$

Similarly, Applying Varignon's theorem about axis OX we can get,

$$\overline{Y} = \frac{(w_1 \times y_1) + (w_2 \times y_2) + (w_3 \times y_3) + ... (w_n \times y_n)}{(w_1 + w_2 + w_3 + .. w_n)} = \sum_{i=1}^{n} \frac{w_i y_i}{w_i}$$

The coordinates $(\overline{X}, \overline{Y})$ of weight W are known as centre of gravity, G.

If the acceleration due to gravity, g, is constant and weights are replaced by product of masses and g then centre of gravity is termed as centre of mass

$$\overline{X} = \frac{(m_1 \times x_1) + (m_2 \times x_2) + (m_3 \times x_3) + ... (m_n \times x_n)}{(m_1 + m_2 + m_3 + .. m_n)} = \sum_{i=1}^{n} \frac{m_i x_i}{m_i}$$

$$\overline{Y} = \frac{(m_1 \times y_1) + (m_2 \times y_2) + (m_3 \times y_3) + ... (m_n \times y_n)}{(m_1 + m_2 + m_3 + .. m_n)} = \sum_{i=1}^{n} \frac{m_i y_i}{m_i}$$

Centroid and Centre of Gravity

If the body is consisting only length then the masses are replaced by length and coordinates $(\overline{X}, \overline{Y})$ of body is again termed as centroid

$$\overline{X} = \left(\frac{l_1 x_1 + l_2 x_2 + l_3 x_3 + l_4 x_4 + l_5 x_5 + l_6 x_6}{l_1 + l_2 + l_3 + l_4 + l_5 + l_6} \right)$$

$$\overline{Y} = \left(\frac{l_1 y_1 + l_2 y_2 + l_3 y_3 + l_4 y_4 + l_5 y_5 + l_6 y_6}{l_1 + l_2 + l_3 + l_4 + l_5 + l_6} \right)$$

If the body is consisting only area then the masses are replaced by area and coordinates $(\overline{X}, \overline{Y})$ of body is termed as centroid

$$\overline{X} = \frac{(a_1 \times x_1) + (a_2 \times x_2) + (a_3 \times x_3) + \dots (a_n \times x_n)}{(a_1 + a_2 + a_3 + \dots a_n)} = \sum_{i=1}^{n} \frac{a_i x_i}{a_i}$$

$$\overline{Y} = \frac{(a_1 \times y_1) + (a_2 \times y_2) + (a_3 \times y_3) + \dots (a_n \times y_n)}{(a_1 + a_2 + a_3 + \dots a_n)} = \sum_{i=1}^{n} \frac{a_i y_i}{a_i}$$

If the body is consisting only volume then the masses are replaced by volume and coordinates $(\overline{X}, \overline{Y})$ of body is again termed as centroid

$$\overline{X} = \frac{(V_1 \times x_1) + (V_2 \times x_2) + (V_3 \times x_3) + \dots (V_n \times x_n)}{(V_1 + V_2 + V_3 + \dots V_n)} = \sum_{i=1}^{n} \frac{V_i x_i}{V_i}$$

$$\overline{Y} = \frac{(V_1 \times y_1) + (V_2 \times y_2) + (V_3 \times y_3) + \dots (V_n \times y_n)}{(V_1 + V_2 + V_3 + \dots V_n)} = \sum_{i=1}^{n} \frac{V_i y_i}{V_i}$$

The formulas discussed earlier can be replaced by integral sign if infinitesimally element of weight dw is considered instead of n number of particles of weights. The centre of gravity is expressed as

$$\overline{X} = \sum_{i=1}^{n} \frac{w_i x_i}{w_i} = \frac{\int dw x^*}{\int dw}$$

$$\overline{Y} = \sum_{i=1}^{n} \frac{w_i y_i}{w_i} = \frac{\int dw y^*}{\int dw}$$

where x^* and y^* are the centre of gravity of elemental weight dw as shown in Fig. 7.2.

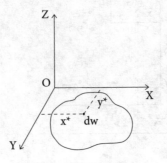

Fig. 7.2

Similarly for line, area and volume, we can express centroid, respectively, as

$$\overline{X} = \frac{\int dl\,x^*}{\int dl}, \overline{X} = \frac{\int dA\,x^*}{\int dA}, \overline{X} = \frac{\int dV\,x^*}{\int dV}$$

$$\overline{Y} = \frac{\int dl\,y^*}{\int dl}, \overline{Y} = \frac{\int dA\,y^*}{\int dA}, \overline{Y} = \frac{\int dV\,y^*}{\int dV}$$

7.3 Centroid of L, C, T and I-Sections

Example: 7.1

Determine the centroid of the given C-section as shown in Fig. 7.3.

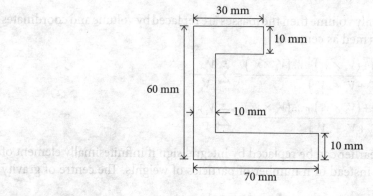

Fig. 7.3

Solution:

Following steps are carried out sequentially. Consider section as shown in Fig. 7.3(a)

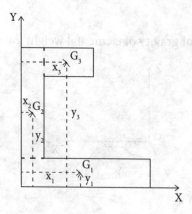

Fig. 7.3 (a)

Centroid and Centre of Gravity

(i) Consider reference axis, i.e. X and Y axis, at the bottom most and the left most, respectively.
(ii) Area selection of different areas, i.e. rectangles in this problem. It is shown by dotted thick lines.
(iii) Mark the individual centroid of each rectangle by G_1, G_2, and G_3.
(iv) Prepare the table accordingly, as given below.

$a_1 = 70 \times 10 \ mm^2$	$x_1 = 35 \ mm$	$y_1 = 5 \ mm$
$a_2 = 50 \times 10 \ mm^2$	$x_2 = 5 \ mm$	$y_2 = 10 + \dfrac{50}{2}$ $= 35 \ mm$
$a_3 = 20 \times 10 \ mm^2$	$x_3 = \left(10 + \dfrac{20}{2}\right)$ $= 20 \ mm$	$y_3 = \left(60 - \dfrac{10}{2}\right)$ $= 55 \ mm$

Thus centroid will be given by

$$\overline{X} = \left(\frac{a_1 x_1 + a_2 x_2 + a_3 x_3}{a_1 + a_2 + a_3}\right) \text{ and } \overline{Y} = \left(\frac{a_1 y_1 + a_2 y_2 + a_3 y_3}{a_1 + a_2 + a_3}\right)$$

$$\overline{X} = \left(\frac{700 \times 35 + 500 \times 5 + 200 \times 20}{700 + 500 + 200}\right), \ \overline{Y} = \left(\frac{700 \times 5 + 500 \times 35 + 200 \times 55}{700 + 500 + 200}\right)$$

$\overline{X} = 22.14 \ mm, \quad \overline{Y} = 22.86 \ mm$

The coordinates of centroid of C-section is (22.14 mm, 22.86 mm)

Example: 7.2

Determine the centroid of L-section as shown in Fig. 7.4.

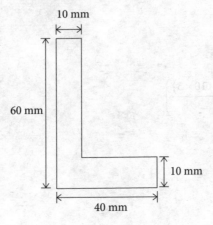

Fig. 7.4

Solution:

First all steps will be carried out as stated in example 7.1.
Consider Fig. 7.4(a)

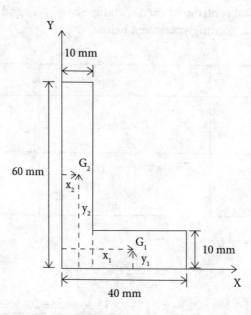

Fig. 7.4 (a)

$a_1 = 30 \times 10 \ mm^2$	$x_1 = \left(10 + \dfrac{30}{2}\right)$ = 25 mm	$y_1 = 5 \ mm$
$a_2 = 60 \times 10 \ mm^2$	$x_2 = 5 \ mm$	$y_2 = 30 \ mm$

The centroid will be

$$\overline{X} = \left(\frac{a_1 x_1 + a_2 x_2}{a_1 + a_2}\right)$$

$$= \left(\frac{(30 \times 10 \times 25) + (60 \times 10 \times 5)}{300 + 600}\right)$$

$$= 11.67 \ mm$$

$$\overline{Y} = \left(\frac{a_1 y_1 + a_2 y_2}{a_1 + a_2}\right)$$

$$= \left(\frac{300 \times 5 + 600 \times 30}{300 + 600}\right)$$

$$= 21.67 \ mm$$

Centroid and Centre of Gravity

Thus the coordinates of centroid of L section is (11.67 mm, 21.67 mm).

Alternative method:

If area of selection is changed then location of centroid of each rectangle is also changed, as shown in Fig. 7.4(b). Thus the table will be

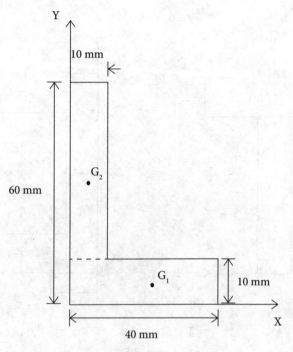

Fig. 7.4 (b)

$a_1 = 400\ mm^2$	$x_1 = 20\ mm$	$y_1 = 5\ mm$
$a_2 = 500\ mm^2$	$x_2 = 5\ mm$	$y_2 = 10 + \dfrac{50}{2}$ $= 35\ mm$

$$\overline{X} = \left(\dfrac{a_1 x_1 + a_2 x_2}{a_1 + a_2}\right) \qquad \overline{Y} = \left(\dfrac{a_1 y_1 + a_2 y_2}{a_1 + a_2}\right)$$

$$= \left(\dfrac{400 \times 20 + 500 \times 5}{400 + 500}\right) \qquad = \left(\dfrac{400 \times 5 + 500 \times 35}{400 + 500}\right)$$

$$\overline{X} = 11.67\ mm \qquad \overline{Y} = 21.67\ mm$$

It can be observed that the centroid has same value if reference axis remains same and intermediate area of selection does not alter the location of centroid.

7.4 Importance of Axis of Symmetry in Centroid and Centre of Gravity

An axis about which a body is split into two equal parts is called as axis of symmetry. If such an axis is considered as reference axis then it reduces the calculation procedure of centroid and centre of gravity. Consider for example T and C section as shown in Figs. 7.5(a) and (b)

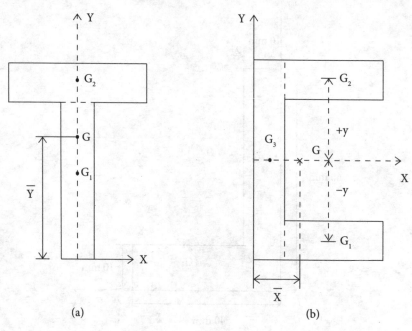

Fig. 7.5

In T section, if Y axis is taken in the middle vertically, it works as axis of symmetry. We observe that the individual centroid G_1 and G_2 of both rectangles lie on axis of symmetry which means that whole centroid of section will lie on Y axis.

As $\overline{X}$ is measured from Y-axis thus $\overline{X}=0$ for T-section and only $\overline{Y}$ is required to be determine.

Similarly for C-section, if X-axis is taken in the middle horizontally, it is called as axis of symmetry. Here centroid G_3 is lying on X-axis, however centroid G_1 and G_2 of equal areas are also lying at equal distances apart from X-axis. Thus algebraic sum of moment of their areas about X-axis will be zero. Thus whole centroid of C-section will lie on X-axis. As $\overline{Y}$ is measured from X-axis thus $\overline{Y}=0$ and only $\overline{X}$ is required to calculate.

Note:
Centroid of a body always lies at a fixed point but its coordinates may vary according to the references axis. For example consider a rectangle where the reference axes are at the bottom most and the left most as shown in Fig. 7.6.

Centroid and Centre of Gravity

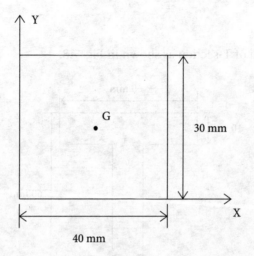

Fig. 7.6

$\overline{X} = 20$ mm
$\overline{Y} = 15$ mm

However if we take axis of symmetry as reference axis, as shown in Fig. 7.7

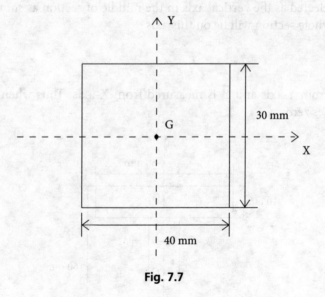

Fig. 7.7

The coordinates of centroid will be $\overline{X} = 0$, $\overline{Y} = 0$

Thus coordinates may have different values but centroid always lies at fixed point.

Example: 7.3

Determine the centroid of T-section as shown in Fig. 7.8.

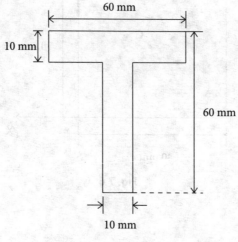

Fig. 7.8

Solution:

If the Y axis is selected as the vertical axis in the middle of section as shown in Fig. 7.8(a), the centroid of whole section will lie on this axis
i.e., $\overline{X} = 0$

Note:
$\overline{X}$ is measured from Y-axis and $\overline{Y}$ is measured from X-axis. Thus when centroid lies on Y-axis, $\overline{X}$ becomes zero.

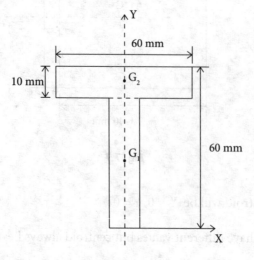

Fig. 7.8 (a)

Centroid and Centre of Gravity

All steps of example 7.1 will be repeated.

$a_1 = 500\ mm^2$	$y_1 = 25\ mm$
$a_2 = 600\ mm^2$	$y_2 = 60 - \dfrac{10}{2}$ $= 55\ mm$

$$\overline{Y} = \left(\dfrac{a_1 y_1 + a_2 y_2}{a_1 + a_2}\right)$$

$$= \left(\dfrac{500 \times 25 + 600 \times 55}{500 + 600}\right)$$

$$\overline{Y} = 41.36\ mm$$

Thus the co-ordinates of T-section will be (0, 41.36 mm)

Example: 7.4

Determine the centroid of the given C-section as shown in Fig. 7.9.

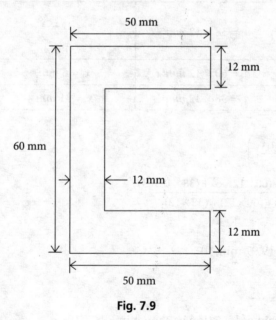

Fig. 7.9

Solution:

Given C-section is symmetrical about X-axis if taken as shown in Fig. 7.9(a)
Thus, $\overline{Y} = 0$

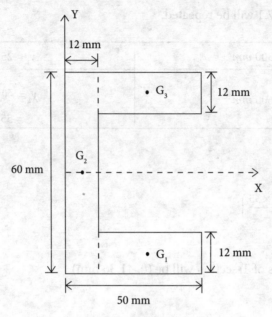

Fig. 7.9 (a)

$a_1 = 38 \times 12\ mm^2$	$x_1 = 12 + \dfrac{38}{2}$ $= 31\ mm$
$a_2 = 60 \times 12\ mm^2$	$x_2 = 6\ mm$
$a_3 = 38 \times 12\ mm^2$	$x_3 = 31\ mm$

$$\overline{X} = \left(\frac{a_1 x_1 + a_2 x_2 + a_3 x_3}{a_1 + a_2 + a_3} \right)$$

$$= \frac{(38 \times 12 \times 31) + (60 \times 12 \times 6) + (38 \times 12 \times 31)}{(38 \times 12) + (60 \times 12) + (38 \times 12)}$$

$$= 19.97\ mm$$

The centroid lies at (19.97 mm, 0)

Example: 7.5

Determine the centroid of I-section as shown in Fig. 7.10.

Centroid and Centre of Gravity

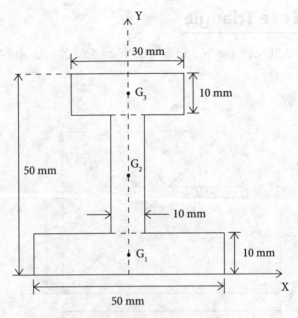

Fig. 7.10

Solution:

As, I-section is symmetrical about Y axis; thus centroid will lie on this axis which results $\overline{X} = 0$

$a_1 = 500\ mm^2$	$y_1 = 5\ mm$
$a_2 = 300\ mm^2$	$y_2 = 10 + \dfrac{30}{2} = 25\ mm$
$a_3 = 300\ mm^2$	$y_3 = 50 - \dfrac{10}{2} = 45\ mm$

$$\overline{Y} = \left(\frac{a_1 y_1 + a_2 y_2 + a_3 y_3}{a_1 + a_2 + a_3}\right)$$

$$= \left(\frac{(500 \times 5) + (300 \times 25) + (300 \times 45)}{500 + 300 + 300}\right)$$

$$= 21.36\ mm$$

The coordinates of centroid is (0, 21.36 mm)

7.5 Centroid of a Triangle

Consider a triangle ABC of base 'b' and height 'h' against X axis and Y axis as shown in Fig. 7.11.

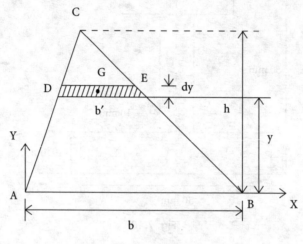

Fig. 7.11

Let an elementary strip DE of width b′ and height dy lies at distance 'y' from X-axis.

Thus $\bar{Y} = \dfrac{\int dA \cdot y^*}{\int dA}$

Here dA = area of elementary strip = $b' \times dy$ ...... (1)

As $\triangle$ CDE and $\triangle$ CAB are similar triangles,

$\dfrac{h-y}{b'} = \dfrac{h}{b}$

i.e. $b' = \dfrac{b(h-y)}{h}$

Substituting b' in equation (1),

$dA = \dfrac{b(h-y)}{h} \times dy$

y^* = centroid of strip DE from X axis

$= y + \dfrac{dy}{2}$

$y^* \approx y$

Centroid and Centre of Gravity

$$\bar{Y} = \frac{\int_0^h \frac{b(h-y)}{h}.dy.y}{\int_0^h \frac{b(h-y)}{h}.dy}$$

$$= \frac{\frac{b}{h}\int_0^h (h.y - y^2).dy}{\frac{b}{h}\int_0^h (h-y).dy}$$

$$= \frac{h\int_0^h y.dy - \int_0^h y^2.dy}{h\int_0^h 1.dy - \int_0^h y.dy}$$

$$= \frac{h.\left[\frac{y^2}{2}\right]_0^h - \left[\frac{y^3}{3}\right]_0^h}{h.\left[y\right]_0^h - \left[\frac{y^2}{2}\right]_0^h}$$

$$= \frac{\frac{h^3}{2} - \frac{h^3}{3}}{h^2 - \frac{h^2}{2}} = \frac{h}{3}$$

Thus centroid of a triangle lies at one third of height from the base. Or, the centroid from the vertex (apex) lies at two third of its height.

7.6 Centroid of a Quarter Circle and Semicircle

Consider a quarter circle of radius R as shown in Fig. 7.12.

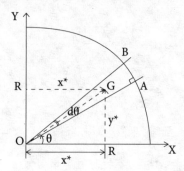

Fig. 7.12

Let an elemental circular sector OAB lies at an angle θ and subtends angle $d\theta$ at the centre 'O',

$$\overline{X} = \frac{\int dA \cdot x^*}{\int dA}$$

where dA = elemental area of sector OAB,

As $\angle AOB = d\theta$ which is very small thus AB may be consider as perpendicular to OA. Thus $\triangle OAB$ may be approached as a right angle triangle.

$$dA = \frac{1}{2} \times OA \times AB \quad \tan d\theta = \frac{AB}{OA}$$

$$= \frac{1}{2} \times R \times R \times d\theta \quad d\theta = \frac{AB}{R}$$

$$dA = \frac{R^2 \cdot d\theta}{2} \quad AB = R \cdot d\theta$$

$$x^* = OG \cos\theta$$

$$= \frac{2}{3} R \cos\theta$$

Note:
OG is the centroidal distance of $\triangle AOB$ from vertex O, which will be two-third of its height i.e., $\frac{2}{3}R$.

$$\overline{X} = \frac{\int_0^{\pi/2} \frac{R^2 \cdot d\theta}{2} \cdot \frac{2}{3} R \cos\theta}{\int_0^{\pi/2} \frac{R^2 \cdot d\theta}{2}}$$

$$= \frac{\frac{R^2}{2} \cdot \frac{2}{3} R \int_0^{\pi/2} \cos\theta \cdot d\theta}{\frac{R^2}{2} \int_0^{\pi/2} d\theta}$$

$$= \frac{\frac{2}{3} R [\sin\theta]_0^{\pi/2}}{[\theta]_0^{\pi/2}}$$

$$= \frac{\frac{2}{3} R [\sin\pi/2 - \sin 0]}{\left(\frac{\pi}{2}\right)}$$

$$\overline{X} = \frac{4R}{3\pi}$$

Centroid and Centre of Gravity

$$\bar{Y} = \frac{\int dA.y^*}{\int dA}$$

$$y^* = OG\sin\theta$$
$$= \frac{2}{3}R\sin\theta$$

$$= \frac{\int_0^{\pi/2} \frac{R^2 d\theta}{2}.\frac{2}{3}R\sin\theta}{\int_0^{\pi/2} \frac{R^2}{2}.d\theta}$$

$$= \frac{\frac{2}{3}R[-\cos\theta]_0^{\pi/2}}{[\theta]_0^{\pi/2}}$$

$$\bar{Y} = \frac{4R}{3\pi}$$

Consider a semicircle lies against X and Y axis as shown in Fig. 7.13. The semicircle is equally divided about OY axis. Thus its centroid will lie on this axis i.e., $\bar{X} = 0$

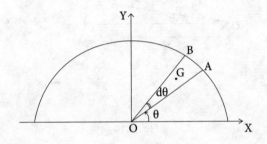

Fig. 7.13

To determine location from OX axis

$$\bar{Y} = \frac{\int dA.y^*}{\int dA}$$

$$dA = \frac{1}{2}.R.R.d\theta$$
$$y^* = OG.\sin\theta$$
$$= \frac{2}{3}R.\sin\theta$$

$$= \frac{\int_0^{\pi} \frac{R^2}{2}.d\theta.\frac{2}{3}R\sin\theta}{\int_0^{\pi} \frac{R^2}{2}.d\theta}$$

$$= \frac{\frac{R^2}{2}.\frac{2}{3}R\int_0^{\pi}\sin\theta.d\theta}{\frac{R^2}{2}\int_0^{\pi}d\theta}$$

$$= \frac{\frac{2}{3}R[-\cos\theta]_0^{\pi}}{[\theta]_0^{\pi}}$$

$$= \frac{\frac{2}{3}R[-\cos\pi + \cos 0]}{\pi}$$

$$= \frac{4R}{3\pi} \quad \text{i.e.} \left(0, \frac{4R}{3\pi}\right)$$

Example: 7.6

Determine centroid of a circular sector of radius R as shown in Fig. 7.14, projecting an angle 2β at its centre.

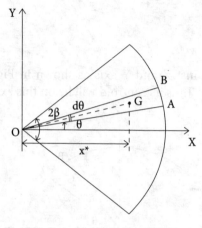

Fig. 7.14

Solution:

Given the circular sector is symmetrical about OX axis, thus the centroid will lie on this axis

$$\bar{Y} = 0$$

$$\bar{X} = \frac{\int dA \cdot x^*}{\int dA}$$

$$\bar{X} = \frac{\int_{-\beta}^{+\beta} \frac{R^2}{2} d\theta \cdot \frac{2}{3} R\cos\theta}{\int_{-\beta}^{+\beta} \frac{R^2}{2} d\theta} \qquad \begin{array}{l} dA = \frac{1}{2} \times R \times R d\theta \\ x^* = OG\cos\theta = \frac{2}{3} R\cos\theta \end{array}$$

$$= \frac{2 \cdot \frac{R^2}{2} \cdot \frac{2}{3} R \int_0^\beta \cos\theta \, d\theta}{2 \cdot \frac{R^2}{2} \cdot \int_0^\beta d\theta} \qquad = \frac{\frac{2}{3} R [\sin\theta]_0^\beta}{[\theta]_0^\beta}$$

$$\bar{X} = \frac{2R\sin\beta}{3\beta}$$

Centroid and Centre of Gravity

Example: 7.7

Determine the centroid of the length of a circular arc of radius R projecting an angle 2β as shown in Fig. 7.15.

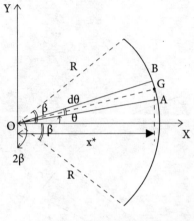

Fig. 7.15

Solution:

As the arc is symmetrical about OX axis, thus $\overline{Y} = 0$

$$\overline{X} = \frac{\int dL \cdot x^*}{\int dL}$$

Note:

The centroid G of elementary arc AB will lie on the middle of arc AB as shown in Fig. 7.15.

$$\overline{X} = \frac{\int_{-\beta}^{+\beta} R \, d\theta \cdot R\cos\theta}{\int_{-\beta}^{+\beta} R \, d\theta} \qquad \begin{array}{l} dl = AB = R \times d\theta \\ x^* = OA \cdot \cos\theta \\ \quad = R \cdot \cos\theta \end{array}$$

$$= \frac{2\int_{0}^{+\beta} R^2 \cos\theta \, d\theta}{2\int_{0}^{\beta} R \, d\theta}$$

$$= \frac{R^2 \cdot [\sin\theta]_0^\beta}{R \cdot [\theta]_0^\beta}$$

$$\overline{X} = \frac{R\sin\beta}{\beta}$$

Example: 7.8

Determine the centroid of a uniform thin wire bent in the form of quarter circle of radius R as shown in Fig. 7.16.

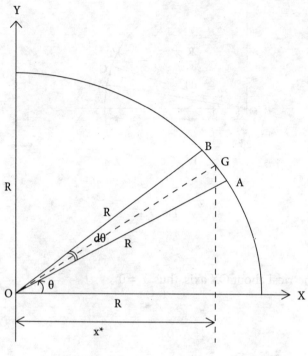

Fig. 7.16

Solution:

$$\overline{X} = \frac{\int dL \cdot x^*}{\int dL}$$

$dL = AB = R \cdot d\theta$

$x^* = OG \cdot \cos\theta = R\cos\theta$

$$\overline{X} = \frac{\int_0^{\pi/2} R \cdot d\theta \cdot R\cos\theta}{\int_0^{\pi/2} R \, d\theta}$$

$$= \frac{R^2 \cdot [\sin\theta]_0^{\pi/2}}{R[\theta]_0^{\pi/2}}$$

$$\overline{X} = \frac{2R}{\pi}$$

Centroid and Centre of Gravity

$$\bar{Y} = \frac{\int dx \cdot y^*}{\int dx}$$

$$y^* = OG \cdot \sin\theta$$
$$= R \cdot \sin\theta$$

$$\bar{Y} = \frac{\int_0^{\pi/2} R \cdot d\theta \cdot R \sin\theta}{\int_0^{\pi/2} R \, d\theta}$$

$$\bar{Y} = \frac{R^2 \left[-\cos\theta\right]_0^{\pi/2}}{R \left[\theta\right]_0^{\pi/2}}$$

$$\bar{Y} = \frac{2R}{\pi}$$

Example: 7.9

Determine the centroid of the area under parabolic equation given by $y = k \cdot x^2$ as shown in Fig. 7.17, where a = 8 m and b = 6 m.

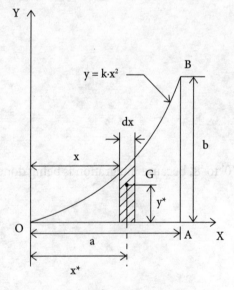

Fig. 7.17

Solution:

Consider an elemental vertical strip of area dA at distance x from OY axis as shown in Fig. 7.17.

Let the strip is of thickness dx

$$\overline{X} = \frac{\int dA \cdot x^*}{\int dA}$$

$dA = dx \times y = k \cdot x^2 \cdot dx$

$x^* = x + \dfrac{dx}{2} \simeq x$

$$\overline{X} = \frac{\int_0^8 k \cdot x^2 dx \cdot x}{\int_0^8 kx^2 dx} = \frac{k \int_0^8 x^3 \cdot dx}{k \int_0^8 x^2 \cdot dx}$$

$$\overline{X} = \frac{\left[\dfrac{x^4}{4}\right]_0^8}{\left[\dfrac{x^3}{3}\right]_0^8}$$

$\overline{X} = 6m$

$$\overline{Y} = \frac{\int dA \cdot y^*}{\int dA}$$

$y^* = \dfrac{y}{2}$

$y^* = \dfrac{1}{2} \cdot kx^2$

$$= \frac{\int_0^8 k \cdot x^2 \cdot dx \cdot \left(\dfrac{1}{2} kx^2\right)}{\int_0^8 kx^2 \cdot dx}$$

Note:
Limits will remain same '0' to '8', because integration is being done with respect to X again.

$$\overline{Y} = \frac{1}{2} \cdot \frac{k^2 \int_0^8 x^4 \cdot dx}{k \int_0^8 x^2 \cdot dx}$$

$$= \frac{1}{2} \cdot \frac{k \cdot \left[\dfrac{x^5}{5}\right]_0^8}{\left[\dfrac{x^3}{3}\right]_0^8}$$

$$= \frac{1}{2} \cdot \frac{k \cdot \dfrac{8^5}{5}}{\dfrac{8^3}{3}}$$

$$= \frac{k}{2} \cdot \frac{3}{5} \cdot 8^2 \quad \ldots\ldots (1)$$

Centroid and Centre of Gravity

At point B, $x = 8$ and $y = 6\ m$,
Substituting these values in parabolic equation,
$y = k.x^2$
$6 = k.8^2$ i.e. $k = 3/32$

Substituting value of 'k' in equation (1),

$$\overline{Y} = \frac{k}{2}\cdot\frac{3}{5}\cdot 8^2 = \frac{1}{2}\cdot\frac{3}{32}\cdot\frac{3}{5}\cdot 8^2$$
$$= 1.8\ m$$

Example: 7.10

Determine the centroid of the area formed by quarter of an ellipse as shown in Fig. 7.18.

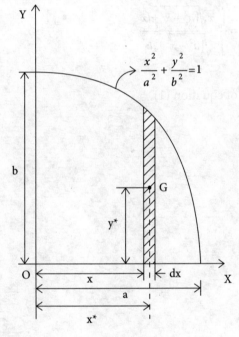

Fig. 7.18

Solution:

Consider an elemental strip of length dx at distance x from axis OY as shown in Fig. 7.18.

$$\overline{X} = \frac{\int dA.x^*}{\int dA}$$

dA = height of strip $\times dx$
   $= y.dx$

Consider equation of ellipse,

$$\frac{x^2}{a^2}+\frac{y^2}{b^2}=1$$

$$y=\frac{b}{a}\sqrt{(a^2-x^2)}$$

thus $dA = \frac{b}{a}\sqrt{(a^2-x^2)}.dx$

and $x^* = x + \frac{dx}{2} \simeq x$

$$\overline{X} = \frac{\int_0^a \frac{b}{a}\sqrt{(a^2-x^2)}dx.x}{\int_0^a \frac{b}{a}\sqrt{(a^2-x^2)}dx}$$

$$= \frac{\frac{b}{a}\int_0^a x\sqrt{(a^2-x^2)}dx}{\frac{b}{a}\int_0^a \sqrt{(a^2-x^2)}.dx} = \frac{\int_0^a x\sqrt{(a^2-x^2)}dx}{\int_0^a \sqrt{(a^2-x^2)}.dx} \qquad \ldots\ldots (1)$$

Consider the numerator of equation (1),

$$\int_0^a x\sqrt{(a^2-x^2)}dx$$

$$= \int_0^a x(a^2-x^2)^{\frac{1}{2}}.dx$$

$$= \left[\frac{x(a^2-x^2)^{3/2}}{\frac{3}{2}(-2x)}\right]_0^a$$

$$= \left[\frac{(a^2-x^2)^{3/2}}{-3}\right]_0^a$$

$$= \frac{a}{3}$$

Consider the denominator of equation (1),

Using integration by parts

$$\int_0^a \sqrt{(a^2-x^2)}.dx = \left[\frac{x\sqrt{(a^2-x^2)}}{2}+\frac{a^2}{2}\sin^{-1}\left(\frac{x}{a}\right)\right]_0^a$$

$$= \frac{a^2}{2}\sin^{-1}(1) = \frac{\pi a^2}{4}$$

Centroid and Centre of Gravity

Substituting values of numerator and denominator in equation (1),

$$\overline{X} = \frac{a^3/3}{\pi a^2/4} = \frac{4a}{3\pi}$$

$$\overline{X} = \frac{4a}{3\pi}$$

$$\overline{Y} = \frac{\int dA \cdot y^*}{\int dA}$$

$$y^* = \frac{y}{2}$$

$$= \frac{b}{2a}\sqrt{(a^2 - x^2)}$$

$$= \frac{\int_0^a \frac{b}{a}\sqrt{(a^2 - x^2)} \cdot dx \cdot \frac{1}{2} \cdot \frac{b}{a}\sqrt{(a^2 - x^2)}}{\int_0^a \frac{b}{a}\sqrt{(a^2 - x^2)} \cdot dx}$$

$$\overline{Y} = \frac{\frac{b^2}{2a^2}\int_0^a (a^2 - x^2) \cdot dx}{\frac{b}{a}\int_0^a \sqrt{(a^2 - x^2)} \cdot dx} = \frac{\frac{b}{2a}\int_0^a (a^2 - x^2) \cdot dx}{\int_0^a \sqrt{(a^2 - x^2)} \cdot dx} \qquad \ldots (2)$$

Considering the numerator of equation (2),

$$= \frac{b}{2a}\left[\int_0^a a^2 \cdot dx - \int_0^a x^2 \cdot dx\right]$$

$$= \frac{b}{2a}\left[a^2 \cdot a - \frac{a^3}{3}\right]$$

$$= \frac{b}{2a}\left[\frac{2a^3}{3}\right]$$

$$= \frac{a^2 b}{3}$$

Substituting the values of numerator and denominator in equation (2),

$$\text{thus } \overline{Y} = \frac{\frac{a^2 b}{3}}{\frac{\pi a^2}{4}}$$

$$\overline{Y} = \frac{4b}{3\pi}$$

Table 7.1: Centroid of some common shapes

Serial No.	Shape	Figure	Length/ Area/ Volume	$\bar{X}$	$\bar{Y}$
1.	Rectangle		$A = b \cdot h$	$b/2$	$h/2$
2.	Un-symmetrical triangle		$A = \dfrac{1}{2}(b \cdot h)$	$(a+b)/3$	$h/3$
3.	Isosceles triangle		$A = \dfrac{1}{2}(b \cdot h)$	$b/2$	$h/3$
4.	Right angled triangle		$A = \dfrac{1}{2}(b \cdot h)$	$b/3$	$h/3$
5.	Circle		$A = \pi R^2$	R	R

Centroid and Centre of Gravity

	Shape	Figure	Area	$\bar{x}$	$\bar{y}$
6.	Semicircle		$A = \dfrac{1}{2}(\pi R^2)$	R	$\left(\dfrac{4R}{3\pi}\right)$
7.	Quarter circle		$A = \dfrac{1}{4}(\pi R^2)$	$\left(\dfrac{4R}{3\pi}\right)$	$\left(\dfrac{4R}{3\pi}\right)$
8.	Circular Sector		$A = \alpha \cdot R^2$	$\left(\dfrac{2R\sin\alpha}{3\alpha}\right)$	0
9.	Semi-Parabola		$A = \left(\dfrac{2}{3}ah\right)$	$\dfrac{3}{8}a$	$\left(\dfrac{3}{5}h\right)$
10.	Parabola		$A = \left(\dfrac{4}{3}ah\right)$	0	$\left(\dfrac{3h}{5}\right)$

#	Shape	Figure	Area/Length	$\bar{x}$	$\bar{y}$
11.	Trapezium		$A = \dfrac{1}{2}(a+b)h$	$\dfrac{a}{2}$	$\left(\dfrac{2b+a}{a+b}\right)\dfrac{h}{3}$
12.	Straight Line		l	$l/2$	0
13.	Inclined line		l	$\dfrac{l}{2}\cos\alpha$	$\dfrac{l}{2}\sin\alpha$
14.	Circular arc		$C = 2\pi R$	R	R
15.	Semi-circular arc		$C = \pi R$	R	$\left(\dfrac{2R}{\pi}\right)$
16.	Quarter circular arc		$C = \left(\dfrac{\pi R}{2}\right)$	$\left(\dfrac{2R}{\pi}\right)$	$\left(\dfrac{2R}{\pi}\right)$

Centroid and Centre of Gravity

17.	Circular Arc		$PQ = 2R\alpha$	$\left(\dfrac{R\sin\alpha}{\alpha}\right)$	0
18.	Right Circular arc		$V = \left(\dfrac{1}{3}\pi R^2 h\right)$	0	$h/4$
19.	Hemisphere		$V = \left(\dfrac{2}{3}\pi R^3\right)$	0	$\dfrac{3R}{8}$

7.7 Centroid of Composite Sections and Bodies

Generally in engineering work, bodies are composed of different cross-section areas. The cross-section areas may be of rectangular, triangular, quarter circular, semi-circular and circular in shape. In composite sections, the given area is divided into suitable cross-section areas whose individual centroid is known. The coordinates $(\overline{X}, \overline{Y})$ of composite section is determined by using

$$\overline{X} = \dfrac{(a_1 \times x_1) + (a_2 \times x_2) + (a_3 \times x_3) + \ldots (a_n \times x_n)}{(a_1 + a_2 + a_3 + \ldots a_3)} = \sum_{i=1}^{n} \dfrac{a_i x_i}{a_i}$$

$$\overline{Y} = \dfrac{(a_1 \times y_1) + (a_2 \times y_2) + (a_3 \times y_3) + \ldots (a_n \times y_n)}{(a_1 + a_2 + a_3 + \ldots a_3)} = \sum_{i=1}^{n} \dfrac{a_i y_i}{a_i}$$

This method can be used to determine centre of gravity and centroid where bodies comprise weights, areas, lines or volumes.

Note:

i. If some part of the section is hollow then whole section is taken first as a solid section and divided into suitable sections. Then hollow section whose centroid is available, is removed from solid section. Due to this, the hollow section is considered with negative values in both numerator and denominator of centroid formula.

ii. If some portion of the section lies opposite to the reference axis then its centroidal distance from reference axis is taken with negative value in numerator of centroid formula. However denominator does not contain negative value for such portion. Denominator contains negative value only for hollow sections.

Example: 7.11

From a circular plate of diameter 100 *mm*, a circular part is cut out whole diameter is 50 *mm*. Find the centroid of the remainder shown in Fig. 7.19.

(UPTU, II Sem, 2002–03)

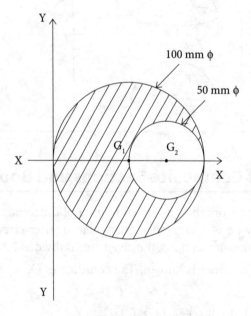

Fig. 7.19

Solution:

The shaded area is symmetrical about X–X axis; hence the centroid lies on this axis which means that
$\overline{Y} = 0$
Now, mark centroid G_1 for full circular plate and G_2 for cut-out circular plate, $R_1 = 50$ *mm*, $R_2 = 25$ *mm*

Centroid and Centre of Gravity

$a_1 = \pi(50)^2$	$x_1 = 50\ mm$
$a_2 = \pi(25)^2$	$x_2 = 50 + 25$ $= 75\ mm$

$$\overline{X} = \left(\frac{a_1 x_1 - a_2 x_2}{a_1 - a_2}\right)$$

$$= \left[\frac{\pi(50)^2 \times 50 - \pi(25)^2 \times 75}{\pi(50)^2 - \pi(25)^2}\right]$$

$$= 41.67\ mm$$

Example: 7.12

Find the centroid of the Fig. 7.20.

UPTU, IInd Sem (2008–2009)

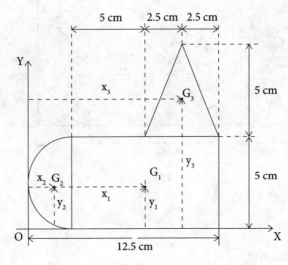

Fig. 7.20

$a_1 = 10 \times 5 = 50\ cm^2$	$x_1 = 7.5\ cm$	$y_1 = 2.5\ cm$
$a_2 = \dfrac{\pi(2.5)^2}{2}$ $= 9.82\ cm^2$	$x_2 = \left(\dfrac{4 \times 2.5}{3\pi}\right)$ $= 1.44\ cm$	$y_2 = 2.5\ cm$
$a_3 = \dfrac{1}{2} \times 5 \times 5$ $= 12.5\ cm^2$	$x_3 = 10\ cm$	$y_3 = 5 + \dfrac{5}{3}$ $= 6.67\ cm$

$$\overline{X} = \left(\frac{a_1 x_1 + a_2 x_2 + a_3 x_3}{a_1 + a_2 + a_3}\right), \quad \overline{Y} = \left(\frac{a_1 y_1 + a_2 y_2 + a_3 y_3}{a_1 + a_2 + a_3}\right)$$

$$\overline{X} = \left(\frac{50 \times 7.5 + 9.82 \times 1.44 + 12.5 \times 10}{50 + 9.82 + 12.5}\right), \quad \overline{Y} = \left(\frac{50 \times 2.5 + 9.82 \times 2.5 + 12.5 \times 6.67}{50 + 9.82 + 12.5}\right)$$

$$\overline{X} = 7.11\ cm, \quad \overline{Y} = 3.22\ cm$$

Example 7.13

Determine the centroid of the given Fig. 7.21.

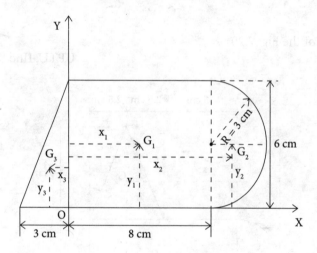

Fig. 7.21

Solution:

The table will be as follows:

$a_1 = 8 \times 6 = 48\ cm^2$	$x_1 = 4\ cm$	$y_1 = 3\ cm$
$a_2 = \dfrac{\pi(6)^2}{8}$ or $\dfrac{\pi(3)^2}{2}$ $= 14.14\ cm^2$	$x_2 = \left(8 + \dfrac{4 \times 3}{3\pi}\right)$ $= 9.27\ cm$	$y_2 = 3\ cm$
$a_3 = \dfrac{1}{2} \times 3 \times 6$ $= 9\ cm^2$	$x_3 = \left(\dfrac{-3}{3}\right)$ $= (-1\ cm)$	$y_3 = \dfrac{1}{3} \times 6$ $= 2\ cm$

Centroid and Centre of Gravity

Note:

(i) x_3 will be negative as it is measured in opposite direction of X-axis.
(ii) y_2 is equal to the radius of semicircle.

$$\overline{X} = \left(\frac{a_1 x_1 + a_2 x_2 + a_3 x_3}{a_1 + a_2 + a_3}\right), \quad \overline{Y} = \left(\frac{a_1 y_1 + a_2 y_2 + a_3 y_3}{a_1 + a_2 + a_3}\right)$$

$$\overline{X} = \left(\frac{48 \times 4 + 14.14 \times 9.27 + 9(-1)}{48 + 14.14 + 9}\right), \quad \overline{Y} = \left(\frac{48 \times 3 + 14.14 \times 3 + 9 \times 2}{48 + 14.14 + 9}\right)$$

$$\overline{X} = 4.41 \, cm, \quad \overline{Y} = 2.87 \, cm$$

Example: 7.14

A thin wire is bent into a closed loop A-B-C-D-E-A as shown in Fig. 7.22, where the portion AB is a circular arc. Determine the centroid of the wire.

(GBTU - 2011)

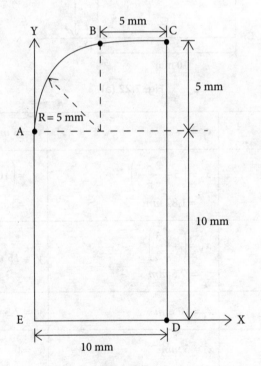

Fig. 7.22

Solution:

Consider reference axis X and Y as the bottom most and the left most as the figure is not symmetrical about any axis. Mark individual centroid in different five lengths as shown in Fig. 7.22(a).

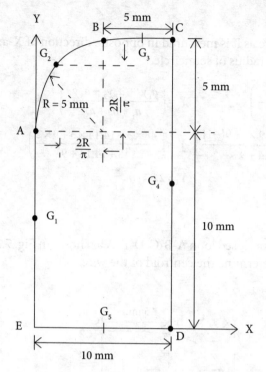

Fig. 7.22 (a)

$l_1 = 10$ mm (AE)	$x_1 = 0$	$y_1 = 5$ mm
$l_2 = \dfrac{2\pi R}{4} = 2.5\pi$ mm (AB) $= 7.85$ mm	$x_2 = \left(5 - \dfrac{2\times 5}{\pi}\right)$ $= 1.82$ mm	$y_2 = \left(10 + \dfrac{2\times 5}{\pi}\right)$ $= 13.18$ mm
$l_3 = 5$ mm (BC)	$x_3 = \left(5 + \dfrac{5}{2}\right)$ $= 7.5$ mm	$y_3 = 15$ mm
$l_4 = 15$ mm (CD)	$x_4 = 10$ mm	$y_4 = 7.5$ mm
$l_5 = 10$ mm (DE)	$x_5 = 5$ mm	$y_5 = 0$

$$\overline{X} = \left(\frac{l_1 x_1 + l_2 x_2 + l_3 x_3 + l_4 x_4 + l_5 x_5}{l_1 + l_2 + l_3 + l_4 + l_5}\right)$$

$$= \left(\frac{(10\times 0) + (7.85\times 1.82) + (5\times 7.5) + (15\times 10) + (10\times 5)}{10 + 7.85 + 5 + 15 + 10}\right)$$

$$= 5.26 \text{ mm}$$

Centroid and Centre of Gravity

$$\bar{Y} = \left(\frac{l_1 y_1 + l_2 y_2 + l_3 y_3 + l_4 y_4 + l_5 y_5}{l_1 + l_2 + l_3 + l_4 + l_5} \right)$$

$$= \left(\frac{(10 \times 5) + (7.85 \times 13.18) + (5 \times 15) + (15 \times 7.5) + (10 \times 0)}{10 + 7.85 + 5 + 15 + 10} \right)$$

$$= 7.13 \ mm$$

The centroid of wire is (5.26 *mm*, 7.13 *mm*).

Example: 7.15

A thin wire is bent into a closed loop A-B-C-D-E-F-A as shown in Fig. 7.23. Determine the centroid of the wire.

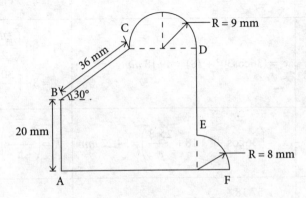

Fig. 7.23

Solution:

Consider reference axis and show centroid of different lengths as shown in Fig. 7.23(a).

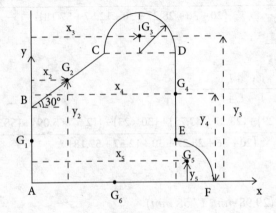

Fig. 7.23 (a)

Here, length of DE segment = (AB + BC sin30° − Radius of quarter circle)
= (20 + 36 sin30° − 8)
DE = 30 mm
Length of AF segment = (BC cos30° + 18 + 8)
= (36cos30° + 26) = 57.18 mm

$l_1(AB) = 20$ mm	$x_1 = 0$	$y_1 = 10$ mm
$l_2(BC) = 36$ mm	$x_2 = 18\cos 30° = 15.59$ mm	$y_2 = (20 + 18\sin 30°) = 29$ mm
$l_3(CD) = \pi(9$ mm$)$ $= 28.27$ mm	$x_3 = (36\cos 30° + 9) = 40.18$ mm	$y_3 = \left(20 + 36\sin 30° + \dfrac{2 \times 9}{\pi}\right)$ $= 43.73$ mm
$l_4(DE) = 30$ mm	$x_4 = (36\cos 30° + 18) = 49.18$ mm	$y_4 = 8 + \left(\dfrac{DE}{2}\right)$ $= 8 + \left(\dfrac{30}{2}\right) = 23$ mm
$l_5(EF) = \dfrac{2\pi(8)}{4}$ $= 12.57$ mm	$x_5 = \left(36\cos 30° + 18 + \dfrac{2 \times 8}{\pi}\right) = 54.27$ mm	$y_5 = \left(\dfrac{2 \times 8}{\pi}\right) = 5.09$ mm
$l_6(AF) = 57.18$ mm	$x_6 = \dfrac{57.18}{2} = 28.59$ mm	$y_6 = 0$

$$\overline{X} = \left(\frac{l_1 x_1 + l_2 x_2 + l_3 x_3 + l_4 x_4 + l_5 x_5 + l_6 x_6}{l_1 + l_2 + l_3 + l_4 + l_5 + l_6}\right)$$

$$= \frac{[(20 \times 0) + (36 \times 15.59) + (28.27 \times 40.18) + (30 \times 49.18) + (12.57 \times 54.27) + (58.18 \times 28.59)]}{(20 + 36 + 28.27 + 30 + 12.57 + 57.18)}$$

$\overline{X} = 29.98$ mm

$$\overline{Y} = \left(\frac{l_1 y_1 + l_2 y_2 + l_3 y_3 + l_4 y_4 + l_5 y_5 + l_6 y_6}{l_1 + l_2 + l_3 + l_4 + l_5 + l_6}\right)$$

$$= \frac{[(20 \times 10) + (36 \times 29) + (28.27 \times 43.73) + (30 \times 23) + (12.57 \times 5.09) + (58.18 \times 0)]}{(20 + 36 + 28.27 + 30 + 12.57 + 57.18)}$$

$= 17.48$ mm

The centroid of wire is (29.98 mm, 17.58 mm).

Centroid and Centre of Gravity

Example: 7.16

A semi-circular area is removed from the trapezoid as shown in Fig. 7.24. Determine the centroid of the remaining area.

(UPTU, Ist Sem, 2009–10)

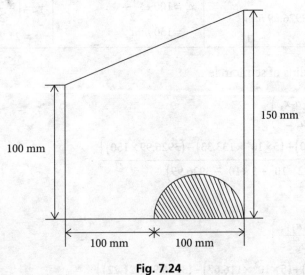

Fig. 7.24

Solution:

Consider Fig. 7.24(a) with reference axis and individual centroid

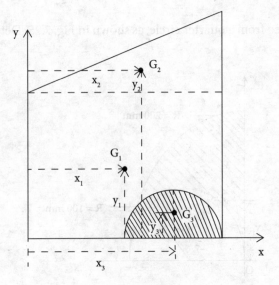

Fig. 7.24 (a)

$a_1 = (200 \times 100)\ mm^2 = 2 \times 10^4\ mm^2$	$x_1 = 100\ mm$	$y_1 = 50\ mm$
$a_2 = \dfrac{1}{2} \times 200 \times 50\ mm^2 = 5 \times 10^3\ mm^2$	$x_2 = \dfrac{2}{3} \times 200$ $= 133.33\ mm$	$y_2 = 100 + \dfrac{50}{3}$ $= 116.67\ mm$
$a_3 = \dfrac{\pi (50)^2}{2}\ mm^2 = 3926.99\ mm^2$	$x_3 = 100 + \dfrac{100}{2}$ $= 150\ mm$	$y_3 = \left(\dfrac{4 \times 50}{3\pi}\right)$ $= 21.22\ mm$

Note: $x_3 = 100 +$ radius of semicircle

$$\overline{X} = \left(\frac{a_1 x_1 + a_2 x_2 - a_3 x_3}{a_1 + a_2 - a_3}\right)$$

$$= \frac{\left[(2 \times 10^4 \times 100) + (5 \times 10^3 \times 133.33) - (3926.99 \times 150)\right]}{(2 \times 10^4 + 5 \times 10^3 - 3926.99)}$$

$= 98.59\ mm$

$$\overline{Y} = \left(\frac{a_1 y_1 + a_2 y_2 - a_3 y_3}{a_1 + a_2 - a_3}\right)$$

$$= \frac{\left[(2 \times 10^4 \times 50) + (5 \times 10^3 \times 116.67) - (3926.99 \times 21.22)\right]}{(2 \times 10^4 + 5 \times 10^3 - 3926.99)}$$

$= 71.18\ mm$

Example: 7.17

A semicircle is removed from a quarter circle, as shown in Fig. 7.25. Determine the centroid of the shaded area.

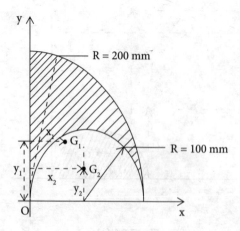

Fig. 7.25

Centroid and Centre of Gravity

Solution:

Mark individual centroid G_1, G_2 as shown in Fig. 7.25.

$a_1 = \dfrac{\pi(200)^2}{4}$ $= 31415.93\ mm^2$	$x_1 = \dfrac{4 \times 200}{3\pi}$ $= 84.88\ mm$	$y_1 = 84.88\ mm$
$a_2 = \dfrac{\pi(100)^2}{2}$ $= 15707.96\ mm^2$	$x_2 = 100\ mm$	$y_2 = \dfrac{4 \times 100}{3\pi}$ $= 42.44\ mm$

$$\overline{X} = \left(\dfrac{a_1 x_1 - a_2 x_2}{a_1 - a_2}\right), \quad \overline{Y} = \left(\dfrac{a_1 y_1 - a_2 y_2}{a_1 - a_2}\right)$$

$$\overline{X} = \left[\dfrac{(31415.93 \times 84.88) - (15707.96 \times 100)}{(31415.93 - 15707.96)}\right] = 69.76\ mm$$

$$\overline{Y} = \left[\dfrac{(31415.93 \times 84.88) - (15707.96 \times 42.44)}{(31415.93 - 15707.96)}\right] = 127.36\ mm$$

Example: 7.18

A triangle and a semicircle are removed from a rectangle as shown in Fig. 7.26. Locate the centroid of the remaining object.

(UPTU, IInd Sem, Special C.O., 2008–09)

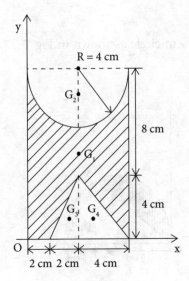

Fig. 7.26

Solution:

$a_1 = 8 \times 12\ cm^2 = 96\ cm^2$	$x_1 = 4\ cm$	$y_1 = 6\ cm$
$a_2 = \dfrac{\pi(4)^2}{2} = 25.13\ cm^2$	$x_2 = 4\ cm$	$y_2 = \left(12 - \dfrac{4 \times 4}{3\pi}\right) = 10.30\ cm$
$a_3 = \dfrac{1}{2} \times 2 \times 4 = 4\ cm^2$	$x_3 = 2 + \left(2 - 2 \times \dfrac{1}{3}\right)$ $= 3.33\ cm$	$y_3 = 1.33\ cm$
$a_4 = \dfrac{1}{2} \times 4 \times 4 = 8\ cm^2$	$x_4 = 4 + 4 \times \dfrac{1}{3}$ $= 5.33\ cm$	$y_4 = 1.33\ cm$

$$\overline{X} = \left(\dfrac{a_1 x_1 - a_2 x_2 - a_3 x_3 - a_4 x_4}{a_1 - a_2 - a_3 - a_4}\right)$$

$$\overline{X} = \left[\dfrac{(96 \times 4) - (25.13 \times 4) - (4 \times 3.33) - (8 \times 5.33)}{(96 - 25.13 - 4 - 8)}\right]$$

$= 3.86\ cm$

$$\overline{Y} = \left(\dfrac{a_1 y_1 - a_2 y_2 - a_3 y_3 - a_4 y_4}{a_1 - a_2 - a_3 - a_4}\right)$$

$$= \left[\dfrac{(96 \times 6) - (25.13 \times 10.3) - (4 \times 1.33) - (8 \times 1.33)}{(96 - 25.13 - 4 - 8)}\right]$$

$= 5.12\ cm$

Example: 7.19

A triangle is removed from a semicircle as shown in Fig. 7.27. Locate the centroid of the remaining object.

(UPTU, Sem1, 2008–09)

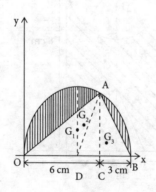

Fig. 7.27

Centroid and Centre of Gravity

Solution:

Radius, $OD = AD = \dfrac{OB}{2}$

$= \dfrac{9}{2} = 4.5\ cm$

$CD = OC - DO$

$= 6 - 4.5\ cm$

$= 1.5\ cm$

$AC = \sqrt{AD^2 - CD^2}$

$= \sqrt{(4.5)^2 - (1.5)^2}$

$AC = 4.24\ cm$

Mark individual centroids G_1, G_2 and G_3 as shown in Fig. 7.27.

$a_1 = \dfrac{\pi(4.5)^2}{2} = 31.81\ cm^2$	$x_1 = 4.5\ cm$	$y_1 = \left(\dfrac{4 \times 4.5}{3\pi}\right)$ $= 1.91\ cm$
$a_2 = \dfrac{1}{2} \times 6 \times 4.24 = 12.72\ cm^2$	$x_2 = \left(6 - \dfrac{1}{3} \times 6\right)$ $= 4\ cm$	$y_2 = \dfrac{1}{3} \times 4.24$ $= 1.41\ cm$
$a_3 = \dfrac{1}{2} \times 3 \times 4.24 = 6.36\ cm^2$	$x_3 = \left(6 + \dfrac{1}{3} \times 3\right)$ $= 7\ cm$	$y_3 = \dfrac{1}{3} \times 4.24$ $= 1.41\ cm$

$\overline{X} = \left(\dfrac{a_1 x_1 - a_2 x_2 - a_3 x_3}{a_1 - a_2 - a_3}\right)$

$= \left[\dfrac{(31.81 \times 4.5) - (12.72 \times 4) - (6.36 \times 7)}{(31.81 - 12.72 - 6.36)}\right]$

$= 3.75\ cm$

$\overline{Y} = \left(\dfrac{a_1 y_1 - a_2 y_2 - a_3 y_3}{a_1 - a_2 - a_3}\right)$

$= \left[\dfrac{(31.81 \times 1.91) - (12.72 \times 1.41) - (6.36 \times 1.41)}{(31.81 - 12.72 - 6.36)}\right]$

$= 2.66\ cm$

Example: 7.20

Three identical boxes, each having length l and weight W, are placed as shown in Fig. 7.28. Find out the maximum possible distance m through which the top box can extend out from the bottom so that there is no possibility of the toppling of the stack.

(UPTU, 2002)

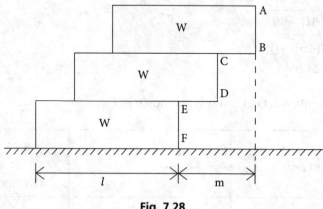

Fig. 7.28

Solution:

For maximum possible distance 'm', the top box can be placed on intermediate box such that its weight W passes through CD and the entire centre of gravity of the top box and intermediate box should pass through the EF as shown in Fig. 7.28(a)

Let the intermediate box is placed at distance 'a' from bottom box, Fig. 7.28(a)

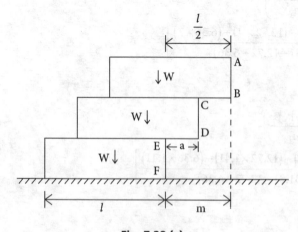

Fig. 7.28 (a)

Centroid and Centre of Gravity

$$\overline{X} = \frac{W_1 \cdot x_1 + W_2 \cdot x_2}{W_1 + W_2} \qquad W_1 = W, \; x_1 = a$$

where $\overline{X} = 0 \qquad W_2 = W, \; x_2 = -\left(\dfrac{l}{2} - a\right)$

$$0 = \frac{W \cdot (a) + W\left[-\left(\dfrac{l}{2} - a\right)\right]}{W_1 + W_2}$$

$$0 = W \cdot a - W\left(\dfrac{l}{2} - a\right)$$

$$0 = a - \dfrac{l}{2} + a$$

$$a = \dfrac{l}{4} = 0.25l$$

thus, $m = a + \dfrac{l}{2}$

$= 0.25l + 0.5l$

$m = 0.75l$

Example: 7.21

Determine the co-ordinate x_c and y_c of the centre of a 100 mm diameter circular hole cut in a thin plate so that this point will be the centroid of the remaining shaded area as shown in Fig. 7.29.

(UPTU, Ist Sem, 2001–02)

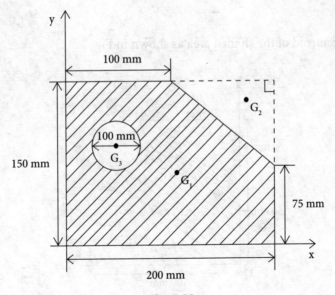

Fig. 7.29

Solution:

Here $\overline{X} = x_c$ and $\overline{Y} = y_c$, i.e., the centre of circle as given in the question. The shaded area can be attained by subtracting a circle and triangle from a rectangle.

$a_1 = 200 \times 150 \ mm^2$	$x_1 = 100 \ mm$	$y_1 = 75 \ mm$
$a_2 = \dfrac{1}{2} \times 100 \times 75 \ mm^2$ $= 50 \times 75 \ mm^2$	$x_2 = \left(100 + \dfrac{2}{3} \times 100\right)$ $= 166.67 \ mm$	$y_2 = \left(75 + 75 \times \dfrac{2}{3}\right)$ $= 125 \ mm$
$a_3 = \pi(50)^2 \ mm^2$	$x_3 = x_c$	$y_3 = y_c$

$$\overline{X} = \left(\frac{a_1 x_1 - a_2 x_2 - a_3 x_3}{a_1 - a_2 - a_3}\right) \quad \text{and} \quad \overline{Y} = \left(\frac{a_1 y_1 - a_2 y_2 - a_3 y_3}{a_1 - a_2 - a_3}\right)$$

$$x_c = \frac{\left[(200 \times 150 \times 100) - (50 \times 75 \times 166.67) - \left(\pi(50)^2 \times x_c\right)\right]}{\left[(200 \times 150) - (50 \times 75) - \pi(50)^2\right]}$$

$x_1 = 90.5 \ mm$

$$\overline{Y} = y_c = \frac{\left[(200 \times 150) \times 75 - (75 \times 50) \times 125 - \left(\pi(50)^2 \times y_c\right)\right]}{\left[(200 \times 150) - (75 \times 50) - \pi(50)^2\right]}$$

$y_c = 67.9 \ mm$

Example: 7.22

Determine the centroid of the shaded area as shown in Fig. 7.30.

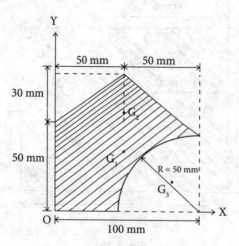

Fig. 7.30

Centroid and Centre of Gravity

Solution:

Here a quarter circle of radius 50 mm is cut from a rectangle. The shaded area can be obtained by subtracting area of quarter circle from area of rectangle and triangle.

$a_1 = 100 \times 50 \ mm^2$ $= 5000 \ mm^2$	$x_1 = 50 \ mm$	$y_1 = 25 \ mm$
$a_2 = \dfrac{1}{2} \times 100 \times 30$ $= 1500 \ mm^2$	$x_2 = 50 \ mm$	$y_2 = 50 + \dfrac{1}{3} \times 30$ $= 60 \ mm$
$a_3 = \dfrac{\pi(50)^2}{4}$ $= 1963.50 \ mm^2$	$x_3 = \left(100 - \dfrac{4 \times 50}{3\pi}\right)$ $= 78.78 \ mm$	$y_3 = \dfrac{4 \times 50}{3\pi}$ $= 21.22 \ mm$

$$\overline{X} = \left(\dfrac{a_1 x_1 + a_2 x_2 - a_3 x_3}{a_1 + a_2 - a_3}\right)$$

$$\overline{X} = \dfrac{[(5000 \times 50) + (1500 \times 50) - (1963.5 \times 78.78)]}{[(5000 + 1500 - 1963.5)]}$$

$$= 37.54 \ mm$$

$$\overline{Y} = \left(\dfrac{a_1 y_1 + a_2 y_2 - a_3 y_3}{a_1 + a_2 - a_3}\right)$$

$$= \dfrac{[(5000 \times 25) + (1500 \times 60) - (1963.5 \times 21.22)]}{[(5000 + 1500 - 1963.5)]}$$

$$= 38.21 \ mm$$

Example: 7.23

A prismatic bar of weight 500 kN and length 20 m is bent to form a regular hexagon with five sides as shown in Fig. 7.31. Determine the distance OG_1.

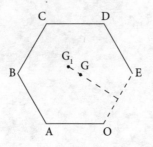

Fig. 7.31

Solution:

The bar is bent into five sides which are equal to each other. Length of each side = 4 m
Thus weight of each side will be 100 kN.

If side OE is assumed to contain equal rod then the centre of gravity of whole hexagon will lie at G but in the given case, the centre of gravity will shift at G_1.

Thus consider Fig. 7.31(a) where the weight of five sides is acting at G_1 and weight of OE side is acting at F.

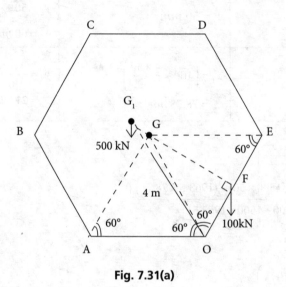

Fig. 7.31(a)

$\Sigma M_G = 0$,
$500 \times GG_1 = 100 \times GF$ ------- (1)
as $\angle GOF = \angle GEF = 60°$
Δ OGE is an equilateral triangle
Thus OG = GE = OE = 4 m, and GF = GE.Sin60° = 4.sin60°

Substituting in equation (1),
$500 \times GG_1 = 100 \times GF$
$500 \times GG_1 = 100 \times 4.\sin60°$
$\quad GG_1 = 0.69$ m
From Fig. 7.31 (a) consider Δ OFG_1,
where OF = 2 m

$OG_1 = \sqrt{OF^2 + FG_1^2}$
$\quad = \sqrt{2^2 + (GF + GG_1)^2}$
$\quad = \sqrt{\left[2^2 + (4\sin60° + 0.69)^2\right]}$
$\quad = 4.61$ m

Centroid and Centre of Gravity

Example: 7.24

A thin wire is bent 45° with horizontal and attached to the hinge 'O' as shown in Fig. 7.32. Determine the suitable length of AB so that the wire AB is remained horizontal.

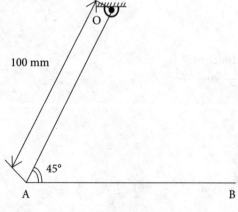

Fig. 7.32

Solution:

The wire AB will remain horizontal when the entire centroid of OA and AB will pass through vertical axis about hinge 'O' i.e. $\overline{X} = 0$ consider Fig. 7.32(a)

Let the length of wire AB is 'a'.

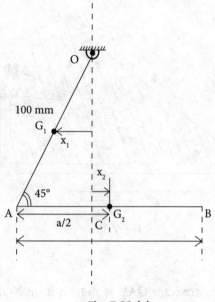

Fig. 7.32 (a)

$l_1 = 100$ mm	$x_1 = -50\cos 45°$
$l_2 = a$	$x_2 = \left(\dfrac{a}{2} - 100\cos 45°\right)$

$$\overline{X} = \left(\frac{l_1 x_1 + l_2 x_2}{l_1 + l_2}\right)$$

$$0 = \frac{100(-50\cos 45°) + a\left(\dfrac{a}{2} - 100\cos 45°\right)}{l_1 + l_2}$$

$$\frac{a^2}{2} - 100 \cdot a \cdot \frac{1}{\sqrt{2}} - \frac{5000}{\sqrt{2}} = 0$$

$$a^2 - 100\sqrt{2} \cdot a - 500\sqrt{2} = 0$$

$$a = 180.58 \text{ mm}$$

Example: 7.25

Determine the centroid of a circular sector of radius R subtending an angle θ as shown in Fig. 7.33.

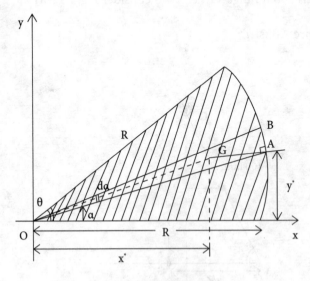

Fig. 7.33

Solution:

Consider an elementary circular sector OAB at angle α from X-axis and making an angle $d\alpha$ as shown in Fig. 7.33.

Centroid and Centre of Gravity

$$\overline{X} = \frac{\int dA \cdot x^*}{\int dA} \text{ and } \overline{Y} = \frac{\int dA \cdot y^*}{\int dA}$$

$$dA = \frac{1}{2} \times OA \times AB = \frac{1}{2} \times R \times R \times d\alpha = \frac{R^2}{2} \cdot d\alpha$$

$$x^* = OG\cos\alpha = \frac{2}{3} R\cos\alpha$$

$$y^* = OG\sin\alpha = \frac{2}{3} R\sin\alpha$$

$$\overline{X} = \frac{\int_0^\theta \frac{R^2}{2} d\alpha \cdot \frac{2}{3} \cdot R\cos\alpha}{\int_0^\theta \frac{R^2}{2} \cdot d\alpha}$$

$$= \frac{\frac{R^2}{2} \times \frac{2}{3} R \int_0^\theta \cos\alpha \cdot d\alpha}{\frac{R^2}{2} \int_0^\theta d\alpha}$$

$$= \frac{\frac{2}{3} R \cdot [\sin\alpha]_0^\theta}{[\alpha]_0^\theta}$$

$$\overline{X} = \frac{2}{3} R \left(\frac{\sin\theta}{\theta} \right)$$

$$\overline{Y} = \frac{\int_0^\theta \frac{R^2}{2} d\alpha \cdot \frac{2}{3} \cdot R\sin\alpha}{\int_0^\theta \frac{R^2}{2} \cdot d\alpha}$$

$$= \frac{\frac{R^2}{2} \times \frac{2}{3} R \int_0^\theta \sin\alpha \cdot d\alpha}{\frac{R^2}{2} \int_0^\theta d\alpha}$$

$$= \frac{\frac{2}{3} R \cdot [-\cos\alpha]_0^\theta}{[\alpha]_0^\theta} = \frac{2R}{3} \frac{[-\cos\theta + \cos 0°]}{(\theta - 0)}$$

$$\overline{Y} = \frac{2R(1-\cos\theta)}{3\theta}$$

Note: if θ is varied from 0° to 90° i.e. $\theta = \pi/2$

then circular sector will become quarter cirlce

$$\overline{X} = \frac{2}{3}R\left[\frac{\sin\theta}{\theta}\right]_0^{\pi/2} = \frac{2}{3}R \cdot \frac{1}{\frac{\pi}{2}}$$

$$= \frac{4R}{3\pi}$$

similarly, $\overline{Y} = \frac{2}{3}R\left[\frac{1-\cos\theta}{\theta}\right]_0^{\pi/2} = \frac{2}{3}R\frac{(1-\cos\pi/2)}{\frac{\pi}{2}}$

$$= \frac{4R}{3\pi}$$

which is the centroid of a quarter circle

Example: 7.26

Determine the centroid of the shaded area as shown in Fig. 7.34.

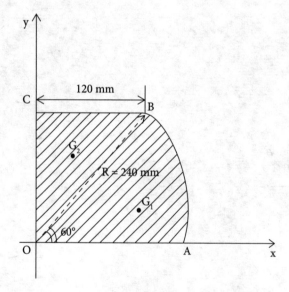

Fig. 7.34

Solution:

∠BOA = 60°
∠COB = 90° − 60°
    = 30°
 OC = OBcos30°
    = 240.cos30°
 OC = 207.85 *mm*

Centroid and Centre of Gravity

The given circular sector is varying from 0° to 60°. Thus the area of circular sector will be

$$= \frac{\pi R^2}{360°} \times 60°$$

$$= \frac{\pi R^2}{6} = \frac{\pi (240)^2}{6}$$

$$= 30159.29 \text{ mm}^2$$

The centroid of sector we have determined in example 7.25 and given by

$$x_1 = \frac{2}{3} R \left(\frac{\sin \theta}{\theta} \right)$$

for $\theta = 60°$, $\quad x_1 = \frac{2}{3} \times 240 \left(\frac{\sin 60°}{\frac{\pi}{3}} \right)$

$$= 132.32 \text{ mm}$$

Similarly $\quad y_1 = \frac{2}{3} R \left[\frac{(1 - \cos \theta)}{\theta} \right]$

$$y_1 = \frac{2}{3} \times 240 \frac{(1 - \cos 60°)}{\frac{\pi}{3}}$$

$$= 76.39 \text{ mm}$$

$a_1 = 30159.29 \text{ mm}^2$	$x_1 = 132.32 \text{ mm}$	$y_1 = 76.39 \text{ mm}$
$a_2 = \frac{1}{2} \times 120 \times 207.85$ $= 12471 \text{ mm}^2$	$x_2 = \frac{1}{3} \times 120$ $= 40 \text{ mm}$	$y_2 = \frac{2}{3} \times 207.85$ $= 138.57 \text{ mm}$

$$\bar{X} = \left(\frac{a_1 x_1 + a_2 x_2}{a_1 + a_2} \right) \quad \text{and} \quad \bar{Y} = \left(\frac{a_1 y_1 + a_2 y_2}{a_1 + a_2} \right)$$

$$\bar{X} = \left[\frac{(30159.29 \times 132.32) + (12471 \times 40)}{(30159.29 + 12471)} \right]$$

$$\bar{X} = 105.31 \text{ mm}$$

$$\bar{Y} = \left[\frac{(30159.29 \times 76.39) + (12471 \times 138.57)}{(30159.29 + 12471)} \right]$$

$$= 94.58 \text{ mm}$$

Example: 7.27

Determine the centroid of the shaded area as shown in Fig. 7.35.

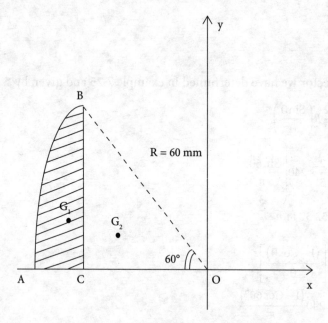

Fig. 7.35

Solution:

OA = OB = 60 mm
BC = OB.sin60° = 60.sin60°
OC = OB.cos60° = 60.cos60°
BC = 51.96 *mm* and OC = 30 *mm*

Area of circular sector from OA and OB is given by

$$= \frac{\pi R^2}{360°} \times 60°$$

$$= \frac{\pi R^2}{6} = \frac{\pi (60)^2}{6}$$

$$= 1884.96 \ mm^2$$

The centroid of circular sector is given by (as solved in example 7.25)

$$x_1^* = -\frac{2}{3} R \left(\frac{\sin \theta}{\theta} \right)$$

Note: Here negative sign is taken as x_1^* is measured opposite to X-axis.

Centroid and Centre of Gravity

for $\theta = 60°$

$$x_1^* = -\frac{2}{3} \times 60 \frac{\sin 60°}{\left(\frac{\pi}{3}\right)}$$

$$= -33.08 \text{ mm}$$

$$y_1^* = +\frac{2}{3}R\left(\frac{1-\cos\theta}{\theta}\right)$$

$$= +\frac{2}{3} \times 60 \frac{(1-\cos 60°)}{\left(\frac{\pi}{3}\right)}$$

$$= 19.10 \text{ mm}$$

$a_1 = 1884.96 \text{ mm}^2$	$x_1 = -33.08 \text{ mm}$	$y_1 = 19.10 \text{ mm}$
$a_2 = \frac{1}{2} \times 51.96 \times 30$	$x_2 = \frac{-2}{3} \times 30$	$y_2 = \frac{1}{3} \times 51.96$
$= 779.4 \text{ mm}^2$	$= -20 \text{ mm}$	$= 17.32 \text{ mm}$

$$\bar{X} = \left(\frac{a_1 x_1 - a_2 x_2}{a_1 - a_2}\right)$$

$$= \left[\frac{1884.96 \times (-33.08) - 779.4(-20)}{(1884.96 - 779.4)}\right]$$

$$= -42.30 \text{ mm}$$

$$\bar{Y} = \left(\frac{a_1 y_1 - a_2 y_2}{a_1 - a_2}\right)$$

$$= \left[\frac{(1884.96 \times 19.10) - (779.4 \times 17.32)}{(1884.96 - 779.4)}\right]$$

$$= 20.35 \text{ mm}$$

7.8 Centre of Gravity of Cone and Hemisphere

Example: 7.28

Determine the centre of gravity of a right circular solid cone of base radius 'R' and height 'h' about the axis of symmetry.

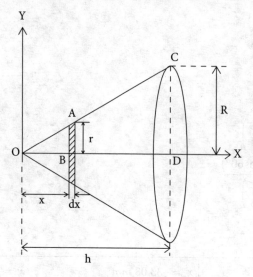

Fig. 7.36

Solution:

Let the density of material of cone is ρ. Consider an elemental plate of radius 'r' and thickness dx at distance 'x' from vertex 'O'. As OX is axis of symmetry thus centre of gravity will lie on axis OX.

Thus $\bar{Y} = 0$

$$\bar{X} = \frac{\int dw \cdot x^*}{\int dw}$$

Weight of elemental plate = ρ × Volume of elemental plate × g
$$= \rho \times dv \times g$$
$$= \rho \times dx \times dA \times g$$
$$dW = \rho \times dx \times \pi r^2 \times g \quad \text{---------(1)}$$

Δ OBA and Δ ODC are similar triangles, thus

$$\frac{r}{x} = \frac{R}{h}$$

$$\therefore r = \frac{R \cdot x}{h} \quad \text{substituting in equation (1)}$$

$$dw = dx \times \rho \left(\frac{R \cdot x}{h}\right)^2 \times g$$

$$x^* = x + \frac{dx}{2} \simeq x$$

Centroid and Centre of Gravity

$$\bar{X} = \frac{\int dW \cdot x^*}{\int dW}$$

$$= \frac{\int_0^h dx \cdot \rho \frac{R^2 \cdot x^2}{h^2} \cdot g \cdot x}{\int_0^h dx \cdot \rho \frac{R^2 \cdot x^2}{h^2} \cdot g}$$

$$= \frac{\frac{\rho R^2}{h^2} g \int_0^h x^3 \, dx}{\frac{\rho R^2}{h^2} g \int_0^h x^2 \, dx}$$

$$= \frac{\left[\frac{x^4}{4}\right]_0^h}{\left[\frac{x^3}{3}\right]_0^h}$$

$$= \frac{\frac{h^4}{4}}{\frac{h^3}{3}}$$

$$= \frac{3h}{4}$$

Thus centre of gravity lies from the vertex of cone at $\frac{3h}{4}$ or lies at $\frac{h}{4}$ from the base of cone.

Example 7.29

Determine the centre of gravity of a frustum of a cone of top diameter 300 mm and bottom diameter 600 mm. Take height of frustum of a cone as 450 mm.

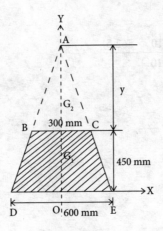

Fig. 7.37

Solution:

Extend CE and BD to y axis which meets at A. Let the A is at distance y from BC.
Δ ABC and Δ ADE are similar triangles,

$$\frac{y}{300} = \frac{y+450}{600}$$

$$2y = y + 450$$
$$y = 450 \text{ mm}$$

As frustum of a cone is symmetrically distributed about axis OA thus $\bar{X} = 0$,

$$\bar{Y} = \left(\frac{w_1 \cdot y_1 - w_2 y_2}{w_1 - w_2}\right)$$

$$= \frac{\rho g}{\rho g}\left(\frac{V_1 \cdot y_1 - V_2 y_2}{V_1 - V_2}\right) \quad \ldots (1)$$

$$V_1 = \frac{1}{3}\pi R_1^2 \cdot h_1 = \frac{1}{3}\pi(300)^2 \times 900$$
$$= 2.7 \times \pi \times 10^7 \text{ mm}^3$$

$$V_2 = \frac{1}{3}\pi R_2^2 \cdot h_2$$
$$= \frac{1}{3}\pi(150)^2 \cdot 450$$
$$= 0.34 \times \pi \times 10^7 \text{ mm}^3$$

$$h_1 = \frac{900}{4} \quad \text{and} \quad h_2 = 450 + \left(\frac{450}{4}\right)$$
$$= 225 \text{ mm} \quad\quad\quad\quad = 562.5 \text{ mm}$$

$$\bar{Y} = \left(\frac{V_1 y_1 - V_2 y_2}{V_1 - V_2}\right)$$

$$= \frac{(2.7\pi \times 10^7 \times 225) - (0.34\pi \times 10^7 \times 562.5)}{(2.7\pi \times 10^7 - 0.34\pi \times 10^7)}$$

$$= 176.38 \text{ mm}$$

Centroid and Centre of Gravity

Example: 7.30

Determine the centre of gravity of a solid hemisphere of radius R from its diametric axis.

Solution:

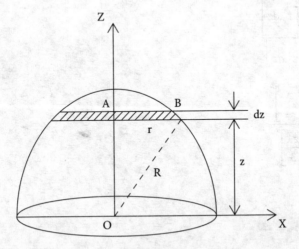

Fig. 7.38

Let the density of material is ρ. The hemisphere is lying on X-Y plane and OZ is the axis of symmetry as shown in fig. The centre of gravity lies on OZ axis thus $\bar{X} = 0$ and $\bar{Y} = 0$,

$$\bar{Z} = \frac{\int dw \cdot z^*}{\int dw}$$

Consider an elemental plate of thickness dz at distance z from OX.

$$dW = \rho \times dv \times g \text{ and } z^* = z + \frac{dz}{2} \simeq z$$
$$dW = \rho \times dz \times dA \times g$$

If radius of elemental plate is r then $dW = \rho \times dz \times \pi r^2 \times g$ ------- (1)

OB is radius of hemisphere 'R'

$$r = \sqrt{(R^2 - Z^2)}$$

Substituting value of r in equation (1),
$$dW = \rho.dz\,\pi(R^2 - Z^2).g$$

$$\bar{Z} = \frac{\int_0^R \rho.dz.\pi(R^2 - z^2)g.z}{\int_0^R \rho.dz.\pi(R^2 - z^2).g}$$

$$\bar{Z} = \frac{\rho.\pi.g\int_0^R (R^2 - z^2)zdz}{\rho.\pi.g\int_0^R (R^2 - z^2)dz}$$

$$= \frac{\left[R^2\int_0^R z.dz - \int_0^R z^3.dz\right]}{\left[R^2\int_0^R dz - \int_0^R z^2.dz\right]}$$

$$= \frac{R^2.\left[\frac{z^2}{2}\right]_0^R - \left[\frac{z^4}{4}\right]_0^R}{R^2.[z]_0^R - \left[\frac{z^3}{3}\right]_0^R}$$

$$= \frac{R^2.\frac{R^2}{2} - \frac{R^4}{4}}{R^3 - \frac{R^3}{3}}$$

$$= \frac{\frac{R^4}{4}}{\frac{2R^3}{3}}$$

$$= \frac{3R}{8}$$

Numerical Problems

N 7.1 Determine centroid of the Z-section as shown in Fig. NP 7.1.

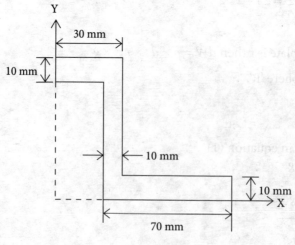

Fig. NP 7.1

Centroid and Centre of Gravity

N 7.2 Determine centroid of the section as shown in Fig. NP 7.2

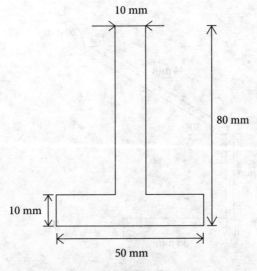

Fig. NP 7.2

N 7.3 Determine the centroid of the shaded area formed by intersection of a cone $y = x^2/2$ and the straight line $y = x$, as shown in Fig. NP 7.3.

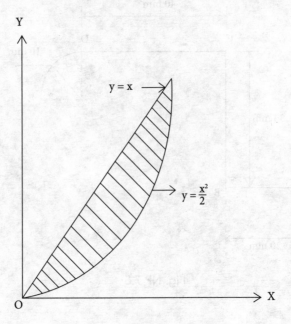

Fig. NP 7.3

N 7.4 A thin wire is bent into AB, BC, CD and DE segments as shown in Fig. NP 7.4; determine the centroid.

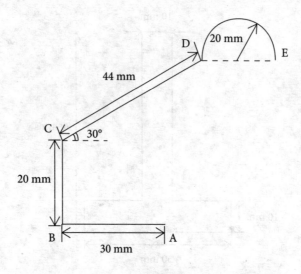

Fig. NP 7.4

N 7.5 Determine the centroid of thin wire as shown in Fig. NP 7.5.

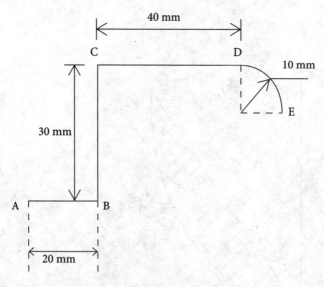

Fig. NP 7.5

Centroid and Centre of Gravity

N 7.6 Determine the centroid of the shaded area as shown in Fig. NP 7.6.

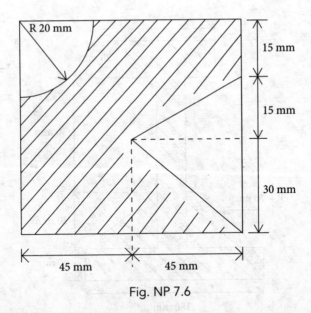

Fig. NP 7.6

N 7.7 Determine the centroid of the Fig. NP 7.7.

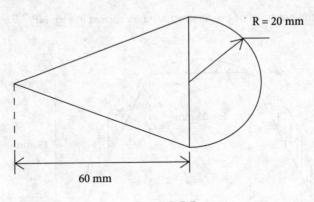

Fig. NP 7.7

N 7.8 Determine centroid of the shaded area as shown in Fig. NP 7.8.

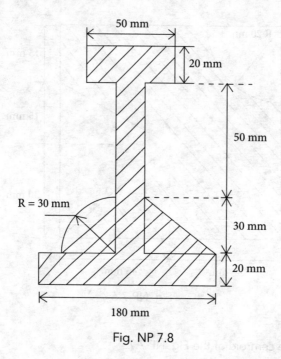

Fig. NP 7.8

N 7.9 Determine the centroid of the shaded area as shown in Fig. NP 7.9.

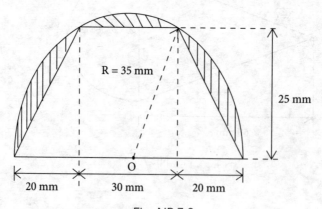

Fig. NP 7.9

Centroid and Centre of Gravity

Multiple Choice Questions

1. Centroid or Centre of gravity can determined by using
   a. Method of moment
   b. Method of Integration
   c. Varignon's theorem
   d. all of these

2. Centroid is not applicable in bodies comprising
   a. length
   b. volume
   c. area
   d. none of these

3. Centre of gravity is applicable in bodies comprising
   a. length
   b. volume
   c. weight
   d. all of these

4. Centre of mass is analogous to centre of gravity if body is subjected to uniform
   a. mass
   b. density
   c. volume
   d. acceleration under gravity

5. A body whose area is symmetric about $\overline{Y}$-axis, shows that
   a. $\overline{X} = 0$
   b. its centroid lie on Y-axis
   c. the first moment of area about Y-axis will be zero
   d. all of these

6. Centroid of triangle from base lies at
   a. One third of its height
   b. two third of its height
   c. three fourth of its height
   d. none of these

7. Centroid of a quarter circle of radius r from its centre lies at
   a. $\left(\frac{4\pi}{3r}, \frac{4\pi}{3r}\right)$
   b. $\left(\frac{4r}{3\pi}, \frac{4r}{3\pi}\right)$
   c. $\left(\frac{4\pi}{3r}, \frac{4r}{3\pi}\right)$
   d. none of these

8. Centroid of a semicircle of radius r varying from 0° to 180° lies at
   a. $\left(0, \frac{4r}{3\pi}\right)$
   b. $\left(0, \frac{2\pi}{3r}\right)$
   c. $\left(0, \frac{2r}{3\pi}\right)$
   d. $\left(0, \frac{4\pi}{3r}\right)$

9. Centre of gravity of a right circular solid cone from its vertex lies at
   a. One fourth of its height
   b. two third of its height
   c. three fourth of its height
   d. none of these

10. A solid hemisphere of radius r is placed on table with its flat surface. The centre of gravity of solid hemisphere from flat surface will lie at
   a. $4r/8$
   b. $3r/8$
   c. $3r/8\pi$
   d. $2r/\pi$

Answers

1. d 2. d 3. c 4. d 5. d 6. a 7. b 8. a 9. c 10. b

Chapter 8

Moment of Inertia

8.1 Moment of Inertia of Plane Area and Mass

We know that inertia deals with resistance to change. However, moment of inertia has two different roles in terms of area moment of inertia and mass moment of inertia. In strength of a material, the area moment of inertia is concerned with resistance of area against bending, buckling of beams and columns, respectively. In dynamics, the mass moment of inertia is concerned with rotatory motion in the same way as mass is concerned with linear motion. That is why mass moment of inertia is considered as analogous to the mass.

In the earlier chapter, we have studied that the first moment of area provides the centroid of the area. If the moment of first moment of area is taken again, it becomes the second moment of area and is called the area moment of inertia. Similarly, in case of mass, the second moment of mass is called the mass moment of inertia.

Consider a rectangular section from pure bending region of beam. Let an elemental area dA is at distance y from neutral axis as shown in Fig. 8.1.

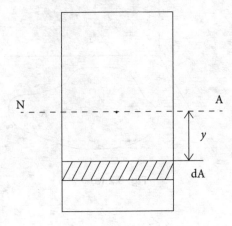

Fig. 8.1

If the elemental area dA offers bending stresses σ against loading then the bending stresses induced are directly proportional to the distance from neutral axis, i.e., σ α y

As the stresses vary linearly across the section, $\sigma = C \times y$
where, C is a constant.
The resistance force or inertia force offered by elemental area dA $= \sigma \times dA$
(representing the first moment of area) $= C \times y \times dA$
The moment of resistance force or inertia force offered by elemental
area dA (representing the second moment of area) $= C \times y \times dA \times y$
The moment of inertia force offered by full cross-section area A $= \int C \times y \times dA \times y$
The moment of inertia force, i.e., moment of inertia of the area $= C \int dA \times y^2$
Thus the moment of inertia of the area from X-axis is given by, I_x $= \int dA \times y^2$
and the moment of inertia of the area from Y-axis is given by, I_y $= \int dA \times x^2$

8.2 Radius of Gyration

If an area A is having moment of inertia I about an axis, then the radius of gyration is such distance at which area A can be concentrated in such a way which will provide the same moment of inertia I, from the same axis.

Let an area A lying in a XY plane is concentrated at distance k_y and k_x from axis OX and OY as shown in Fig. 8.2 (a) and (b).

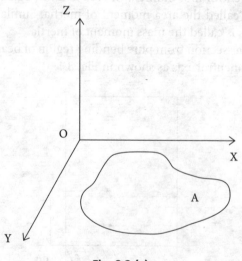

Fig. 8.2 (a)

Moment of Inertia

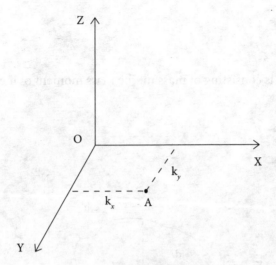

Fig. 8.2 (b)

If the moment of inertia from axis OX and OY is I_x and I_y, respectively, the moment of inertia of the area from OX-axis,

$$I_x = A \times K_y^2$$

The radius of gyration from OX-axis,

$$k_y = \sqrt{\frac{I_x}{A}}$$

Similarly, the radius of gyration from OY-axis,

$$k_x = \sqrt{\frac{I_y}{A}}$$

If mass moment of inertia of a mass m is I from an axis then radius of gyration from such axis will be given by

$$k = \sqrt{\frac{I}{m}}$$

8.3 Parallel Axis Theorem and its Significance

The parallel axis theorem states that moment of inertia of a plane area about an axis, parallel to its centroidal axis is given by the summation of two terms: first, the moment of inertia about its centroidal axis and second, the product of area with square of distance between two parallel axes. Consider an area A lying in a plane and having centroidal axes XX and YY as shown in Fig. 8.3. As per this theorem, the moment of inertia of such area about axis LM and NP, which are parallel to its centroidal axes XX and YY at distances h_y and h_x' will be given by:

$$I_{LM} = (I_G)_{XX} + A \cdot h_y^2$$
$$I_{NP} = (I_G)_{YY} + A \cdot h_x'^2$$

Similarly, if the body is consisting of mass m, the mass moment of inertia will be given by

$$I_{LM} = (I_G)_{XX} + m \cdot h_y^2$$
$$I_{NP} = (I_G)_{YY} + m \cdot h_x'^2$$

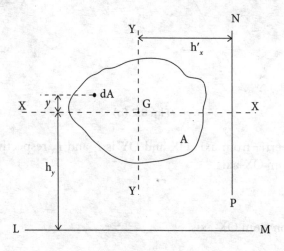

Fig. 8.3

Proof:

Let an elemental area dA is at distance y from centroidal axis XX as shown in Fig. 8.3. The moment of inertia of area about horizontal centroidal axis will be

$$(I_G)_{xx} = \int dA \times y^2$$

According to the definition, the moment of inertia of the elemental area dA from parallel axis LM will be

$$I_{LM} = \int dA \times (y + h_y)^2$$
$$I_{LM} = \int dA \times y^2 + h_y^2 \int dA + 2h_y \int dA \times y$$
$$I_{LM} = (I_G)_{XX} + A \cdot h_y^2 + 0$$
$$I_{LM} = (I_G)_{XX} + A \cdot h_y^2$$

Similarly, the moment of inertia of area about its vertical centroidal axis will be

$$I_{NP} = (I_G)_{YY} + A \cdot h_x'^2$$

Moment of Inertia

Note:

i. Term $\int dA = A$
ii. Term $\int dA \times y$ shows the first moment of full area from its centroidal axis XX i.e. zero as whole area is concentrated there.

This theorem is useful to determine the moment of inertia of any composite section. The composite section comprises different combination of cross-section areas which may be of rectangular, triangular, quarter circular, semi-circular and circular sections. In such cases, the moment of inertia of composite section can be determined about its both horizontal and vertical centroidal axis by using the parallel axis theorem. The moment of inertia of composite section is computed to meet the designing requirements.

8.4 Perpendicular Axis Theorem

The perpendicular axis theorem is used to determine the moment of inertia about polar axis. The polar axis is an axis which lies at intersection of two perpendicular axis lying in a common plane. According to this theorem, the moment of inertia of an area about polar axis is given by the summation of moment of inertia of both centroidal axes i.e.,

$$I_{ZZ} = (I_G)_{XX} + (I_G)_{YY}$$

where, I_{ZZ} is called as polar moment of inertia

Consider an area A lying in a XY plane where an elemental area dA is at distance x^*, y^* and z^* from the axes OY, OX and OZ, respectively, as shown in Fig. 8.4.

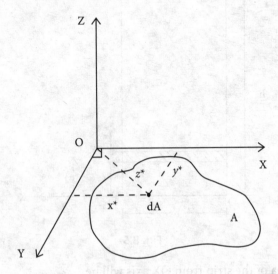

Fig. 8.4

According to the definition, the moment of inertia of the elemental area dA from axis OZ will be

$I_{oz} = I_{zz} = \int dA \times (z^*)^2$

As $(z^*)^2 = (x^*)^2 + (y^*)^2$

$I_{zz} = \int dA \times ((x^*)^2 + (y^*)^2)$

$I_{zz} = \int dA \times (x^*)^2 + \int dA \times (y^*)^2$

$I_{zz} = (I_G)_{YY} + (I_G)_{XX}$

Or,

$I_{zz} = (I_G)_{XX} + (I_G)_{YY}$

This theorem is used to determine the polar moment of inertia which gives a measure of resistance to twist about polar axis.

8.5 Moment of Inertia of a Rectangle

Consider a rectangle of base b and height h as shown in Fig. 8.5. Let an elemental strip parallel to base and having area dA is at distance y from X axis. If the thickness of the strip is dy,

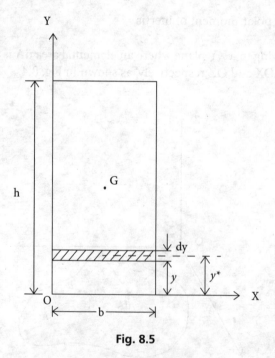

Fig. 8.5

The moment of inertia of the strip from OX axis will be
$= dA \cdot (y^*)^2$

The moment of inertia of rectangle from OX axis will be

Moment of Inertia

$$I_{OX} = I_{Base} = \int dA \cdot (y^*)^2$$

where $dA = b \times dy$

and $y^* = y + \dfrac{dy}{2}$

$\simeq y$

$$I_{Base} = \int_0^h b \cdot dy \cdot y^2$$
$$= b \int_0^h y^2 \cdot dy$$

$$I_{OX} = I_{Base} = \dfrac{bh^3}{3}$$

Similarly, the moment of inertia of the rectangle from OY axis will be,

$$I_{OY} = \left(\dfrac{b^3 h}{3}\right)$$

The moment of inertia of the rectangle from the horizontal centroidal axis can be determined by using the parallel axis theorem,

$$I_{LM} = (I_G)_{XX} + A \cdot h_y^2$$

Here $I_{LM} = I_{OX} = \dfrac{bh^3}{3}$

and $h_y = \dfrac{h}{2}$, $A = b \cdot h$

Substituting all values in the theorem,

$$\left(\dfrac{bh^3}{3}\right) = (I_G)_{XX} + (bh) \cdot \left(\dfrac{h}{2}\right)^2$$

$$(I_G)_{XX} = \dfrac{bh^3}{12}$$

Similarly, the moment of inertia of rectangle from the vertical centroidal axis will be

$$(I_G)_{YY} = \dfrac{b^3 h}{12}$$

8.6 Moment of Inertia of a Triangle

Let the base and height of the triangle is b and h, respectively. Consider an elemental strip DE at a height y from base AB. If the area of elemental strip is dA and the width and thickness are a and dy, respectively, as shown in Fig. 8.6,

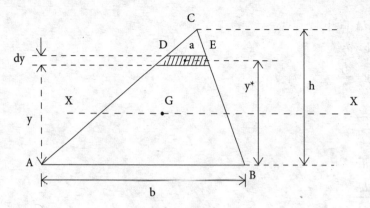

Fig. 8.6

the moment of inertia of strip from AB will be

$= dA \cdot (y^*)^2$

The moment of inertia of triangle from base, AB will be

$I_{AB} = \int dA \cdot (y^*)^2$

where, $dA = a \times dy$

Since $\triangle$ ABC and $\triangle$ DEC are similar triangles,

$$\frac{a}{(h-y)} = \frac{b}{h}$$

i.e. $a = \frac{b(h-y)}{h}$

and $y^* = y + \frac{dy}{2}$

$\simeq y$

$I_{AB} = I_{Base} = \int_0^h \frac{b(h-y)}{h} \cdot dy \cdot y^2$

$I_{AB} = \frac{b}{h}\left[h\int_0^h y^2 \cdot dy - \int_0^h y^3 \cdot dy\right]$

$= \frac{b}{h}\left[\frac{h^4}{3} - \frac{h^4}{4}\right]$

$I_{AB} = \frac{b \cdot h^3}{12}$ and $I_{perpendicular\ to\ AB} = \frac{b^3 \cdot h}{12}$

The moment of inertia from the horizontal centroidal axis will be,

$I_{LM} = (I_G)_{XX} + A \cdot h_y^2$

Here, $I_{LM} = I_{AB} = \frac{bh^3}{12}$

and $h_y = \frac{h}{3}$, $A = \left(\frac{1}{2} \times b \times h\right)$

Moment of Inertia

Substituting all values in the theorem,

$$\left(\frac{bh^3}{12}\right) = (I_G)_{XX} + \left(\frac{1}{2} \times b \times h\right)\left(\frac{h}{3}\right)^2$$

$$(I_G)_{XX} = \frac{bh^3}{36}$$

Similarly, the moment of inertia of the triangle from the vertical centroidal axis will be,

$$(I_G)_{YY} = \frac{b^3 \cdot h}{36}$$

8.7 Moment of Inertia of a Circle, a Quarter Circle and a Semicircle

Let a circle of radius R is lying in XY plane where XX and YY are the horizontal and vertical centroidal axis, respectively, as shown in Fig. 8.7.

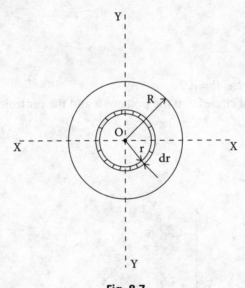

Fig. 8.7

The polar axis ZZ is acting at O perpendicular to the plane. Consider an elemental ring of radius r from polar axis.

The moment of inertia of ring from polar axis will be $= dA \cdot r^2$

The moment of inertia (M.I.) of the circle will be,

$$I_{ZZ} = \int_0^R dA \cdot r^2$$

where, $dA = 2\pi r \cdot dr$

$$I_{ZZ} = 2\pi \int_0^R r^3 \cdot dr$$

$$I_{ZZ} = \frac{2\pi \cdot R^2}{4}$$

$$I_{ZZ} = \frac{\pi R^4}{2}$$

Using the perpendicular axis theorem,

$$I_{ZZ} = (I_G)_{XX} + (I_G)_{YY}$$

$$\frac{\pi R^4}{2} = (I_G)_{XX} + (I_G)_{YY}$$

as $(I_G)_{XX} = (I_G)_{YY}$ due to symmetry for circle,

$$2(I_G)_{XX} = \frac{\pi R^4}{2}$$

$$(I_G)_{XX} = \frac{\pi R^4}{4}$$

$$(I_G)_{YY} = \frac{\pi R^4}{4}$$

If d is the diameter of circle, $R = d/2$

The moment of inertia of circle from the polar axis and the centroidal axis will be,

$$I_{ZZ} = \frac{\pi \cdot d^4}{32}$$

$$I_{XX} = I_{YY} = \frac{\pi d^4}{64}$$

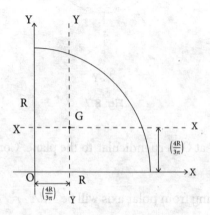

Fig. 8.8

Moment of Inertia

A quarter circle is one fourth of a circle.

Thus $I_{OX} = \frac{1}{4} \cdot (I_G)_{XX}$ of the circle

$$I_{OX} = \frac{\pi R^4}{16}$$

or, $I_{OX} = I_{OY} = \frac{\pi R^4}{16}$

The moment of inertia of a quarter of a circle can be determined from the centroidal axis by using the parallel axis theorem,

$$I_{LM} = (I_G)_{XX} + A \cdot h_y^2$$

Here, $I_{OX} = I_{LM} = \frac{\pi R^4}{16}$

$A = \left(\frac{1}{4} \cdot \pi R^2\right)$

and $h_y = \left(\frac{4R}{3\pi}\right)$

Substituting all values in the theorem,

$$\frac{\pi R^4}{16} = (I_G)_{XX} + \left(\frac{\pi R^2}{4}\right) \cdot \left(\frac{4R}{3\pi}\right)^2$$

$(I_G)_{XX} = 0.055 R^4$

Similarly,

$(I_G)_{YY} = 0.055 R^4$

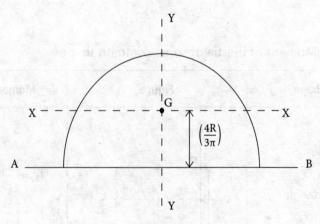

Fig. 8.9

Like quarter circle, the moment of inertia of a semicircle can be determined:

$$I_{AB} = \frac{1}{2}(I_G)_{XX}$$

$$I_{AB} = \frac{\pi R^4}{8}$$

The moment of inertia of semicircle will be double that of quarter circle about the XX axis.

Thus $(I_G)_{XX} = 2 \times (I_G)_{XX}$ of quarter circle

$$(I_G)_{XX} = 0.11 R^4$$

Alternative Method:

Using the parallel axis theorem,

$$I_{LM} = (I_G)_{XX} + A \cdot h_y^2$$

where $I_{LM} = I_{AB} = \left(\dfrac{\pi R^4}{8}\right)$

$$A = \frac{\pi R^2}{2}$$

and $h_y = \left(\dfrac{4R}{3\pi}\right)$

Substituting all values in the theorem,

$$\left(\frac{\pi R^4}{8}\right) = (I_G)_{XX} + \left(\frac{\pi R^4}{2}\right)\left(\frac{4R}{3\pi}\right)^2$$

$$(I_G)_{XX} = 0.11 R^4$$

Table 8.1: Moment of inertia of some common shapes

Serial No.	Shape	Figure	Moment of Inertia
1.	Rectangle		$I_{AB} = \left(\dfrac{bh^3}{3}\right)$, $I_{AD} = \left(\dfrac{b^3 h}{3}\right)$ $(I_G)_{XX} = \left(\dfrac{bh^3}{12}\right)$, $(I_G)_{YY} = \left(\dfrac{b^3 h}{12}\right)$

2.	Right angled Triangle	*(figure: right-angled triangle with legs b and h, centroid G, axes X-X through G and O-A along base, Y-Y through G and O-B along vertical leg)*	$I_{OX} = I_{OA} = I_X = \left(\dfrac{bh^3}{12}\right)$ $I_{OY} = I_{OB} = I_Y = \left(\dfrac{b^3 h}{12}\right)$ $(I_G)_{XX} = \left(\dfrac{bh^3}{36}\right)$ $(I_G)_{YY} = \left(\dfrac{b^3 h}{36}\right)$
3.	Circle	*(figure: circle of radius R with X-X and Y-Y axes through center)*	$I_{XX} = \left(\dfrac{\pi R^4}{4}\right)$ $I_{YY} = \left(\dfrac{\pi R^4}{4}\right)$ $I_{ZZ} = \left(\dfrac{\pi R^4}{2}\right)$
4.	Semicircle	*(figure: semicircle of radius R with base along O-X, centroid G, X-X axis through G)*	$I_{OX} = I_X = \left(\dfrac{\pi R^4}{8}\right)$ $I_{OY} = (I_G)_{YY} = \left(\dfrac{\pi R^4}{8}\right)$ $(I_G)_{XX} = 0.11 R^4$
5.	Quarter circle	*(figure: quarter circle of radius R in first quadrant with centroid G)*	$I_{OX} = I_X = \left(\dfrac{\pi R^4}{16}\right)$ $I_{OY} = I_Y = \left(\dfrac{\pi R^4}{16}\right)$ $(I_G)_{XX} = (I_G)_{YY} = 0.055 R^4$
6.	Slender rod	*(figure: slender rod AB of length L with centroid G at midpoint, Y-Y axis through G)*	$I_A = \dfrac{M \cdot L^2}{3}$ $(I_G)_{YY} = \dfrac{ML^2}{12}$

7.	Rectangular plate	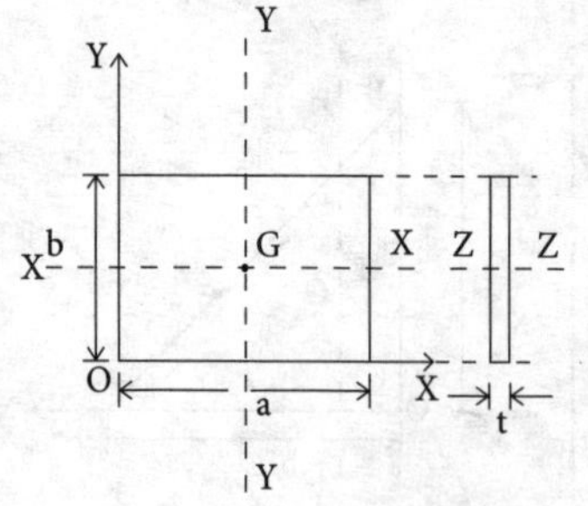	$I_{XX} = \dfrac{Mb^2}{12}$, $I_{YY} = \dfrac{M \cdot a^2}{12}$ $I_{ZZ} = \dfrac{M(a^2 + b^2)}{12}$
8.	Circular plate	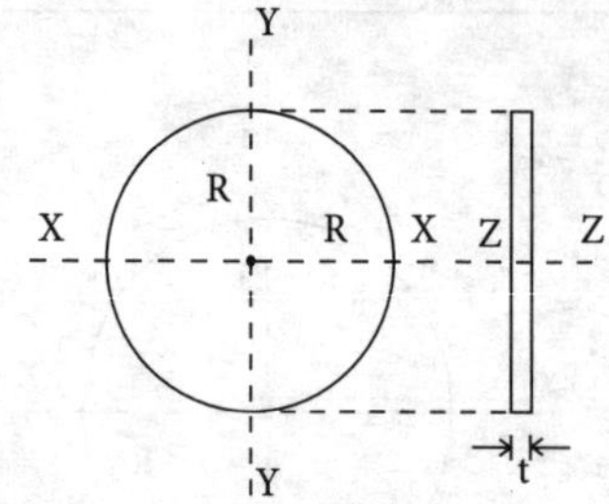	$I_{XX} = \dfrac{MR^2}{4}$ $I_{YY} = \dfrac{MR^2}{4}$ $I_{ZZ} = \dfrac{MR^2}{2}$
9.	Right Circular cone	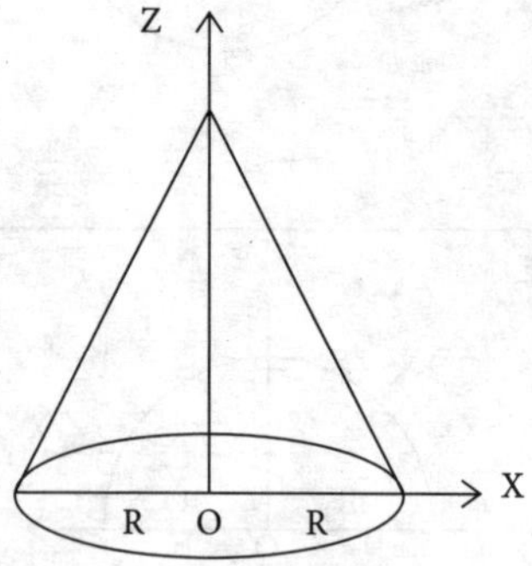	$I_{OZ} = \dfrac{3}{10} MR^2$
10.	Sphere	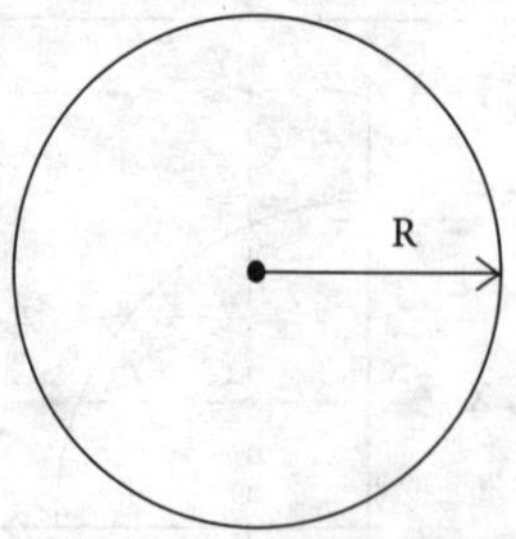	$I_{XX} = I_{YY} = I_{ZZ} = \dfrac{2}{5} MR^2$

8.8 Moment of Inertia of Composite Sections and Bodies

Example: 8.1

Find the moment of inertia of I-section as shown in Fig. 8.10 about *X-X* and *Y-Y* axis.

(UPTU, Ist sem, 2000–2001)

Moment of Inertia

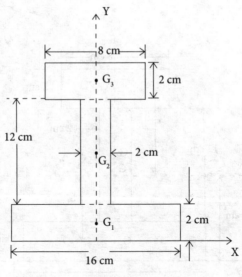

Fig. 8.10

Solution:

First we will determine centroid of given I-section. As section is symmetrical about the *y* axis, its centroid will lie on this axis,

Hence $\overline{X} = 0$

Prepare the table.

$a_1 = 16 \times 2 \ cm^2$	$y_1 = 1 \ cm$
$a_2 = 12 \times 2 \ cm^2$	$y_2 = 2 + \dfrac{12}{2} = 8 \ cm$
$a_3 = 8 \times 2 \ cm^2$	$y_3 = 2 + 12 + \dfrac{2}{2} = 15 \ cm$

$$\overline{Y} = \left(\frac{a_1 y_1 + a_2 y_2 + a_3 y_3}{a_1 + a_2 + a_3} \right)$$

$$= \frac{(16 \times 2 \times 1) + (12 \times 2 \times 8) + (8 \times 2 \times 15)}{(16 \times 2) + (12 \times 2) + (8 \times 2)}$$

$$= 6.44 \ cm$$

To determine moment of inertia, consider Fig. 8.10(a) where *XX* and *YY* axis are centroidal axis of I–section.

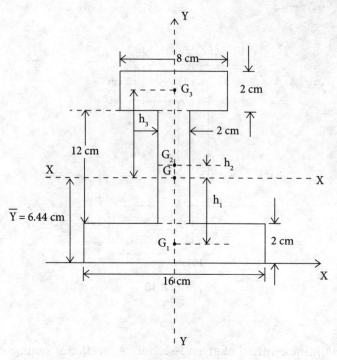

Fig. 8.10 (a)

$I_{XX} = (I_1 + I_2 + I_3)_{\text{about } XX \text{ axis}}$ ..... (1)

$h_1 = \overline{Y} - \dfrac{2}{2} = 6.44 - 1 = 5.44 \ cm,$

$h_2 = y_2 - \overline{Y} = 8 - 6.44 = 1.56 \ cm,$

$h_3 = y_3 - \overline{Y} = 15 - \overline{Y} = 15 - 6.44 = 8.56 \ cm.$

$I_1 \text{ about } XX = \left(I_{G_1}\right)_{XX} + A_1 \cdot h_1^2$

$= \left(\dfrac{16 \times 2^3}{12}\right) + (16 \times 2) \cdot (6.44 - 1)^2$

$= 957.67 \ cm^4$

$I_2 \text{ about } XX = \left(I_{G_2}\right)_{XX} + A_2 \cdot h_2^2$

$= \left(\dfrac{2 \times 12^3}{12}\right) + (2 \times 12) \cdot (8 - 6.44)^2$

$= 346.41 \ cm^4$

$I_3 \text{ about } XX = \left(I_{G_3}\right)_{XX} + A_3 \cdot h_3^2$

$= \left(\dfrac{8 \times 2^3}{12}\right) + (8 \times 2) \cdot (15 - 6.44)^2$

$= 1177.71 \ cm^4$

Moment of Inertia

From equation (1),

I_{xx} = 957.67 + 346.41 + 1177.71

= 2481.79 cm^4

$I_{YY} = (I_1 + I_2 + I_3)_{\text{about YY axis}}$

I_1 about yy axis $= (I_{G1})_{YY} + A_1 (h_1')^2$

As the axis YY passes through all centroids of individual areas, h_1', h_2', $h_3' = 0$

thus $I_{YY} = (I_{G_1})_{YY} + (I_{G_2})_{YY} + (I_{G_3})_{YY}$

$= \left(\dfrac{16^3 \times 2}{12}\right) + \left(\dfrac{2^3 \times 12}{12}\right) + \left(\dfrac{8^3 \times 2}{12}\right)$

$= 776 \text{ cm}^4$

Example: 8.2

Find the moment of inertia of ISA 100 × 75 × 6 about the centroidal XX and YY axis.

(UPTU IInd sem, 2001–02)

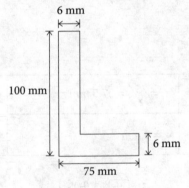

Fig. 8.11

Solution:

Following steps are taken:

(i) First, take reference axis i.e. X-axis at bottom end and Y-axis at left end of the figure as shown in Fig. 8.11(a)
(ii) Second select the area in a suitable manner. Here the given figure consists of two rectangles which may be of size 75 mm × 6 mm and 94 mm × 6 mm or size 69 mm × 6 mm and 100 mm × 6 mm. These sizes depended upon whether dash line is drawn parallel to 75 mm or parallel to 100 mm.
Here the dotted line is drawn parallel to 75 mm.
(iii) Mark centroids of different areas by G_1 and G_2.

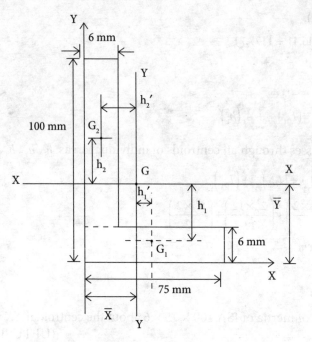

Fig. 8.11 (a)

Prepare the table,

$a_1 = 75 \times 6\ mm^2$	$x_1 = 75/2 = 37.5\ mm$	$y_1 = 3\ mm$
$a_2 = 94 \times 6\ mm^2$	$x_2 = 3\ mm$	$y_2 = \left(6 + 94/2\right) = 53\ mm$

$$\overline{X} = \left(\frac{a_1 x_1 + a_2 x_2}{a_1 + a_2}\right) \qquad \overline{Y} = \left(\frac{a_1 y_1 + a_2 y_2}{a_1 + a_2}\right)$$

$$= \frac{(75 \times 6 \times 37.5) + (94 \times 6 \times 3)}{(75 \times 6) + (94 \times 6)}, \quad \overline{Y} = \frac{(75 \times 6 \times 3) + (94 \times 6 \times 53)}{(75 \times 6) + (94 \times 6)}$$

$\overline{X} = 18.31\ mm$ and $\overline{Y} = 30.81\ mm$

$h_1 = \overline{Y} - \dfrac{6}{2} = 30.81 - 3 = 27.81\ mm$

$h_2 = y_2 - \overline{Y} = 53 - 30.81 = 22.19\ mm$

$h_1' = x_2 - \overline{X} = \dfrac{75}{2} - 18.31 = 19.19\ mm$

$h_2' = \overline{X} - x_2 = 18.31 - 3 = 15.31\ mm$

Moment of Inertia

$I_{XX} = (I_1 + I_2)$ about X-axis

$= \left[(I_{G_1})_{XX} + A_1 h_1^2 \right] + \left[(I_{G_2})_{XX} + A_2 \cdot h_2^2 \right]$

$= \left[\left(\dfrac{75 \times 6^3}{12} \right) + (75 \times 6) \cdot (27.81)^2 \right] + \left[\left(\dfrac{6 \times 94^3}{12} \right) + (6 \times 94) \cdot (22.19)^2 \right]$

$= 10.42 \times 10^5 \, mm^4$

$I_{YY} = (I_1 + I_2)$ about Y-axis

$= \left[(I_{G_1})_{YY} + A_1 (h_1')^2 \right] + \left[(I_{G_2})_{YY} + A_2 (h_2')^2 \right]$

$= \left[\left(\dfrac{75^3 \times 6}{12} \right) + (75 \times 6)(19.19)^2 \right] + \left[\left(\dfrac{6^3 \times 94}{12} \right) + (6 \times 94)(15.31)^2 \right]$

$= 5.11 \times 10^5 \, mm^4$

Example: 8.3

Determine the moment of inertia of channel, C-section as shown in Fig. 8.12 about the centroidal X-X axis. (UPTU Ist sem. Co, 2003)

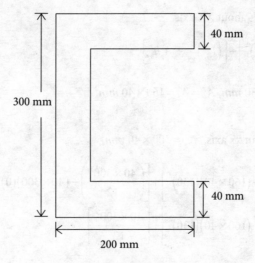

Fig. 8.12

Solution:

The centroidal axis of the given C-section will lie as shown in Fig. 8.12(a) as discussed in Section 7.4.

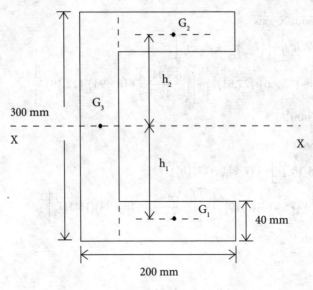

Fig. 8.12 (a)

The moment of inertia about section XX will be,

$I_{XX} = I_1 + I_2 + I_3$ about X-axis.

$\quad = 2 I_1 + I_3 \quad$ as $I_1 = I_2$ about XX axis.

$\quad = 2\left[(I_{G_1})_{XX} + A_1(h_1)^2\right] + \left[(I_{G_3})_{XX} + A_3 h_3^2\right]$

here $h_1 = 150 - \dfrac{40}{2} = 130$ mm, $A_1 = A_2 = 160 \times 40$ mm²

$h_2 = 130$ mm

$h_3 = 0$ as G_3 lies an xx axis., $A_3 = 300 \times 40$ mm²

$I_{XX} = 2\left[\left(\dfrac{160 \times 40^3}{12}\right) + (160 \times 40)(130)^2\right] + \left[\left(\dfrac{40 \times 300^3}{12}\right) + (40 \times 300)(0)^2\right]$

$\quad = 2\left[\left(\dfrac{160 \times 40^3}{12}\right) + (160 \times 40)(130)^2\right] + \left[\dfrac{40 \times 300^3}{12}\right]$

$\quad = 3.08 \times 10^8$ mm⁴

Example: 8.4

Determine the moment of inertia of T-section as shown in Fig. 8.13 about both the centroidal axes.

Moment of Inertia

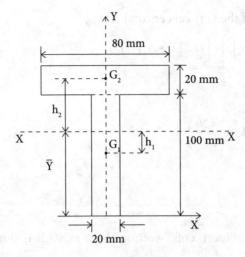

Fig. 8.13

Solution:

As discussed earlier, the centroid of T-section will lie on Y axis, thus $\overline{X} = 0$

Prepare table.

$a_1 = 100 \times 20 \ mm^2$	$y_1 = 50 \ mm$
$a_2 = 80 \times 20 \ mm^2$	$y_2 = 100 + \dfrac{20}{2} = 110 \ mm$

$$\overline{Y} = \left(\dfrac{a_1 y_1 + a_2 y_2}{a_1 + a_2}\right)$$

$$= \dfrac{(100 \times 20 \times 50) + (80 \times 20 \times 110)}{(100 \times 20) + (80 \times 20)}$$

$\overline{Y} = 76.67 \ mm$

$h_1 = \overline{Y} - y_1 = 76.67 - 50 = 26.67 \ mm$
$h_2 = y_2 - \overline{Y} = 110 - 76.67 = 33.33 \ mm$

$$I_{XX} = \left[(I_{G_1})_{XX} + A_1 h_1^2\right] + \left[(I_{G_2})_{XX} + A_2 h_2^2\right]$$

$$= \left[\left(\dfrac{20 \times 100^3}{12}\right) + (20 \times 100)(26.67)^2\right] + \left[\left(\dfrac{80 \times 20^3}{12}\right) + (80 \times 20)(33.33)^2\right]$$

$$= 4.92 \times 10^6 \ mm^4$$

Moment of inertia about the vertical centroidal axis,

$$I_{YY} = \left[(I_{G_1})_{YY} + A_1(h_1')^2\right] + \left[(I_{G_2})_{YY} + A_2(h_2')^2\right]$$

as $h_1' = h_2' = 0$

$$I_{YY} = (I_{G_1})_{YY} + (I_{G_2})_{YY}$$

$$= \left(\frac{20^3 \times 100}{12}\right) + \left(\frac{80^3 \times 20}{12}\right)$$

$$= 0.92 \times 10^6 \, mm^4$$

Example: 8.5

Determine the moment of inertia of T-section about axis AB as shown in Fig. 8.14.

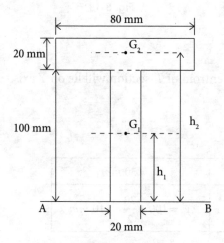

Fig. 8.14

Solution:

$I_{AB} = (I_1 + I_2)$ about X-axis

$$= \left[(I_{G_1})_{XX} + A_1 h_1^2\right] + \left[(I_{G_2})_{XX} + A_2 h_2^2\right]$$

$h_1 = 50 \, mm,$

$h_2 = 100 + \dfrac{20}{2}$

$= 110 \, mm$

$$= \left[\left(\frac{20 \times 100^3}{12}\right) + (20 \times 100)(50)^2\right] + \left[\left(\frac{80 \times 20^3}{12}\right) + (80 \times 20)(110)^2\right]$$

$$= 26.08 \times 10^6 \, mm^4$$

Moment of Inertia

Note: AB is the base of Ist rectangle, 20 mm × 100 mm

Alternate Method

$$I_{AB} = \left[(I_1)_{AB}\right] + \left[(I_{G_2})_{XX} + A_2 h_2^2\right]$$

$$= \left[\left(\frac{20 \times 100^3}{3}\right)\right] + \left[\left(\frac{80 \times 20^3}{12}\right) + (80 \times 20)(110)^2\right]$$

$$= 26.08 \times 10^6 \, mm^4$$

Example: 8.6

Determine the moment of inertia of given arrow section about vertical centroidal axis.

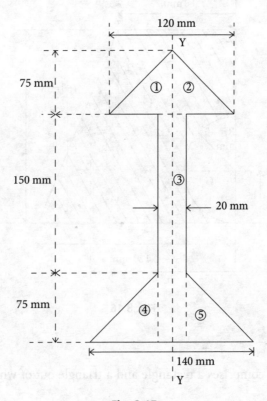

Fig. 8.15

Solution:

The given section is symmetrical about the vertical axis YY; thus it will be the centroidal axis.

The given section can be divided into four triangles 1, 2, 4 and 5 of equal size as marked in figure with base 60 mm and height 75 mm and one rectangle of size 20 mm × 225 mm.

$$I_{YY} = 4(I_1)_{YY} + (I_3)_{YY}$$
$$= 4\left[\frac{75 \times 60^3}{12}\right] + \left[\frac{225 \times 20^3}{12}\right]$$
$$= 55.5 \times 10^5 \, mm^4$$

Example: 8.7

For the shaded area as shown in Fig. 8.16, find the moment of inertia about the lines A-A and B-B.

(UPTU, Ist sem, 2003–04)

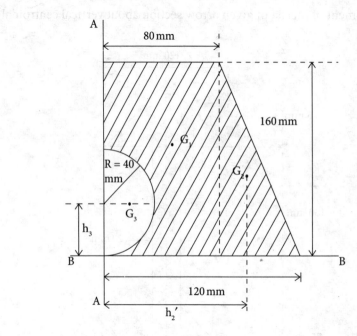

Fig. 8.16

Solution:

Here the shaded area comprises a rectangle and a triangle out of which a semicircle is cut out.

$$I_{A-A} = (I_{rectangle})_{y-axis} + (I_{Triangle})_{y-axis} - I_{semicircle}$$

$$= \left(\frac{160 \times 80^3}{3}\right) + \left[\left(\frac{160 \times 40^3}{36}\right) + \left(\frac{1}{2} \times 160 \times 40\right)(h_2')^2\right] - \left[\frac{\pi(40)^4}{8}\right]$$

Here, $h_2' = \left(80 + \frac{40}{3}\right) = 93.33 \, mm$

Moment of Inertia

$$I_{A-A} = \left(\frac{160 \times 80^3}{3}\right) + \left[\left(\frac{160 \times 40^3}{36}\right) + \left(\frac{1}{2} \times 160 \times 40\right)(93.33)^2\right] - \left[\frac{\pi(40)^4}{8}\right]$$

$$= 54.46 \times 10^6 \, mm^4$$

$$I_{B-B} = \left(I_{rectangle}\right)_{x-axis} + \left(I_{Triangle}\right)_{x-axis} - \left(I_{semicircle}\right)_{x-axis}$$

$$= \left(\frac{80 \times 160^3}{3}\right) + \left(\frac{40 \times 160^3}{12}\right) - \left[(I_{G3})_{XX} + A_3 h_3^2\right]$$

here, $(I_{G3})_{XX} = \frac{\pi R^4}{8} = \frac{\pi(40)^4}{8}$

$$A_3 = \frac{\pi R^2}{2} = \frac{\pi(40)^2}{2}$$

$$h_3 = 40 \, mm$$

$$I_{B-B} = \left(\frac{80 \times 160^3}{3}\right) + \left(\frac{40 \times 160^3}{12}\right) - \left[\frac{\pi(40)^4}{8} + \frac{\pi(40)^2}{2}(40)^2\right]$$

$$= 109.23 \times 10^6 + 13.65 \times 10^6 - 5.03 \times 10^6$$

$$= 117.85 \times 10^6 \, mm^4$$

Example: 8.8

Determine the moment of inertia of the given Fig. 8.17 about both centroidal axes.

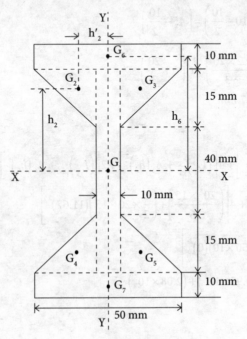

Fig. 8.17

Solution:

The given Fig. 8.17 is symmetrical about both the horizontal and vertical axes if taken at the middle.

Thus $\overline{X} = 0, \overline{Y} = 0$ and XX and YY axis will be the horizontal and vertical centroidal axis, respectively.

This figure comprises a rectangle (10 mm × 70 mm), four triangles (20 mm × 15 mm) and two rectangles (50 mm × 10 mm).

$$I_{XX} = \left[(I_{G_1})_{XX} + A_1 h_1^2\right]_{\text{rectangle}} + 4\left[(I_{G_2})_{XX} + A_2 h_2^2\right]_{\text{Triangle}} + 2\left[(I_{G_6})_{XX} + A_6 \cdot h_6^2\right]_{\text{rectangle}}$$

$$I_{XX} = \left[\left(\frac{10 \times 70^3}{12}\right) + (10 \times 90)(0)^2\right] + 4\left[\left(\frac{20 \times 15^3}{36}\right) + \left(\frac{1}{2} \times 20 \times 15\right)\left(35 - \frac{15}{3}\right)^2\right]$$

$$+ 2\left[\left(\frac{50 \times 10^3}{12}\right) + (50 \times 10)\left(45 - \frac{10}{2}\right)^2\right]$$

$$= (2.86 \times 10^5) + (5.48 \times 10^5) + (16.08 \times 10^5)$$

$$= 24.42 \times 10^5 \, mm^4$$

Note: $h_2 = \left(20 + 15 - \frac{15}{3}\right) = \left(35 - \frac{15}{3}\right)$

$h_6 = \left(20 + 15 + 10 - \frac{10}{2}\right) = \left(45 - \frac{10}{2}\right)$

$h_1' = 0, \quad h_2' = \left(25 - \frac{2 \times 20}{3}\right)$ or $\left(5 + \frac{20}{3}\right)$

$h_2' = 11.67 \, mm$

and $h_6' = 0$

$$I_{YY} = \left[(I_{G_1})_{YY} + A_1 (h_1')^2\right] + 4\left[(I_{G_2})_{YY} + A_2 (h_2')^2\right] + 2\left[(I_{G_6})_{YY} + A_6 (h_6')^2\right]$$

$$= \left[\left(\frac{10^3 \times 70}{12}\right) + 0\right] + 4\left[\left(\frac{20^3 \times 15}{36}\right) + \left(\frac{1}{2} \times 20 \times 15\right)(11.67)^2\right]$$

$$+ 2\left[\left(\frac{50^3 \times 10}{12}\right) + (50 \times 10)(0)^2\right]$$

$$= (0.058 \times 10^5) + (0.95 \times 10^5) + (2.08 \times 10^5)$$

$$= 3.09 \times 10^5 \, mm^4$$

Moment of Inertia

Example: 8.9

Determine the moment of inertia of the shaded area about axis AB and OC as shown in Fig. 8.18.

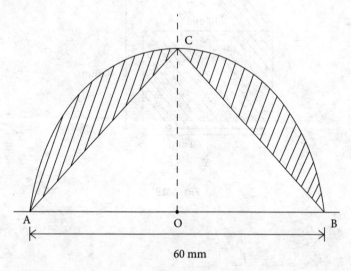

Fig. 8.18

Solution:

In this figure, a triangle of size (60 mm × 30 mm) is cut out from a semicircle of radius 30 mm.

$$I_{AB} = \left[I_{semicircle} - I_{triangle} \right]_{along\ x-axis}$$

$$= \left[\frac{\pi (30)^4}{8} \right] - \left[\frac{60 \times 30^3}{12} \right]$$

$$= 3.18 \times 10^5 - 1.35 \times 10^5$$

$$I_{AB} = 1.83 \times 10^5\ mm^4$$

$$I_{OC} = \left[I_{semicircle} - I_{triangleOAC} - I_{triangel\ OBC} \right]_{along\ y-axis}$$

$$= \left[\frac{\pi (30)^4}{8} \right] - 2 \left[\frac{30 \times 30^3}{12} \right]$$

$$= 3.18 \times 10^5 - 1.35 \times 10^5$$

$$= 1.83 \times 10^5\ mm^4$$

Example: 8.10

Determine the moment of inertia of the shaded area about vertical centroidal axis as shown in Fig. 8.19.

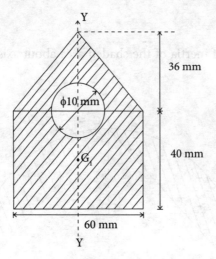

Fig. 8.19

Solution:

$$I_{YY} = \left[I_{rectangle} + 2.I_{right\ angle\ triangle} - I_{circle} \right]_{along\ y-axis}$$

$$= \left(\frac{40 \times 60^3}{12} \right) + 2 \left(\frac{36 \times 30^3}{12} \right) - \left(\frac{\pi (5)^4}{4} \right)$$

$$= (7.2 \times 10^5) + (1.62 \times 10^5) - (0.00491 \times 10^5)$$

$$= 8.82 \times 10^5\ mm^4$$

Example: 8.11

Determine the moment of inertia of Fig. 8.20 about both centroidal axes.

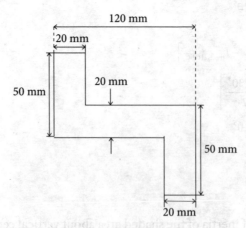

Fig. 8.20

Moment of Inertia

Solution:

The given figure is symmetrical about both horizontal and vertical axis, thus centroid of this section will lie at their intersection i.e. G_2

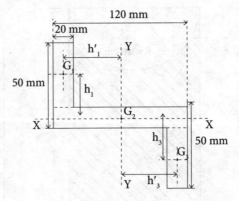

Fig. 8.20 (a)

Consider Fig. 8.20 (a) which comprises three rectangles of areas 20 mm × 30 mm, 120 mm × 20 mm and 20 mm × 30 mm, respectively.

$$h_1 = h_3 = \left(\frac{20}{2} + \frac{50-20}{2}\right), \quad h'_1 = h'_3 = \left(\frac{120}{2} - \frac{20}{2}\right)$$

$$= 25 \text{ mm} \quad\quad\quad = 50 \text{ mm}$$

$h_2 = 0$ and $h'_2 = 0$

Here two rectangles have identical moment of inertia about both axes.

$$I_{XX} = (I_1 + I_2 + I_3)_{\text{along x-axis}}$$

$$= \left[(I_{G1})_{XX} + A_1 h_1^2\right] + \left[(I_{G2})_{XX} + A_2 h_2^2\right] + \left[(I_{G3})_{XX} + A_3 h_3^2\right]$$

$$I_{XX} = 2\left[(I_{G1})_{XX} + A_1 h_1^2\right] + \left[(I_{G2})_{XX} + 0\right]$$

$$= 2\left[\left(\frac{20 \times 30^3}{12}\right) + (20 \times 30)(25)^2\right] + \left[\frac{120 \times 20^3}{12}\right]$$

$$= 9.2 \times 10^5 \text{ mm}^4$$

$$I_{YY} = (I_1 + I_2 + I_3)_{\text{along y-axis}}$$

$$= \left[(I_{G_1})_{YY} + A_1 (h'_1)^2\right] + \left[(I_{G_2})_{YY} + A_2 (h'_2)^2\right] + \left[(I_{G_3})_{YY} + A_3 (h'_3)^2\right]$$

$$= 2\left[(I_{G_1})_{YY} + A_1 (h'_1)^2\right] + \left[(I_{G_2})_{YY} + 0\right]$$

$$= 2\left[\left(\frac{20^3 \times 30}{12}\right) + (20 \times 30)(50)^2\right] + \left[\frac{120^3 \times 20}{12}\right]$$

$$= 59.2 \times 10^5 \text{ mm}^4$$

Example: 8.12

Determine the moment of inertia of the shaded area about axis AB as shown in the Fig. 8.21.

(MTU 2013–14, Ist sem)

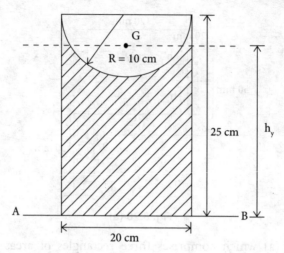

Fig. 8.21

Solution:

$$I_{AB} = \left[(I)_{rectangle} - I_{semicircle}\right]_{along\ AB} \quad \ldots (1)$$

$$(I_{rectangle})_{AB} = \left(\frac{20 \times 25^3}{3}\right) = 1.04 \times 10^5\ cm^4$$

$$(I_{semicricle})_{AB} = \left[(I_G)_{XX} + A.h^2\right]$$

where, $h_y = \left[25 - \left(\frac{4R}{3\pi}\right)\right]$

$$= \left[25 - \frac{40}{3\pi}\right]$$

$$= 20.76\ cm$$

$$(I_{semicricle})_{AB} = \left[0.11R^4 + \frac{\pi R^2}{2}(20.76)^2\right]$$

$$= \left[0.11(10)^4 + \frac{\pi(10)^2}{2}(20.76)^2\right]$$

$$= 0.69 \times 10^5\ cm^4$$

Substituting values in equation (1),

$$I_{AB} = 1.04 \times 10^5 - 0.69 \times 10^5$$

$$= 0.35 \times 10^5\ cm^4$$

Moment of Inertia

Example: 8.13

Determine moment of inertia of the shaded area in Fig. 8.22 about:

(i) Axis AB
(ii) Vertical centroidal axis

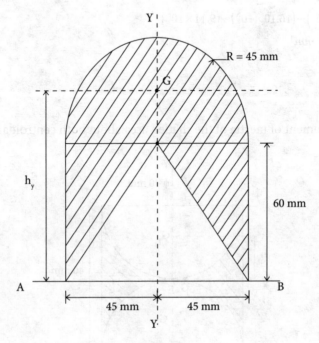

Fig. 8.22

Solution:

Here, $h_y = \left(60 + \dfrac{4 \times 45}{3\pi}\right)$
$= 79.1\ mm$

The moment of inertia of shaded area can be determined by considering rectangle (90 mm × 60 mm), semicircle of radius 50 mm and subtracting a triangle (90 mm × 60 mm)

(i)

$I_{AB} = \left[I_{rectangle} + I_{semicircle} - I_{Triangle}\right]_{\text{along x-axis}}$

$= \left[\dfrac{90 \times 60^3}{3}\right] + \left[0.11(45)^4 + \dfrac{\pi(45)^2}{2} \cdot (79.1)^2\right] - \left[\dfrac{90 \times 60^3}{12}\right]$

$= 64.8 \times 10^5 + 203.53 \times 10^5 - 16.2 \times 10^5$

$= 252.13 \times 10^5\ mm^4$

(ii) The moment of inertia of the shaded area about vertical centroidal axis will be given by

$$I_{YY} = \left[I_{rectangle} + I_{semicircle} - I_{triangle} \right]_{along\ y\text{-}axis}$$

$$I_{YY} = \left[\frac{90^3 \times 60}{12} \right] + \left[\frac{\pi(45)^4}{8} \right] - 2\left[\frac{60 \times 45^3}{12} \right]$$

$$= (36.45 \times 10^5) + (16.10 \times 10^5) - (9.11 \times 10^5)$$

$$I_{YY} = 43.44 \times 10^5\ mm^4$$

Example: 8.14

Determine the moment of inertia of the shaded area about both centroidal axis of Fig. 8.23.

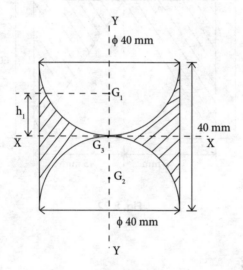

Fig. 8.23

Solution:

Here, $\bar{X} = 0$ and $\bar{Y} = 0$

$$I_{XX} = \left[I_{square} - 2.I_{semicircle} \right]_{along\ x\text{-}axis}$$

$$= \left(\frac{40 \times 40^3}{12} \right) - 2\left[(I_{G_1})_{XX} + A_1 h_1^2 \right]$$

$$= \left(\frac{40 \times 40^3}{12} \right) - 2\left[0.11(20)^4 + \frac{\pi(20)^2}{2} \left(20 - \frac{4 \times 20}{3\pi} \right)^2 \right]$$

$$= 2.13 \times 10^5 - 2.02 \times 10^5$$

$$= 0.11 \times 10^5\ mm^4$$

$$I_{YY} = \left[I_{square} - 2.I_{semicircle} \right]_{along\ y-axis}$$
$$= \left(\frac{40^3 \times 40}{12} \right) - 2\left(\frac{\pi(20)^4}{8} \right)$$
$$= 2.13 \times 10^5 - 1.26 \times 10^5$$
$$= 0.87 \times 10^5\ mm^4$$

Example: 8.15

Determine the M.I. of the shaded area about both the centroidal axes.

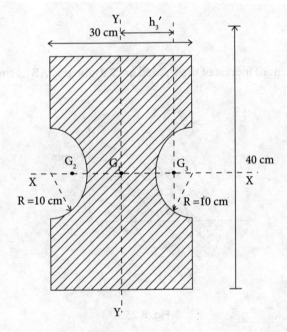

Fig. 8.24

Solution:

$\overline{X} = 0, \overline{Y} = 0$

Here G_1, G_2 and G_3 are centroids of all areas are lying on XX axis.

Thus $h_1 = h_2 = h_3 = 0$

However,

$$h_1' = 0 \text{ and } h_2' = h_3' = \left(15 - \frac{4 \times 10}{3\pi} \right)$$
$$= 10.76\ cm$$

$$I_{XX} = \left[I_{rectangle} - 2 \cdot I_{semicircle} \right]_{\text{along x-axis}}$$

$$= \left[\frac{30 \times 40^3}{12} \right] - 2 \cdot \left[\frac{\pi(10)^4}{8} \right]$$

$$= 1.52 \times 10^5 \, cm^4$$

$$I_{YY} = \left[I_{rectangle} - 2 \cdot I_{semicircle} \right]_{\text{along y-axis}}$$

$$= \left[\frac{40 \times 30^3}{12} \right] - 2 \cdot \left[0.11(10)^4 + \frac{\pi(10)^2}{2}(10.76)^2 \right]$$

$$= 0.514 \times 10^5 \, cm^4$$

Example: 8.16

Determine the moment of inertia of the shaded area about axis AB as shown in the Fig. 8.25.

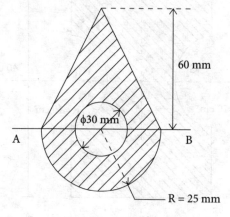

Fig. 8.25

Solution:

Here given figure comprises an isosceles triangle and a semicircle out of which a circle is cut-out.

Here AB is the base of the triangle and it is diameter of the both circular laminas,

$$I_{AB} = I_{triangle} + I_{semicircle} - I_{circle}$$

$$= \left(\frac{50 \times 60^3}{12} \right) + \left[\frac{\pi(25)^4}{4 \times 2} \right] - \left[\frac{\pi(15)^4}{4} \right]$$

$$= 9 \times 10^5 + 1.53 \times 10^5 - 0.398 \times 10^5$$

$$= 10.13 \times 10^5 \, mm^4$$

Moment of Inertia

Example: 8.17

A girder is composed of a web plate of size 500 mm × 10 mm, cover plates of size 400 mm × 10 mm and four angles of size 100 mm × 10 mm as shown in Fig. 8.26. Determine the M.I. of girder about both the centroidal axes.

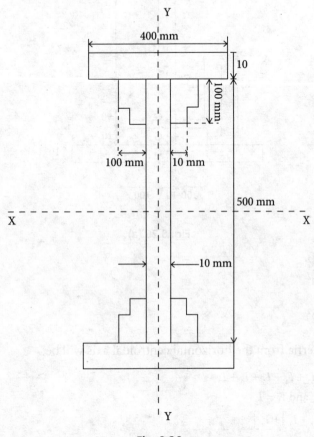

Fig. 8.26

Solution:

Here $\overline{X} = 0, \overline{Y} = 0$

The given figure can be considered as seven rectangles with different centroids as shown in Fig. 8.26 (a).

Here the plates 1, 2, 3 are identical with the plates 5, 6 and 7, respectively.

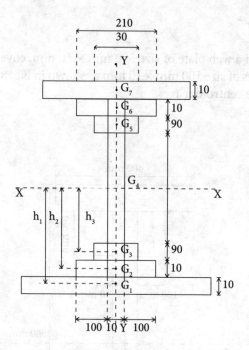

Fig. 8.26 (a)

$A_1 = 400 \times 10$ mm²
$A_2 = 210 \times 10$ mm²
$A_3 = 30 \times 90$ mm²
$A_4 = 300 \times 10$ mm²

The moment of inertia from the horizontal centroidal axis will be,

$I_{XX} = I_1 + I_2 + I_3 + I_4 + I_5 + I_6 + I_7$
as $I_1 = I_7, I_2 = I_6$ and $I_3 = I_5$

$I_{XX} = \left[2[I_1 + I_2 + I_3] + I_4 \right]_{\text{along x-axis}}$

$= 2\left[(I_{G_1})_{XX} + A_1 h_1^2 \right] + 2\left[(I_{G_2})_{XX} + A_2 h_2^2 \right] + 2\left[(I_{G_3})_{XX} + A_3 h_3^2 \right] +$
$\left[(I_{G_4})_{XX} + A_4 \cdot h_3^2 \right]$

Here, $h_1 = \left(\dfrac{520}{2} - \dfrac{10}{2} \right) = 255$ mm, $\quad h_1' = 0$

$h_2 = \left(\dfrac{520}{2} - 10 - \dfrac{10}{2} \right) = 245$ mm, $\quad h_2' = 0$

$h_3 = \left(\dfrac{520}{2} - 10 - 10 - \dfrac{90}{2} \right) = 195$ mm, $h_3' = 0$

and $h_4 = 0, \quad h_4' = 0$

Moment of Inertia

$$I_{XX} = 2\left[\left(\frac{400 \times 10^3}{12}\right) + (400 \times 10)(255)^2\right] + 2\left[\left(\frac{210 \times 10^3}{12}\right) + (210 \times 10)(245)^2\right]$$

$$+ 2\left[\left(\frac{30 \times 90^3}{12}\right) + (30 \times 90)(195)^2\right] + \left[\left(\frac{10 \times 300^3}{12}\right)\right]$$

$$I_{XX} = 10.04 \times 10^8 \, mm^4$$

$$I_{YY} = 2\left[\left(I_{G_1}\right)_{YY} + \left(I_{G_2}\right)_{YY} + \left(I_{G_3}\right)_{YY}\right] + \left(I_{G_4}\right)_{YY}$$

$$= 2\left[\left(\frac{400^3 \times 10}{12}\right) + \left(\frac{210^3 \times 10}{12}\right) + \left(\frac{30^3 \times 90}{12}\right)\right] + \left[\left(\frac{10^3 \times 300}{12}\right)\right]$$

$$= (1.23 \times 10^8) + (0.00025 \times 10^8) = 1.23 \times 10^8 \, mm^4$$

Example: 8.18

Determine the moment of inertia of shaded area about both centroidal axis for given Fig. 8.27.

(MTU 2011–12)

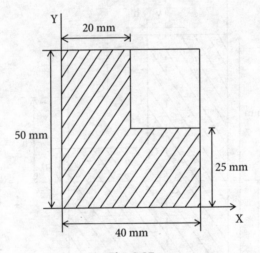

Fig. 8.27

Solution:

The centroid of shaded area will be determined first.

$a_1 = 40 \times 50 \, mm^2$	$x_1 = 20 \, mm$	$y_1 = 25 \, mm$
$a_2 = 20 \times 25 \, mm^2$	$x_2 = \left(20 + \dfrac{20}{2}\right)$ $= 30 \, mm$	$y_2 = \left(25 + \dfrac{25}{2}\right)$ $= 37.5 \, mm$

$$\overline{X} = \left(\frac{a_1 x_1 - a_2 x_2}{a_1 - a_2} \right)$$

$$= \left[\frac{(40 \times 50 \times 20) - (20 \times 25 \times 30)}{(40 \times 50) - (20 \times 25)} \right]$$

$= 16.67 \; mm$

$$\overline{Y} = \left(\frac{a_1 y_1 - a_2 y_2}{a_1 - a_2} \right)$$

$$= \left[\frac{(40 \times 50 \times 25) - (20 \times 25 \times 37.5)}{(40 \times 50) - (20 \times 25)} \right]$$

$= 20.83 \; mm$

Centroidal axis XX and YY are shown in Fig. 8.27(a).

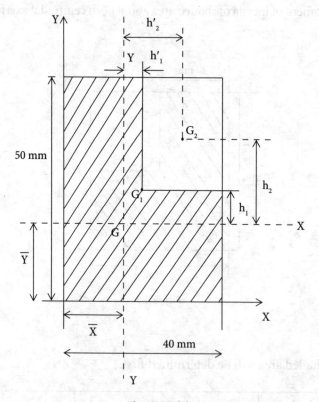

Fig. 8.27 (a)

$$I_{XX} = (I_1 - I_2)_{\text{along x-axis.}}$$

$$= \left[(I_{G_1})_{XX} + A_1 h_1^2 \right] - \left[(I_{G_2})_{XX} + A_2 h_2^2 \right]$$

here $h_1 = (25 - \overline{Y}) = (25 - 20.83)$
$$= 4.17 \, mm$$
$$h_2 = \left(50 - \overline{Y} - \frac{25}{2}\right) = 16.67 \, mm$$
$$= \left[\left(\frac{40 \times 50^3}{12}\right) + (40 \times 50)(4.17)^2\right] - \left[\left(\frac{20 \times 25^3}{12}\right) + (20 \times 25)(16.67)^2\right]$$
$$I_{XX} = 2.86 \times 10^5 \, mm^4$$
$$I_{YY} = (I_1 - I_2)_{\text{along y-axis}}$$
$$= \left[(I_{G_1})_{YY} + A_1 (h'_1)^2\right] - \left[(I_{G_2})_{YY} + A_2 (h'_2)^2\right]$$
$$h'_1 = (20 - \overline{X}) = (20 - 16.67) = 3.33 \, mm$$
$$h'_2 = \left(40 - \overline{X} - \frac{20}{2}\right) = (40 - 16.67 - 10) = 13.33 \, mm$$
$$I_{YY} = \left[\left(\frac{40^3 \times 50}{12}\right) + (40 \times 50)(3.33)^2\right] - \left[\left(\frac{20^3 \times 25}{12}\right) + (20 \times 25)(13.33)^2\right]$$
$$= 1.83 \times 10^5 \, mm^4$$

Example: 8.19

Determine the moment of inertia of the area under the curve as shown in Fig. 8.28 about both axes.

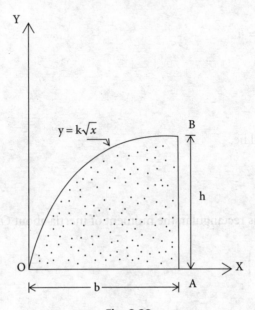

Fig. 8.28

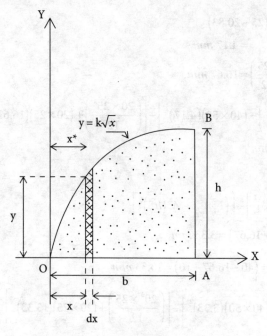

Fig. 8.28 (a)

Solution:

Consider an elemental strip of width dx, at distance x from OY axis as shown in Fig. 8.28 (a)
as $y = k\sqrt{x}$

at point B, $x = b$ and $y = h$

Thus,

$h = k\sqrt{b}$

or, $k = h/\sqrt{b}$

The curve equation will be,

$y = \dfrac{h}{\sqrt{b}}(x)^{1/2}$

As the elemental strip is rectangular, the moment of inertia about OX axis will be,

$d(I_{ox}) = dI_x = \left(\dfrac{dx \cdot y^3}{3}\right)$

or,

Moment of Inertia

$$I_{OX} = I_X = \int_0^b \frac{dx \cdot h^3}{3(b)^{3/2}} (x)^{3/2}$$

$$I_X = \frac{h^3}{3b^{3/2}} \int_0^b x^{3/2} \cdot dx$$

$$= \frac{h^3}{3b^{3/2}} \left[\frac{x^{5/2}}{5/2} \right]_0^b$$

$$I_X = \frac{2}{15} bh^3$$

The moment of inertia about OY axis will be,

$$d(I_{OY}) = dI_y = dA \cdot (x^*)^2$$

where $dA = y \cdot dx = \frac{h}{\sqrt{b}} \cdot x^{1/2} \cdot dx$

and $x^* = x + \frac{dx}{2} \simeq x$

$$I_{OY} = I_Y = \int_0^b dA \cdot (x^*)^2$$

$$= \int_0^b \frac{h}{\sqrt{b}} \cdot x^{1/2} dx \cdot x^2$$

$$I_y = \frac{h}{\sqrt{b}} \int_0^b x^{5/2} \cdot dx$$

$$= \frac{h}{\sqrt{b}} \cdot \left[\frac{x^{7/2}}{7/2} \right]_0^b$$

$$I_Y = \frac{2}{7} \cdot b^3 \cdot h$$

Note: Integration limit in both cases is 0 to b as integration is with respect to differential element dx.

Example: 8.20

Determine the moment of inertia of the area under the curve as shown in Fig. 8.29 about both axes.

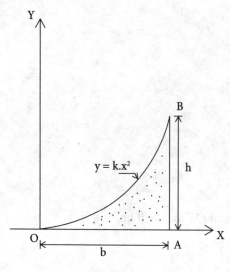

Fig. 8.29

Solution:

For point B,

$y = k \cdot x^2$

$h = k \cdot b^2$

i.e., $k = h/b^2$

The equation will become, $y = \dfrac{h}{b^2} \cdot x^2$

Consider an elemental strip of width dx at distance x from OY axis as shown in Fig. 8.29 (a).

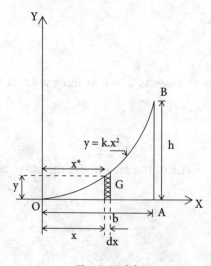

Fig. 8.29 (a)

Moment of Inertia

The moment of inertia of the strip from OX axis will be,

$$d(I_{OX}) = d(I_X) = \left(\frac{dx \cdot y^3}{3}\right)$$

$$I_{OX} = I_X = \int_0^b \frac{dx}{3} \cdot \left(\frac{h}{b^2} \cdot x^2\right)^3$$

$$= \frac{h^3}{3b^6} \int_0^b x^6 \cdot dx$$

$$I_X = \frac{h^3}{3b^6} \left[\frac{x^7}{7}\right]_0^b$$

$$I_X = \left(\frac{b \cdot h^3}{21}\right)$$

The moment of inertia of the strip from OY axis will be,

$$d(I_{OY}) = dI_y = dA \cdot (x^*)^2$$

where $dA = y \cdot dx$

$$= \frac{h}{b^2} \cdot x^2 \cdot dx$$

and $x^* = x + \dfrac{dx}{2}$

$$\simeq x$$

$$I_Y = \int_0^b dA \cdot (x^*)^2$$

$$= \int_0^b \frac{h}{b^2} \cdot x^2 dx \cdot x^2$$

$$= \frac{h}{b^2} \int_0^b x^4 \cdot dx$$

$$= \frac{h}{b^2} \left[\frac{x^5}{5}\right]_0^b$$

$$I_Y = \left(\frac{b^3 h}{5}\right)$$

8.9 Mass Moment of Inertia of Prismatic Bar, Rectangular Plate, Circular Disc, Solid Cone and Sphere about Axis of Symmetry

Example: 8.21

Determine the mass moment of inertia of a prismatic bar of mass M and length L about Y axis from its:

(i) one end
(ii) centre of mass

Solution:

(i) M.I. About Y-axis from end A–A,

Consider an elementary mass dm of length dx at distance x from end AA as shown in Fig. 8.30. Let the density of material of rod is ρ and cross-section area is A.

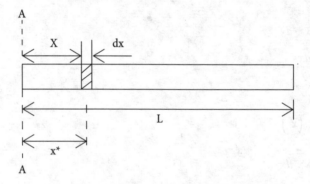

Fig. 8.30

The mass moment of inertia of elemental mass dm from end AA will be

$$I_{A\text{-}A} = dm \cdot (x^*)^2$$

where $dm = \rho \times dv = \rho \times A \times dx$

$$x^* = x + \frac{dx}{2} \simeq x$$

thus $I_{AA} = \int_0^L \rho \cdot A \cdot dx \cdot x^2$

$$= \rho \cdot A \cdot \left[\frac{x^3}{3}\right]_0^L$$

$$I_{AA} = \frac{\rho \cdot A \cdot L^3}{3}$$

Moment of Inertia

The mass of the rod, $M = \rho \cdot A \cdot L$

$$I_{AA} = \frac{\rho \cdot A \cdot L}{3} \times L^2$$

$$I_{AA} = \frac{M \cdot L^2}{3}$$

(ii) M.I. about YY axis from the centre of mass,

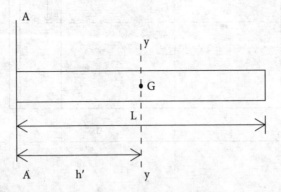

Fig. 8.30 (a)

Using the parallel axis theorem,

$$I_{AA} = (I_G)_{YY} + M(h')^2$$

$$h' = L/2$$

$$\left(\frac{ML^2}{3}\right) = (I_G)_{YY} + M \cdot \left(\frac{L}{2}\right)^2$$

$$(I_G)_{YY} = \frac{ML^2}{12}$$

Example: 8.22

Determine mass moment of inertia of a rectangular plate of mass M, size $a \times b$ and thickness t about its:

(i) Centroidal axis
(ii) Polar axis

Solution:

(i) Consider an elemental strip of mass dm, width dy at a distance y from XX axis as shown in Fig. 8.31. Let the thickness of plate is t.

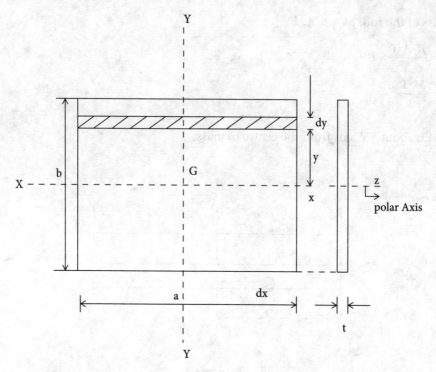

Fig. 8.31

$$I_{XX} = \int_{-b/2}^{+b/2} dm \cdot (y^*)^2$$

where $dm = \rho \times dv = \rho \times a \cdot dy \cdot t$

and $y^* = y + \dfrac{dy}{2} \simeq y$

$$I_{XX} = \int_{-b/2}^{+b/2} \rho \cdot a \cdot dy \cdot t \cdot y^2$$

$$= \rho \cdot a \cdot t \left[\dfrac{y^3}{3} \right]_{-b/2}^{b/2}$$

$$= \dfrac{\rho \cdot a \cdot t}{3} \left[\dfrac{b^3}{8} + \dfrac{b^3}{8} \right]$$

$$I_{XX} = \dfrac{\rho \cdot a \cdot t \cdot b^3}{12} = \left(\dfrac{\rho \cdot at \cdot b}{12} \right) \cdot b^2$$

$$I_{XX} = \dfrac{M \cdot b^2}{12}$$

Similarly, if the elemental strip is taken parallel to Y axis,

Moment of Inertia

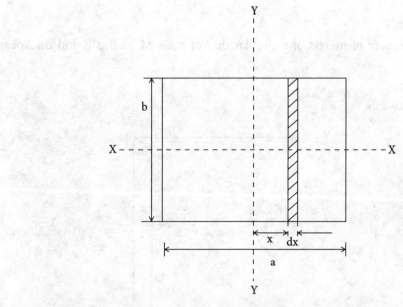

Fig. 8.31 (a)

$$I_{YY} = \int_{-a/2}^{a/2} dm \cdot (x^*)^2$$

$$dm = \rho \cdot b \cdot dx \cdot t$$

$$x^* = x + \frac{dx}{2}$$

$$\simeq x$$

$$I_{YY} = \int_{-a/2}^{a/2} \rho \cdot b \cdot dx \cdot t (x)^2$$

$$I_{YY} = \rho \cdot b \cdot t \left[\frac{x^3}{3}\right]_{-a/2}^{+a/2}$$

$$= \frac{\rho \cdot bt \cdot a^3}{12} = \left(\frac{\rho abt}{12}\right) \cdot a^2$$

$$I_{YY} = \frac{M \cdot a^2}{12}$$

(ii) M.I. about polar axis can be determined by using the perpendicular axis theorem,

$$I_{ZZ} = I_{XX} + I_{YY}$$

$$I_{ZZ} = \frac{M}{12}(a^2 + b^2)$$

Example: 8.23

Determine the moment of inertia of a circular disc of mass M, radius R and thickness t about:

(i) the polar axis
(ii) the centroidal axis

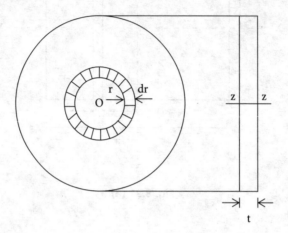

Fig. 8.32

Solution:

(i) About the polar axis:

Consider an elemental ring of mass 'dm' at radius 'r' from the polar axis as shown in Fig. 8.32. Let the width of elemental ring is dr and thickness of the plate is t.

The moment of inertia of elemental ring dm will be

$$I_{zz} = \int_0^R dm \cdot r^2$$

Here the limits vary from 0 to R as the ring varies from centre of disc to the circumference. Let the density of material of disc is ρ,

$$dm = \rho \times dv = \rho \times dA \times t = \rho \times 2\pi r \times dr \times t$$

$$I_{zz} = \int_0^R \rho \cdot 2\pi r dr t \cdot r^2$$

$$= 2\pi \rho t \int_0^R r^3 \cdot dr = 2\pi \rho t \left[\frac{r^4}{4} \right]_0^R$$

$$I_{zz} = 2\pi \rho t \left[\frac{R^4}{4} \right]$$

$$I_{zz} = \frac{2\pi \rho t R^4}{4}$$

Moment of Inertia

But the mass of disc, $M = \rho \times V = \rho \times A \times t = \rho \cdot \pi R^2 t$

$$I_{ZZ} = \frac{2(\pi \rho t R^2) \cdot R^2}{4}$$

$$I_{ZZ} = \frac{MR^2}{2}$$

(ii) About the centroidal axis:
We know that

$$I_{XX} + I_{YY} = I_{ZZ}$$

$$I_{XX} + I_{YY} = \frac{MR^2}{2}$$

Since the disc is symmetrical about XX and YY axis,

$$I_{XX} = I_{YY}$$

thus $2I_{XX} = \dfrac{MR^2}{2}$

$$I_{XX} = I_{YY} = \frac{MR^2}{4}$$

Example: 8.24

Determine the mass moment of inertia of a right circular solid cone of height h and base radius R about axis of symmetry or rotation.

Solution:

Consider an elemental disc of mass dm and radius r is lying at distance x from the vertex is as shown in Fig. 8.33.

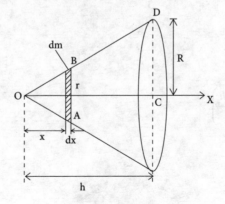

Fig. 8.33

The moment of inertia of the elemental disc about the polar axis is given by

$$\left(\frac{dm \cdot r^2}{2}\right)$$

where $dm = \rho \times dv = \rho \times dx \times dA$
$= \rho \times dx \times \pi r^2$

thus, $I_{xx} = \int_0^h \dfrac{\rho \cdot dx \cdot \pi r^4}{2}$

$$I_{xx} = \frac{\rho \cdot \pi}{2} \int_0^h r^4 \cdot dx \qquad \ldots\ldots(1)$$

Δ OAB and Δ OCD are similar triangle,

$$\frac{AB}{OA} = \frac{CD}{OC}$$

$$\frac{r}{x} = \frac{R}{h}$$

$$r = \left(\frac{R \cdot x}{h}\right)$$

Substituting value of r in equation (1),

$$I_{xx} = \frac{\rho \pi}{2} \int_0^h \left(\frac{R \cdot x}{h}\right)^4 \cdot dx$$

$$= \frac{\rho \pi}{2} \cdot \frac{R^4}{h^4} \left[\frac{x^5}{5}\right]_0^h$$

$$I_{xx} = \frac{\rho \pi R^4}{2h^4} \times \frac{h^5}{5}$$

Mass of solid cone is given by $M = \rho \times \dfrac{1}{3}\pi R^2 h$

$$I_{xx} = \frac{\rho \pi R^4 h^5}{10 h^4} \times \left(\frac{M}{M}\right)$$

$$= \frac{\rho \pi R^4 h}{10} \times \frac{M}{\left(\dfrac{\rho \pi R^2 h}{3}\right)}$$

$$I_{xx} = \frac{3}{10} MR^2$$

Moment of Inertia

Example: 8.25

Determine the mass moment of inertia of a solid sphere of mass M and radius R about its axis of rotation.

Solution:

Consider an elemental disc of radius r at a distance y. Let the thickness of the disc is dy as shown in Fig. 8.34. Here all the axis *XX*, *YY* and *ZZ*, are axis of rotation or diametrical axis or polar axis. If the mass of elemental disc is dm and density is ρ then moment of inertia from *YY* axis will be,

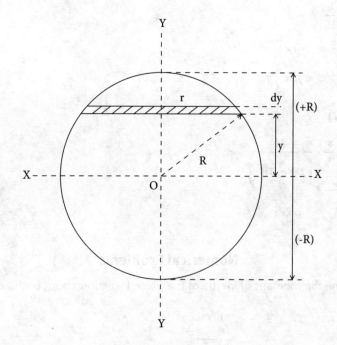

Fig. 8.34

$$I_{YY} = \int \left(\frac{dm \cdot r^2}{2} \right)$$

$$= \int_{-R}^{+R} \frac{\rho \cdot dy \pi r^4}{2}$$

$$= 2 \int_{-R}^{+R} \frac{\rho \cdot dy \pi (R^2 - y^2)^2}{2}$$

$$I_{YY} = \frac{2\rho \cdot \pi}{2} \int_0^{+R} \left(R^4 + y^4 - 2R^2 \cdot y^2\right) dy$$

$$= \rho\pi \left[R^4 \int_0^{+R} dy + \int_0^{+R} y^4 \cdot dy - 2R^2 \int_0^{+R} y^2 dy \right]$$

$$= \rho\pi \left[R^4 [y]_0^R + \left[\frac{y^5}{5}\right]_0^R - 2R^2 \left[\frac{y^3}{3}\right]_0^R \right]$$

$$= \rho\pi \left[R^5 + \frac{R^5}{5} - \frac{2R^5}{3} \right]$$

$$= \rho\pi R^5 \left[\frac{15 + 3 - 10}{15}\right]$$

$$I_{YY} = \rho\pi R^5 \times \frac{8}{15} \qquad \ldots (1)$$

where $dm = \rho \times dv$
$= \rho \cdot dy \times dA$
$= \rho \cdot dy \times \pi r^2$
and $R^2 = y^2 + r^2$
$r^2 = (R^2 - y^2)$

mass of sphere, $M = \rho \times V$

$$= \rho \times \frac{4}{3}\pi R^3$$

From equation (1),

$$I_{YY} = \rho\pi R^5 \times \frac{8}{15} \times \frac{M}{\left(\rho \times \frac{4}{3}\pi R^3\right)}$$

$$I_{YY} = \frac{2}{5} MR^2$$

Numerical Problems

N 8.1 Determine the moment of inertia of the given L-section about both centroidal axes.

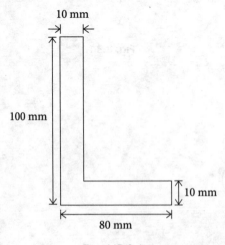

Fig. NP 8.1

Moment of Inertia

N 8.2 Determine the moment of inertia of T-section about –
  (i) AB as shown in the figure
  (ii) both the centroidal axes.

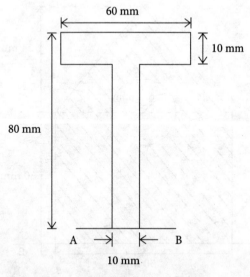

Fig. NP 8.2

N 8.3 Determine the moment of inertia of the I-section about both the centroidal axes.

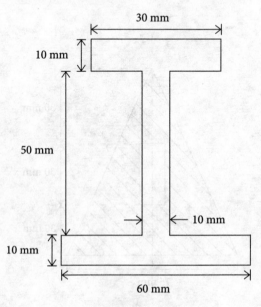

Fig. NP 8.3

N 8.4 Determine the moment of inertia of the shaded section about:
(i) AB as shown in Fig. NP 8.4
(ii) both the centroidal axes

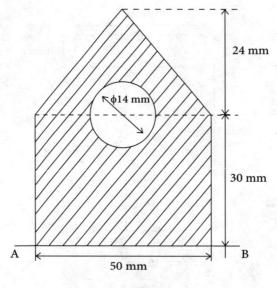

Fig. NP 8.4

N 8.5 Determine the moment of inertia of shaded area about the vertical centroidal axis for Fig. NP 8.5.

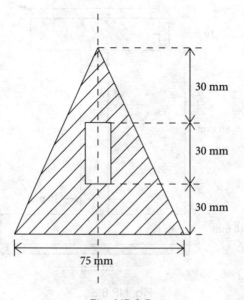

Fig. NP 8.5

Moment of Inertia

N 8.6 Determine the moment of inertia of shaded section about AB as shown in Fig. NP 8.6.

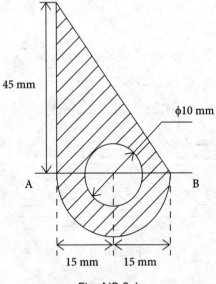

Fig. NP 8.6

N 8.7 Determine the M.I. of Fig. NP 8.7 about both the centroidal axes.

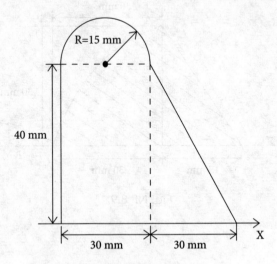

Fig. NP 8.7

N 8.8 Determine M.I. of the given shaded area about AB as shown in Fig. NP 8.8.

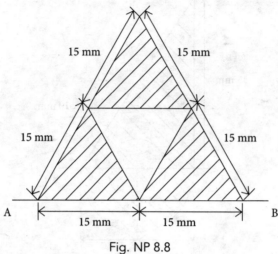

Fig. NP 8.8

N 8.9 Determine M.I. of the shaded area about AB as shown in Fig. NP 8.9.

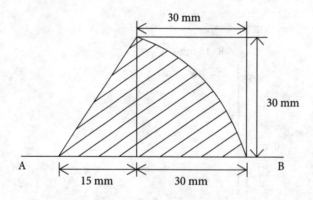

Fig. NP 8.9

Moment of Inertia

N 8.10 Determine the moment of inertia of the area under the curve given by equation, $y = h \cdot \sin\left(\dfrac{\pi x}{b}\right)$ from X-axis as shown in Fig. NP 8.10.

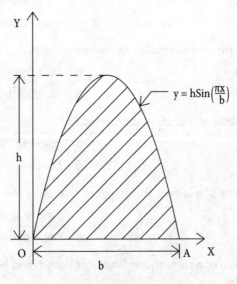

Fig. NP 8.10

N 8.11 Determine moment of inertia of the area under the curve given by equation, $x = k \cdot y^2$ from X-axis as shown in Fig. NP 8.11.

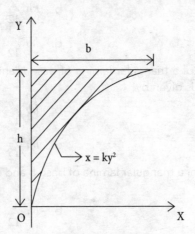

Fig. NP 8.11

Multiple Choice Questions

1. Area moment of inertia of a lamina has always
   a. Positive value
   b. Negative value
   c. zero
   d. none of these

2. Area moment of inertia of a lamina is not concerned with
   a. bending
   b. buckling
   c. twisting
   d. none of these

3. Mass moment of inertia of a body is concerned with
   a. twisting
   b. buckling
   c. motion
   d. bending

4. The moment of inertia of a lamina about axis of symmetry remains
   a. Positive
   b. Negative
   c. zero
   d. none of these

5. The moment of inertia of a rectangular lamina of base b and height h from its horizontal centroidal axis will be
   a. $bh^3/3$
   b. $bh^3/12$
   c. $bh^3/36$
   d. $hb^3/12$

6. Moment of inertia of a rectangular lamina of base b and height h from its base will be
   a. $bh^3/3$
   b. $bh^3/12$
   c. $bh^3/36$
   d. $hb^3/12$

7. The moment of inertia of a triangular lamina of base b and height h from its centroidal axis parallel to its base is given by
   a. $bh^3/3$
   b. $hb^3/12$
   c. $bh^3/36$
   d. $bh^3/12$

8. The moment of inertia of a triangular lamina of base b and height h from its base is given by
   a. $bh^3/3$
   b. $bh^3/12$
   c. $bh^3/36$
   d. $hb^3/12$

Moment of Inertia

9. The polar moment of inertia of a circular area of radius *r* is given by
   a. $\pi r^4 / 2$
   b. $\pi r^4 / 4$
   c. $\pi r^4 / 8$
   d. $\pi r^4 / 16$

10. The moment of inertia of a quarter circular area of radius *r* from its centroidal axis parallel to the diametrical axis is given by
    a. $\pi r^4 / 16$
    b. $\pi r^4 / 8$
    c. $0.11 r^4$
    d. $0.055 r^4$

Answers

1. a 2. d 3. c 4. a 5. b 6. a 7. c 8. b 9. a 10. d

Chapter 9

Shear Force and Bending Moment Diagrams

9.1 Beams

Beams are widely used as a structural member to support heavy transverse load throughout its length. Generally beams are used with rectangular cross section and I-section out of circular, square and T-section. Beam consists of very large length as compared to width and depth and supported by columns at its ends. Beams are made of timber, reinforced concrete and steel. Skyscrapers are constructed with the help of beams and columns.

9.2 Types of Beams

Cantilever Beam: The beam which is fixed at one end and remains free at other end called cantilever beam as shown in Fig. 9.1 (a).
Propped Cantilever Beam: The beam which is fixed at one end and remains simply supported at other end called propped cantilever beam as shown in Fig. 9.1 (b).

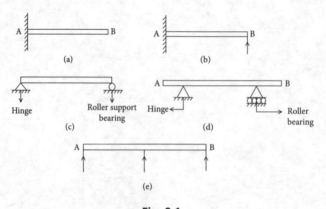

Fig. 9.1

Simply Supported Beam: The beam which is simply supported at both ends called simply supported beam as shown in Fig. 9.1 (c). The supports used are combination of hinge and roller bearings.

Overhanging Beam: The beam which extends beyond its one or two supports is called overhanging beam as shown in Fig. 9.1 (d). The supports used are combination of hinge and roller bearings.

Continuous Beam: The beam which is supported throughout its length by more than two supports called continuous beam as shown in Fig. 9.1 (e).

9.3 Types of Loads and Beams

There are four types of loading acts on beams. The beam may be subjected to any type of single loading or in combination of two or more. The different types of loading are described below:

Point load or Concentrated Load: This kind of load is concentrated at a point or acts at a point hence called as Point load or Concentrated load. It is represented by an arrow. Consider Fig. 9.2 (a) where W kN Point load is acting on a simply supported beam.

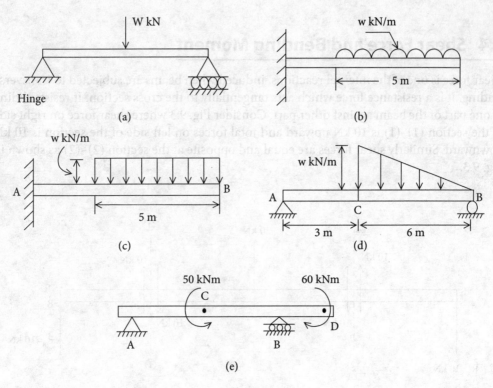

Fig. 9.2

Uniformly Distributed Load: It is also abbreviated as UDL. This kind of load is distributed over certain distance with constant intensity. It can be represented by two ways as shown in Fig. 9.2 (b) and (c) where load of constant intensity w kN/m is distributed over certain distance 5 m on a cantilever beam. For analysis purpose total load is computed i.e., area of rectangle (w*5) and assumed that acting on the middle of length of its length (Centroid of rectangle) i.e., 2.5 m from end B.

Uniformly Varying Load: It is also abbreviated as UVL. This kind of load is distributed over certain distance with uniformly varying intensity. Consider Fig. 9.2 (d) where load of uniformly varying intensity w kN/m is distributed over certain distance 6 m on a simply supported beam. For analysis purpose total load is computed i.e., area of triangle $\left(\frac{1}{2}*w*6\right)$ and assumed that acting on the Centroid of triangle i.e., 5 m from end A $\left(3+\frac{1}{3}*6=5\right)$, 4 m from end $\left(\frac{2}{3}*6=4\right)$ B and 2 m from $C\left(\frac{1}{3}*6=2\right)$.

Applied External Moment or Couple: This kind of loading tends to rotate the beam as shown in Fig. 9.2 (e) where two moments 50 kNm and 60 kNm are acting at points C and D anticlockwise and clockwise respectively an on an overhanging beam.

9.4 Shear Force and Bending Moment

Shear force is one of the internal reactions, induced when beams are subjected to transverse loading. It is a resistance force which acts tangentially to the cross section. It resists sliding of one part of the beam against other part. Consider Fig. 9.3 where shear force on right side of the section (1)–(1) is 10 kN upward and total forces on left side of the section is 10 kN downward. Similarly shear forces are equal and opposite at the section (2)–(2) as shown in Fig. 9.3.

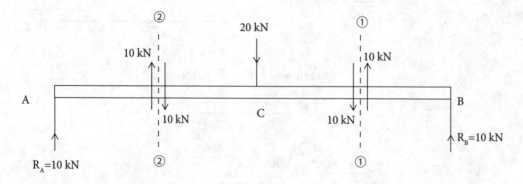

Fig. 9.3

Shear Force and Bending Moment Diagrams

Bending moment is one of the internal reactions, induced when beams are subjected to transverse loading. It acts along the longitudinal axis and resist bending due to transverse loading. Bending moment is of two types:

(i) Sagging moment
(ii) Hogging moment

Sagging moment is one where the upper fibre of the beam becomes concave and the Hogging moment is one where the upper fibre of the beam becomes convex. The sagging moment and hogging moment are shown in Fig. 9.4.

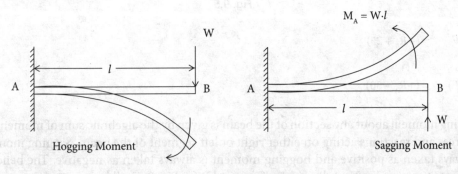

Fig. 9.4

9.5 Shear Force Diagram (SFD) and Bending Moment Diagram (BMD)

These diagrams show the variation of shear force and bending moment throughout the length of the beam and called SFD (also S.F.D.) and BMD (also B.M.D.), respectively. Such diagrams play crucial role in the safe designing of beams. The value and shape of shear force and bending moment in SFD and BMD give necessary information to the designer to design the beam efficiently and economically.

9.6 Sign Convention of Shear Force and Bending Moment in SFD and BMD

Shear force about any section of the beam is given by the algebraic sum of all forces and reactions acting on either right or left segment of the beam. Generally about a section, downward forces acting on right segment of the beam are taken as positive and vice versa. Consider simple supported beam carrying two point loads and reactions R_A and R_B as shown in Fig. 9.5. The shear force at section X_1–X_1 at distance x_1 from end B will be

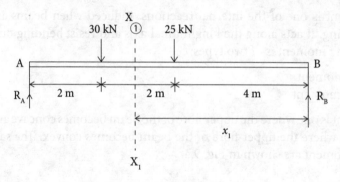

Fig. 9.5

$(SF)_{x_1 x_1} = (-R_B + 25)$
or,
$(SF)_{x_1 x_1} = (+R_A - 30)$

Bending moment about any section of the beam is given by the algebraic sum of moments of all forces and reactions acting on either right or left segment of the beam. Sagging moment is always taken as positive and hogging moment is always taken as negative. The bending moment at section X_1–X_1 at distance x_1 from end B for Fig. 9.5 will be

$(BM)_{x_1 x_1} = (+R_B \cdot x_1 - 25(x_1 - 4))$
or
$(BM)_{x_1 x_1} = +R_A \cdot (8 - x_1) - 30(6 - x_1)$

9.7 Relationship between Load Intensity (w), Shear Force (S) and Bending Moment (M)

The load intensity (w), shear force (S) and bending moment (M) is given by:

$$w = \frac{dS}{dx} \text{ and } S = \frac{dM}{dx}$$

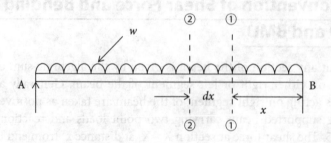

Fig. 9.6

Shear Force and Bending Moment Diagrams

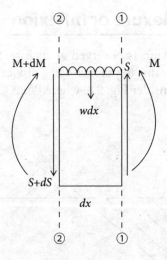

Fig. 9.7

Proof:

Consider Fig. 9.6 where a simple supported beam is supporting a uniformly distributed load of intensity 'w' throughout its entire span. Let the elemental length dx lies at distance x from end B. If the shear force and bending moment at distance x are S and M, respectively and at distance $(x + dx)$ are $(S + dS)$ and $(M + dM)$, respectively, as shown in Fig. 9.7.

Applying equations of equilibrium on F.B.D. of elemental length dx,
$\Sigma Y = 0$, $S + dS + w \cdot dx - S = 0$

$$w = -\frac{dS}{dx}$$

where $\frac{dS}{dx}$ represents the slope of SFD; thus the slope of SFD depends upon the intensity of loading acting on the beam.

$(\Sigma M)_{\text{about section (2)-(2)}} = 0$,

$$-(M + dM) + M + S \cdot dx - w dx \cdot \frac{dx}{2} = 0$$

$-dM + S \cdot dx = 0,$ where $w \cdot dx \cdot \frac{dx}{2}$ will be negligeble being a very-2 small value

$$S = \frac{dM}{dx}$$

where $\frac{dM}{dx}$ represents the slope of BMD; thus the slope of BMD depends upon the intensity of shear force acting on the beam.

9.8 Point of Contraflexure or Inflexion

Point of contraflexure or inflexion is observed at such sections where bending moment gradually varies from negative value to positive value or vice versa. At such point the value of bending moment is zero. Consider Fig. 9.8, where Point C is the point of contraflexure.

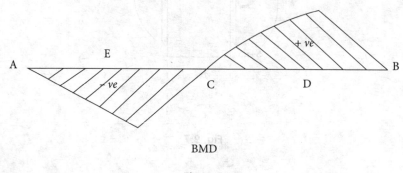

Fig. 9.8

9.9 Characteristics of SFD and BMD

SFD and BMD diagrams show some unique characteristics for different types of beam and loading. Some of the characteristics are given below:

a. SFD contains a straight line for span containing point load, a linear or tapered line for span containing UDL and a parabolic curve for span containing UVL.
b. BMD contains a linear or tapered line for span containing point load and a parabolic curve for span containing UDL or UVL.
c. Concentrated load at its point of application, sharply affects the SFD by its magnitude however its effect is directly proportional to the distance in BMD.
d. Moment or couple acting on a point does not affect SFD however, affects the BMD sharply by its magnitude at its point of application.
e. The bending moment is observed maximum where SFD changes its sign gradually from negative value to positive value or vice versa.
f. The bending moment is always observed zero at the free end of a cantilever and zero at both ends of a simple supported or overhanging beam.
g. The value of shear force and bending moment is always observed maximum at the fixed end of a cantilever beam.
h. The slope of SFD directly depends on the intensity of loading.
i. The slope of BMD directly depends on the value of shear force.

Shear Force and Bending Moment Diagrams

Example: 9.1

Draw S.F.D. and B.M.D. for simple supported beam as shown in Fig. 9.9.

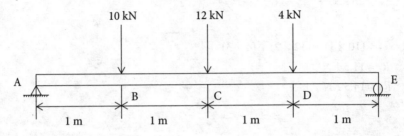

Fig. 9.9

Solution:

First, support reactions are to be determined. Consider F.B.D. of the simple supported beam AE.

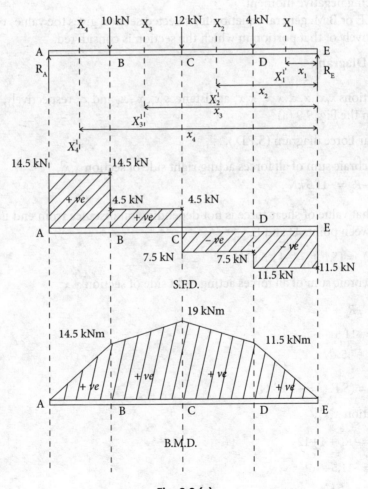

Fig. 9.9 (a)

$\Sigma Y = 0$,

$$R_A + R_E = 10 + 12 + 4$$
$$= 26 \text{ kN} \quad \quad \ldots\ldots (1)$$

$\Sigma M_A = 0$

$$R_E \times 4 = (10 \times 1) + (12 \times 2) + (4 \times 3)$$
$$R_E = 11.5 \text{ KN}$$
$$R_A = 14.5 \text{ KN}$$

Note:

(i) For S.F.D., all downward forces acting on right side of a selected section are taken as positive while upward forces are taken as negative.

(ii) In B.M.D., force causing concavity of beam (Sagging) are considered as producing positive moment while force causing convex shape of beam are considered as producing negative moment.

(iii) Each S.F. or B.M. general equation for selected section gives the values of S.F. or B.M. respectively of that portion in which the section is considered.

Shear Force Diagram:

Consider sections $x_1 x_1^1, x_2 x_2^1, x_3 x_3^1, x_4 x_4^1$ at distances x_1, x_2, x_3 and x_4, respectively, from the end E as shown in the Fig. 9.9 (a)

To draw Shear Force diagram (S.F.D.),

Consider algebraic sum of all forces acting right side of section $x_1 x_1^1$,

$$(S.F)_{x_1 x_1^1} = -R_E = -11.5 \text{ kN}$$

This shows that value of shear force is not depending on distance from end E and remains constant between points D and E.

Thus $(S.F.)_E = (S.F.)_D = -11.5 \text{ kN}$

Consider algebraic sum of all forces acting right side of section $x_2 x_2^1$,

$$(S.F.)_{x_2 x_2^1} = -R_E + 4$$
$$= -11.5 + 4$$
$$= -7.5 \text{ kN}$$

Thus, $(S.F.)_D = (S.F.)_C = -7.5 \text{ kN}$

Consider section $x_3 x_3^1$,

$$(S.F.)_{x_3 x_3^1} = -R_E + 4 + 12$$
$$= -11.5 + 16$$
$$= +4.5 \text{ kN}$$

$$(S.F)_C = (S.F.)_B = +4.5 \text{ kN}$$

Similarly, section $x_4 x_4^1$,

$$(S.F.)_{x_4 x_4^1} = -11.5 + 4 + 12 + 10$$
$$= 14.5\, kN$$
$$(SF)_B = (S.F.)_A = +14.5\, kN$$

Bending Moment Diagram:

To draw B.M.D., moments will be taken from axes $X_1 X_1^1, X_2 X_2^1, X_3 X_3^1$ and $X_4 X_4^1$, respectively for all forces acting before corresponding section form the point E,

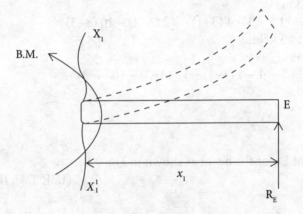

Fig. 9.9 (b)

$$(B.M.)_{X_1 X_1^1} = +R_E \times x_1$$
$$= +11.5 x_1$$

We can observe that this equation is dependent of distance x_1 from end E. B.M. is taken as positive as it is sagging moment.

Thus, $(BM)_E = 0$, $(B.M.)_D = +11.5$ kNm

Similarly,

$$(B.M)_{X_2 X_2^1} = +R_E \cdot x_2 - 4(x_2 - 1)$$

Here B.M. due to 4 kN will negative as it is hogging moment.

For B.M. at D put $x_2 = 1$ m,
$(B.M.)_D = +11.5 \times 1 - 4(1-1)$
$= +11.5$ kNm (Unchanged)

For B. M. at C put $x_2 = 2$ m,
$(B. M.)_C = +11.5 \times 2 - 4(2-1)$
$= 19$ kNm

$$(B.M)_{X_3,X_3^1} = +R_E \times x_3 - 4(x_3-1) - 12(x_3-2)$$
$$= 11.5x_3 - 4(x_3-1) - 12(x_3-2)$$

For (B.M.) at C put $x_3 = 2$ m,
  $(B.M.)_C = 11.5 \times 2 - 4(2-1) - 12(2-2)$
  $= 19$ kNm (Unchanged)
  at $x_3 = 3$ m,
  $(B.M.)_B = +11.5 \times 3 - 4(3-1) - 12(3-2)$
  $= +14.5$ kNm

$$(B.M)_{X_4 X_4^1} = 11.5x_4 - 4(x_4-1) - 12(x_4-2) - 10(x_4-3)$$

  at $x_4 = 3$ m,
  $(B.M.)_B = +11.5 \times 3 - 4(3-1) - 12(3-2) - 10(3-3)$
  $= 14.5$ kNm
  at $x_4 = 4$ m,
  $(B.M.)_B = 11.5 \times 4 - 4(4-1) - 12(4-2) - 10(4-3)$
  $= 0$

Example: 9.2

Draw S.F.D and B.M.D. for the beam as shown in Fig. 9.10.

(U. P. T. U. IInd Sem, 2001–02)

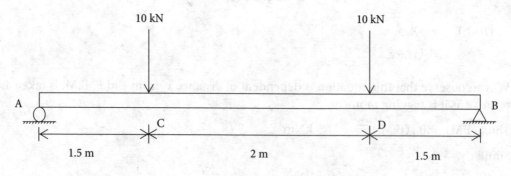

Fig. 9.10

Solution:

Consider F.B.D. of the beam AB, as shown in Fig. 9.10 (a)

Shear Force and Bending Moment Diagrams

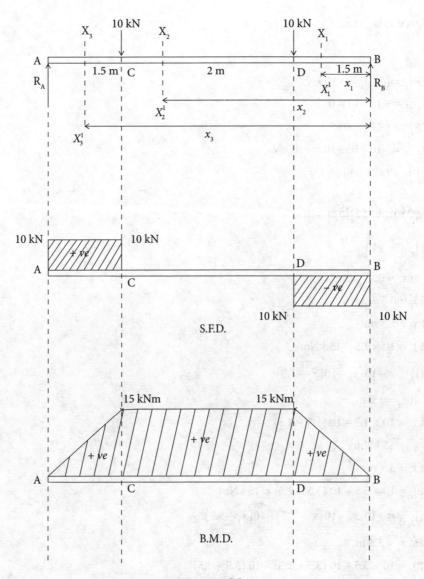

Fig. 9.10 (a)

$\Sigma Y = 0$,

$R_A + R_B = 10 + 10$

$\quad\quad\quad = 20$ kN ..... (1)

$\Sigma M_A = 0$,

$R_B \times 5 = 10 \times 1.5 + 10 \times 3.5$

$R_B = 10$ kN

$R_A = 10$ kN

Shear Force Diagram:

$(SF)_{X_1 X_1^1} = -10 \, kN$

$(SF)_B = (S.F)_D = -10 \, kN$

$(S.F)_{X_2 X_2^1} = -10 + 10 = 0$

$(S.F)_D = (S.F)_C = 0$

$(SF)_{X_3 X_3^1} = -10 + 10 + 10 = +10 \, kN$

$(SF)_C = (SF)_A = +10 \, kN$

Bending Moment Diagram:

$(B.M)_{X_1 X_1^1} = +10 \cdot x_1$

at $x_1 = 0$,

(B. M.)$_B$ = 0

at $x_1 = 1.5$ m,

(B. M.)$_D$ = 10 × 1.5 = 15 kNm

$(B.M)_{X_2 X_2^1} = +10 \cdot x_2 - 10(x_2 - 1.5)$

at $x_2 = 1.5$ m,

(B. M.)$_D$ = 10 × 1.5 − 10 (1.5 − 1.5)

= 15 kNm

at $x_2 = 3.5$ m,

(B. M.)$_C$ = 10 × 3.5 − 10 (3.5 − 1.5) = 15 kNm

$(B.M)_{X_3 X_3^1} = +10 \cdot x_3 - 10(x_3 - 1.5) - 10(x_3 - 3.5)$

at $x_2 = 3.5$ m,

(B. M.)$_C$ = 10 × 3.5 − 10 (3.5 − 1.5) − 10 (3.5 − 3.5)

= 15 kNm

at $x_2 = 5$ m,

(B. M.)$_A$ = 10 × 5 − 10 (5 − 1.5) − 10 (5 − 3.5)

= 50 − 35 − 15

= 0

Example: 9.3

Draw SFD and BMD for a simple supported beam carrying point load 'W' at its mid-span of length 'l' as shown in Fig. 9.11.

Shear Force and Bending Moment Diagrams

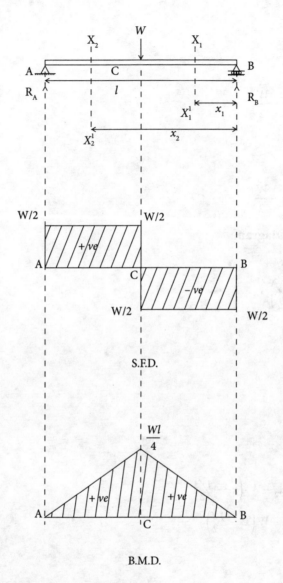

Fig. 9.11

Solution:

$\Sigma Y = 0$, $R_A + R_B = W$

As the beam is symmetrically loaded, $R_A = R_B = W/2$

Shear Force Diagram:

Considering a section $X_1 X_1^1$ at distance x_1 from the end B,

$$(SF)_{X_1X_1^1} = -R_B$$
$$= -\frac{W}{2}$$
$$(SF)_B = (SF)_C = -\frac{W}{2}$$
$$(SF)_{X_2X_2^1} = -\frac{W}{2} + W$$
$$= +\frac{W}{2}$$
$$(SF)_C = (SF)_A = +\frac{W}{2}$$

Bending Moment Diagram:

$$(BM)_{X_1X_1^1} = +R_B \cdot x_1$$
$$= \left(\frac{W}{2} \cdot x_1\right)$$

$(BM)_B = 0$
  at $x_1 = 0$,

$$(BM)_C = \left(\frac{W \cdot l}{4}\right)$$
at $x_1 = \frac{l}{2}$

$$(BM)_{X_2X_2^1} = +R_B \cdot x_2 - W \cdot \left(x_2 - \frac{l}{2}\right)$$
$$(BM)_C = \frac{W}{2} \cdot x_2 - W\left(x_2 - \frac{l}{2}\right)$$
at $x_2 = \frac{l}{2}$
$$(BM)_C = +\frac{Wl}{4}$$

$(BM)_A = 0$
  at $x_2 = l$

Example: 9.4

Draw the shear force and bending moment diagrams for a simple supported beam of span 'l' carrying a couple M at a distance 'a' from the left.

(UPTU, IInd Sem, 2009)

Shear Force and Bending Moment Diagrams

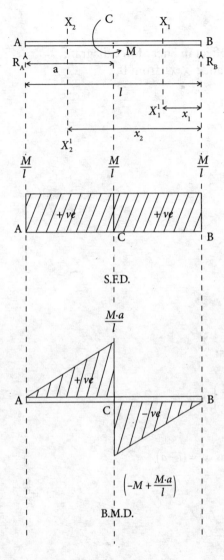

Fig. 9.12

Solution:

$\Sigma Y = 0$,

$R_A + R_B = 0$ ..... (1)

$\Sigma M_A = 0, R_B \times l + M = 0$

$R_B = -\dfrac{M}{l}$

$R_A = +\dfrac{M}{l}$

Shear Force Diagram:

Couple or moment has no effect in Shear force diagram
Consider the section at distance $X_1 X_1^1$ from B,

$$(S.F.)_{X_1 X_1^1} = -R_B$$
$$= +\frac{M}{l}$$
$$(SF)_B = (SF)_C = +\frac{M}{l}$$

Similarly,

$$(S.F.)_{X_2 X_2^1} = -R_B$$
$$= +\frac{M}{l}$$
$$(SF)_C = (SF)_A = +\frac{M}{l}$$

Bending Moment Diagram:

$$(BM)_{X_1 X_1^1} = +R_B \cdot x_1 = \left(\frac{-M}{l}\right) \cdot x_1$$

$$(B \cdot M)_B = 0 \text{ as } x_1 = 0$$

$$(BM)_C = \frac{-M}{l} \times (l-a) \text{ as } x_1 = (l-a)$$
$$= -M + M \cdot a/l$$

$$(BM)_{X_2 X_2^1} = +R_B \cdot x_2 + M$$
$$= \frac{-M}{l} \cdot x_2 + M$$

$$(B \cdot M)_C = \frac{-M}{l}(l-a) + M = -M + \frac{Ma}{l} + M$$

$$(BM)_A = \frac{-M}{l} \times l + M = 0$$

Example: 9.5

Draw SFD and BMD for a simple supported beam carrying a UDL of intensity w on its entire length l.

Shear Force and Bending Moment Diagrams

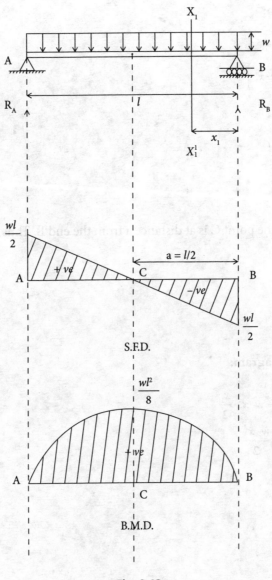

Fig. 9.13

Solution:

$\Sigma Y = 0$,
$R_A + R_B = w \cdot l$
As the beam is symmetrically loaded,

$$R_A = R_B = \frac{w \cdot l}{2}$$

Shear Force Diagram:

$$(SF)_{X_1 X_1^1} = -R_B + w \cdot x_1$$
$$= \left(\frac{-wl}{2} + w \cdot x_1\right)$$
$$(SF)_B = \frac{-wl}{2}$$
$$(SF)_A = \frac{-wl}{2} + w \cdot l$$
$$= +\frac{wl}{2}$$

(SF) at C is zero. Let the point C is at distance a from the end B. Thus,

$$(SF)_C = \frac{-wl}{2} + w \cdot a$$
$$0 = \frac{-wl}{2} + w \cdot a$$
i.e. $a = l/2$

Bending Moment Diagram:

$$(BM)_{X_1 X_1^1} = +R_B \cdot x_1 - w \cdot x_1 \cdot \frac{x_1}{2}$$
$$= \frac{wl}{2} \cdot x_1 - \frac{w \cdot x_1^2}{2}$$

at $x_1 = 0$,

$(BM)_B = 0$

at $x_1 = l$,
$(BM)_A = 0$
and $x_1 = l/2$,
$(BM)_C = wl^2/8$

Example: 9.6

Draw S.F.D and B.M.D. for the simple supported beam as shown in Fig. 9.14.

(UPTU 2006)

Shear Force and Bending Moment Diagrams

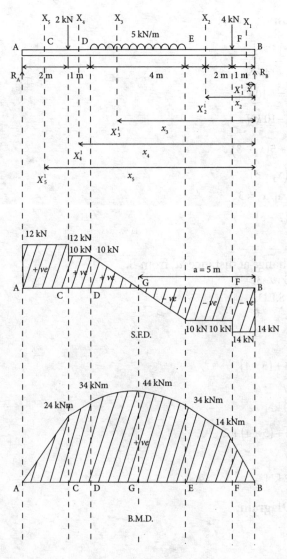

Fig. 9.14

Solution:

$\Sigma Y = 0$, $R_A + R_B = 4 + (5 \times 4) + 2$
$\qquad\qquad\qquad = 26$ ..... (1)

$\Sigma M_A = 0$,
$R_B \times 10 = 4 \times 9 + (5 \times 4) \times 5 + 2 \times 2$
$\qquad\quad\ = 36 + 100 + 4$
$\quad R_B = 14$ kN

from equation (1), $R_A = 12$ kN

Shear Force Diagram:

$(SF)_{X_1 X_1^1} = -R_B = -14 \ kN$

$(SF)_B = (SF)_F = -14 \ kN$

$(SF)_{X_2 X_2^1} = -14 + 4 = -10 \ kN$

$(SF)_F = (SF)_E = -10 \ kN$

$(SF)_{X_3 X_3^1} = -14 + 4 + 5(x_3 - 3)$

$\qquad = -10 + 5(x_3 - 3)$

$(SF)_E = -10 \ kN$ at $x_3 = 3$

$(SF)_D = -10 + 5(7 - 3)$

$\qquad = +10 \ kN$

Let the shear force change at distance 'a' from B,
$\quad (SF)_G = -10 + 5(a-3)$
But $(SF)_G = 0$ from S.F.D.
$\quad 5(a-3) = 10$
$\qquad a = 5 \ m$

$(SF)_{X_4 X_4^1} = -14 + 4 + (5 \times 4)$

$\qquad = +10 \ kN$

$(SF)_D = (SF)_C = +10 \ kN$

$(SF)_{X_5 X_5^1} = -14 + 4 + (5 \times 4) + 2$

$\qquad = +12 \ kN$

$(SF)_C = (SF)_A = +12 \ kN$

Bending Moment Diagram:

$(BM)_{X_1 X_1^1} = +R_B \cdot x_1$

$\qquad = +14 \cdot x_1$

$(BM)_B = 0$, $(BM)_F = 14 \times 1 = 14 \ kNm$,

$(BM)_{X_2 X_2^1} = 14 \cdot x_2 - 4(x_2 - 1)$

$(BM)_F = 14 \ kNm$, $(BM)_E = 14 \times 3 - 4(3-1) = 42 - 8 = 34 \ kNm$

$(BM)_{X_3 X_3^1} = 14(x_3) - 4(x_3 - 1) - 5(x_3 - 3) \times \dfrac{(x_3 - 3)}{2}$

$(BM)_E = 14 \times 3 - 4 \times 2 - 0 = 34 \ kNm$

$(BM)_D = 14 \times 7 - 4 \times 6 - 5 \times 4 \times \dfrac{4}{2} = 34 \ kNm$

$(BM)_G = 14 \times 5 - 4 \times 4 - 5 \times 2 \times \dfrac{2}{2} = 44 \ kNm$

$(B.M.)_{X_4 X_4^1} = +14 \cdot x_4 - 4(x_4 - 1) - 20(x_4 - 5)$

$(BM)_D = 14 \times 7 - 4 \times 6 - 20 \times 2$
$= 34$ kNm

$(BM)_{X_5 X_5^1} = +14 \cdot x_5 - 4(x_5 - 1) - 20(x_5 - 5) - 2(x_5 - 8)$

$(BM)_C = 14 \times 8 - 4(8-1) - 20(8-5) - 2(8-8)$
$= 24$ kNm

$(BM)_A = 14 \times 10 - 4(10-1) - 20(10-5) - 2(10-8)$
$= 0$

Example: 9.7

Find the positions of maximum shear force for the beam as shown in Fig. 9.15.

(UPTU, IInd Sem, 2003–04)

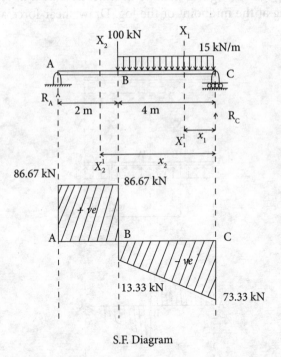

S.F. Diagram

Fig. 9.15

Solution:

$\Sigma Y = 0, R_A + R_C = 100 + (15 \times 4)$
$= 160$ kN ..... (1)

$\Sigma M_A = 0, R_C \times 6 = 100 \times 2 + (15 \times 4) \times 4$

$R_C = 73.33$ kN and $R_A = 86.67$ kN

Shear Force Diagram:

$(S.F.)_{x_1 x_1^1} = -R_c + 15 \cdot x_1$

$\qquad = -73.33 + 15\, x_1$

$(SF)_C = -73.33$ kN i.e., at $x_1 = 0$

$(SF)_B = -73.33 + 15 \times 4 = -13.33$ kN at $x_1 = 4$ m

$(S.F.)_{x_2 x_2^1} = -R_c + (15 \times 4) + 100$

$\qquad = -73.33 + 60 + 100$

$\qquad = +86.67$ kN

As it is independent of x_2 distance, thus $(SF)_B = (SF)_A = +86.67$ kN

Example: 9.8

A log of wood of negligible weight is floating in water as shown in Fig. 19.16. A person of weight W is standing at the midpoint of the log. Draw shear force and bending moment diagram of the log.

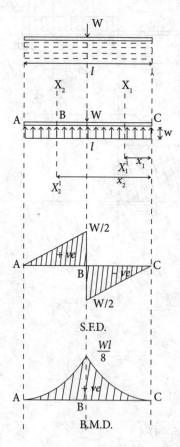

Fig. 9.16

Solution:

Let the length of log is 'l'. Since the wooden log is floating on water. Thus total downward weight, W will be equal to buoyant force acting upward over entire length of the log. Let the intensity of buoyant force per unit length is w. Thus under equilibrium,

$W = w.l$

i.e.,

$w = \dfrac{W}{l}$

Shear Force Diagram:

Considering $X_1X_1^1$ and $X_2X_2^1$ at distances x_1 and x_2, respectively from C,

$(SF)_{X_1X_1^1} = -w \cdot x_1 = \dfrac{-W}{l} \cdot x_1$

$(SF)_C = 0$,   when $x_1 = 0$

and $(SF)_B = \dfrac{-W}{l} \cdot \dfrac{l}{2} = -W/2$

$(SF)_{X_2X_2^1} = -w \cdot x_2 + W$

$= \dfrac{-W}{l} \cdot x_2 + W$

$(SF)_B = \dfrac{-W}{l} \times \dfrac{l}{2} + W$

$= +\dfrac{W}{2}$

$(SF)_A = \dfrac{-W}{l} \cdot l + W$

$= 0$

Bending Moment Diagram:

$(BM)_{X_1X_1^1} = w \cdot x_1 \times \dfrac{x_1}{2}$

$= \dfrac{W}{l} \times \dfrac{x_1^2}{2}$

$(BM)_C = 0$

$(BM)_B = \dfrac{W}{l} \cdot \dfrac{1}{2} \cdot \left(\dfrac{l}{2}\right)^2$

$= \dfrac{W \cdot l}{8}$

$$(BM)_{X_2X_2^1} = +wx_2 \cdot \frac{x_2}{2} - W\left(x_2 - \frac{l}{2}\right)$$

$$= \frac{W}{l} \times \frac{x_2^2}{2} - W\left(x_2 - \frac{l}{2}\right)$$

$$(BM)_B = \frac{W}{l} \cdot \frac{l}{2} \cdot \frac{l}{4} - W(0)$$

$$(BM)_B = \frac{Wl}{8}$$

$$(BM)_A = \frac{W}{l} \cdot l \cdot \frac{l}{2} - W\left(l - \frac{l}{2}\right)$$

$$= \frac{Wl}{2} - \frac{Wl}{2}$$

$$= 0$$

Note: The slope of BMD at point B will be maximum as the value of shear force is maximum at point B.

Example: 9.9

A simple supported beam is shown in the Fig. 9.17. Draw shear force and bending moment diagram.

Solution:

$\Sigma y = 0$,

$R_A + R_B = 20 \times 18$

$ = 360$ kN ..... (1)

$\Sigma M_A = 0$,

$R_B \times 18 = (20 \times 18) \times \dfrac{18}{2}$

$R_B = 180$ kN
$R_A = 180$ kN

Note: As it is symmetrically loaded beam thus we can say $R_A = R_B = 180$ kN without taking moment.

Shear Force Diagram:

Considering $X_1 X_1^1$ at distance 'x_1' from the end B,

$(SF)_{X_1 X_1^1} = -R_B + 20 \cdot x_1$

$\phantom{(SF)_{X_1 X_1^1}} = -180 + 20 \cdot x_1$

Shear Force and Bending Moment Diagrams

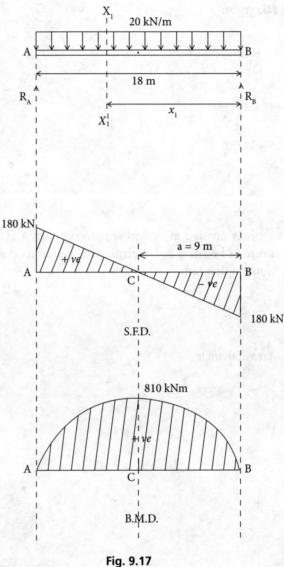

Fig. 9.17

$(SF)_B = -180$ kN,
$(SF)_A = -180 + 20 \times 18$
$= +180$ kN

It is to be noted that at point 'C' the shear face becomes zero gradually from negative value to positive value. Let 'C' is at distance 'a' from the end B,

$(SF)_C = -180 + 20 \cdot a$

$0 = -180 + 20 \cdot a$

$a = 9$ m i.e., at centre of beam

Bending Moment Diagram:

$$(BM)_{X_1 X_1^1} = +R_B \times x_1 - 20 \cdot x_1 \cdot \frac{x_1}{2}$$

$(BM)_B = 0$
$(BM)_A = 180 \times 18 - 10\,(18)^2$
$\quad\quad\quad = 0$
$(BM)_C = 180 \times 9 - 10\,(9)^2$
$\quad\quad\quad = 810 \text{ kNm}$

Example: 9.10

Figure 9.18 shows a beam pivoted at A and simple supported at B and carrying a load varying from zero at A to 12 kN/m at B. Determine the reactions at A and B, draw the shear force and bending moment diagram.

(U.P.T.U. Ist Sem, 2002–03)

Solution:

$\Sigma Y = 0$, $R_A + R_B$ = Area of triangle

$$= \frac{1}{2} \times 3 \times 12$$

$$= 18 \quad\quad\quad\quad\quad \ldots\ldots (1)$$

$\Sigma M_A = 0$,

$$R_B \times 3 = 18 \times \frac{2}{3} \times 3$$

$R_B = 12 \text{ kN}$
$R_A = 6 \text{ kN}$

In this question we have to consider section $X_1 X_1^1$ from left side A at distance x_1. The reason is if section XX is considered from right side the remaining area will be trapezium whose centroid would be difficult to consider.

Shear Force Diagram:

$$(S \cdot F.)_{X_1 X_1^1} = +R_A - \text{Load of UVL upto } x_1 \text{ from A}$$

$$= 6 - \text{ Area of triangle up to } x_1$$

$$= \left(6 - \frac{1}{2} \times x_1 \times h\right)$$

Shear Force and Bending Moment Diagrams

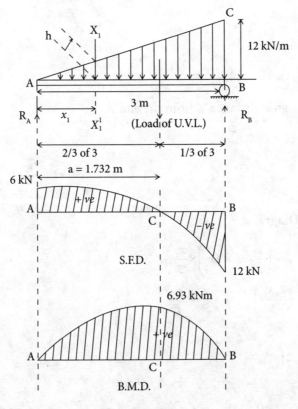

Fig. 9.18

Now height, h of UVL at x_1 distance can be determined by three ways:

(i) By law of similar triangles: $\dfrac{h}{x_1} = \dfrac{12}{3}$ i.e., $h = 4x_1$

(ii) By trigonometry: if $\angle CAB = \alpha$

then $\tan\alpha = \dfrac{h}{x_1}$ and $\tan\alpha = \dfrac{12}{3}$

i.e., $\dfrac{h}{x_1} = \dfrac{12}{3}$ i.e., $h = 4x_1$

(iii) By method of proportionality:

The intensity of load at 3 m from A to B = 12 kN

thus the intensity of load at x_1 distance, i.e., $h = \left(\dfrac{12 \times x_1}{3}\right)$

Substituting value of h in equation (1),

$$(SF)_{X_1X_1} = 6 - 2x_1^2$$

$$(SF)_A = +6 \text{ kN}$$

$$(SF)_B = 6 - \frac{1}{2} \times 3 \times 4 \times 3 = -12 \text{ kN}$$

Let S.F. changes its sign at distance 'a' from the end A,

$$(SF)_C = 6 - \frac{1}{2} \times a \times 4 \times a$$

$$0 = 6 - 2a^2$$

$$a = 1.732 \text{ m}$$

Bending Moment Diagram:

$$(BM)_{X_1X_1} = +R_A \times x_1 - \left(\frac{1}{2} \times x_1 \times 4x_1\right) \times \frac{x_1}{3}$$

$$= \left(6 \cdot x_1 - \frac{2 \cdot x_1^3}{3}\right)$$

$$(BM)_A = 0$$

$$(BM)_B = 6 \times 3 - \frac{2}{3}(3)^3$$

$$= 0$$

$$(BM)_C = 6 \times 1.732 - \frac{2(1.732)^3}{3}$$

$$= 6.93 \text{ kNm}$$

Example: 9.11

The intensity of loading on simple supported beam of spam 10 m increases uniformly from 10 kN/m at the left support to 20 kN/m at the right support. Find the position and magnitude of maximum bending moment.

(U.P.T.U., Ist Sem, 2004–05)

Solution:

The load lying over simple supported beam is a combination of U.D.L and U.V.L.

$\Sigma Y = 0$,

$R_A + R_B$ = Downward load of U.D.L + Downward load of U.V.L.

$$= (10 \times 10) + \left(\frac{1}{2} \times 10 \times (20 - 10)\right)$$

$$= 100 + 50$$

$$R_A + R_B = 150 \qquad \ldots\ldots (1)$$

Shear Force and Bending Moment Diagrams

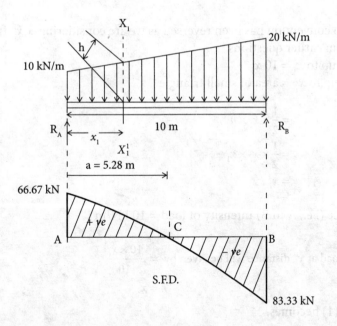

Fig. 9.19

$\Sigma M_A = 0$,

$$R_B \times 10 = (10 \times 10) \times \frac{10}{2} + \left(\frac{1}{2} \times 10 \times 10\right) \times \left(\frac{2}{3} \times 10\right)$$

$R_B = 83.33$ kN
$R_A = 66.67$ kN

Shear Force Diagram:

$(S.F.)x_1 x_1^1 = +R_A - \text{Load of U.D.L up to } x_1 - \text{Load of U.V.L up to } x_1$ ..... (1)

Note: Here sign convention has been reversed as we are considering $x_1 x_1^1$ from left side i.e., reverse side from earlier questions.

Load of U.D.L. up to $x_1 = 10 \cdot x_1$

Load of U.V.L. up to x_1 = area of small triangle

$$= \frac{1}{2} \times x_1 \times h$$

$$= \frac{1}{2} \times x_1 \times x_1$$

$$= x_1^2 / 2$$

At 10 m distance (i.e., A to B) intensity of load = 10 kN/m

intensity of load at x_1 distance will be given by $= \dfrac{10 \times x_1}{10}$

$$h = x_1$$

Thus equation (1) becomes,

$$(SF) x_1 x_1^1 = 66.67 - 10 x_1 - \frac{x_1^2}{2}$$

$$(SF)_A = +66.67 \text{ kN}$$

$$(SF)_B = 66.67 - 10 \times 10 - \frac{(10)^2}{2}$$

$$= -83.33 \text{ kN}$$

Let the shear force changes its sign at distance 'a' from end A.

$$(S.F.)_c = 66.67 - 10a - \frac{a^2}{2}$$

$$0 = 66.67 - 10a - \frac{a^2}{2}$$

$$a^2 + 20a - 133.34 = 0$$

$$a = \frac{-20 \pm \sqrt{(20)^2 - 4.1.(-133.4)}}{2 \times 1}$$

$$a = \frac{-20 \pm 30.55}{2} \quad \text{i.e. } a = -25.28 \text{ m or } 5.28 \text{ m}$$

$a = 5.28\ m$ as negative value does not lie on the beam.

Bending Moment Diagram:

$$(B.M.)_{x_1 x_1^1} = +R_A \times x_1 - 10 \times x_1 \times \frac{x_1}{2} - \left(\frac{1}{2} \times x_1 \times x_1\right) \times \frac{x_1}{3}$$

$$= 66.67 x_1 - 5x_1^2 - \frac{x_1^3}{6}$$

$(BM)_A = 0$

$(BM)_B = 66.67 \times 10 - 5(10)^2 - \frac{(10)^3}{6} = 0$

$(BM)_C = 66.67 \times 5.28 - 5(5.28)^2 - \frac{(5.28)^3}{6}$

$= 188.09 \, kNm$

Example: 9.12

Draw S.F.D and B.M.D for cantilever beam subjected to U.D.L of intensity 3 kN/m and point load 10 kN as shown in Fig. 9.20. Also determine the supporting reaction and moment offered by wall on beam.

Solution:
To maintain the beam in equilibrium, wall will offer an upward force R_A and moment in anticlockwise direction as shown in Fig. 9.20 at point A.

Shear Force Diagram:

$(S.F.)_{x_1 x_1^1} = +3.x_1$

$(SF)_D = 0, (SF)_C = 3 \times 4 = +12 \, kN$

$(SF)_{x_2 x_2^1} = +3 \times 4 = +12 \, kN$

$(SF)_C = (SF)_B = +12 \, kN$

$(SF)_{x_3 x_3^1} = +12 + 10 = 22 \, kN$

$(SF)_B = (SF)_A = +22 \, kN$

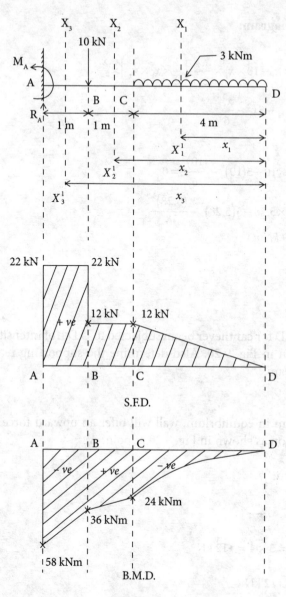

Fig. 9.20

Bending Moment Diagram:

$$(BM)_{x_1 x_1^1} = -3x_1 \times x_1/2$$

$$(BM)_D = 0, \ (BM)_C = -3 \times 4 \times \frac{4}{2} = -24 \text{ kNm}$$

$$(BM)_{x_2 x_2^1} = -(3 \times 4) \times (x_2 - 2)$$

Shear Force and Bending Moment Diagrams

$(BM)_C = -12\,(4-2)$
$= -24$ kNm
$(BM)_B = -12\,(5-2) = -36$ kNm

$(BM)_{x_3 x_3^1} = -(3 \times 4)(x_3 - 2) - 10(x_3 - 5)$

$(BM)_B = -3 \times 4\,(5-2) - 0 = -36$ kNm

$(BM)_A = -3 \times 4\,(6-2) - 10\,(6-5) = -58$ kNm

For support reaction and moment,

$\Sigma Y = 0,\; R_A = 10 + (3 \times 4) = 22$ kN

$\Sigma M_A = 0,\; -M_A + 10 \times 1 + (3 \times 4) \times 4 = 0$

$M_A = 58$ kNm

Note: Here full load of U.D.L (3×4) is acting at mid i.e. at 2 m from D thus from $X_2 X_2^1$ the distance of load will be $(x_2 - 2)$.

Example: 9.13

Draw SFD and BMD for the cantilever beam as shown in Fig. 9.21.

Solution:

Shear Force Diagram:

$(SF)_{X_1 X_1^1} = +\dfrac{1}{2} \times x_1 \times h$

The intensity of load at 6 m is = 24 kN/m

The intensity of load at x_1 will be, $h = \dfrac{24 \times x_1}{6} = 4 \cdot x_1$

$(SF)_{X_1 X_1^1} = +\dfrac{1}{2} \times x_1 \times 4x_1$

$= 2 \cdot x_1^2$

$(SF)_B = 0$
$(SF)_A = 2\,(6)^2$
$= 72$ kN

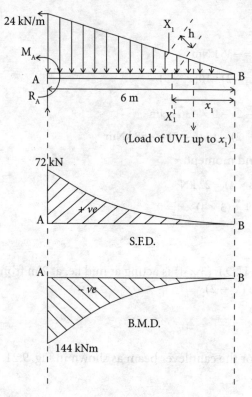

Fig. 9.21

Bending Moment Diagram:

$$(BM)_{X_1 X_1^1} = -\left(\frac{1}{2} + x_1 \times 4x_1\right) \times \frac{x_1}{3}$$

$(BM)_B = 0$

$$(BM)_A = -\frac{1}{2} \times 6 \times 4 \times 6 \times \frac{6}{3}$$

$$= -144 \text{ kNm}$$

Example: 9.14

Vertical forces 20 kN, 40 kN and uniformly distributed load of 20 kN/m is acting in 3 m length as shown in Fig. 9.22. Find the resultant force of the system and draw S.F.D and B.M.D.

(U.P.T.U., Ist Sem, 2001–2002)

Solution:

$\Sigma Y = 0$, $R_c = 20 + 40 + (20 \times 3)$
$= 120$ kN

$\Sigma M_c = 0$, $-20 \times 3 - 40 \times 2 - (20 \times 3) \times 1.5 + M_c = 0$
$M_c = 230$ kNm

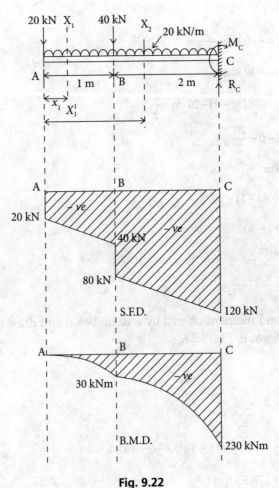

Fig. 9.22

Shear Force Diagram:

Consider sections $X_1 X_1^1$ and $X_2 X_2^1$ from left side i.e., at x_1 and x_2, respectively. Thus sign convention will get changed for shear force diagram.

$(SF)_{X_1 X_1^1} = -20 - 20 x_1$

$(SF)_A = -20$ kN, $(SF)_B = -20 - 20 \times 1 = -40$ kN

$(SF)_{X_2 X_2^1} = -20 - 40 - 20 x_2$

$(SF)_B = -20 - 40 - 20 = -80$ kN
$(SF)_C = -20 - 40 - 20 \times 3$
$= -120$ kN

Bending Moment Diagram:

$$(BM)_{X_1 X_1^1} = -20 \cdot x_1 - 20 \cdot x_1 \cdot \frac{x_1}{2}$$

$$(BM)_A = 0, \quad (BM)_B = -20 - \frac{20 \times 1 \times 1}{2}$$

$$= -30 \text{ kNm}$$

$$(BM)_{X_2 X_2^1} = -20 \cdot x_2 - 40(x_2 - 1) - 20 \cdot x_2 \cdot \frac{x_2}{2}$$

$$(BM)_B = -20 \times 1 - 0 - \frac{20}{2}$$

$$= -30 \text{ kNm}$$

$$(BM)_C = -20 \times 3 - 40(3 - 1) - 20 \cdot 3 \cdot \frac{3}{2}$$

$$= -60 - 80 - 90$$

$$= -230 \text{ kNm}$$

Example: 9.15

Determine reaction and moment offered by wall on beam and draw S.F.D. and B.M.D. for cantilever beam as shown in Fig. 9.23.

Solution:

$\Sigma Y = 0, R_A = 9 + (5 \times 3) + 16$
$= 40$ kN

$\Sigma M_A = 0, \quad -M_A + 9 \times 1 + 6 + (5 \times 3)\left(3 + \frac{3}{2}\right) + 16 \times 6 = 0$

$M_A = 178.5$ kNm

Shear Force Diagram:

$$(SF)_{X_1 X_1^1} = 16 + 5 \cdot x_1$$

$(SF)_E = 16$ kN, $(SF)_D = 16 + 5 \times 3$
$= 31$ kN

$$(SF)_{X_2 X_2^1} = 16 + (5 \times 3) = 31 \text{ kN}$$

$(SF)_D = (SF)_C = 31$ kN

Shear Force and Bending Moment Diagrams

(couple or moment has no effect in shear force diagram)

$(SF)_{X_3 X_3^1} = 16 + (5 \times 3) = 31 \ kN$

$(SF)_C = (SF)_B = 31 \ kN$

$(SF)_{X_4 X_4^1} = 16 + (5 \times 3) + 9 = 40 \ kN$

$(SF)_B = (SF)_A = 40 \ kN$

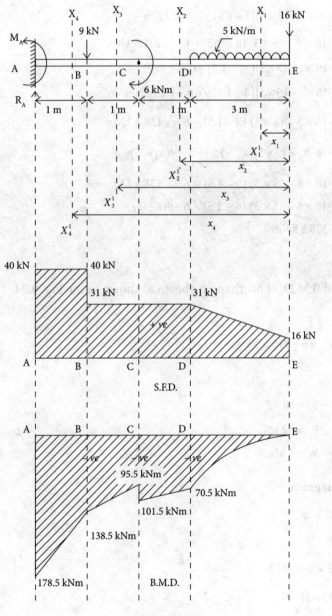

Fig. 9.23

Bending Moment Diagram:

$(BM)_{X_1 X_1^1} = -16x_1 - 5x_1 \times \dfrac{x_1}{2}$

$(BM)_E = 0$

$(BM)_D = -16 \times 3 - 5 \times 3 \times \dfrac{3}{2} = -70.5 \ kNm$

$(BM)_{X_2 X_2^1} = -16x_2 - (5 \times 3)(x_2 - 1.5)$

$(BM)_D = -16 \times 3 - (5 \times 3)(3 - 1.5) = -70.5 \ kNm$

$(BM)_C = -16 \times 4 - (5 \times 3)(4 - 1.5) - 6 = -101.5 \ kNm$

$(BM)_{X_3 X_3^1} = -16 \cdot x_3 - (5 \times 3)(x_3 - 1.5) - 6$

$(BM)_C = -16 \times 4 - (5 \times 3)(4 - 1.5) - 6 = -95.5 \ kNm$

$(BM)_B = -16 \times 5 - (5 \times 3)(5 - 1.5) - 6 = -138.5 \ kNm$

$(BM)_{X_4 X_4^1} = -16 \cdot x_4 - (5 \times 3)(x_4 - 1.5) - 6 - 9(x_4 - 5)$

$(BM)_B = -16 \times 5 - (5 \times 3)(5 - 1.5) - 6 = -138.5 \ kNm$

$(BM)_A = -16 \times 6 - (5 \times 3)(6 - 1.5) - 6 - 9(6 - 5)$

$\qquad = -178.5 \ kNm$

Example: 9.16

Draw S.F.D. and B.M.D. of overhanging beam as shown in the Fig. 9.24.

Solution:

$\Sigma Y = 0, \ R_B + R_C = 6 + (15 \times 2)$

$\qquad \qquad \qquad = 36$ ..... (1)

$\Sigma M_B = 0,$

$R_C \times 4 + 6 \times 2 + 10 = 30 \times 5$

$\qquad \qquad R_C = 32 \ kN$

$\qquad \qquad R_B = 4 \ kN$

Shear Force Diagram:

$(SF)_{X_1 X_1^1} = +15 \times x_1$

$(SF)_D = 0$

$(SF)_C = 15 \times 2 = 30 \ kN$

$(SF) x_2 x_2^1 = (15 \times 2) - R_C$

$\qquad \qquad = 30 - 32 = -2 \ kN$

Shear Force and Bending Moment Diagrams

$(SF)_{X_2X_2^1} = -2 \text{ kN} \quad (SF)_C = (SF)_E = -2 \text{ kN}$

$(SF)_{X_3X_3^1} = (15 \times 2) - R_C$

$(SF)_E = (SF)_B = -2 \text{ kN}$

$(SF)_{X_4X_4^1} = (15 \times 2) - R_C - R_B$

$\qquad = 30 - 32 - 4$

$\qquad = -6 \text{ kN}$

$(SF)_B = (SF)_A = -6 \text{ kN}$

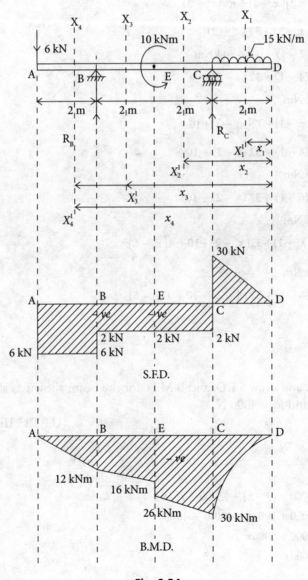

Fig. 9.24

Bending Moment Diagram:

$$(BM)_{X_1X_1^1} = -15x_1 \times \frac{x_1}{2}$$

$$(BM)_D = 0$$

$$(BM)_C = -15 \times 2 \times \frac{\cancel{2}}{\cancel{2}}$$

$$= -30 \text{ kNm}$$

$$(BM)_{X_2X_2^1} = -(15 \times 2)(x_2 - 1) + R_C(x_2 - 2)$$

$$= -30(x_2 - 1) + 32(x_2 - 2)$$

$(BM)_C = -30(2 - 1) + 0 = -30$ kNm

$(BM)_E = -30(4 - 1) + 32(4 - 2)$

$\qquad = -26$ kNm

$$(BM)_{X_3X_3^1} = -30(x_3 - 1) + 32(x_3 - 2) + 10$$

$(BM)_E = -30(4 - 1) + 32(4 - 2) + 10$

$\qquad = -16$ kNm

$(BM)_B = -30(6 - 1) + 32(6 - 2) + 10$

$(BM)_B = -12$ kNm

$$(BM)_{X_4X_4^1} = -30(x_4 - 1) + 32(x_4 - 2) + 10 + 4(x_4 - 6)$$

$(BM)_B = -12$ kNm

$(BM)_A = -30(8 - 1) + 32(8 - 2) + 10 + 4(8 - 6)$

$\qquad = 0$

Example: 9.17

Find the value of x and draw S.F.D. and B.M.D. for the beam shown as shown in Fig. 9.25. Take $R_A = 1000$ N and $R_B = 4000$ N.

(U.P.T.U. IInd Sem, 2000–01)

Solution:

$\Sigma M_A = 0$,

$R_B \times (2 + x) = 1000 \times (3 + x) + (2000 \times 2)(x + 1)$

$4000(2 + x) = 3000 + 1000x + 4000x + 4000$

$8000 + 4000x = 7000 + 5000x$

$\qquad 1000 = 1000x$

$\qquad\qquad x = 1 \ m$

Shear Force Diagram:

$(SF)_{X_1 X_1^1} = +1000$ N

$(SF)_D = (SF)_B = +1000$ N

$(SF)_{x_2 x_2^1} = +1000 - 4000 + 2000(x_2 - 1)$

$(SF)_B = -3000$ N

$(SF)_C = 1000 - 4000 + 2000 (3 - 1)$

$\quad\quad\quad = 1000$ N

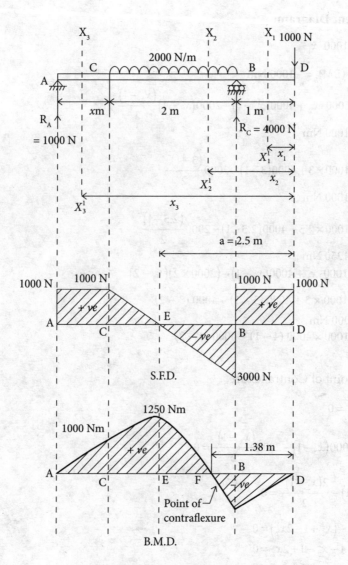

Fig. 9.25

$(SD)_{x_3 x_3^1} = +1000 - 4000 + (2000 \times 2)$
$\qquad = +1000 \text{ N}$
$(SF)_C = (SF)_A = +1000 \text{ N}$

Note in portion BC, (S.F.) changes its sign and let it is zero at point E which is at distance 'a' from point D. Using equation $(SF)_{x_2 x_2^1}$,

$(SF)_E = 1000 - 4000 + 2000(a - 1)$
$0 = -3000 + 2000(a - 1)$
$a = 2.5 \text{ m}$

Bending Moment Diagram:

$(BM)_{x_1 x_1^1} = -1000 \cdot x_1$
$(BM)_D = 0, (BM)_B = -1000 \text{ Nm}$

$(BM)_{x_2 x_2^1} = -1000 \cdot x_2 + 4000(x_2 - 1) - 2000(x_2 - 1)\dfrac{(x_2 - 1)}{2}$

$(BM)_B = -1000 \text{ Nm}$

$(BM)_C = -1000 \times 3 + 4000(3 - 1) - 2000\dfrac{(3-1)^2}{2}$
$\qquad = +1000 \text{ Nm}$

$(BM)_E = -1000 \times 2.5 + 4000(2.5 - 1) - 2000\dfrac{(2.5 - 1)}{2}$
$\qquad = +1250 \text{ Nm}$

$(BM)_{x_3 x_3^1} = -1000 \cdot x_3 + 4000(x_3 - 1) - (2000 \times 2)(x_3 - 2)$

$(BM)_C = -1000 \times 3 + 4000(3 - 1) - 4000(3 - 2)$
$\qquad = 1000 \text{ Nm}$

$(BM)_A = -1000 \times 4000(4 - 1) - 4000(4 - 2)$
$\qquad = 0.$

To determine Point of Contraflexure,

Put $(BM)_{x_2 x_2^1} = 0$

$-1000 \cdot x_2 + 4000(x_2 - 1) - 2000\dfrac{(x_2 - 1)^2}{2} = 0$

$-x_2 + 4(x_2 - 1) - \dfrac{2(x_2 - 1)^2}{2} = 0$

$-x_2 + 4x_2 - 4 - (x_2^2 + 1 - 2x_2) = 0$
$+3x_2 - 4 - x_2^2 - 1 + 2x_2 = 0$
$-x_2^2 + 5x_2 - 5 = 0$
$x_2^2 - 5x_2 + 5 = 0$

Shear Force and Bending Moment Diagrams

$$x_2 = \frac{+5 \pm \sqrt{(+5)^2 - 4 \times 1 \times 5}}{2},$$

$$x_2 = \frac{5 \pm \sqrt{5}}{2}$$

x_2 = 3.62 m and 1.38 m where first value does not lie for x_2 as shown in BMD, thus
$x_2 = 1.38\ m$

Example: 9.18

Draw S.F.D. and B.M.D. of an overhanging beam as shown in Fig. 9.26.

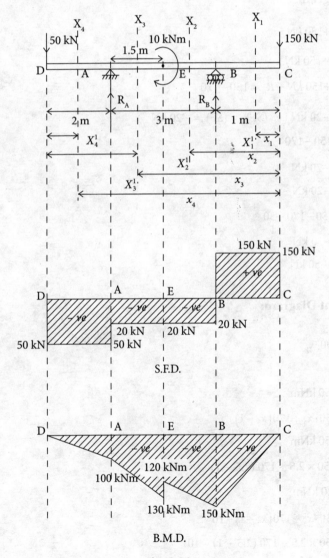

Fig. 9.26

Solution:

$\Sigma Y = 0$, $R_A + R_B = 150 + 50$

$\quad = 200$ kN ..... (1)

$\Sigma M_A = 0$,

$R_B \times 3 + 50 \times 2 = 10 + 150 \times 4$

$\quad\quad\quad = 610$

$\quad\quad R_B = 170$ kN

$\quad\quad R_A = 30$ kN

Shear Force Diagram:

$(SF)_{X_1 X_1^1} = +150$ kN

$(SF)_C = (SF)_B = +150$ kN

$(SF)_{X_2 X_2^1} = +150\ kN - R_B = 150 - 170$

$\quad\quad = -20$ kN i.e., $(SF)_B = (SF)_E = -20$ kN

$(SF)_{X_3 X_3^1} = 150 - 170$

$\quad\quad = -20$ kN

$(SF)_E = (SF)_A = -20$ kN

$(SF)_{X_4 X_4^1} = 150 - 170 - 30$

$\quad\quad = -50$ kN

$(SF)_A = (SF)_D = -50$ kN

Bending Moment Diagram:

$(BM)_{X_1 X_1^1} = -150 \cdot x_1$

$(BM)_C = 0$

$(BM)_B = -150$ kNm

$(BM)_{X_2 X_2^1} = -150 \cdot x_2 + 170(x_2 - 1)$

$(BM)_B = -150$ kNm

$(BM)_E = -150 \times 2.5 + 170(2.5 - 1)$

$\quad\quad = -120$ kNm

$(BM)_{X_3 X_3^1} = -150 \cdot x_3 + 170(x_3 - 1) - 10$

$(BM)_E = -150 \times 2.5 + 170(2.5 - 1) - 10$

$\quad\quad = -130$ kNm

$(BM)_A = -150 \times 4 + 170(4-1) - 10$

$\quad = -100 \text{ kNm}$

$(BM)_{X_4 X_4^1} = -150 \cdot x_4 + 170(x_4 - 1) - 10 + 30(x_4 - 4)$

$(BM)_A = -150 \times 4 + 170x(4-1) - 10$

$\quad = -100 \text{ kNm}$

$(BM)_D = -150 \times 6 + 170(6-1) - 10 + 30(6-4)$

$\quad = 0$

Example: 9.19

Draw SFD and BMD for an overhanging beam as shown in Fig. 9.27.

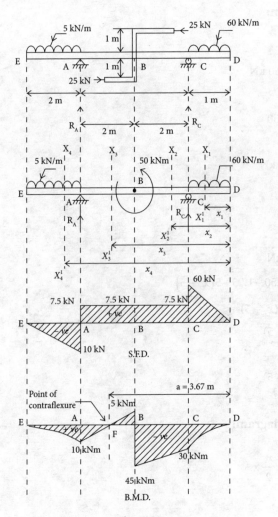

Fig. 9.27

Solution:

The figure can be modified about point B where Couple = 25 × 2 = 50 kNm (anticlockwise) act.

$\Sigma Y = 0, R_A + R_C = (5 \times 2) + (60 \times 1)$
$= 70 \text{ kN}$ ..... (1)

$\Sigma M_A = 0, R_C \times 4 + 50 + (5 \times 2) \times 1 = 60 \times 1 \times (4 + 0.5)$
$4 R_C + 60 = 270$
$R_C = 52.50 \text{ kN}$
$R_A = 17.50 \text{ kN}$

Shear Force Diagram:

$(SF)_{X_1 X_1^1} = +60 \cdot x_1$

$(SF)_D = 0, (SF)_C = +60 \text{ kN}$

$(SF)_{X_2 X_2^1} = +60 \times 1 - R_C$ ..... (2)
$= 60 - 52.50$
$= 7.5 \text{ kN}$

$(SF)_C = (SF)_B = +7.5 \text{ kN}$

$(SF)_B = 7.5 \text{ kN}.$

$(SF)_{X_3 X_3^1} = 60 - 52.5$
$= 7.5 \text{ kN}$

$(SF)_B = (SF)_A = 7.5 \text{ kN}$

$(SF)_{X_4 X_4^1} = +60 - 52.50 - 17.5 + 5(x_4 - 5)$
$= -10 + 5 (x_4 - 5)$

$(SF)_A = -10 + 5 (5 - 5)$
$= -10 \text{ kN}$

$(SF)_E = -10 + 5 (7 - 5)$
$= 0.$

Bending Moment Diagram:

$(BM)_{X_1 X_1^1} = -60 x_1 \times \dfrac{x_1}{2}$

$(BM)_D = 0$

$$(BM)_C = -60 \times \frac{1 \times 1}{2} = -30 \, kNm$$

$$(BM)_{X_2X_2^1} = -60(x_2 - 0.5) + 52.50(x_2 - 1)$$

$(BM)_C = -60(1 - 0.5) + 0 = -30$ kNm

$(BM)_B = -60(3 - 0.5) + 52.50(3 - 1)$

$\quad = -45$ kNm

$$(BM)_{X_3X_3^1} = -60(x_3 - 0.5) + 52.50(x_3 - 1) + 50$$

$(BM)_B = -60(3 - 0.5) + 52.5(3 - 1) + 50$

$\quad = +5$ kNm

$(BM)_A = -60(5 - 0.5) + 52.5(5 - 1) + 50$

$\quad = -10$ kNm

$$(BM)_{X_4X_4^1} = -60(x_4 - 0.5) + 52.5(x_4 - 1) + 50 + 17.50(x_4 - 5) - 5(x_4 - 5)(x_4 - 5)/2$$

$(BM)_A = -10$ kNm

$(BM)_E = 0$

To determine point of Contraflexure put $(B.M.)_{X_3X_3^1} = 0$ and $x_3 = a$

$(BM)_F = 0 = -60(a - 0.5) + 52.5(a - 1) + 50$

$\quad 0 = -60 \cdot a + 30 + 52.5a - 52.5 + 50$

$\quad a = 3.67 \, m$

Theoretical Problems

T 9.1 Define beam and its different types with suitable examples.
T 9.2 Discuss different types o loads to which beams are subjected.
T 9.3 Briefly explain the role of couple in shear force diagram and bending moment diagram.
T 9.4 What do you mean by shear force diagram and bending moment diagram?
T 9.5 Illustrate sagging and hogging moments with near and clean figure.
T 9.6 Explain point of contraflexure with neat and clean figure.
T 9.7 Derive the relation between intensity of loading, shear force and bending moment

$$w = \frac{dF}{dx} \text{ and } F = \frac{dM}{dx}$$

where w = intensity of loading, F = Shear force and M = Bending Moment

Numerical Problems

N 9.1 Draw SFD and BMD for the simple supported beam as given in Fig. NP 9.1.

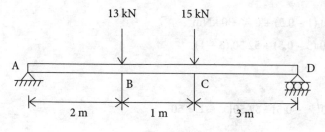

Fig. NP 9.1

N 9.2 Draw SFD and BMD for simple supported beam as given in Fig. NP 9.2.

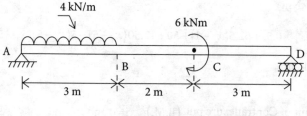

Fig. NP 9.2

N 9.3 Draw SFD and BMD for Cantilever beam as shown in Fig. NP 9.3.

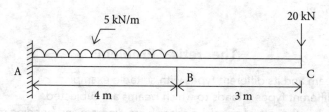

Fig. NP 9.3

N 9.4 Draw SFD and BMD for Cantilever beam as shown in Fig. NP 9.4.

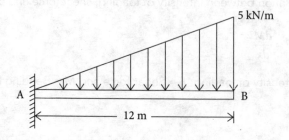

Fig. NP 9.4

N 9.5 Draw SFD and BMD for overhanging beam as shown in Fig. NP 9.5.

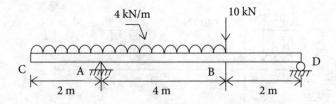

Fig. NP 9.5

N 9.6 Draw SFD and BMD for simple supported beam carrying point load and moments as shown in Fig. NP 9.6.

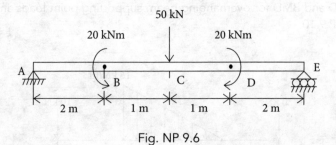

Fig. NP 9.6

N 9.7 Draw SFD and BMD Carrying UVL and UDL as shown in Fig. NP 9.7.

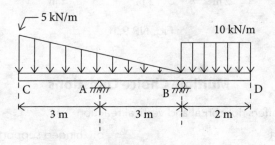

Fig. NP 9.7

N 9.8 Draw SFD and BMD for Cantilever beam as shown in Fig. NP 9.8.

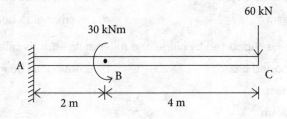

Fig. NP 9.8

N 9.9 Draw SFD and BMD for overhanging beam as shown in Fig. NP 9.9.

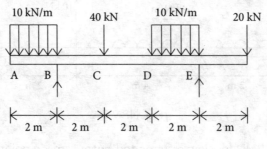

Fig. NP 9.9

N 9.10 Draw SFD and BMD for overhanging beam supporting point loads and UVL as shown in Fig. NP 9.10.

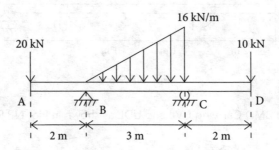

Fig. NP 9.10

Multiple Choice Questions

1. Which support offers horizontal and vertical reaction
   a. roller support
   b. hinged support
   c. fixed support
   d. all of these

2. The beam whose one end remains fixed and other end is supported by roller, called as
   a. simple supported beam
   b. cantilever beam
   c. propped beam
   d. overhanging beam

3. The beam which is supported by hinge and roller at its ends, is called as
   a. simple supported beam
   b. continuous beam
   c. propped beam
   d. cantilever beam

Shear Force and Bending Moment Diagrams

4. Which is statically determinate beam
   a. simply supported beam
   b. cantilever beam
   c. overhanging beam
   d. all of these

5. The load which is distributed over entire length of the beam with uniform intensity is called as
   a. UVL
   b. UDL
   c. concentrated load
   d. all of these

6. Which quantity does not affect SFD
   a. point load
   b. UDL
   c. UVL
   d. Couple

7. In SFD, if shear force changes its sign gradually at a point then its bending moment diagram at that point will be
   a. point of Contraflexure
   b. maximum bending moment
   c. zero bending moment
   d. none of these

8. A point load W is acting at midpoint of simple supported beam. If the span of the beam is L then the maximum bending moment will be
   a. $WL/4$
   b. $WL/8$
   c. $WL/2$
   d. none of these

9. A UDL of load intensity ω is acting on entire span of a simple supported beam. If the span of the beam is L then the maximum bending moment will be
   a. $\omega L^2/4$
   b. $\omega L^2/8$
   c. $\omega L^2/2$
   d. none of these

10. In a numerical problem, if bending moment is constant in some region then its shear force will be
    a. increasing
    b. decreasing
    c. zero
    d. none of these

11. The relationship between Load intensity (ω), Shear Force (S) and Bending Moment (M) is given by
    a. $\omega = dS/dx$ and $S = dM/dx$
    b. $\omega = dM/dx$ and $S = dS/dx$
    c. $S = d\omega/dx$ and $M = dS/dx$
    d. none of these

Answers
1. b 2. c 3. a 4. d 5. b 6. d 7. b 8. a 9. b 10. c 11. a

Chapter 10

Kinematics: Rectilinear Motion of Particles

10.1 Introduction

Previous chapters were concerned with the forces and reactions under which a body was held stationary; that branch of mechanics is called Statics. However, if the forces remained unbalanced, the body moves; this branch of mechanics is called Dynamics. It is further divided into two branches:

a. **Kinematics** that deals with the analysis of displacement, time, velocity and acceleration of a motion without reason of motion i.e., forces.
b. **Kinetics** that deals with the analysis of displacement, time, velocity, acceleration of a motion along with the forces causing motion. It also deals with the mass of the body and relationships between the forces acting on a body.

A body under motion can be idealized as a particle which is concentrated about its centre of gravity. It is assumed that a particle does not have rotational movement. However, if rotational movement of the body cannot be neglected then such body cannot be treated as particle.

10.2 Displacement, Velocity and Acceleration

Displacement is the minimum distance or smallest route travelled by a body from a starting point to the destination point along a straight line. It is a vector quantity as it is concerned with both magnitude and direction of motion.

Kinematics: Rectilinear Motion of Particles

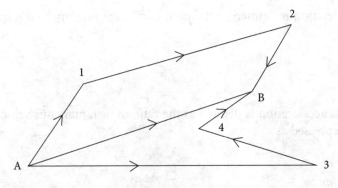

Fig. 10.1

Consider a particle moves from origin point A to destination point B via different paths A-1-2-B, A-B and A-3-4-B as shown in Fig. 10.1. Here the smallest route A-B travelled along a straight line will be displacement. However the other route traversed by the body will be called distance which depends upon magnitude only and called a scalar quantity. Being vector quantity, the displacement can have positive or negative value but distance always has a positive quantity.

Velocity of a body is described as the rate of change of displacement with respect to time. Velocity can be average velocity, instantaneous velocity and uniform velocity depending upon the variation of displacement and time.

Average velocity: In the previous example if body travels displacement Δs during time interval Δt via path A-B then the average velocity is given by

$$v_{avg} = \frac{\Delta s}{\Delta t}$$

Instantaneous velocity: At a particular instant, the velocity of a body is known as Instantaneous velocity. If displacement Δs and time interval Δt are reduced considerably then the instantaneous velocity is given by

$$v_{inst.velocity} = \lim_{t \to 0} \frac{\Delta s}{\Delta t} \text{ or } \frac{ds}{dt}$$

Uniform velocity: If body travels equal displacements in equal time intervals then the velocity of a body is known as uniform velocity. Under uniform velocity, the body does not have acceleration or retardation as it moves with constant velocity.

Acceleration: Acceleration of a body is described as increase in the rate of change of velocity or variation of velocity with respect to time. If rate of change of velocity decreases with respect to time, it is known as retardation or deceleration. Acceleration can be average acceleration or instantaneous acceleration depending upon the type of velocity and time.

The average acceleration is defined as the ratio of average velocity to the time internal Δt and expressed as

$$a_{avg} = \frac{v_{avg}}{\Delta t}$$

The instantaneous acceleration is defined as the ratio of instantaneous velocity to the time interval Δt and expressed as

$$\lim_{t \to 0} \frac{v_{inst.velocity}}{\Delta t} \text{ or } \frac{dv}{dt}$$

10.3 Rectilinear Motion

During motion, a particle can have three types of motion i.e., translational, rotational or general plane motion. In this chapter, we will study translational motion. Rotational motion and general plane motion are discussed in chapter 15. In translational motion, the particle travels along a straight line path without rotation; it is also known as rectilinear motion. In this type of motion, all particles of a body travel along a straight line only i.e., bears only one dimensional motion. The rectilinear motion may lie along X-axis or Y-axis. The examples of rectilinear motion along X-axis are motion of vehicles on a straight track (road without speed breaker), motion of piston in a horizontal cylinder. However, the motion of lift in a building and falling of a stone vertically are examples of rectilinear motion along Y-axis.

10.4 Rectilinear Motion in Horizontal Direction (X-axis)

A particle moving in a straight line can have either uniform or variable acceleration. Generally the particles consist of variable acceleration in actual practice. The motion of a body under **variable acceleration** can be analyzed by using differential and integral equations of motion. The motion under uniform acceleration is analyzed by using equations of motion as discussed in next section (10.4.2).

10.4.1 Motion with variable acceleration

Differential equations of motion: These equations are used to analyze the motion of a body having variable acceleration. The equations for four parameters of motion i.e., displacement, time, velocity and acceleration are given by

$$v = \frac{dx}{dt}$$

$$a = \frac{dv}{dt} \text{ or } \frac{d^2x}{dt^2} \text{ or } v\frac{dv}{dx}$$

Kinematics: Rectilinear Motion of Particles

In some cases variable motion is also defined as a function of time. Here the displacement travel by particle is a function of time and expressed as

$$x = f(t)$$

In such cases the velocity and acceleration can be obtained by differentiating the function with respect to time as given below:

$$v = f(t) = \frac{dx}{dt}$$

$$\text{and } a = f(t) = \frac{d^2x}{dt^2}$$

Integral equations of motion: These equations are used to analyze the parameters of motion like displacement, time or velocity of a body where acceleration is described as a function of time.

If $a = f(t)$

or $a = \dfrac{dv}{dt} = f(t)$

$dv = f(t).dt$

$v = \int f(t).dt + C_1$

where, C_1 is a constant.
As we know that

$$v = \frac{dx}{dt}$$

$$dx = v.dt$$

Further integration after substituting the value of v in the above equation, we get the displacement with respect to time,

$$\int dx = \int v.dt$$

or $\int dx = \int f(t).dt + C_2$
where, C_2 is a constant.

Example: 10.1

The distance of a particle moving in a straight with variable acceleration is given by, $s = 15t + 3t^2 - t^3$ in which, 's' is the distance measured in meter and the time 't' is measured in seconds. Calculate:

(i) the velocity and acceleration at start
(ii) the time, at which the particle attains its maximum velocity
(iii) the maximum velocity of the particle

Solution:

The distance is given by equation

$$s = 15t + 3t^2 - t^3$$

Since $v = \dfrac{ds}{dt}$ and $a = \dfrac{dv}{dt}$, the equation will differentiated with respect to time. The first and second differential will provide velocity and acceleration respectively.

$$v = \dfrac{ds}{dt} = 15 + 6t - 3t^2 \qquad \ldots (1)$$

$$a = \dfrac{dv}{dt} = 6 - 6t \qquad \ldots (2)$$

(i) When particle start, $t = 0$
$V = 15$ m/sec and $a = 6$ m/sec²

(ii) When particle attains maximum velocity, its acceleration becomes zero.
i.e., $a = 0$
From equation (2),
$6 - 6t = 0$
$t = 1$ sec

(iii) The maximum velocity will take place after 1 sec.
thus from equation (1)
$v = 15 + 6 - 3$
$v_{max} = 18$ m/sec

Example: 10.2

The rectilinear motion of particle is expressed by relation $x = t^3 - 3t^2 + 15t - 8$, where 'x' is the displacement measured in meter and the time 't' is measured in seconds. Calculate:

(i) The velocity and acceleration after 1 second and 2 second.
(ii) The average velocity and acceleration after 1 second and 2 second.

Solution:

The displacement of particle is given by

$$x = t^3 - 3t^2 + 15t - 8$$

(i) The velocity will be given by,

$$v = \frac{dx}{dt} = 3t^2 - 6t + 15 \quad \ldots (1)$$

The acceleration will be given by,

$$a = \frac{dv}{dt} = 6t - 6 \quad \ldots (2)$$

Using equation (1) velocity after 1 second and 2 second will be

$$v_{t=1\text{ sec}} = 3(1)^2 - 6(1) + 15 = 12 \text{ m/sec}$$
$$v_{t=2\text{ sec}} = 3(2)^2 - 6(2) + 15 = 15 \text{ m/sec}$$

Using equation (2), acceleration after 1 second and 2 second will be,

$$a_{t=1\text{ sec}} = 6(1) - 6 = 0$$
$$a_{t=2\text{ sec}} = 6(2) - 6 = 6 \text{ m/sec}^2$$

(ii) since, the average velocity $= \dfrac{\text{Net displacement}}{\text{time duration}} \quad \ldots (3)$

Net displacement after 1 second and 2 second will be determined by using equation,

$$x = t^3 - 3t^2 + 15t - 8$$
$$x_{t=1\text{ sec}} = (1)^3 - 3(1)^2 + 15(1) - 8 = 5 \text{ m}$$
$$x_{t=2\text{ sec}} = (2)^3 - 3(2)^2 + 15(2) - 8 = 18 \text{ m}$$

Using equation (3),
average velocity after 1 second and 2 second

$$= \left(\frac{18-5}{2-1}\right)$$
$$= 13 \text{ m/second}$$

since, the average acceleration $= \dfrac{\text{change in velocity}}{\text{time duration}} \quad \ldots (4)$

Change in velocity = velocity at 2 second − velocity at 1 second
$$= 15 - 12$$
$$= 3 \text{ m/sec}$$

Using equation (4),

The average acceleration $= \left(\dfrac{3}{2-1}\right)$

$$= 3 \text{ m/sec}^2$$

Example: 10.3

The rectilinear motion of a particle is expressed by equation, $a = t^3 - 4t^2 + 12$

where 'a' is the acceleration in m/sec² and 't' is the time in seconds. If particle attains velocity of 10 m/sec and displacement 24 m after 2 seconds then calculate velocity and displacement after 3.6 seconds.

Solution:

The variable acceleration is given by equation,

$$a = t^3 - 4t^2 + 12$$

as $a = \dfrac{dv}{dt} = t^3 - 4t^2 + 12$

$$dv = (t^3 - 4t^2 + 12).dt$$

$$\int dv = \int (t^3 - 4t^2 + 12).dt$$

$$v = \dfrac{t^4}{4} - \dfrac{4t^3}{3} + 12t + C_1 \qquad \ldots (1)$$

Given that after $t = 2$ second,
$v = 10$ m/sec and $x = 24$ m
Using equation (1) by substituting values of v and t,

$$10 = \dfrac{2^4}{4} - \dfrac{4(2)^3}{3} + 12(2) + C_1$$

$$C_1 = -7.33$$

Thus equation (1) will become,

$$v = \dfrac{t^4}{4} - \dfrac{4}{3}t^3 + 12t - 7.33 \qquad \ldots (2)$$

Velocity after 3.6 second,

$$v_{t=3.6\ sec} = \dfrac{(3.6)^4}{4} - \dfrac{4}{3}(3.6)^3 + 12(3.6) - 7.33$$

$$v_{t=3.6\ sec} = 15.65\ m/sec$$

Kinematics: Rectilinear Motion of Particles

Equation (2) can be expressed as,

$$v = \frac{dx}{dt}$$

i.e. $dx = v.dt$

$$x = \int v.dt$$

$$x = \int \left(\frac{t^4}{4} - \frac{4}{3}t^3 + 12t - 7.33 \right) dt$$

$$x = \frac{t^5}{20} - \frac{t^4}{3} + 6t^2 - 7.33t + C_2 \quad \ldots\ldots (3)$$

as, $x = 24$ m after 2 second.

$$24 = \frac{2^5}{20} - \frac{2^4}{3} + 6(2)^2 - 7.33(2) + C_2$$

$$C_2 = 18.39$$

Substituting values of C_2 and $t = 3.6$ seconds in equation (3),

$$x_{t=3.6\ sec} = \left[\frac{(3.6)^5}{20} - \frac{(3.6)^4}{3} + 6(3.6)^2 - 7.33(3.6) + 18.39 \right]$$

$$= 44\ m.$$

Example: 10.4

A particle under rectilinear motion has an acceleration given by $a = 3(v)^{1/2}$ where 'v' is the velocity expressed in m/sec. During motion it has velocity 64 m/sec and the displacement 108 m at time 't' = 6 sec. Determine at time 't' = 10 sec, the velocity, acceleration and displacement of the particle.

Solution:

Given: $v = 64$ m/sec, $s = 108$ m at $t = 6$ sec

and $a = 3\sqrt{v}$

$$v^{-1/2} dv = 3.dt$$

$$\int v^{-1/2}.dv = 3\int dt$$

$$2\sqrt{v} = 3t + C_1 \quad \ldots\ldots (1)$$

Substituting values of v and t in equation (1),

$$2\sqrt{64} = 3 \times 6 + C_1$$

$$C_1 = -2$$

Substituting values of C_1 and $t = 10$ sec in equation (1),

$$2\sqrt{v} = 3 \times 10 - 2$$
$$v = 196 \ m/sec$$

thus acceleration will be,

$$a = 3\sqrt{v} = 3\sqrt{196} = 42 \ m/sec^2$$

we know that,

$$a = \frac{v.dv}{ds}$$
$$3\sqrt{v} = \frac{v.dv}{ds}$$
$$3.ds = \sqrt{v}.dv$$
$$3\int ds = \int v^{1/2}.dv$$
$$3s = \frac{2v^{3/2}}{3} + C_2 \quad \ldots (2)$$

Substituting values of v and s at 6 sec in equation (2),

$$3 \times 108 = \frac{2}{3}(64)^{3/2} + C_2$$
$$C_2 = -17.33$$

at 10 sec, substituting values of v and C_2 in equation (2),

$$3s = \frac{2}{3}(196)^{3/2} - 17.33$$
$$s = 604 \ m$$

Example: 10.5

A rectilinear motion of motor car starting from rest is governed by the equation $a = \left(\frac{8}{1.5v + 2}\right)$, where '$a$' is the acceleration in m/sec² and v is velocity in m/sec at any instant.

Find the distance moved and the time taken by the car to attain a velocity of 8 m/sec.

(UPTU Ist Sem 2012–13)

Solution:

Given: $a = \left(\frac{8}{1.5v + 2}\right), v = 0$

$$a = \frac{dv}{dt}$$

thus $\dfrac{dv}{dt} = \dfrac{8}{1.5v+2}$

$(1.5v + 2).dv = 8.dt$

when $v = 0$, $t = 0$

$$\int (1.5v+2)dv = 8\int dt$$

$$\frac{1.5v^2}{2} + 2v = 8t + C$$

Substituting values of v and t
$C = 0$

thus, $\dfrac{1.5v^2}{2} + 2v = 8t$

as $v = 8$ m/sec

$$\frac{1.5(8)^2}{2} + 2 \times 8 = 8t$$

$t = 8$ sec

Using equation,

$$a = \frac{v.dv}{ds}$$

$$\left(\frac{8}{1.5v+2}\right) = \frac{v.dv}{ds}$$

$$8.ds = 1.5v^2.dv + 2vdv$$

$$8\int ds = 1.5\int v^2.dv + 2\int v.dv$$

$$8.s = \frac{1.5.v^3}{3} + \frac{2.v^2}{2} + C_2$$

At $t = 0$. $v = 0$, $s = 0$
$C_2 = 0$

$$8.s = \frac{1.5v^3}{3} + \frac{2.v^2}{2}$$

at $v = 8$ m/sec

$$8.s = \frac{1.5(8)^3}{3} + \frac{2(8)^2}{2}$$

$s = 40$ m

10.4.2 Motion with uniform acceleration

Let a body accelerates with uniform acceleration 'a' from initial velocity 'u' to final 'v' and travels distance 's' during time 't', then the relation between parameters of motion are described by following three equations:

as $a = \dfrac{dv}{dt}$

$dv = a.dt$

Using integration limits,

$$\int_u^v dv = \int_0^t a.dt$$

$$\int_u^v dv = a\int_0^t dt$$

$v - u = a.t$

$v = u + a.t$

We know that $v = \dfrac{dx}{dt}$

i.e., $v = \dfrac{dx}{dt} = u + at$

$dx = u.dt + a.t.dt$

Using integration limits,

$$\int_0^s dx = \int_0^t u.dt + a\int_0^t t.dt$$

$$s = ut + \dfrac{1}{2}at^2$$

We know that acceleration is also given by,

$a = v\dfrac{dv}{dx}$

$v\,dv = a\,dx$

Using integration limits,

$$\int_u^v v\,dv = \int_0^s a.dx$$

$\dfrac{v^2 - u^2}{2} = a.s$

$v^2 - u^2 = 2as$

$v^2 = u^2 + 2as$

Kinematics: Rectilinear Motion of Particles

Thus there are three equations of motion when particle moves with uniform acceleration in X-direction.

1. $v = u + a.t$
2. $s = ut + \frac{1}{2}at^2$
3. $v^2 = u^2 + 2as$

Distance covered by a particle during n^{th} second

If a particle covered distance s_n during n second and s_{n-1} during $(n-1)$ second then the distance covered during n^{th} second will be given by

$$s_{n^{th}} = s_n - s_{n-1}$$
$$= \left\{ un + \frac{1}{2}an^2 \right\} - \left\{ u(n-1) + \frac{1}{2}a(n-1)^2 \right\}$$
$$s_{n^{th}} = u + \frac{a}{2}(2n-1)$$

Example: 10.6

A bus accelerates from rest with uniform acceleration of 0.10 m/sec² for 18 sec. Determine the distance covered by bus and its final velocity.

Solution:

Given: $u = 0$, $a = 0.10$ m/sec², $t = 18$ sec.
We know that,

$$s = ut + \frac{1}{2}at^2$$
$$= 0 + \frac{1}{2} \times 0.10 \times (18)^2$$
$$s = 16.2 \, m$$

Using equation,

$v^2 = u^2 + 2as$
$v^2 = 0 + 2 \times 0.1 \times 16.2$
$v = 1.8$ m/sec

Example: 10.7

An electric train is moving with constant velocity of 90 km/hr. The train driver applies emergency brake and stops the train in 15 sec. Determine:

(i) The deceleration of the train
(ii) Distance travelled by train after the brakes are applied.

Solution:

Given: $u = 90$ km/hr $= \dfrac{90 \times 5}{18} = 25$ m/sec

$v = 0, t = 15$ sec

(i) $v = u + at$
$0 = 25 + a \times 15$
$a = -1.67$ m/sec^2

(ii) Using equation,
$v^2 = u^2 + 2as$
$0 = (25)^2 + (-1.67) \times s$
$s = 374.25$ m

Example: 10.8

A truck moving on a straight highway covered its half journey with speed 50 km/hr and another half with a speed of 80 km/hr. Determine the average speed of the truck during journey.

Solution:

Given, $v_1 = 50$ km/hr, $v_2 = 80$ km/hr
Let the track travelled distance 'x' during time 't'

Average velocity $= \dfrac{x}{t}$ ..... (1)

If truck take time t_1 and t_2 during two halves of the journey then,

$$50 = \dfrac{x/2}{t_1} \text{ and } 80 = \dfrac{x/2}{t_2}$$

$$t_1 = \dfrac{x}{100} \text{ and } t_2 = \dfrac{x}{160}$$

thus, $t = t_1 + t_2 = \dfrac{x}{100} + \dfrac{x}{160} = \dfrac{26.x}{1600}$

Kinematics: Rectilinear Motion of Particles

Substitute 't' in equation (1),

$$\text{average velocity} = \frac{x}{\left(\frac{26.x}{1600}\right)}$$

$$= \frac{1600}{26}$$

$$= 61.54 \text{ km/hr}$$

Example: 10.9

A bike rider is moving with uniform speed of 36 km/hr and he realizes that he will reach late to railway station by 40 seconds. He further accelerates at a constant rate of 8 m/sec² to reach the railway station by right time. Calculate:

(i) The time taken to reach the railway station after acceleration
(ii) The distance travelled after accelerating the bike

Solution:

Given: $u = 36$ km/hr $= \frac{36 \times 5}{18} = 10$ m/sec

Let the time taken to reach the railway station without delay is 't' and station is at distance 'x'.

(i) Before acceleration, as the bike is moving with uniform speed, $a = 0$ and time will be $(t + 40)$

$$x = ut + \frac{1}{2}at^2$$

$$x = ut + 0$$

$$x = 10(t + 40) \quad \quad \ldots (1)$$

After the acceleration,

$$x = ut + \frac{1}{2}at^2$$

$$= 10.t + \frac{1}{2} \times 8.t^2 \quad \quad \ldots (2)$$

Equating equation (1) and (2),

$$10(t + 40) = 10.t + \frac{1}{2}.8t^2$$

$$10t + 400 = 10t + 4t^2$$

$$t = 10 \text{ sec}$$

Substituting values of t in equation (1),

$x = 10(10 + 40)$

$= 500$ m

Example: 10.10

A train moving on a straight track, decelerates uniformly until comes to rest. At the beginning of deceleration, it travels 24 m in first three seconds and further 39 m in next 6 seconds. Determine:

(i) The initial velocity and uniform deceleration
(ii) The total distance travelled by train until comes to rest

Solution:

(i) Let the initial velocity of the train is 'u' m/sec just before deceleration. Consider Fig. 10.2 drawn as per given data.

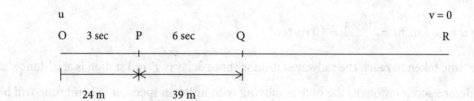

Fig. 10.2

Using equation of motion for path OP,

$$s = ut - \frac{1}{2}at^2$$

$$24 = u \times 3 - \frac{1}{2} \times a(3)^2$$

$$8 = u - 1.5a \quad \quad \ldots\ldots (1)$$

Similarly for path OQ, $S = 63$ m and $t = 9$ sec

$$63 = u \times 9 - \frac{1}{2}a(9)^2$$

$$7 = u - 4.5a \quad \quad \ldots\ldots (2)$$

from equation (1) and (2),
$a = 0.33$ m/sec^2 and $u = 8.5$ m/sec

Kinematics: Rectilinear Motion of Particles

(ii) Let the total distance travelled by train is 'x'.
i.e., OR = x, as $u = 8.5$ m/sec and $a = 0.33$ m/sec^2 and $v = 0$
Using equation,
$v^2 = u^2 - 2as$
$0 = (8.5)^2 - 2 \times 0.33 \times x$
$x = 109.47$ m

Example: 10.11

A biker is driving his bike at 72 km/hr. He watches a traffic light 240 m away becomes red. If the light takes 20 sec to turn green, determine the required deceleration to cross the light without stopping the bike. Determine the speed of the bike when he passes the traffic light.

Solution:

Given Data:

$u = 72$ km/hr $= 72 \times \dfrac{5}{18} = 20$ m/sec

$s = 240$ m

$t = 20$ sec

here the bike rider would cover 240 m in 20 sec so that he can move further after turning green light.

We know that,

$s = ut + \dfrac{1}{2}at^2$

$240 = 20 \times 20 + \dfrac{1}{2} \times a(20)^2$

$a = -0.8$ m/sec^2

The speed of bike when it passes the traffic light will be given by,

$v = u + at$
$ = 20 + (-0.8) \times 20$
$v = 4$ m/sec

Or

$v^2 = u^2 + 2as$
$v^2 = (20)^2 + 2(-0.8) \times 240$
$v = 4$ m/sec

10.5 Graphical Method for Motion Curves

The motion curves for displacement, velocity and acceleration can be plotted with respect to time in graphical method. The different types of curves are discussed below:

Displacement v/s Time graph:

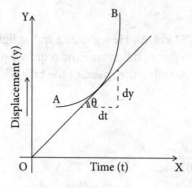

Fig. 10.3

In this curve, time and displacement are taken along X-axis and Y-axis respectively. Consider Fig. 10.3 where a particle travels displacement dy during time dt, then slope of the curve θ provides the velocity i.e.,

$$\theta = \frac{dy}{dt} \text{ or } v = \frac{dy}{dt}$$

Velocity v/s Time graph:

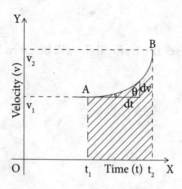

Fig. 10.4

Kinematics: Rectilinear Motion of Particles

In this curve, time and velocity are taken along X-axis and Y-axis, respectively. Consider Fig. 10.4 where a particle travels with velocity dv during time dt, then slope of the curve provides the acceleration i.e.,

$$\theta = \frac{dv}{dt} \text{ or } a = \frac{dv}{dt}$$

However the earlier equation from Displacement v/s Time graph can also be considered as $dy = v.dt$

or, displacement $= \int v.dt$

i.e., area under the velocity and time graph provides displacement shown by hatching area in Fig. 10.4.

Acceleration v/s Time graph:

In this curve, time and acceleration are taken along X-axis and Y-axis respectively. Consider Fig. 10.5.

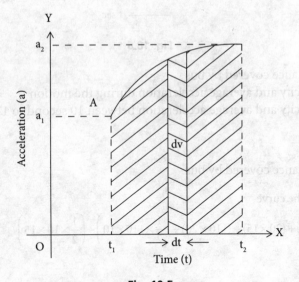

Fig. 10.5

If a particle travels with velocity dv during time dt, then

$$a = \frac{dv}{dt}$$
$$dv = a.dt$$

or, velocity, $v = \int a.dt$

i.e., area under the acceleration and time graph provides velocity.

Example: 10.12

The velocity–time curve of a bus moving in a rectilinear path is shown in Fig. 10.6. Determine:

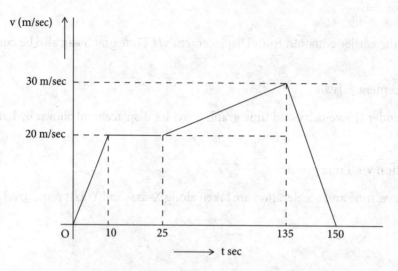

Fig. 10.6

(i) The total distance covered by bus
(ii) Average velocity and average acceleration during the motion
(iii) Average velocity and average acceleration between 10 second to 135 seconds

Solution:

(i) The total distance covered by bus

= area under the curve

$$= \left(\frac{1}{2} \times 10 \times 20\right) + (20 \times 15) + (110 \times 20) + \left(\frac{1}{2} \times 110 \times 10\right) + \left(\frac{1}{2} \times 10 \times 15\right)$$

$$= 3225 \; m$$

(ii) Average velocity during the motion will be $= \dfrac{\text{total distance}}{\text{total time}}$

$$= \frac{3225}{150}$$

$$= 21.5 \; m/sec$$

Average acceleration during the motion $= \dfrac{\text{change in velocity}}{\text{total time}}$

$$= \frac{(30-20)}{150}$$

$$= 0.067 \; m/sec^2$$

(iii) Average velocity between 10 sec to 135 seconds will be given by

$$= \frac{\text{Distance travelled between 10 sec to 135 sec}}{\text{time interval}}$$

$$= \frac{\left[(20 \times 15) + \left(\frac{1}{2} \times 110 \times 10\right) + (20 \times 110)\right]}{(135 - 10)}$$

$$= 24.4 \ m/sec$$

The average acceleration between 10 sec to 135 seconds will be given by,

$$= \frac{\text{change in velocity}}{\text{time interval}}$$

$$= \frac{(30 - 20)}{(135 - 10)}$$

$$= 0.08 \ m/sec^2$$

Example: 10.13

A vehicle starts from rest with a linear motion and comes to rest after 25 seconds. The velocity–time graph is shown in Fig. 10.7. Calculate the total distance travelled by the vehicle.

(UPTU, Ist Sem, 2000–2001)

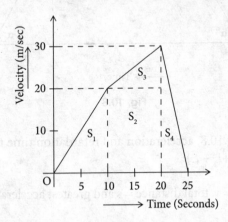

Fig. 10.7

Solution:

The total distance will be given by area under the curve. i.e., $s = s_1 + s_2 + s_3 + s_4$

$$s = \left(\frac{1}{2} \times 10 \times 20\right) + (10 \times 20) + \left(\frac{1}{2} \times 10 \times 10\right) + \left(\frac{1}{2} \times 5 \times 30\right)$$

$$= 425 \ m$$

Example: 10.14

The greatest possible acceleration or deceleration that a train may have is α and its maximum speed is v. Find the minimum time in which the train can get from one station to the next. If stations are s distance apart.

(UPTU, Ist Sem, 2001–2002)

Solution:

For minimum time, train will have to accelerate and decelerates as much as possible. As the greatest possible acceleration and deceleration is same thus α the slope, in velocity-time curve will be identical as shown in Fig. 10.8.

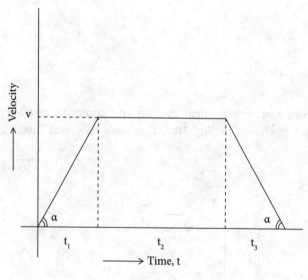

Fig. 10.8

It can be observed from fig.10.8, acceleration and retardation time t_1 and t_3 will be equal. Thus minimum time = $(t_1 + t_2 + t_3)$

Given: maximum velocity = v, total distance = s and greatest acceleration or deceleration = α

s = area under the curve

$$= \left(\frac{1}{2} \times v.t_1\right) + \left(v.t_2\right) + \left(\frac{1}{2} \times v.t_3\right)$$

$$s = \frac{v}{2}(t_1 + 2t_2 + t_3)$$

$$(t_1 + 2t_2 + t_3) = \frac{2s}{v} \qquad \qquad \ldots (1)$$

Kinematics: Rectilinear Motion of Particles

α = slope of acceleration or deceleration

$$\alpha = \frac{v}{t_1} \text{ or } \frac{v}{t_3}$$

thus, $t_1 = t_3 = \left(\frac{v}{\alpha}\right)$

Substituting value of t_1 and t_3 in equation (1),

$$\frac{v}{\alpha} + 2.t_2 + \frac{v}{\alpha} = \frac{2s}{v}$$

$$2\left(\frac{v}{\alpha} + t_2\right) = \frac{2s}{v}$$

$$t_2 = \left(\frac{s}{v} - \frac{v}{\alpha}\right)$$

Substituting t_1, t_2 and t_3 in equation (1),

$$(t_1 + t_2 + t_3) = \frac{v}{\alpha} + \left(\frac{s}{v} - \frac{v}{\alpha}\right) + \frac{v}{\alpha}$$

Minimum time for train to reach next station = $(t_1 + t_2 + t_3) = \left(\frac{s}{v} + \frac{v}{\alpha}\right)$

Example: 10.15

A metro train moves between two stations 4.2 km apart. Train starts from first station at rest and uniform accelerated to a speed of 72 km/hr in 24 seconds. The speed remains constant for some time and finally it is brought to rest at second station by deceleration of 0.75 m/sec². Determine the total time taken by train to reach second station.

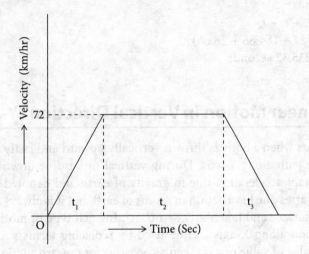

Fig. 10.9

Solution:

Consider motion curve for metro train,
Given,
$u = 0$, $s = 4.2 \times 10^3$ m
$t_1 = 24$ seconds
$v = 72$ km/hr

$$= 72 \times \frac{5}{18}$$

$v = 20$ m/sec

The total distance,

$$s = \left[\frac{1}{2}(v.t_1)\right] + \left[(v.t_2)\right] + \left[\frac{1}{2} \times t_3 \times v\right]$$

$$4.2 \times 10^3 = \left(\frac{1}{2} \times 20 \times 24\right) + (20 \times t_2) + \left(\frac{1}{2} \times 20 \times t_3\right)$$

$$396 = 2t_2 + t_3 \qquad \qquad \ldots (1)$$

During deceleration, $a = 0.75$ m/sec²
$u = v = 20$ m/sec, final velocity $v = 0$
$v = u + at$
$0 = 20 - 0.75 \times t_3$
$t_3 = 26.66$ seconds
Substituting value of t_3 in equation (1),
$396 = 2t_2 + 26.66$
$t_2 = 184.66$ seconds
Thus total time $= t_1 + t_2 + t_3$
$\qquad \qquad = (24 + 184.66 + 26.66)$
$\qquad \qquad = 235.32$ seconds

10.6 Rectilinear Motion in Vertical Direction (Y-axis)

This motion appears when a body is thrown vertically upward and lastly comes downward by following same path along Y-axis. During vertical upward or downward motion, the body bears a constant acceleration due to gravity of earth and denoted by symbol g. The value of g slightly varies due to variation in radius of earth but it is always assumed constant all over the earth surface and taken as 9.81 m/sec². In such type of motion, the equations of rectilinear motion along X-axis can be used by replacing terms s and a by h and g, respectively. The value of value of g is taken as negative for upward motion and positive for downward motion.

Kinematics: Rectilinear Motion of Particles

Thus when particle moves under gravity in Y-direction, the equations of motion will be given by:

1. $v = u \pm g.t$
2. $h = ut \pm \dfrac{1}{2} gt^2$
3. $v^2 = u^2 \pm 2gh$

Example: 10.16

A particle is dropped from the top of a tower 100 m high. After one second another particle is projected upwards from the foot of the tower which meets the first particle at a height of 18 m. Find the velocity with which the second particle was projected.

(UPTU, IInd Sem, 2001–02)

Solution:

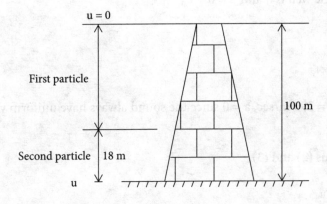

Fig. 10.10

Let the first particle meet second particle after 't' seconds when dropped from the top of the tower.

Thus first particle will fall down by = 100 – 18 = 82 m

We know,

$$h = ut + \dfrac{1}{2} gt^2$$

$$82 = 0 + \dfrac{1}{2} \times 9.81 . t^2$$

$$t = 4.09 \text{ seconds}$$

As second particle is projected after one second, it will go up 18 m in 3.09 seconds. Let the velocity of projection is 'u'

$$h = ut - \frac{1}{2} \times gt^2$$

$$18 = u(3.09) - \frac{1}{2} \times 9.81(3.09)^2$$

$$u = 20.98 \; m/\sec$$

Example: 10.17

A boy dropped his ball into a well and heard the sound of strike the water in 5.6 seconds. If the velocity of second is 290 m/sec then determine the depth of the well.

Solution:

If ball takes time t_1 to hit water and sound takes time t_2 to reach the boy.

$t_1 + t_2 = 5.6$ seconds ..... (1)

Let the depth of the well is h and u = 0

$$h = ut + \frac{1}{2}gt^2$$

$$h = 0 + \frac{1}{2} \times 9.81 \times t_1^2 \qquad \qquad ..... (2)$$

velocity of sound = 290 m/sec, a = 0 since the sound always have uniform velocity

$h = 290 \times t_2 + 0$ ..... (3)

Equating equations (2) and (3),

$$\frac{1}{2} \times 9.81 t_1^2 = 290 t_2$$

$$t_1^2 - 59.12 t_2 = 0$$

Substituting $t_2 = (5.6 - t_1)$
$t_1^2 - 59.12(5.6 - t_1) = 0$
$t_1^2 + 59.12 t_1 - 331.1 = 0$
$t_1 = 5.15$ sec

Substituting value of t_1 in equation (2),

$$h = \frac{1}{2} \times 9.81 (5.15)^2$$

$$h = 130.09 \; m$$

Example: 10.18

A cage descends in a mine shaft with an acceleration of 0.6 m/sec². After the cage has travelled 30 m, a stone is dropped from the top of the shaft. Determine:

(i) The time taken by the stone to hit the cage
(ii) Distance travelled by the cage before impact

(UPTU, Ist Sem, 2003–04)

Solution:

Given: $u = 0$, $a = 0.6$ m/sec^2, $h = 30$ m
Time taken by cage to descend $= t$

We know that $\quad h = ut + \dfrac{1}{2} \cdot a \cdot t^2$

$$30 = 0 + \dfrac{1}{2} \times 0.6 \times t^2$$

$$t = 10 \text{ sec}$$

(i) Let the stone takes time t' seconds to hit the cage after traveling distance y. Here stone will fall under acceleration, $g = 9.81$ m/sec^2 however cage is descending under acceleration, $a = 0.6$ m/sec^2 and will take time $(t' + 10)$ to travel same distance y.

Distance travelled by stone,

$$y = ut + \dfrac{1}{2} gt^2$$

$$y = 0 + \dfrac{1}{2} \times 9.81 \times (t')^2 \qquad \ldots (1)$$

And distance travelled by cage,

$$y = ut + \dfrac{1}{2} at^2$$

$$y = 0 + \dfrac{1}{2} \times 0.6 (t' \times 10)^2 \qquad \ldots (2)$$

Equalising equations (1) and (2)

$$\dfrac{1}{2} \times 9.81 (t')^2 = \dfrac{1}{2} \times 0.6 (t' + 10)^2$$

$$9.81 (t')^2 = 0.6 \left((t')^2 + 20t' + 100 \right)$$

$$9.81 (t')^2 - 12t' - 60 = 0$$

$$3.07 (t')^2 - 4t' - 20 = 0$$

$$t' = 3.29 \; sec$$

(ii) Distance travelled by cage before impact, using equation (1),

$$y = \dfrac{1}{2} \times 9.81 (3.29)^2$$

$$y = 53.09 \; m$$

Example: 10.19

Two persons are having balls, one at the top of a multi-storey building and another one 32 m below the top of the building. When ball is dropped from the top of the building by first person for height 15 m, the second person also dropped his ball. If both balls hit the ground at the same time then determine:

(i) Height of the building
(ii) Velocity of both stones when hit the ground
(iii) Time taken by both stones to hit the ground

Solution:

Let the height of the building is h,

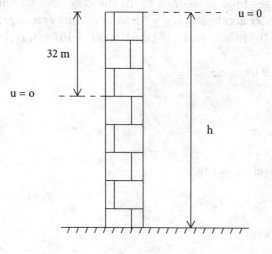

Fig. 10.11

For first person ball,
$u = 0$, $h' = 15$ m
$v^2 = u^2 + 2gh'$
$v^2 = 0 + 2 \times 9.81 \times 15$
$v = 17.16$ m/sec i.e., the velocity of first ball. After this, let both balls take time 't' to hit the ground.

First ball will travel by = $(h - 15)$ and second ball will travel by = $(h - 32)$

i.e., $(h-15) = 17.16t + \dfrac{1}{2} \times 9.81(t)^2$ ..... (1)

$(h-32) = 0 + \dfrac{1}{2} \times 9.81(t)^2$ ..... (2)

Subtracting equation (2) from equation (1)
$h - 15 - h + 32 = 17.16t$
$t = 0.99$ sec

(i) Height of the building,

$$(h - 32) = \frac{1}{2} \times 9.81 (0.99)^2$$
$$h - 32 = 4.81$$
$$h = 36.81 \; m$$

(ii) Velocity of first stone when hit the ground,
$v^2 = u^2 + 2gh$
$v^2 = 0 + 2 \times 9.81 \times 36.81$
$v = 26.87$ m/sec

Velocity of second stone,
$v^2 = 0 + 2 \times 9.81 \times 4.81$
$v = 9.71$ m/sec

(iii) Time taken by first stone to hit the ground,
$v = u + gt$
$26.87 = 0 + 9.81 \times t_1$
$t_1 = 2.74$ sec
Time taken by second stone to hit the ground,
$t_2 = 0.99$ sec

Theoretical Problems

T 10.1 Define the difference between statics and dynamics.
T 10.2 Define the difference between kinematics and kinetics.
T 10.3 Discuss the average velocity and instantaneous velocity.
T 10.4 Differentiate the distance and displacement during motion of a body.
T 10.5 What do you mean by particle?
T 10.6 Define rectilinear motion of a body.
T 10.7 Discuss the differential equations of motion used under variable acceleration.
T 10.8 Discuss the different types of motion curves.

Numerical Problems

N 10.1 The equation of motion of a particle moving in a straight line is given by:

$$s = 18t + 3t^2 - 2t^3$$

where s is the total distance covered from the starting point in meters at the end of t seconds. Find out:

(i) velocity and acceleration at the start
(ii) the time when the particle reaches its maximum velocity

[UPTU 2013–14]

N 10.2 The acceleration of a particle moving along a straight line is given by the relation $a = 3(v)^{2/3}$. When $t = 3$ sec, its displacement is $s = 37.516$ m and velocity $v = 42.87$ m/sec. Determine the displacement, velocity and acceleration when $t = 5$ sec

N 10.3 The vehicle starting from rest is accelerated along straight line where acceleration is governed by relation, $a = \left(\dfrac{10}{2v+3}\right)$, where 'a' is the acceleration in m/sec² and v is the velocity in m/sec. Determine:
(i) time taken by vehicle to reach velocity of 12 m/sec
(ii) distance moved during that time

N 10.4 A scooterist is travelling at 45 km/hr in a straight road observes a traffic signal light 250 m ahead of him turning red. The light is timed to stay red for 15 seconds. If the scooterist wishes to pass the signal without stopping, just as it turn green, find:
(i) the required uniform deceleration of scooter.
(ii) the speed of scooter as it passes the signal.

N 10.5 A truck uniformly accelerates to 0.6 m/sec² for 10 sec. If truck started from rest, determine the distance travel for this duration and its final velocity also.

N 10.6 An automobile starts from rest with a linear motion and comes to rest after 60 seconds. The velocity-time graph is shown in Fig. NP. 10.1. Determine:

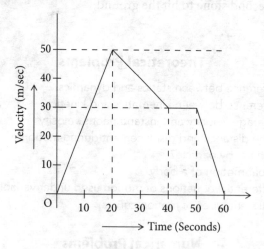

Fig. NP 10.1

(i) The total distance travelled by the vehicle
(ii) Average velocity and average acceleration during the motion

Kinematics: Rectilinear Motion of Particles 527

N 10.7 A boy dropped a stone from the top of a building 88 m high. Another boy is standing at foot of the building projects stone upward after 1.2 sec. If both stones meet at a height of 14 m, determine the projection velocity of second stone.

N 10.8 A car starts from rest, accelerated to an acceleration of 0.8 m/sec². Afterwards it comes to rest with deceleration of 1.2 m/sec² after travelling a distance of 1.5 km. Determine:
  (i) The distance travelled by car during acceleration
  (ii) Maximum velocity of the car

N 10.9 A bike is moving under uniform acceleration, travels 40 m in tenth second and 60 m in 14 seconds. Determine the initial velocity at start and uniform acceleration.

N 10.10 Three persons P, Q, R are standing 80 m apart on a straight highway. A bike starting from rest moves with uniform acceleration passes person P and then takes 8 seconds to reach person Q and further 6 seconds to reach last person R.
Determine:
  (i) The velocity of the bike when crosses person P and Q
  (ii) Acceleration of the bike
  (iii) Distance travelled by bike when it crosses person R

Multiple Choice Questions

1. Kinematics problems are related with the following parameters
   a. displacement, acceleration, velocity and time
   b. force, displacement, acceleration, velocity and time
   c. displacement, acceleration and velocity
   d. all of these

2. During motion a body can be idealized as a particle if it does not consist
   a. translatory motion
   b. rotary motion
   c. either (a) or (b)
   d. none of these

3. In rectilinear motion, determine the correct relation between distance and displacement
   a. displacement ≥ distance
   b. displacement ≤ distance
   c. displacement = distance
   d. none of these

4. State the scalar quantity out of following quantities
   a. displacement
   b. velocity
   c. acceleration
   d. distance

5. If a body travels equal displacements in equal time intervals, the motion is called as
   a. uniform velocity
   b. instantaneous velocity
   c. average velocity
   d. none of these

6. If a body travels with uniform velocity then it contains
   a. constant acceleration
   b. instantaneous acceleration
   c. average acceleration
   d. zero acceleration

7. The variable acceleration in differential equations of motion is given by
   a. $\dfrac{dv}{dt}$
   b. $v\dfrac{dv}{dx}$
   c. $\dfrac{d^2x}{dt^2}$
   d. all of these

8. The equation of motion $v^2 = (u^2 \pm 2\,as)$ is valid if body moves with
   a. constant acceleration
   b. instantaneous acceleration
   c. average acceleration
   d. none of these

9. The distance covered by a particle during n^{th} second is given by
   a. $u + \dfrac{a}{2}(n-2)$
   b. $u + \dfrac{a}{2}(2n-1)$
   c. $2u + \dfrac{a}{2}(2n-1)$
   d. none of these

10. In a Displacement v/s Time graph, the slope of curve provides
    a. acceleration
    b. velocity
    c. average acceleration
    d. none of these

11. State area under which graph provides velocity
    a. acceleration v/s time graph
    b. average velocity v/s time graph
    c. displacement v/s time graph
    d. none of these

12. State area under which graph provides displacement
    a. acceleration v/s time graph
    b. average velocity v/s time graph
    c. displacement v/s time graph
    d. none of these

Answers

1. a   2. b   3. b   4. d   5. a   6. d   7. d   8. a   9. b   10. b   11. a   12. b

Chapter 11

Kinematics: Curvilinear Motion of Particles

11.1 Introduction

When the motion of a body takes place along a curved path, it is called curvilinear motion. The curvilinear motion is known as a two-dimensional motion as it takes place in the X–Y plane. The basic difference between rectilinear and curvilinear motion is that in a rectilinear motion, all particles of a body move parallel to each other along a straight line but in a curvilinear motion all particles move parallel to each other along a curved path in a common plane. Some common examples of curvilinear motion are as follows:

a. A vehicle passing over a speed breaker.
b. Turning of a vehicle along a curved path.
c. Motion of a ball during lofted six in cricket.
d. Motion of missile fired from a fighter plane.
e. Oscillatory motion of a pendulum.

11.2 Rectangular Coordinates

This system is used to analyse the curvilinear motion of a particle moving in X–Y plane. Here the velocity and acceleration are resolved in two perpendicular components along X and Y axis. The resultant value of velocity and acceleration can be determined by using vector approach or combining its components as discussed below:

Velocity: Consider a ball during lofted six, it consists two types of motions, one rotary motion and another curvilinear motion. Here the rotary motion is neglected and assumed that all particles of ball will travel with same displacement, velocity and acceleration. Consider a displacement-time graph where a particle is moving along a curved path in X–Y plane. Let at any instant t, particle has position A. If particle moves from point A to B by displacement Δs during time interval Δt as shown in Fig. 11.1.

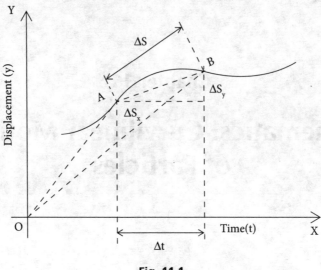

Fig. 11.1

where $\Delta s = \Delta s_x + \Delta s_y$
We know that the instantaneous velocity is given by

$$v_{inst.velocity} = \lim_{\Delta t \to 0} \frac{\Delta s}{\Delta t}$$

$$v = \lim_{\Delta t \to 0} \frac{\Delta s_x + \Delta s_y}{\Delta t}$$

$$v = \lim_{\Delta t \to 0} \frac{\Delta s_x}{\Delta t} + \lim_{\Delta t \to 0} \frac{\Delta s_y}{\Delta t}$$

$$v = v_x + v_y$$

The above equation provides instantaneous velocity by using vector approach. It can also obtained from its components,

$$v = \sqrt{v_x^2 + v_y^2} \text{ and } \theta = \tan^{-1}\left(\frac{v_y}{v_x}\right)$$

Acceleration: Consider velocity–time graph where a particle is moving along a curved path in the X–Y plane. Let at any instant t, particle has velocity v. During motion from point A to B, if velocity of particle increases by Δv in time interval Δt as shown in Fig. 11.2.

Kinematics: Curvilinear Motion of Particles

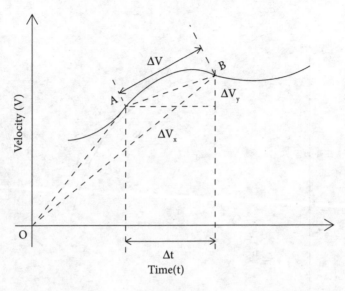

Fig. 11.2

where $\Delta v = \Delta v_x + \Delta v_y$
We know that the instantaneous acceleration is given by

$$a_{inst.acceleration} = \lim_{\Delta t \to 0} \frac{\Delta v}{\Delta t}$$

$$a = \lim_{\Delta t \to 0} \frac{\Delta v_x + \Delta v_y}{\Delta t}$$

$$a = \lim_{\Delta t \to 0} \frac{\Delta v_x}{\Delta t} + \lim_{\Delta t \to 0} \frac{\Delta v_y}{\Delta t}$$

$$a = a_x + a_y$$

The above equation provides instantaneous acceleration by using vector approach. It can also obtained by combining its components,

$$a = \sqrt{a_x^2 + a_y^2} \text{ and } \theta = \tan^{-1}\left(\frac{a_y}{a_x}\right)$$

11.3 Tangential and Normal Components of Acceleration

In a curvilinear motion the velocity of particle remains always parallel to its path but acceleration lies along the direction of change in velocity. Thus acceleration is used to describe motion via tangential and normal to the path of a curvilinear motion. Let the acceleration along the tangent and normal to the path are denoted by a_t and a_n respectively. Consider Fig. 11.3 (a) where a particle moves from point A to point A_1 in time interval Δt and radius of curvature of curve is r.

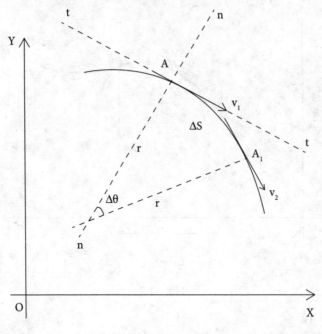

Fig. 11.3 (a)

If particle has velocity v_1, v_2 at point A and A_1, respectively, a velocity vector diagram will be drawn as shown in Fig. 11.3 (b). Such a diagram is further used to determine the tangential and normal acceleration as discussed below:

First draw vector v_1 and v_2 by vector $\overrightarrow{ab}$ and vector $\overrightarrow{ac}$. The change in velocity Δv can be determined by closing side by $\overrightarrow{bc}$. Further $\overrightarrow{bc}$ can be resolved along tangent to the path as $\overrightarrow{bd}$ and normal to the path as $\overrightarrow{dc}$ which can be represented by components Δv_t and Δv_n. It can be expressed as

$$\Delta v = \Delta v_t + \Delta v_n$$

We know that the instantaneous acceleration is given by,

$$a_{inst.acceleration} = \lim_{\Delta t \to 0} \frac{\Delta v}{\Delta t}$$

$$a = \lim_{\Delta t \to 0} \frac{\Delta v_t + \Delta v_n}{\Delta t}$$

$$a = \lim_{\Delta t \to 0} \frac{\Delta v_t}{\Delta t} + \lim_{\Delta t \to 0} \frac{\Delta v_n}{\Delta t}$$

$$a = a_t + a_n$$

Kinematics: Curvilinear Motion of Particles

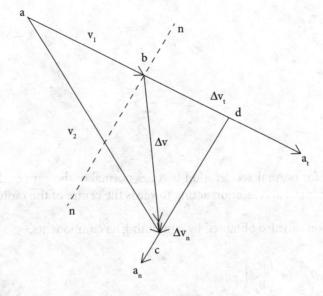

Fig. 11.3 (b)

where a_t lies tangent to the path and a_n lies normal to the path. When Δt tends to zero, the point A_1 approaches to point A and the a_t lies tangent to the path and a_n lies normal to the path about point A.

Further $\overline{bd}$ represents changes in magnitude of velocity in tangential direction, thus acceleration in tangential direction will be

$$a_t = \lim_{\Delta t \to 0} \frac{(v_2 - v_1)}{\Delta t}$$

$$a_t = \lim_{\Delta t \to 0} \frac{\Delta v}{\Delta t}$$

$$a_t = \frac{dv}{dt}$$

It is the expression for tangential acceleration lies along the path of the particle.

Further $\overline{dc}$ represents changes in magnitude of velocity in normal direction, thus acceleration in normal direction will be

$$a_n = \lim_{\Delta t \to 0} \frac{\Delta v_n}{\Delta t}$$

where $\Delta v_n = \overline{dc} = v\Delta\theta$

Thus $a_n = \lim_{\Delta t \to 0} \dfrac{v\Delta\theta}{\Delta t}$

where $\Delta\theta = \dfrac{\Delta s}{r}$

$$a_n = \lim_{\Delta t \to 0} \left(\frac{v \Delta s}{r \Delta t} \right)$$

$$a_n = \frac{v}{r} \lim_{\Delta t \to 0} \left(\frac{\Delta s}{\Delta t} \right)$$

$$a_n = \frac{v}{r} \left(\frac{ds}{dt} \right)$$

$$a_n = \frac{v^2}{r}$$

It is the expression for normal acceleration that lies normal to the path of the particle. It is also known as centripetal acceleration acting towards the centre of the radius of curvature.

The total acceleration can also obtained by combining its components,

$$a = \sqrt{a_x^2 + a_y^2} \text{ and } \theta = \tan^{-1}\left(\frac{a_y}{a_x}\right)$$

Example: 11.1

The curvilinear motion of particle is defined along X and Y direction by following equations:

$x = (3 + 4t^2), y = (5 + 2t^3)$

Determine the position, velocity and acceleration of the particle at 4 sec.

Solution:

Given: $x = (3 + 4t^2), y = (5 + 2t^3)$
The position of particle after 4 sec will be,

$x = 3 + 4(4)^2$

$x = 67 \, m$

and $y = (5 + 2(4)^3)$

$= 133 \, m$

Thus magnitude of position,

$\Delta s = \sqrt{x^2 + y^2}$

$= \sqrt{(67)^2 + (133)^2}$

$\Delta s = 148.92 \, m$

Kinematics: Curvilinear Motion of Particles

and direction,

$$\theta_s = \tan^{-1}\left(\frac{y}{x}\right)$$

$$= \tan^{-1}\left(\frac{133}{67}\right)$$

$$\theta_s = 63.26°$$

The velocity of the particle can be determined by differentiating position of particle,

i.e. $v_x = \dfrac{dx}{dt} = 8t$

v_x at 4 sec = 32 m/sec

and $v_y = \dfrac{dy}{dt} = 6t^2$

v_y at 4 sec = $6(4)^2 = 96$ m/sec

Thus velocity of particle at 4 sec will be,

$$v = \sqrt{v_x^2 + x_y^2}$$

$$= \sqrt{32^2 + 96^2}$$

$$v = 101.19 \ m/sec$$

and direction,

$$\theta_v = \tan^{-1}\left(\frac{v_y}{v_x}\right)$$

$$= \tan^{-1}\left(\frac{96}{32}\right)$$

$$\theta_v = 71.57°$$

Similarly, acceleration will be determined by differentiating velocity of particle in x and y direction,

$$a_x = \frac{dv_x}{dt} = 8 \quad \text{and} \quad a_y = \frac{dv_y}{dt} = 12t$$

At 4 sec,
$a_x = 8 \ m/sec^2$, $a_y = 48 \ m/sec^2$

The acceleration of the particle at 4 sec will be,

$$a = \sqrt{a_x^2 + a_y^2}$$

$$= \sqrt{8^2 + 48^2}$$

$$a = 48.66 \ m/sec^2$$

and direction,

$$\theta_a = \tan^{-1}\left(\frac{a_y}{a_x}\right)$$
$$= \tan^{-1}\left(\frac{48}{8}\right)$$
$$\theta_a = 80.54°$$

Example: 11.2

The motion of a particle lies along path $y^2 = 1.2x^5$, where x and y are expressed in metres. The position of particle in 'x' direction with respect to time is given by relation $x = t/6$.

Determine the position, velocity and acceleration of the particle along y-direction at $x = 2\ m$.

Solution:

Given: $y^2 = 1.2x^5$

i.e. $y = \sqrt{1.2}\,x^{5/2} = 1.10 x^{5/2}$

as $x = t/6$

thus, $y = 1.10\dfrac{t^{5/2}}{(6)^{5/2}}$

$y = 0.012 t^{5/2}$ ..... (1)

at $x = 2\ m$,

$2 = \dfrac{t}{6}$

$t = 12$ sec

Substituting value of 't' in equation (1),

$y = 0.012(12)^{5/2}$

$y = 5.99\ m$

The velocity of particle can be obtained by differentiating equation (1),

$v_y = \dfrac{dy}{dt} = 0.012 \times \dfrac{5}{2} t^{3/2}$ ..... (2)

$v_y = 0.012 \times \dfrac{5}{2}(12)^{3/2}$

$v_y = 1.25\ m/\sec$

The acceleration of particle can be obtained by differentiating equation (2),

$$a_y = \frac{dv_y}{dt} = 0.012 \times \frac{5}{2} \times \frac{3}{2} t^{1/2}$$

$$a_y = 0.012 \times \frac{5}{2} \times \frac{3}{2} (12)^{1/2}$$

$$a_y = 0.16 \, m/\sec^2$$

Example: 11.3

The curvilinear motion of a particle is given by equation $y^2 = \frac{5}{3}.x$

where x and y are expressed in metres.
The position of particle in 'x' direction with respect to time is given by relation, $x = t^2/2$. Determine the position, velocity and acceleration of the particle along y-direction when particle has position $x = 8$ m.

Solution:

Given: The equation of motion is

$$y^2 = \frac{5}{3}.x \quad \ldots (1)$$

The position of particle in x-direction with respect to time is

$$x = t^2/2 \quad \ldots (2)$$

From equation (1) and (2),

$$y^2 = \frac{5}{3} \cdot \frac{t^2}{2}$$

$$y^2 = \frac{5t^2}{6}$$

i.e. $y = 0.91 \, t$ ..... (3)

At $x = 8$ m the time, t will be given by equation (2),

$$8 = \frac{t^2}{2}$$

$$t = 4 \, \sec$$

The position of particle at $x = 8$ m,
Using equation (1),

$$y^2 = \frac{5}{3} \times 8$$

$$y = 3.65 \, m$$

Or, using equation (3), at $t = 4$ sec

$y = 0.91t$

$y = 0.91(4)$

$y = 3.64\ m$

The differentiating equation (3) will provide velocity along y-direction,

$$v_y = \frac{dy}{dt} = 0.91\ m/sec$$

Similarly, differentiating equation (3) twice will provide acceleration along y-direction,

$$a_y = \frac{dv_y}{dt} = 0$$

Example: 11.4

In a bike race, a bike rider is going at a speed of 90 *km/hr* along a curved road of radius 290 *m*. Determine the tangential, normal and total acceleration of the bike. Further, if the bike is accelerated to a speed of 144 *km/hr* in 6 seconds, then determine the tangential, normal and total acceleration.

Solution:

Given: Constant speed of bike = 90 *km/hr*

$$= 90 \times \frac{5}{18}$$

i.e., $v = 25\ m/sec$

$r = 290\ m$

We know that, tangential acceleration $= \frac{dv}{dt}$

$a_t = 0$ as $dv = 0$

The normal acceleration, $= \frac{v^2}{r}$

$$a_n = \frac{(25)^2}{290}$$

$= 2.16\ m/sec^2$

Thus total acceleration of bike,

$a = \sqrt{a_t^2 + a_n^2}$

$= \sqrt{0 + (2 \cdot 16)^2}$

$= 2.16\ m/sec$

Kinematics: Curvilinear Motion of Particles

Further, after acceleration,

$$v = 144 \ km/hr$$
$$= 144 \times \frac{5}{18}$$
$$v = 40 \ m/\sec$$

and t = 6 sec

The tangential acceleration,

$$= \frac{dv}{dt} = \frac{\text{Change in velocity}}{\text{time interval}}$$
$$= \frac{(40-25)}{6}$$
$$a_t = 2.5 \ m/\sec^2$$

The normal acceleration,

$$a_n = \frac{v^2}{r}$$
$$= \frac{(40)^2}{290}$$
$$= 5.52 \ m/\sec^2$$

Total acceleration,

$$a = \sqrt{a_t^2 + a_n^2}$$
$$= \sqrt{(2\cdot 5)^2 + (5\cdot 52)^2}$$
$$= 6.06 \ m/\sec^2$$

11.4 Projectile

When a particle is thrown up from ground at any angle α, (0° < α < 90°), it traverses a curved path called projectile. In projectile motion the particle bears simultaneously the effect of both horizontal motion as well as vertical motion. In the absence of air resistance, the horizontal component of projectile motion remains constant as the acceleration does not act in horizontal direction. However, the vertical component of projectile motion varies as it is subjected to gravitational acceleration or deceleration.

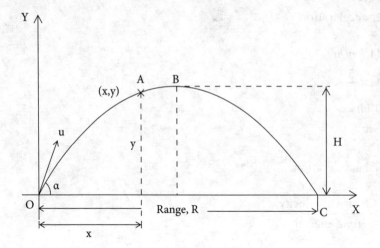

Fig. 11.4

Consider Fig. 11.4, where a particle traverses a curved path/projectile. The various terms related to a projectile are as follows:

Velocity of projection is the initial velocity with which particle is thrown in the atmosphere and represented by **u**.

Angle of projection is the angle α at which a particle is thrown. Its value lies in the range (0° < α < 90°).

Trajectory is the path followed by a particle during its flight.

Range, R is the horizontal distance covered by a particle during its flight.

Time of flight, T is the time taken by a particle to complete the trajectory.

Projectile Motion: Consider Fig. 11.4, where a particle is thrown upward from origin point 'O' with initial velocity 'u' at an angle of projection 'α' on a horizontal plane. As discussed earlier, the particle experiences combination of horizontal motion and vertical motion during its projectile motion. Thus horizontal motion includes, the horizontal component of initial velocity = $u\cos\alpha$ which remains constant as acceleration remains zero in horizontal direction. If the total time taken by particle during flight is T,

the distance along horizontal plane = $u\cos\alpha \times T$

i.e., Range, R = $u\cos\alpha \times T$ ..... (1)

Let the particle reaches the point A(x, y) after 't' seconds as shown in Fig. 11.4, the values of the coordinates of A will be,

Kinematics: Curvilinear Motion of Particles

$$x = u\cos\alpha \times t \quad \ldots (2)$$

$$y = \left(u\sin\alpha \times t - \frac{1}{2}gt^2\right) \quad \ldots (3)$$

From equation (2),

$$t = \left(\frac{x}{u\cos\alpha}\right)$$

Substituting value of 't' in equation (3),

$$y = u\sin\alpha\left(\frac{x}{u\cos\alpha}\right) - \frac{1}{2}g\left(\frac{x}{u\cos\alpha}\right)^2$$

$$y = \left[x \cdot \tan\alpha - \frac{1}{2}g \cdot \frac{x^2}{u^2\cos^2\alpha}\right] \quad \ldots (4)$$

which is called the equation of projectile motion.

Since this equation is in the form of $y = [A.x - B.x^2]$ which represents parabola, thus equation (4) confirms that the path of a projectile is parabola.

Now different terms associated with the projectile can also be determined.

Maximum height (H): Consider Fig. 11.4, when particles reach at point B, the vertical component of initial velocity will become zero. If particle reaches at maximum height H, using equation of motion,

$$v^2 = u^2 - 2gh$$

$$0 = (u\sin\alpha)^2 - 2g \cdot H$$

$$H = \frac{u^2\sin^2\alpha}{2g} \quad \ldots (5)$$

Time of flight (T): If particle takes time t^1 to reach at position B, then using equation of motion

$$v = u - gt$$

$$0 = u\sin\alpha - g \cdot t^1$$

Thus time taken by particle to reach at maximum height or position B, $t^1 = \left(\dfrac{u\sin\alpha}{g}\right)$

Thus time of flight will be

$$T = 2t^1$$

$$T = \left(\frac{2u\sin\alpha}{g}\right) \quad \ldots (6)$$

Range (R): The horizontal distance covered by particle is OC
i.e., R = $u \cos \alpha \times$ T from equation (1),
Substituting value of T,

$$R = u \cos \alpha \left(\frac{2u \sin \alpha}{g} \right)$$

$$R = \left(\frac{u^2 \cdot \sin 2\alpha}{g} \right) \quad\quad\quad ..... (7)$$

Since $\sin 2\alpha$ will have same value as $\sin(180 - 2\alpha)$, there will be two angles of projection which will provide same range, i.e., α and $(90 - \alpha)$.

Maximum Range (R): A projectile can have maximum range if $\sin 2\alpha$ becomes maximum.

The maximum value of $\sin 2\alpha = 1$
$\quad\quad\quad$ i.e., $\alpha = 45°$
from equation (7),

$$R_{max} = \left(\frac{u^2}{g} \right) \quad\quad\quad ..... (8)$$

Example: 11.5

A player kicks the football with a velocity of 24 *m/sec* at an angle of 36° from the ground. Determine for this projectile:

(i) The total time of flight
(ii) The maximum height attained by football
(iii) The velocity of football after 1.2 sec and 2 sec
(iv) The range of football
(v) The maximum range of football it could attain

Solution:

Given: $u = 24$ *m/sec*, $\alpha = 36°$

(i) The total time of flight,

$$T = \left(\frac{2u \sin \alpha}{g} \right)$$
$$= \frac{2 \times 24 \sin 36°}{9.81}$$
$$= 2.88 \text{ sec}$$

Kinematics: Curvilinear Motion of Particles

(ii) The maximum height attained by football,

$$H = \frac{u^2 \sin^2 \alpha}{2g}$$
$$= \frac{(24)^2 \cdot \sin^2 36°}{2 \times 9 \cdot 81}$$
$$= 10.14 \, m$$

(iii) Since total time of flight is 2.88 sec which means football will reach its maximum height in 1.44 sec. Thus in 1.2 sec it will travel less than of its maximum height. Let the football reach at point 'A' with velocity V_A as shown in Fig. 11.5.

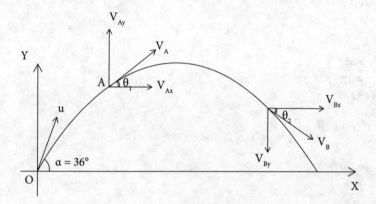

Fig. 11.5

where,

$$v_{Ax} = u \cos\alpha$$
$$= 24 . \cos 36°$$
$$= 19.42 \, m/sec$$

$$v_{Ay} = u \sin\alpha - gt$$
$$= 24 . \sin 36° - 9.81 \times 1.2$$
$$= 2.33 \, m$$

$$v_A = \sqrt{v_{Ax}^2 + v_{Ay}^2} = \sqrt{(19.42)^2 + (2.33)^2}$$
$$= 19.56 \, m/sec$$

$$\tan\theta_1 = \frac{V_{Ay}}{V_{Ax}} = \frac{2.33}{19.42}$$

$$\theta_1 = 6.84°$$

For $t = 2$ sec, the football will be ahead of its maximum height. Let the football reach the point B with velocity v_B,

where $v_{Bx} = 24.\cos 36°$

$= 19.42$ m/sec

$v_{By} = u\sin\alpha - gt$

$= 24.\sin 36° - 9.81 \times 2$

$= -5.51$ m/sec

$v_B = \sqrt{v_{Bx}^2 + v_{By}^2}$

$= \sqrt{(19.42)^2 + (-5.51)^2}$

$v_B = 20.19$ m/sec

$\tan\theta_2 = \dfrac{V_{By}}{V_{Bx}}$

$= \dfrac{5.51}{19.42}$

$\theta_2 = 15.84°$

(iv) Range of football,

$R = \left(\dfrac{u^2 \sin 2\alpha}{g}\right)$

$= \left(\dfrac{(24)^2 \cdot \sin 72°}{9.81}\right)$

$= 55.84$ m

(v) The maximum range of football will be when

$\sin 2\alpha = 1 = \sin 90°$

i.e., $\alpha = 45°$

$R_{max} = \dfrac{u^2}{g} = \dfrac{(24)^2}{9.81}$

$= 58.72$ m

Example: 11.6

In a cricket match, a fielder wants to throw the ball to the wicket keeper who is standing at 48 m apart. If the fielder throws the ball with velocity of 32 m/sec, determine:

(i) The suitable angle of projection so that the ball reaches to the wicket keeper as early as possible.
(ii) The time taken by ball to reach the wicket keeper.

Kinematics: Curvilinear Motion of Particles

Solution:

Assume that both cricketers have height.
Given, R = 48 m, u = 32 m/sec

(i) Since range,

$$R = \frac{u^2 \sin 2\alpha}{g}$$

$$48 = \frac{(32)^2 \cdot \sin 2\alpha}{9.81}$$

$$\sin 2\alpha = 0.46$$

i.e., sin 2α = sin(180 − 2α) = 0.46
thus, 2α = 27.38° and sin(180 − 2α) = 27.38°
α = 13.69° and 76.31°

Thus, to perform the task, the player may throw at 13.69° or 76.31° but the suitable angle of projection will be 13.69° as it will take lesser time for the ball to reach the wicket keeper.

(ii) Time taken by ball,

$$T = \frac{2u \sin \alpha}{g}$$

for α = 13·69°

$$T = \frac{2 \times 32 \sin 13 \cdot 69}{9 \cdot 81} = 1.54 \text{ sec}$$

for α = 76·31°

$$T = \frac{2 \times 32 \sin 76 \cdot 31}{9 \cdot 81} = 6.34 \text{ sec}$$

Thus ball will take 1.54 sec to reach the wicket keeper for suitable projection 13.69°.

Example: 11.7

An athlete throws an iron ball with initial velocity of 8 m/sec in a shot put game. If he releases the ball from height 1.0 m from the ground with angle of projection 30°, determine the maximum distance covered by ball.

Solution:

Given, y = −1.0 m, u = 8 m/sec, α = 30°
Consider Fig. 11.6.

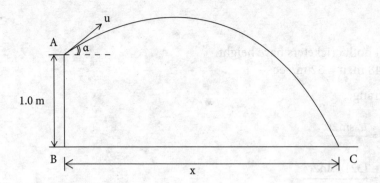

Fig. 11.6

Using the equation of projectile motion,

$$y = x\tan\alpha - \frac{1}{2} \cdot \frac{gx^2}{u^2 \cos^2\alpha}$$

$$-1 \cdot 0 = x\tan 30° - \frac{1}{2} \times \frac{9.81 x^2}{8^2 \cdot \cos^2 30°}$$

$$x^2 - 5 \cdot 65x - 9 \cdot 79 = 0$$

$$x = 7 \cdot 04 \; m$$

Alternative Method:

If the ball takes time 't' from A to C then range, $x = u \cos\alpha \times t$

$$x = 8 \cdot \cos 30° \times t \qquad \qquad \ldots\ldots (1)$$

Using equation,

$$h = ut - \frac{1}{2}gt^2$$

$$-1 \cdot 0 = u \sin\alpha \cdot t - \frac{1}{2} \times 9.81 t^2$$

$$-1 \cdot 0 = 8 \sin 30° \cdot t - \frac{9.81 t^2}{2}$$

$$9.81 t^2 - 8t + 2 = 0$$

$$t = 1 \cdot 02 \; sec$$

Substituting value of 't' in equation (1),
$x = 7.06 \; m$

Example: 11.8

A football player kicks the football at an angle of 25° from the ground. Determine the range and time of flight if the initial velocity of the ball is 20 m/sec.

Kinematics: Curvilinear Motion of Particles

Solution:

Given, α = 25°, u = 20 m/sec
Range,

$$R = \left(\frac{u^2 \sin 2\alpha}{g}\right)$$

$$= \frac{(20)^2 \sin 50°}{9.81}$$

$$= 31.24 \ m$$

Time of flight,

$$T = \left(\frac{2u \sin \alpha}{g}\right)$$

$$= \left(\frac{2 \times 20 \sin 25°}{9.81}\right)$$

$$= 1.72 \ sec$$

Example: 11.9

In a construction site, a labourer at ground wants to throw bricks to a person standing on the roof of a building. If the total height at which brick is to be collect 15 m and the distance between labourer and building is 9 m, determine the suitable velocity with which brick should be thrown. Take angle of projection as 70°.

Solution:

Consider Fig. 11.7. Assume that both labourer have same height thus ignored here

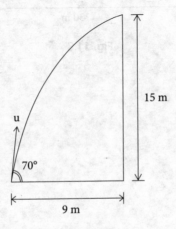

Fig. 11.7

Using equation of projectile motion,

$$y = x\tan\alpha - \frac{1}{2}g \cdot \frac{x^2}{u^2\cos^2\alpha}$$

$$15 = 9\tan 70° - \frac{1}{2} \times \frac{9.81 \times (9)^2}{u^2\cos^2 70°}$$

$$u = 18.66 \ m/sec$$

Example: 11.10

During a Holi festival, a boy throws a water balloon from the roof of a building. If the height of boy and building is 1.5 m and 11 m, respectively, determine the initial velocity by which the balloon should be thrown so that it could hit a person sitting on ground at 30 m away from building. Also determine the time taken by balloon to hit the person. Take angle of projection as 30°.

Solution:

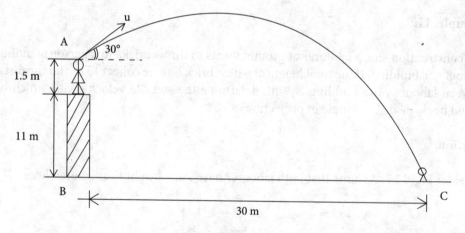

Fig. 11.8

Consider Fig. 11.8.

Given, $y = -(11 + 1.5)$ m

$= -12.5$ m

$x = 30$ m, $\alpha = 30°$

Kinematics: Curvilinear Motion of Particles

Using equation of projectile motion,

$$y = x\tan\alpha - \frac{1}{2} \cdot \frac{g \cdot x^2}{u^2 \cos^2\alpha}$$

$$-12.5 = 30\tan 30° - \frac{1}{2} \times \frac{9.81 \times (30)^2}{u^2 \cos^2 30°}$$

$$u = 14.05 \, m/sec$$

If balloon takes time t to hit the person

Range = $u \cos\alpha \times t$

$30 = 14.05 \cos 30° \times t$

$t = 2.47$ sec

Example: 11.11

A commando wants to jump from tower '1' to tower '2' as shown in Fig. 11.9. The distance between the towers is 5.50 m and heights of towers are 36 m and 30 m, respectively. Determine the minimum velocity by which he should jump so that he can just reach tower '2'.

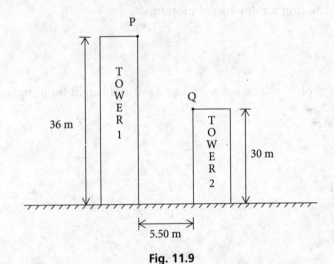

Fig. 11.9

Solution:

When the commando jumps from tower '1' to tower '2', for minimum velocity he runs and then jump from point P with horizontal velocity 'u' to reach at point Q of tower '2' as shown in Fig. 11.9 (a). If the time during jump is 't' then,

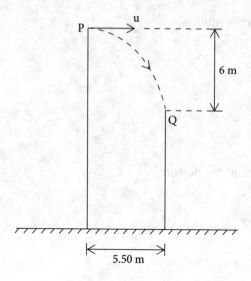

Fig. 11.9 (a)

Distance between towers = u × t

i.e., 5.50 = u × t   ..... (1)

Using equation of motion for downward motion,

$$h = ut + \frac{1}{2}gt^2$$

Vertical component of 'u' will be u cos 90° i.e., u = 0 and height for jump, h = 6 m

$$6 = 0 + \frac{1}{2} \times 9.81 \times t^2$$
$$t = 1.11 \text{ sec}$$

From equation (1),

5.5 = u × 1.11

u = 4.95 m/sec

Alternative Method:

Using the equation of projectile motion,

$$y = x \tan\alpha - \frac{1}{2}\frac{g \cdot x^2}{u^2 \cos^2\alpha}$$

where α = 0°, y = −6 m as jumping downward, x = 5.50 m

$$-6 = 5.50 \tan 0° - \frac{1}{2} \times \frac{9.81 \times (5.50)^2}{u^2 \cos^2 0°}$$

$$-6 = 0 - \frac{1}{2} \times \frac{9.81 \times (5.50)^2}{u^2}$$

$$u = 4.97 \ m/sec$$

Example: 11.12

In a war, a pilot of a fighter plane wishes to hit the target at 800 m away on the ground in 6 seconds. If the plane is moving horizontally, determine the minimum velocity of plane and the height of the plane to hit the target.

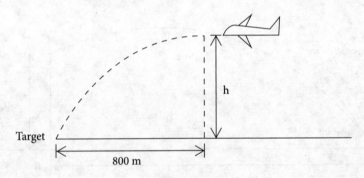

Fig. 11.10

Solution:

Given $x = 800$ m, $t = 6$ sec

$$x = u \times t$$
$$800 = u \times 6$$
$$u = 133.33 \ m/sec \ or \ 480 \ kmph$$

Let the plane is at height 'h'.

$$h = ut + \frac{1}{2}gt^2$$
$$= 0 + \frac{1}{2} \times 9.81 \times (6)^2$$
$$h = 176.58 \ m$$

This can also be determined by using equation of projectile motion,

$$y = x \tan \alpha - \frac{1}{2} \frac{g \cdot x^2}{u^2 \cos^2 \alpha}$$

where, x = 800 m, α = 0° and y = −h

$$-h = 800 \cdot \tan 0° - \frac{1}{2} \times \frac{9.81 \times (800)^2}{(133.33)^2 \cos^2 0}$$

$$h = 176.59 \; m$$

Example: 11.13

In a cricket match, a fielder wants to catch a lofted shot played by a batsman. The ball moves with a velocity of 36 m/sec at 30° from the ground level. If the fielder is standing at 108 m away from the batsman and can take catch up to 3 m high, determine the distance by which fielder should move away to take the catch.

Solution:

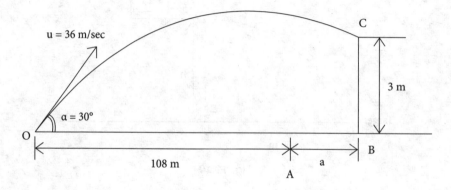

Fig. 11.11

Consider Fig. 11.11. Let the fielder moves away by distance 'a' and takes the catch by jumping at height 3 m.

Using the equation of projectile motion,

$$y = x \tan \alpha - \frac{1}{2} \cdot \frac{g \cdot x^2}{u^2 \cos^2 \alpha} \qquad \ldots (1)$$

where y = 3 m, x = (108 + a), u = 36 m/sec, α = 30°

Using equation (1),

$$3 = (108 + a) \cdot \tan 30° - \frac{1}{2} \times \frac{9.81(108 + a)^2}{(36)^2 \cdot \cos^2 30°}$$

$$(108 + a) = 109.42$$

$$a = 1.42 \; m$$

Kinematics: Curvilinear Motion of Particles

Example: 11.14

A fielder 'P' throws a ball to another fielder 'Q' at distance 'R'. If the ball is thrown two times with same velocity at different angles of projection to reach another fielder 'Q', determine:

(i) The relation between their distances, maximum heights of both projectiles.
(ii) The relation between angle of projection and maximum heights.

Solution:

Let the maximum heights of both projectiles are h_1 and h_2 as shown in Fig. 11.12. Assume that the height of both fielders is same and ignored in figure.

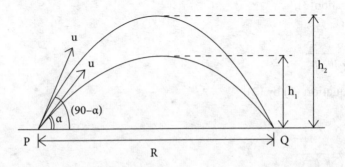

Fig. 11.12

Since the distance between two fielder is same for both projectiles thus angle of projection will be α and $(90-\alpha)$.

(i) The maximum height for both projections will be,

$$h_1 = \frac{u^2 \sin^2 \alpha}{2g} \quad \ldots (1)$$

$$h_2 = \frac{u^2 \cdot \sin^2(90-\alpha)}{2g}$$
$$= \frac{u^2 \cdot \cos^2 \alpha}{2g} \quad \ldots (2)$$

The distance between fielders = range = $R = \left(\dfrac{u^2 \cdot \sin 2\alpha}{g}\right) \quad \ldots (3)$

From equation (1) and (2),

$$h_1 \cdot h_2 = \left(\frac{u^2}{2g} \cdot \sin^2\alpha\right)\left(\frac{u^2 \cdot \cos^2\alpha}{2g}\right)$$

$$= \left(\frac{u^2}{2g}\right)^2 \cdot (\sin\alpha\cos\alpha)^2$$

$$= \left(\frac{u^2}{2g}\right)^2 \cdot \frac{(2\sin\alpha\cos\alpha)^2}{4}$$

$$h_1 \cdot h_2 = \frac{1}{4\times 4}\left(\frac{u^2}{g} \cdot \sin 2\alpha\right)^2$$

from equation (3),

$$h_1 \cdot h_2 = \frac{1}{16} \times R^2$$

or $R = 4\sqrt{h_1 \cdot h_2}$

(ii) Divide equation (1) by (2),

$$\frac{h_1}{h_2} = \tan^2\alpha$$

$$\alpha = \tan^{-1}\left(\sqrt{\frac{h_1}{h_2}}\right)$$

Example: 11.15

A batsman plays a shot in such a way that the maximum height of the ball becomes half of the range. Determine the time of flight of the ball if it is projected with velocity of 24 m/sec.

Solution:

Given, $u = 24$ m/sec

$H = R/2$

$$\frac{u^2 \sin^2\alpha}{2g} = \frac{1}{2}\left(\frac{u^2 \sin 2\alpha}{g}\right)$$

$\sin^2\alpha = 2\sin\alpha.\cos\alpha$

$\sin\alpha(\sin\alpha - 2\cos\alpha) = 0$

$\alpha \neq 0$

$\sin\alpha - 2\cos\alpha = 0$

$\tan\alpha = 2$

$\alpha = 63.43°$

Time of flight,

$$T = \left(\frac{2u\sin\alpha}{g}\right)$$

$$= \left(\frac{2 \times 24 \sin 63.43°}{9.81}\right)$$

$$= 4.38 \text{ sec}$$

Example: 11.16

Two archers 'P' and 'Q' shot arrows at each other simultaneously as shown in Fig. 11.13. If archer 'P' shots arrow with velocity of 36 m/sec at 40° angle of projection and both arrows hit each other after 4 sec. Determine:

(i) The height at which both arrow will hit each other
(ii) The angle of projection with which archer 'Q' should shot arrow with velocity of 28 m/sec
(iii) The distance between two archers

Solution:

Let both archers have same heights on the ground thus height is ignored in the given figure. If both arrows hit each other after time 't' then vertical height 'y' of both arrows will be same but horizontal distances on the ground will be different. Let horizontal distances are x_1 and x_2 as shown in Fig. 11.13.

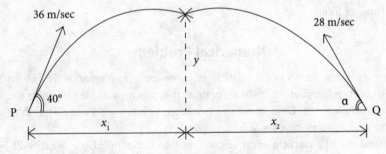

Fig. 11.13

(i) The vertical distance travel by arrow of archer P will be,

$$y = u\sin\alpha \cdot t - \frac{1}{2} \cdot g \cdot t^2$$

$$= 36\sin 40° \times 4 - \frac{1}{2} \times 9.81(4)^2$$

$$y = 14.08 \; m$$

The height at which both arrow will hit each other is 14.08 m

(ii) The vertical distance travel by arrow of archer Q will be,

$$y = 28\sin\alpha \times 4 - \frac{1}{2} \times 9.81(4)^2$$

$$14.08 = 28\sin\alpha \times 4 - \frac{1}{2} \times 9.81(4)^2$$

$$\alpha = 55.73°$$

(iii) The total distance between two archers = $(x_1 + x_2)$

where, $x_1 = 36\cos 40° \times t$
    $= 36\cos 40° \times 4$

and $x_2 = 28\cos\alpha \times t = 28\cos 55.73° \times 4$

thus, total distance = $(36\cos 40° \times 4) + (28\cos 55.73° \times 4)$
    $= 173.38$ m

Theoretical Problems

T 11.1 Differentiate between rectilinear and curvilinear motion.
T 11.2 Discuss curvilinear motion with examples.
T 11.3 Discuss the normal and tangential components of acceleration.
T 11.4 Derive the equation of the path of a projectile.
T 11.5 Determine the following relations for a projectile motion:
   (i) Range
   (ii) Maximum range
   (iii) Maximum height
   (iv) Time of flight

Numerical Problems

N 11.1 The curvilinear motion of a particle is given by equation $y^2 = 1.6 x^5$, where x and y are expressed in meters. In 'x' direction, the position of particle with respect to time is given by x = t/8. Determine the position, velocity and acceleration of the particle along y-direction when x = 2.5 m.

N 11.2 The velocity of a particle in curvilinear motion is defined by $v_x = (3t^3 + 2t^2 - t + 4)$ and $v_y = (2t^3 - t^2 + 2t - 1)$. Determine the position, velocity and acceleration of the particle after 2.5 sec form the beginning of motion.

N 11.3 A car starts from rest and moves along a curved path with uniform acceleration until attains a velocity of 60 km/hr at the end of the fourth minute. If the radius of the curved path is 650 m, determine the tangential, normal and total accelerations in m/sec² of the car at the end of third minute.

N 11.4 A batsman hits the ball in such a way that it moves with a velocity of 32 m/sec at an angle of 32° from the ground. Determine:
   (i) Maximum height attained by ball
   (ii) Total time of flight
   (iii) Range of ball

Kinematics: Curvilinear Motion of Particles

N 11.5 A boy of height 1.1 m throws a stone with initial velocity of 18 m/sec. Determine the maximum distance covered by stone and time of flight.

N 11.6 In a projectile motion, if particle attains maximum height which is one fourth of its range then determine the angle of projection of the particle.

N 11.7 In a basketball game, a player shoots the ball at 30° from height 2 m into the basket, If the distance between the basket and player is 3.6 m, determine the required velocity of the ball to enter into the basket.

N 11.8 If a particle is thrown in such a way that the maximum height becomes double of the range. Determine the angle of projection and time of flight. Take initial velocity of particle as 18 m/sec.

N 11.9 An archer wishes to hit the target on horizontal plane. When arrow is shot at an angle of 18°, it remains short 15 m from target, however when another arrow is shoot at 45°, it remains 20 m ahead of target. Determine:
  (i) Suitable angle of projection to hit the target
  (ii) The distance between archer and target
  (iii) Time required hitting the target

N 11.10 In a trajectory, a particle is projected with initial velocity of 15 m/sec. If particle attains maximum height equal to its horizontal range, Determine:
  (i) Angle of projection
  (ii) Time of flight
  (iii) Maximum height
  (iv) Maximum range

Multiple Choice Questions

1. In curvilinear motion, the rotary motion of a particle is treated as
   a. negligible
   b. considerable
   c. constant
   d. none of these

2. In rectangular coordinates system, the curvilinear motion can be analysed by
   a. vector approach
   b. combining components in x and y axis
   c. both (a) and (b)
   d. none of these

3. In curvilinear motion, it is assumed that all particles of the body travel with same
   a. displacement
   b. velocity
   c. acceleration
   d. all of these

4. The normal acceleration of a particle moving in curvilinear motion is given by
   a. $\dfrac{dv}{dt}$
   b. $\dfrac{v^2}{r}$
   c. $r\omega$
   d. none of these

5. The normal acceleration of a particle moving in curvilinear motion acts
   a. towards to the centre of the radius of curvature
   b. away from the centre of the radius of curvature
   c. both (a) and (b)
   d. none of these

6. The normal acceleration of a particle moving in curvilinear motion is also called as
   a. centripetal acceleration
   b. centrifugal acceleration
   c. both (a) and (b)
   d. none of these

7. The equation of projectile motion is analogous to
   a. spiral
   b. parabola
   c. hyperbola
   d. none of these

8. In projectile motion, the angle of projection (α) must be
   a. $0°$
   b. $0° < \alpha < 90°$
   c. $90°$
   d. none of these

9. In projectile motion, the maximum height travelled by a particle is given by
   a. $\dfrac{2u \sin\alpha}{g}$
   b. $\dfrac{u^2 \sin 2\alpha}{g}$
   c. $\dfrac{u^2 \sin^2 \alpha}{2g}$
   d. none of these

10. In projectile motion, time of flight is given by
    a. $\dfrac{2u \sin\alpha}{g}$
    b. $\dfrac{u^2 \sin 2\alpha}{g}$
    c. $\dfrac{u^2 \sin^2 \alpha}{2g}$
    d. $\dfrac{u \sin\alpha}{2g}$

11. In projectile motion, the range of a particle is given by
    a. $\dfrac{2u \sin\alpha}{g}$
    b. $\dfrac{u^2 \sin 2\alpha}{g}$
    c. $\dfrac{u^2 \sin^2 \alpha}{2g}$
    d. none of these

12. In projectile motion, the maximum range travelled by a particle is given by
    a. $\dfrac{2u}{g}$
    b. $\dfrac{u^2 \sin 2\alpha}{g}$
    c. $\dfrac{u^2 \sin^2 \alpha}{2g}$
    d. $\dfrac{u^2}{g}$

13. In projectile motion, the maximum range can be obtained if angle of projection is
    a. 30°
    b. 45°
    c. 60°
    d. none of these

14. In projectile motion, what combination of angles of projection (α) will produce equal range
    a. 30°, 60°
    b. 20°, 70°
    c. 40°, 50°
    d. all of these

Answers

1. a   2. c   3. d   4. b   5. c   6. c   7. b   8. b   9. c   10. a   11. b   12. d   13. b   14. d

Chapter 12
Kinetics of Particles

12.1 Introduction

In engineering problems, it becomes essential for design engineers to consider the cause of motion along with the analysis of displacement, time, velocity and acceleration of a motion. Such branch of dynamics is known as Kinetics. It also deals with the mass of the body and relationships between the forces, moments or their combination acting on a body. Such motion can be analyzed by using Newton's laws of motion and D'Alembert's principle which are detailed in the upcoming sections.

12.2 Laws of Motion

Sir Isaac Newton (1642–1727) was the first one who stated the three laws of motion in his treatise Principle. These laws govern the motion of particle and demonstrate their validity. These laws are as follows:

(i) **Newton's First law:** A particle always continues to remain at rest or in uniform motion along a straight line in the absence of applied force. This law is also known as law of inertia.
(ii) **Newton's Second law:** If a particle is subjected to force, the magnitude of the acceleration will be directly proportional to the magnitude of the force and indirectly proportional to the mass and lies in the direction of the force.
(iii) **Newton's Third law:** Every action is encountered with equal and opposite reaction, or two interacting bodies have forces of action and reaction collinearly equal in magnitude but opposite in direction.

Basically these laws were presented by Newton in succession of tremendous work done by various philosophers, mainly Aristotle and later improvised by Galileo.

Kinetics of Particles

12.3 D'Alembert's Principle

This principle was introduced by French mathematician Jean le Rond d'Alembert in eighteenth century. He considered Newton's law in another fashion and stated that if the net resultant force causing motion of a body is included with inertia force of body then such body can be treated as in dynamic equilibrium. Further, the motion can be analysed by using equations of static equilibrium i.e.,

$\Sigma X = 0$ and $\Sigma Y = 0$

Consider a body of mass 'm' at rest subjected to various forces P, Q and S as shown in Fig. 12.1. If the net resultant force is R_F which accelerates the body by acceleration 'a' in horizontal direction and inertia force is R_I then according to D'Alembert's Principle,
Net resultant force + inertia force = 0

$R_F + R_I = 0$

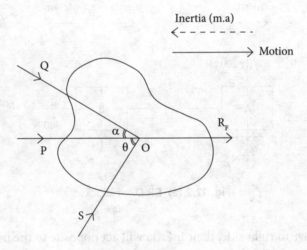

Fig. 12.1

Since inertia force acts in the opposite direction of accelerated motion, $R_I = -ma$
Thus the equation for dynamic equilibrium will be
$R_F + (-ma) = 0$
$R_F = ma$ which is nothing but Newton's second law of motion.

Using this principle we can determine acceleration of the body which could be used to define the displacement and velocity of a body as functions of time.

Example 12.1

Two bodies A and B are connected by a thread and move along a rough horizontal plane ($\mu = 0.3$) under the action of a force 400 N applied to body B as shown in Fig. 12.2. Find the acceleration of the two bodies and tension in the thread using D'Alembert's principle.

(U.P.T.U, Ist Sem 2013–14)

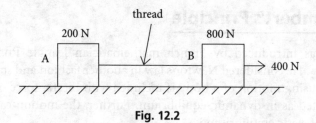

Fig. 12.2

Solution:

Given $u = 0$, $\mu = 0.3$

The free body diagram of both bodies A and B will be as shown in Fig. 12.2 (a). Let blocks A and B are accelerated by acceleration a_A and a_B, respectively.

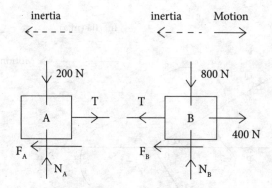

Fig. 12.2 (a) F.B.D. of blocks

As both bodies move to right side, their inertia will act opposite to the motion as shown in Fig. 12.2 (a).

Using D'Alembert's principle for body B,

Net force causing motion + inertia of body = 0

Net force causing motion = $(400 - F_B - T)$

$$\text{Inertia force of body} = \left(-\frac{800}{g} \cdot a_B\right)$$

$$(400 - F_B - T) - \frac{800}{g} \cdot a_B = 0 \qquad \ldots (1)$$

and, $N_B = 800$ N

Kinetics of Particles

Thus equation (1) may be written as,

$$400 - \mu N_B - T = \frac{800}{g} \cdot a_B$$

$$400 - 0.3 \times 800 - T = \frac{800}{g} \cdot a_B$$

$$160 - T = \frac{800}{g} \cdot a_B \qquad \ldots (2)$$

Similarly using D'Alembert's principle for block A,

$$(T - F_A) - \frac{200}{g} \cdot a_A = 0$$

$$T - 0.3 N_A = \frac{200}{g} \cdot a_A$$

and, $N_A = 200\ N$

$$T - 60 = \frac{200}{g} \cdot a_A \qquad \ldots (3)$$

Since both blocks moves together, $a_A = a_B = a$; adding equation (2) and (3),

$$160 - T + T - 60 = \frac{800}{g} \cdot a + \frac{200}{g} \cdot a$$

$$100 = \frac{1000}{g} \cdot a$$

$$a = \frac{g}{10}$$

$$a = 0.981\ m/sec^2$$

Substituting the value of 'a' in equation (3),

$$T - 60 = \frac{200}{9.81} \times 0.981$$

$$T = 80\ N$$

Example: 12.2

A block of 10 kg mass rests on a rough horizontal surface, whose coefficient of kinetic friction is 0.2. It is being pulled by a constant force of 50 N as shown in Fig. 12.3 determine the velocity and distance travelled by the block after 5 seconds.

(U.P.T.U IInd Sem, 2013–14)

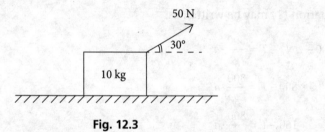

Fig. 12.3

Solution:

Given, m = 10 kg, u = 0, µ = 0.2, t = 5 sec
Consider the free body diagram of the block as shown in Fig. 12.3 (a)

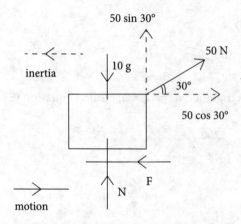

Fig. 12.3 (a) F.B.D. of block

Net pulling force on the block,

$\quad = (50 \cos 30° - F)$ ..... (1)

Inertia of the block = $-m \cdot a$

$\quad = -10 \cdot a$ ..... (2)

Using D'Alembert's principle,

Net force causing motion + inertia of body = 0

$\quad = (50 \cos 30° - F) - 10 \cdot a = 0$ ..... (3)

Kinetics of Particles

Resolving forces along vertical axis,

$$50 \sin 30° + N = 10g$$
$$N = (10 \times 9.81 - 50 \sin 30°)$$
$$N = 73.1 \, N$$

Substitute in equation (3)

$$50 \cos 30° - \mu \cdot N - 10 \cdot a = 0$$
$$50 \cos 30° - 0.2 \times 73.1 - 10 \cdot a = 0$$
$$a = 2.86 \, m/sec^2$$

Using equation,

$$v = u + at$$
$$= 0 + 2.86 \times 5$$
$$v = 14.3 \, m/sec$$

The distance travelled by the block will be,

$$s = ut + \frac{1}{2}at^2$$
$$= 0 + \frac{1}{2} \times 2.86(5)^2$$
$$s = 35.75 \, m$$

Example 12.3

In a mine shaft, an elevator cage is used to ascend or descend the workers. The cage is accelerated to 1.4 m/sec² by means of a wire rope. If the maximum capacity of the elevator cage is to carry 10 persons at a time then determine the range of tension produced in the wire. Take mass of the cage as 600 kg and average mass of a person as 65 kg.

Solution:

Given: a = 1.4 m/sec², u = 0

$$m_{cage} = 600 \, kg$$
$$m_{person} = 65 \times 10 = 650 \, kg$$

Total mass, $m = m_{cage} + m_{pesons}$

$$= 600 + 650$$
$$= 1250 \, kg$$

Let T_1 tension is produced when elevator cage ascends.

Using D'Alembert's principle,

$$(T_1 - 1250g) - 1250 \cdot a = 0$$
$$T_1 = 1250 \cdot g + 1250 \cdot a$$
$$= (1250 \times 9.81) + (1250 \times 1.4)$$
$$T_1 = 14012.5 \ N$$

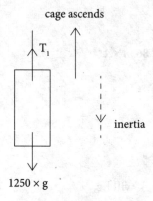

Fig. 12.4 F.B.D. of Cage

If tension is produced T_2 when elevator cage descends, using D'Alembert's principle,

$$(1250.g - T_2) - 1250 \cdot a = 0$$
$$T_2 = 1250(g - a)$$
$$= 1250(9.81 - 1.4)$$
$$T_2 = 10512.5 \ N$$

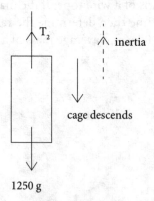

Fig. 12.4 (a) F.B.D. of Cage

Thus the tension produced in wire for range 10512.5 N to 14012.5 N.

Kinetics of Particles

Example: 12.4

A person of mass 70 kg uses a lift. If the acceleration of lift during ascending or descending motion is 1.8 m/sec² then determine the reaction of the floor of the lift during ascending and descending of the lift.

Solution:

Given, m = 70 kg

$\quad$ a = 1.8 m/sec²

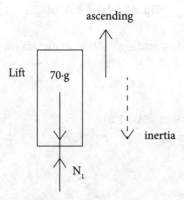

Fig. 12.5 F.B.D. of Lift

Let the reaction of the floor of lift is N_1 when lift is ascending.
$\quad$ Net upward force = $(N_1 - 70 \cdot g)$
$\quad$ The inertia force = $-m \cdot a$
$\quad\quad\quad\quad\quad\quad\quad$ = $-70 \cdot a$

As per D'Alembert's principle,

Net force causing ascending motion + inertia force = 0

$$(N_1 - 70 \cdot g) - 70 \cdot a = 0$$
$$N_1 = 70(g + a)$$
$$= 70(9.81 + 1.8)$$
$$N_1 = 812.7 \, N$$

When lift descends, let the reaction of the floor of lift is N_2.

As per D'Alembert's principle,

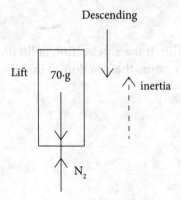

Fig. 12.5 (a) F.B.D. of Lift

Net downward force causing descending motion + inertia force = 0

$$(70 \cdot g - N_2) - m \cdot a = 0$$
$$(70 \cdot g - N_2) - 70 \times 1.8 = 0$$
$$N_2 = (70 \times 9.81 - 70 \times 1.8)$$
$$= 560.7 \ N$$

Example: 12.5

A 750 N crate rests on a 500 N cart, the co-efficient of friction between the crate and the cart is 0.3 and the road is 0.2. If the cart is pulled by force P such that the crate does not slip, determine:

(i) The maximum allowable magnitude of P
(ii) The corresponding acceleration of the cart.

(U.P.T.U., Ist Sem, 2001–02)

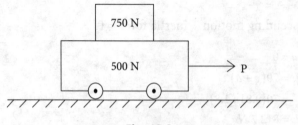

Fig. 12.6

Solution:

Consider the free body diagrams of crate and cart as shown in Fig. 12.6 (a).

Kinetics of Particles

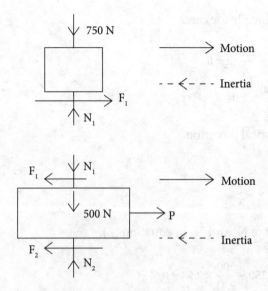

Fig. 12.6 (a) F.B.D. of Crate and Cart

Here, when cart is pulled by force P, the friction acts on its wheel left side which causes rotation of wheel in clockwise direction so that cart moves right side. However at the same time, the crate tends to impend left side but restricted due to the friction imposed by cart on right side and does not allow it to slip. As the cart applied friction F_1 to crate on right side, the same equal and opposite force is applied by crate on cart as shown in Fig. 12.6 (a).

Let the corresponding acceleration of cart is 'a'.

Using D'Alembert's principle for crate,

Net force acting on right side + inertia force = 0

$$F_1 - \frac{750}{g} \cdot a = 0 \quad \ldots (1)$$

Resolving forces in vertical direction,

$N_1 = 750$ N

Substituting the value of N_1 in equation (1),

$$\mu_1 N_1 - \frac{750}{g} \cdot a = 0$$

$$0.3 \times 750 - \frac{750}{9.81} \cdot a = 0$$

$$a = 2.94 \ m/sec^2$$

Using D'Alembert's principle for cart,

$$(P - F_1 - F_2) - \frac{500}{g} \cdot a = 0$$

$$P - \mu_1 N_1 - \mu_2 N_2 - \frac{500}{g} \times 2.94 = 0 \qquad \ldots\ldots (2)$$

Resolving forces in vertical direction,

$$N_2 = N_1 + 500$$
$$= 750 + 500$$
$$= 1250 \ N$$

Substituting the values of N_1 and N_2 in equation (2),

$$P - 0.3 \times 750 - 0.2 \times 1250 - \frac{500}{9.81} \times 2.94 = 0$$

$$P = 624.85 \ N$$

Example: 12.6

Two blocks of weight 500 N and 600 N are connected by a rope passing over an ideal pulley as shown in Fig. 12.7.

Determine:

(i)   Acceleration of blocks
(ii)   Tension induced in the string

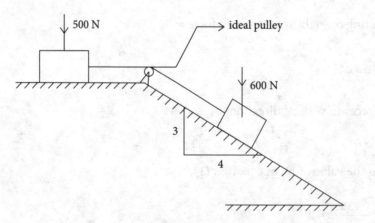

Fig. 12.7

Take co-efficient of friction between the block and the horizontal plane is 0.28 and that between the block and the inclined plane is 0.24.

Kinetics of Particles

Solution:

Let the tension in the string is T as shown in Fig. 12.7 (a) and angle of inclination is θ.

$\tan\theta = 3/4$

$\theta = 36.87°$

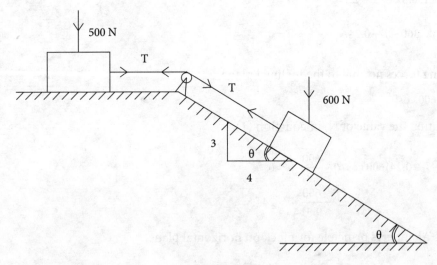

Fig. 12.7 (a)

Consider free body diagram of both blocks as shown in Fig. 12.7 (b). Let the block 600 N accelerates by amount 'a' in downward direction then block 500 N accelerates by same amount in rightward direction.

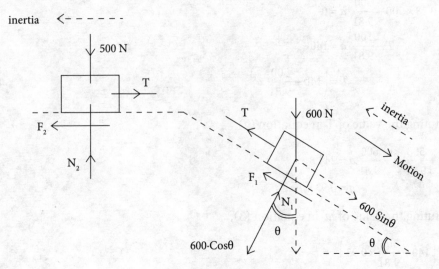

Fig. 12.7 (b) F.B.D. of blocks

Using D'Alembert's principle for block on incline,

Net downward force causing motion on incline + inertia force = 0

$$(600 - \sin\theta - T - F_1) - \frac{600}{g} \cdot a = 0$$

$$600 \sin 36.87° - T - \mu_1 N_1 - \frac{600}{9.81} \cdot a = 0$$

$$360 - T - 0.24 N_1 - \frac{600}{9.81} \cdot a = 0 \quad \ldots\ldots (1)$$

Resolving forces normal to the inclined plane,

$N_1 = 600.\cos\theta$

Substituting the value of N_1 in equation (1),

$$360 - T - 0.24(600\cos\theta) - \frac{600}{9.81} \cdot a = 0$$

$$T + \frac{600}{9.81} \cdot a = 244.8 \quad \ldots\ldots (2)$$

Using D'Alembert's principle for block on horizontal plane,

$$(T - F_2) - \frac{500}{g} \cdot a = 0$$

and $\quad N_2 = 500\ N$

$$T - \mu_2 N_2 - \frac{500}{g} \cdot a = 0$$

$$T - 0.28 \times 500 - \frac{500}{9.81} \cdot a = 0$$

$$T - \frac{500}{9.81} \cdot a = 140$$

$$T = \left(140 + \frac{500}{9.81} \cdot a\right) \quad \ldots\ldots (3)$$

Substituting the value of T in equation (2),

$$140 + \frac{500}{9.81} \cdot a + \frac{600}{9.81} \cdot a = 244.8$$

$$a = 0.93\ m/\sec^2$$

Substituting the value of 'a' in equation (3),

$$T = \left(140 + \frac{500}{9.81} \times 0.93\right)$$

$$T = 187.40\ N$$

Kinetics of Particles

Example: 12.7

Two equal weights of 1000 N each are lying on two identical planes connected to a string passing over a frictionless pulley as shown in Fig. 12.8.

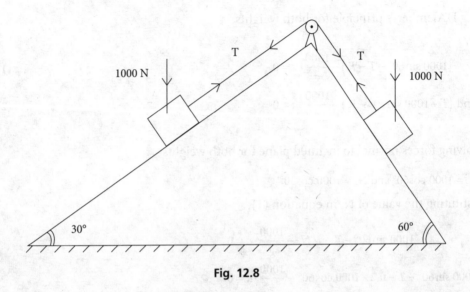

Fig. 12.8

Using D'Alembert's principle, find the acceleration of weights and tension in the string. The coefficient of friction between the planes and weights is 0.2.

(M.T.U. 2010–2011)

Solution:

Consider free body diagrams of both weights as shown in Fig. 12.8 (a)

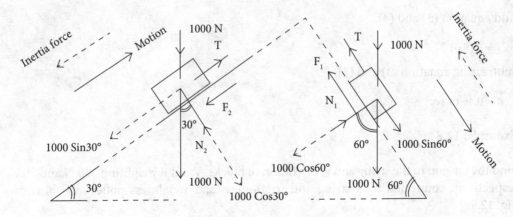

Fig. 12.8 (a) F.B.D. of blocks

We can observe that, magnitude wise the component of weight parallel to the inclined plane is more on 60° incline as compared to 30° incline (1000 sin60° > sin30°) thus 1000 N weight on 60° incline will move downward and another one will move upward. Let the weights are accelerated by amount 'a' and tension induced in the string is T.

Using D'Alembert's principle for both weights,

$$\left(1000 \sin 60° - T - F_1\right) - \frac{1000}{g} \cdot a = 0 \qquad \ldots (1)$$

and $\left(T - 1000 \sin 30° - F_2\right) - \frac{1000}{g} \cdot a = 0 \qquad \ldots (2)$

Resolving forces normal to inclined plane for both weights,

$N_1 = 1000 \cos 60°$ and $N_2 = 1000 \cos 30°$

Substituting the value of N_1 in equation (1),

$$1000 \sin 60° - T - \mu_1 N_1 - \frac{1000}{g} \cdot a = 0$$

$$1000 \sin 60° - T - 0.2 \times 1000 \cos 60° - \frac{1000}{g} \cdot a = 0$$

$$T + \frac{1000}{g} \cdot a = 766 \qquad \ldots (3)$$

Similarly, substitute the value of N_2 in equation (2),

$$T - 1000 \sin 30° - 0.2 \times 1000 \cos 30° - \frac{1000}{g} \cdot a = 0$$

$$T - \frac{1000}{g} \cdot a = 673.21 \qquad \ldots (4)$$

Add equation (3) and (4),

$T = 719.61$ N

Subtracting equation (3) and (4),

$a = 0.46$ m/sec²

Example: 12.8

Find the tension in the string and acceleration of blocks A and B weighting 200 N and 50 N, respectively, connected by a string and frictionless and weightless pulleys as shown in Fig. 12.9.

(UPTU, 2002–03)

Kinetics of Particles

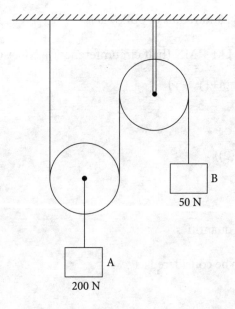

Fig. 12.9

Solution:

Here both pulleys are identical, frictionless and weightless. Thus tension 'T' will remain same in the whole string. Let acceleration of blocks A and B are a_1 and a_2, respectively. Thus from the given Fig. 12.9, we can observe that the relation between acceleration of blocks A and B will be,

$$a_2 = 2 \cdot a_1 \qquad \ldots (1)$$

However equation (1) can be derived from Fig. 12.9 (a) given below.

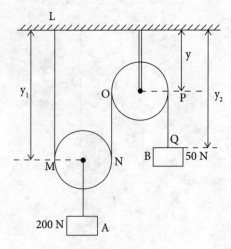

Fig. 12.9 (a)

Let the length of string is l and radius of pulley is r.

Thus length of string, l = LM + MN (half circumference) + NO + OP + PQ

$l = y_1 + \pi r + (y_1 - y) + \pi r + (y_2 - y)$

$l = 2y_1 + 2\pi r - 2y + y_2$

$y_2 = l - 2y_1 - 2\pi r + 2y$ ..... (2)

Differentiating equation (2),

$\dfrac{dy_2}{dt} = \dfrac{-2 \cdot dy_1}{dt}$ ..... (3)

As l, r and y are constant quantities.

The equation (3) can also be considered as

$v_2 = -2 \cdot v_1$

If the equation is differentiated again,

$\dfrac{d^2 y_2}{dt^2} = \dfrac{-2 \cdot d^2 y_1}{dt^2}$

$a_2 = -2 \cdot a_1$

which confirms equation (1), and negative sign shows that both blocks are moving in the opposite direction.

Consider free body diagrams of blocks as shown in Fig. 12.9 (b)

Using D'Alembert's principle for blocks A and B, respectively,

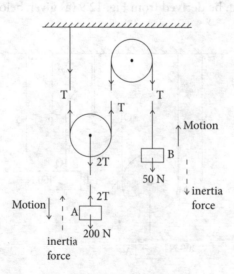

Fig. 12.9 (b) F.B.D. of blocks

Kinetics of Particles

$$(200 - 2T) - \frac{200}{g} \cdot a_1 = 0 \qquad \ldots\ldots (4)$$

$$(T - 50) - \frac{50}{g} \cdot a_2 = 0 \qquad \ldots\ldots (5)$$

Substituting the value of T and a_2 in equation (4),

$$T = \left(50 + \frac{50}{g} \cdot a_2\right)$$

$$200 - 2\left(50 + \frac{50}{g} \cdot (2a_1)\right) - \frac{200}{g} \cdot a_1 = 0$$

$$a_1 = 2.45 \; m/sec^2$$

Substituting the value a_1 in equation (1) and (5), a_2 = 4.9 m/sec² and T = 74.97 N

Example: 12.9

Two blocks 40 kg and 24 kg are connected by inextensible string and supported as shown in Fig. 12.8. Determine tension in string and acceleration of each block. The coefficient of friction between block and horizontal plane is 0.25. Consider pulleys as frictionless and weightless.

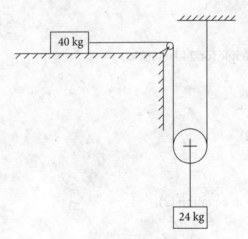

Fig. 12.10

Solution:

Let the tension in the string is 'T' and acceleration of bock 24 kg is 'a' in downward direction. Thus acceleration of block 40 kg will be '2a' towards right side.

Consider free body diagram of both blocks as shown in Fig. 12.10 (a).

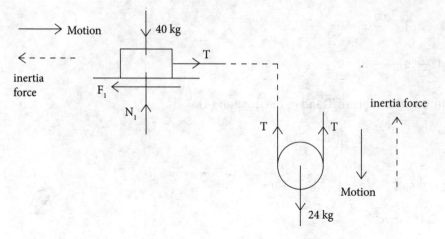

Fig. 12.10 (a) F.B.D. of blocks

Using D'Alembert's Principle for 40 kg block,

$$(T - F_1) - 40(2a) = 0$$
$$\text{and } N_1 = 40g$$
$$T - \mu_1 N_1 - 40(2a) = 0$$
$$T - 0.25 \times 40 \times g - 80a = 0$$
$$T - 10g = 80a \qquad \ldots (1)$$

Using D'Alembert's Principle for 24 kg blocks,

$$(24g - 2T) - 24 \cdot a = 0$$
$$12g - T = 12 \cdot a \qquad \ldots (2)$$

Add equation (1) and (2),

2g = 92.a
a = 0.21 m/sec²

Substitute the value of a in equation (2),

T = 540.48 N

The acceleration of 24 kg block will be 0.21 m/sec²
The acceleration of 40 kg block will be 0.42 m/sec²

Example: 12.10

Neglecting the frictional and inertial effect of a two-step pulley arrangements, determine the acceleration of both masses as shown in the Fig. 12.11. Also determine tension in the strings.

Kinetics of Particles

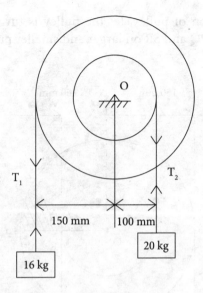

Fig. 12.11

Solution:

Here mass 16 kg will move downward as turning effect about O will be more due to large radius as compared to 20 kg. Let mass 16 kg and 20 kg are accelerated by a_1 and a_2, respectively and have tensions T_1 and T_2 as shown in the free body diagram of both blocks.

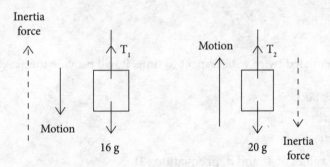

Fig. 12.11 (a) F.B.D. of blocks

Using D'Alembert's principle,

$(16g - T_1) - 16 \cdot a_1 = 0$ ..... (1)

$(T_2 - 20g) - 20 \cdot a_2 = 0$ ..... (2)

Consider free body diagram of two-step pulley. Taking moment about centre O,

$T_1 \times 150 = T_2 \times 100$

$T_2 = 1.5 T_1$ ..... (3)

Consider the rotary motion of pulley, let the pulley is turned by an angle α. Where strings move for distance PQ and SR on larger and smaller pulley respectively as shown in Fig. 12.11 (b).

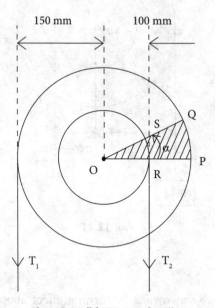

Fig. 12.11 (b) F.B.D. of Pulley

$$\alpha = \frac{PQ}{150} = \frac{SR}{100}$$
$$PQ = 1.5 \cdot SR$$

If arcs are differentiated twice with respect to time, it will provide the acceleration,

$$\frac{d^2(PQ)}{dt^2} = \frac{1.5 \, d^2(SR)}{dt^2}$$
$$a_1 = 1.5 \cdot a_2 \qquad \ldots (4)$$

Substituting the values of T_2 and a_2 in equation (2),

$$1.5 T_1 - 20g - 20\left(\frac{a_1}{1.5}\right) = 0$$

$$T_1 - \frac{20}{1.5} \cdot g - \frac{20}{2.25} \cdot a_1 = 0 \qquad \ldots (5)$$

Add equation (1) and (5),

$$16g - T_1 - 16 \cdot a_1 + T_1 - \frac{20}{1.5} g - \frac{20}{2.25} \cdot a_1 = 0$$

$$a_1 = 1.05 \ m/sec^2$$

Kinetics of Particles

From equation (4),

$$a_2 = \frac{a_1}{1.5}$$
$$a_2 = 0.7 \ m/sec^2$$

Substituting the values of a_1 and a_2 in equation (1) and (2), the tension will be
  $T_1 = 140.16 \ N$
  $T_2 = 210.2 \ N$

Example: 12.11

Determine the acceleration of each block and the tension in the string passing over ideal pulleys as shown in Fig. 12.12. The coefficient of friction for blocks 50 kg and 70 kg is 0.24 and 0.28, respectively. Consider the weight of pulley supporting 60 kg as negligible.

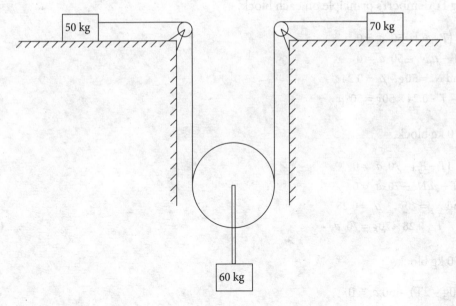

Fig. 12.12

Solution:

Let the tension induced in the string is T and acceleration of 50 kg, 70 kg and 60 kg blocks are a_1, a_2 and a_3, respectively. Consider free body diagram as shown Fig. 12.12 (a).

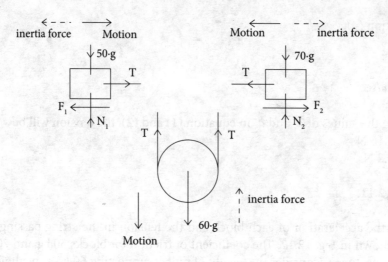

Fig. 12.12 (a) F.B.D. of blocks

Using D'Alembert's principle on each block,

$(T - F_1) - 50 \cdot a_1 = 0$
$T - \mu_1 N_1 - 50 \cdot a_1 = 0$
and $N_1 = 50g$, $\mu_1 = 0.24$
$T - 0.24 \times 50g = 50 \cdot a_1$ ..... (1)

For 70 kg block,

$(T - F_2) - 70 \cdot a_2 = 0$
$T - \mu_2 N_2 - 70 \cdot a_2 = 0$
and $N_2 = 70g$, $\mu_2 = 0.28$
$T - 0.28 \times 70g = 70 \cdot a_2$ ..... (2)

For 60 kg block,

$(60g - 2T) - 60 \cdot a_3 = 0$

The 60 kg block will accelerate downward by average of acceleration of 50 kg and 70 kg block.

i.e. $a_3 = \left(\dfrac{a_1 + a_2}{2}\right)$

$60g - 2T - 60\left(\dfrac{a_1 + a_2}{2}\right) = 0$
$60g - 2T - 30(a_1 + a_2) = 0$
$60g - 2T = 30(a_1 + a_2)$ ..... (3)

Kinetics of Particles

Add equation (1), (2) and (3),

$$60g - 0.24 \times 50g - 0.28 \times 70g = 50a_1 + 70a_2 + 30a_1 + 30a_2$$
$$28.4g = 80a_1 + 100a_2 \qquad \ldots (4)$$

Subtracting equation (1) from equation (2),

$$-7.6g = -50a_1 + 70a_2 \qquad \ldots (5)$$

Solving equation (4) and (5) we get,

$$a_1 = 2.54 \ m/sec^2$$
$$a_2 = 0.75 \ m/sec^2$$

Substituting the value of a_1 in equation (1),

T = 244.72 N

Example: 12.12

Three blocks are suspended by inextensible strings with the help of two identical, frictionless and weightless pulleys as shown in Fig. 12.13. Determine tension in both strings and acceleration of each block.

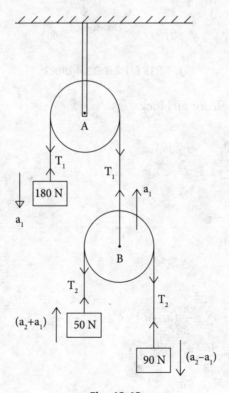

Fig. 12.13

Solution:

Let the tension in the strings over pulley A and B are T_1 and T_2. Here, if block 180 N accelerates by acceleration 'a_1' towards ground, the pulley B also accelerated by same amount 'a_1' towards up side. However, block 90 N comes downward and 50 N moves upward. If block 90 N and 50 N have accelerations 'a_2' with respect to pulley B then net acceleration of block 90 N will be $(a_2 - a_1)$ and net acceleration of block 50 N will be $(a_2 + a_1)$. Consider free body diagrams of each block as shown in Fig. 12.13 (a).

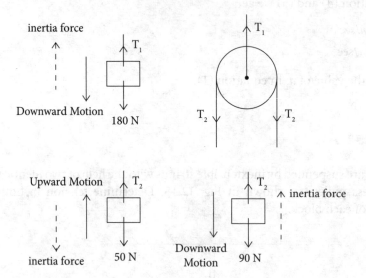

Fig. 12.13 (a) F.B.D. of blocks

Using D'Alembert's principle for all blocks,

$$(180 - T_1) - \frac{180}{g} \cdot a_1 = 0 \qquad \ldots (1)$$

$$T_1 = 2T_2 \qquad \ldots (2)$$

$$(T_2 - 50) - \frac{50}{g}(a_2 + a_1) = 0 \qquad \ldots (3)$$

$$(90 - T_2) - \frac{90}{g}(a_2 - a_1) = 0 \qquad \ldots (4)$$

Add equation (3) and (4),

$$T_2 - 50 + 90 - T_2 - \frac{50}{g}(a_2 + a_1) - \frac{90}{g}(a_2 - a_1) = 0$$

$$40 - \frac{140}{g} \cdot a_2 + \frac{40}{g} \cdot a_1 = 0$$

$$2 - \frac{7a_2}{g} + \frac{2}{g} \cdot a_1 = 0 \qquad \ldots (5)$$

Kinetics of Particles

Substituting the value of T_1 in equation (1),

$$180 - 2T_2 - \frac{180}{g} \cdot a_1 = 0$$

$$90 - T_2 - \frac{90}{g} \cdot a_1 = 0$$

$$(90 - T_2) = \frac{90}{g} \cdot a_1 \qquad \text{substitute in equation (4),}$$

$$\frac{90}{g} \cdot a_1 - \frac{90}{g}(a_2 - a_1) = 0$$

$$a_1 - a_2 + a_1 = 0$$

$$a_1 = a_2/2 \qquad \text{substitute in equation (5),}$$

$$2 - \frac{7 \cdot a_2}{g} + \frac{2}{g}\left(\frac{a_2}{2}\right) = 0$$

$$a_2 = g/3$$

$$= 3.27 \; m/sec^2$$

$$\text{thus } a_1 = 1.64 \; m/sec^2$$

Substituting the value of a_1 in equation (1),

$$(180 - T_1) - \frac{180 \times 1.64}{9.81} = 0$$

$$T_1 = 149.91 \; N$$

From equation (2),

$$T_2 = \frac{T_1}{2}$$

$$T_2 = 74.95 \; N$$

acceleration of block 180 N $= a_1 = 1.64 \; m/sec^2$

acceleration of block 90 N $= (a_2 - a_1) = 1.63 \; m/sec^2$

acceleration of block 50 N $= (a_2 + a_1) = 4.91 \; m/sec^2$

Example: 12.13

The system of blocks is initially at rest as shown in Fig. 12.14. The weight of blocks A and B are 240 N and 120 N, respectively. Find the velocity of blocks when the block A moves by 1.6 m. Assume both pulleys are frictionless and weightless.

(U.PTU, 2003–04)

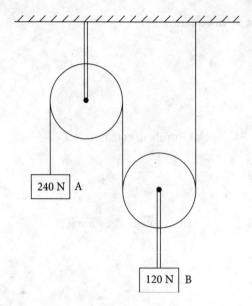

Fig. 12.14

Solution:

Given, $u = 0$, $s = 1.6$ m

Like earlier example 12.8, the acceleration of block A will be twice of that of B. If block A and B are accelerated by a_1 and a_2, respectively, $a_1 = 2.a_2$ ..... (1)

The free body diagram of blocks is shown in Fig. 12.14 (a),

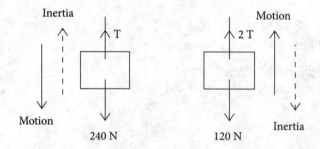

Fig. 12.14 (a) F.B.D. of blocks

Using D'Alembert's principle for all blocks,

$$(240 - T) - \frac{240}{g} \cdot a_1 = 0 \qquad \ldots (2)$$

$$(2T - 120) - \frac{120}{g} \cdot a_2 = 0 \qquad \ldots (3)$$

Kinetics of Particles

Substituting the value of a_1 in equation (2) and then solving equation (2) and (3) we get,

$a_2 = 3.27\ m/sec^2$, $a_1 = 6.54\ m/sec^2$ and $T = 80\ N$

Let velocities of blocks A and B are V_A and V_B, respectively,

The velocity of block A when moves by 1.6 m

$$V_A^2 = u^2 + 2as$$
$$= 0 + 2 \cdot a_1 s$$
$$= 2 \times 1.6 \times 6.54$$
$$V_A = 4.57\ m/sec$$

as $a_2 = \dfrac{a_1}{2}$

thus, $V_B = \dfrac{V_A}{2}$
$$= \dfrac{4.57}{2}$$
$$= 2.29\ m/sec$$

Theoretical problems

T 12.1 Discuss Newton's laws of motion.
T 12.2 Explain D'Alembert's principle and its significance.
T 12.3 State the difference between D'Alembert's principle and Newton's second law of motion.

Numerical Problems

N 12.1 A block of weight 20 kN is lying on a horizontal plane as shown in Fig. NP 12.1

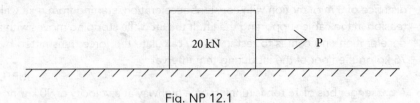

Fig. NP 12.1

If the co-efficient of kinetic friction between the contact surface is 0.22 then determine force pull 'P' to accelerate the block by 4.2 m/sec² using D'Alembert's principle, if:
(i) Force P is applied parallel to the horizontal plane
(ii) Force P is applied at 35° from the horizontal plane

N 12.2 A block of 2 kN is placed on an inclined plane and pulled by a force P as shown in the Fig. NP 12.2. If block attains a velocity of 6 m/sec in 4 sec and kinetic friction between contact surfaces is 0.24 then determine force P causing motion by using D'Alembert's principle.

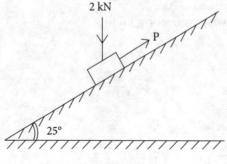

Fig. NP 12.2

N 12.3 Find the acceleration of bodies and tension in the string joining blocks A and B as shown in Fig. NP 12.3.

[UPTU 2002–03]

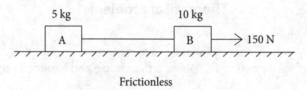

Frictionless

Fig. NP 12.3

N 12.4 A vertical lift of total mass 500 kg acquires an upward velocity of 2 m/sec over a distance of 3 m of motion with constant acceleration, starting from rest. Calculate the tension in the cable supporting the lift. If the lift while stopping moves with a constant deceleration and comes to rest in 2 sec, calculate the force transmitted by a man of 75 kg on the floor of the lift during that interval.

[UPTU 2002–03]

N 12.5 A passenger bus of 16 ton is running on a highway at a velocity of 90 km/hr. Suddenly, one rear wheel got punctured and bus driver stops the bus in 10 seconds.
Determine:
(i) The distance covered by bus during that time
(ii) Braking force applied by driver to stop the bus

N 12.6 An effect of 200 N is required to just move a certain body up on inclined plane of 15°; the force is acting parallel to the plane. If the angle of inclination of the plane is made 20°, the effort required being again parallel to the plane, is found to be 230 N, find the weight of the body and co-efficient of friction.

(U.P.T.U., C.O, IInd Sem, 2012–13)

N 12.7 A body of 12 kN is kept on a rough incline. Determine the velocity and distance covered by body in 4 seconds. Take angle of incline as 30° and coefficient of friction between the contact surfaces as 0.28.

N 12.8 Two blocks of weight 120 N and 100 N are suspended as shown in Fig. NP 12.4. If the pulleys are weightless and frictionless, determine tension in the string and acceleration of each block by using D'Alembert's principle.

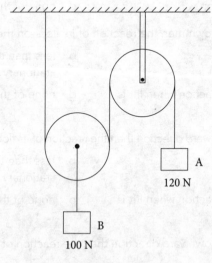

Fig. NP 12.4

Multiple Choice Questions

1. D' Alembert's principle is useful where kinetic problems are mainly concerned with
   a. acceleration
   b. time
   c. displacement
   d. velocity

2. Mathematically, D' Alembert's principle is based on which Newton's law
   a. First law
   b. Second law
   c. Third law
   d. none of these

3. Kinetic problems relate the following parameters
   a. acceleration, displacement and time
   b. force, displacement, acceleration, velocity and time
   c. displacement, acceleration and velocity
   d. none of these

4. Newton's first law is the law of
   a. displacement
   b. velocity
   c. inertia
   d. none of these

5. Which quantity is analogous to moment of inertia
   a. velocity
   b. acceleration
   c. mass
   d. force

6. Inertia force becomes zero, when body moves with
   a. uniform velocity
   b. uniform acceleration
   c. uniform deceleration
   d. none of these

7. If a lift suddenly falls down then the reaction of its floor on the passengers will be
   a. zero
   b. less than the reaction when lift is stationary
   c. greater than the reaction when lift is stationary
   d. none of these

8. If a lift accelerates in upward direction then the reaction of its floor on the passengers will be
   a. zero
   b. less than the reaction when lift is stationary
   c. greater than the reaction when lift is stationary
   d. none of these

9. If a lift accelerates in downward direction then the reaction of its floor on the passengers will be
   a. zero
   b. less than the reaction when lift is stationary
   c. greater than the reaction when lift is stationary
   d. none of these

10. If a lift decelerates in upward direction then the reaction of its floor on the passengers will be
    a. zero
    b. less than the reaction when lift is stationary
    c. greater than the reaction when lift is stationary
    d. none of these

11. If a lift decelerates in downward direction then the reaction of its floor on the passengers will be
    a. zero
    b. less than the reaction when lift is stationary
    c. greater than the reaction when lift is stationary
    d. none of these

Answers

1. a   2. b   3. b   4. c   5. c   6. a   7. a   8. c   9. b   10. b   11. c

Chapter 13

Work and Energy

13.1 Introduction

In this chapter we will study two principles, i.e., the work and energy principle and the principle of conservation of energy. Earlier, we have used D' Alembert's principle to analyse the kinetic problems. The work and energy principle is another alternative method to analyse the kinetic problems. The basic difference in its application is that the work and energy principle is preferred when problems deal especially with velocity. However D' Alembert's principle is preferred when problems deal with acceleration.

13.2 Work Done by a Force

Work is done by a force when a force displaces a particle by some displacement in its direction. It is a scalar quantity. Consider a particle of mass 'm' kg lying on a smooth horizontal surface as shown in Fig. 13.1. If the particle is pushed by a constant force 'p' by displacement 'd' then the work done is given by

$W = p \times d$

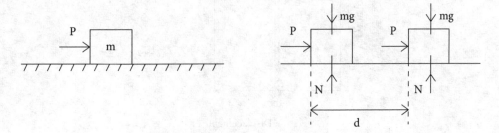

Fig. 13.1

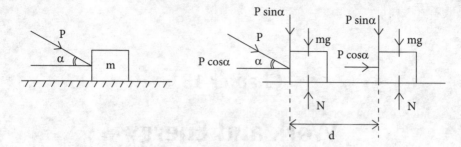

Fig. 13.2

However, if the particle is pushed for displacement 'd' by a constant force 'p' applied at an angle 'α' from the horizontal surface as shown in Fig. 13.2, the work done is given by

W = *Force in the direction of displacement* × d

W = p *cos α* × d

If α = 0° then it belongs to the earlier case as discussed.

If α = 90° then it shows that work done will be equal to zero as force is not causing displacement.

For example consider Fig. 13.2, where normal reaction 'N' of mass 'm' is not producing displacement due to which work done by normal reaction will be zero.

Work done will be negative if force and displacement are opposite in direction. For example work done by frictional force is always negative.

It is to be noted that force and displacement both are vector quantities but work is a scalar quantity. The unit of work is Nm or Joule in S.I. unit.

13.3 Work Done by a Variable Force

Consider a particle displaced by variable force as shown in Fig. 13.3.

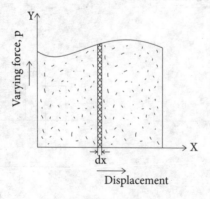

Fig. 13.3

Work and Energy

If a particle is displaced by displacement dx under variable force p then the work done will be given by $\delta W = p\, dx$

The total work done by a variable force will be given by

$$W = \int p\, dx$$

i.e., area under the curve provides the work done by variable force.

13.4 Energy

Energy is the capacity of a body to conduct work. In other words it is equivalent to the work. It is also a scalar quantity like work. A particle can have various types of energies like electrical energy, chemical energy, heat energy and mechanical energy. Out of these energies only mechanical energy is considered in engineering mechanics which includes mainly kinetic energy (KE) and potential energy (PE).

If a particle of mass 'm' is moving with a velocity 'v' at height 'h' then the kinetic energy of the particle is given by

$$K.E. = \frac{1}{2}mv^2$$

and the potential energy is given by

$$P.E. = mgh$$

13.5 Work–Energy Principle

Consider a particle of mass 'm' kg is moving on a smooth plane surface and a pushing force 'F' changes its velocity from initial velocity v_1 to final velocity v_2 as shown in Fig. 13.4. If the particle is infinitesimally displaced by displacement 'ds' during time 'dt' then the work done by pushing force will be given by

$$\delta w = F.ds$$
$$\delta w = m.a.ds \quad \ldots\ldots (1)$$

where $a = \dfrac{dv}{dt}$

Substituting the value of a in equation (1),

$$\delta w = m.\frac{dv}{dt}.ds$$
$$\delta w = m.\frac{ds}{dt}.dv$$
$$\delta w = m.v.dv$$

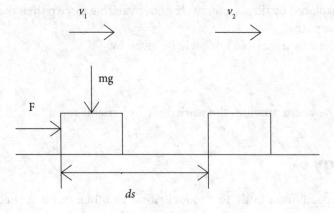

Fig. 13.4

If the particle is finally displaced by distance 's' from position 1 to position 2 (not shown), then work done can be determine by integrating the above equation,

$$\int_1^2 \delta w = \int_1^2 m.v.dv$$

$$\int_1^2 \delta w = m\int_1^2 v.dv$$

$$W = \frac{1}{2}m(v_2^2 - v_1^2)$$

Work done = Change in kinetic energy of the particle

Thus according to this principle, the work done on a moving particle is equal to the change in its kinetic energy. This principle is useful when kinetic problems are required to be exclusively dealt with in variation of velocities.

13.6 Power

Power is a term which designates the work performed with minimum time. For example, a person wants to travel a distance of 5 km. Suppose he travels the same distance by three different modes, i.e., by walking, bicycle and motor cycle, and takes time 70 min, 25 min, and 8 min, respectively. Thus the work performed in the third case is most meaningful as it is conducted with the least time. Hence power is the capacity of a machine to perform work with respect to time or it is the rate at which work is performed. If a machine performs work W in total time duration t then, the average power of such machine will be given by

$$P_{avg} = \frac{W}{t}$$

Work and Energy

If a machine performs work δw for instant time dt then the instantaneous power of machine will be given by

$$P_{inst} = \frac{\delta w}{dt}$$

as $\delta w = F.ds$

$$P_{inst} = \frac{F.ds}{dt}$$
$$P_{inst} = F.dv$$

Hence, Power $P = F.\, v$

Power is expressed in watt or joule per second in S.I. units. However, in electrical and mechanical equipment devices like motor and engine, the power is defined in terms of horsepower (hp).

The relation between horsepower and watt in Metric units is given by

$1\ hp = 735.5\ watt$

The relation between horsepower and watt in British units is given by

$1\ hp = 746\ watt$

13.7 Principle of Conservation of Energy

According to this principle, the total energy of a body remains constant if there is no transaction of energy allowed between the body and its surrounding. In other words, energy cannot be produced or destroyed; however, its existence varies in different forms.

Consider a stone of mass 'm' kg is lying at height h and dropped under gravity from the rest as shown in Fig. 13.5. If the total energy of stone is E then it will be given by

$E = $ kinetic energy + potential energy
$E = 0 + mgh$
$E = mgh$

Let the stone reaches a position A-A after falling to height h_a then its velocity will be given by

$v^2 = u^2 + 2gh_a$
$v^2 = 0 + 2gh_a$
$v^2 = 2gh_a$

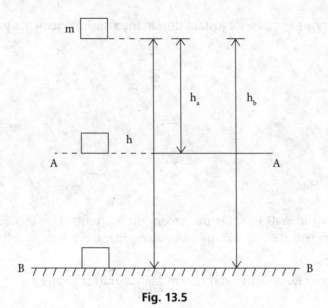

Fig. 13.5

If the total energy of stone at position A-A is E_a then it will be given by

$$E_a = \frac{1}{2}mv^2 + mg(h - h_a)$$

$$E_a = \frac{1}{2}m(2gh_a) + mg(h - h_a)$$

$$E_a = mgh$$

Finally, the stone reaches ground at position B-B after falling through the height h_b, then its velocity will be given by

$$v^2 = 2gh_b$$

If the total energy of stone at position B-B is E_b then it will be given by

$$E_b = \frac{1}{2}mv^2 + mg(h - h_b) \quad \text{where } h = h_b$$

$$E_b = \frac{1}{2}m(2gh_b) + 0$$

$$E_b = mgh$$

We observe that during falling of stone, its potential energy is continuously changing into kinetic energy but total energy of a body remains constant. Thus, it can be concluded that total energy of a body remains constant and its existence varies from one form to another form at different locations.

Work and Energy

Example: 13.1

A block of mass 10 kg rests on a rough surface. The block is pulled by a force of 50 N acting parallel to the plane. Determine the velocity of block after it travels 24 m, starting from the rest. If the force of 50 N is then removed, determine the distance traverse by block until it stops. Take co-efficient of friction as 0.24.

Solve the problem by using:

(i) Work–energy principle
(ii) D' Alembert's principle

Solution:

Given: $m = 10$ kg, $P = 50$ N
$v_1 = 0$, $s = 24$ m
$\mu = 0.24$

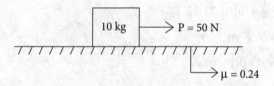

Fig. 13.6

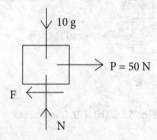

Fig. 13.6 (a) F.B.D. of block

(i) Using work–energy principle,
Net work done = Change in K.E.
Net force causing motion × displacement = Change in K. E. of block

$$(P - F)s = \frac{1}{2}m\left(V_2^2 - V_1^2\right) \quad \ldots\ldots (1)$$

As the block is moving on surface, its acceleration along vertical axis will be zero thus $\Sigma Y = 0$

i.e., $N = 10 \cdot g$ ..... (2)

Consider equation (1),

$$(P - \mu N)s = \frac{1}{2}m(V_2^2 - V_1^2)$$

Substituting the value of N,

$$(50 - 0.24 \times 10 \times g)24 = \frac{1}{2} \times 10(V_2^2 - 0)$$

$$V_2 = 11.27 \ m/sec$$

When the pull force is removed, let it traverse distance 'x' m where only frictional force acts and block comes to rest from final velocity, i.e., $v_2 = 0$ and $v_1 = 11.27$ m/sec

Using work–energy principle,

Net W. D. = Change in K.E. of block

Net force causing motion × distance = Change in K.E. of block

$$(0 - F)x = \frac{1}{2}m(v_2^2 - v_1^2)$$

$$(0 - 0.24 \times 10 \times g)x = \frac{1}{2} \times 10(0 - 11.27^2)$$

$$x = 26.97 \ m$$

(ii) Using D' Alembert's principle,

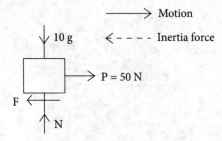

Fig. 13.6 (b) F.B.D. of block

Let the body is accelerated by acceleration 'a',

Net force causing motion + Inertia force = 0

$(P - F) + (-m \cdot a) = 0$

$(P - \mu \cdot N) - m \cdot a = 0$

$(50 - 0.24 \times 10 \times g) - 10 \cdot a = 0$

$$a = 2.65 \ m/sec^2$$

as, $u = v_1 = 0$

$$v^2 = u^2 + 2as$$

$$= 0 + 2 \times 2.65 \times 24$$

$$v = 11.27 \ m/sec$$

Work and Energy

When the pull force is removed, only frictional and inertia force acts. Let the body has deceleration a_1.

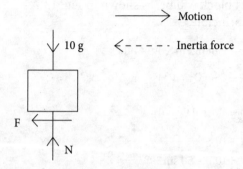

Fig. 13.6 (c) F.B.D. of block

Using D' Alembert's principle,

(Net force causing motion) + Inertia force = 0

$$(0 - F) + (- m \cdot a_1) = 0$$
$$- \mu \cdot N - m \cdot a_1 = 0$$
$$- 0.24 \times 10 \times g - 10 \times a_1 = 0$$
$$a_1 = - 2.35 \text{ m/sec}^2$$
$$v^2 = u^2 + 2as$$
$$0 = (11.27)^2 - 2 \times 2.35 \times x,$$
$$x = 27 \text{ m}$$

Example: 13.2

If pull 50 N is applied at an angle of 30° to the horizontal in the previous question, then solve the same by using work–energy principle.

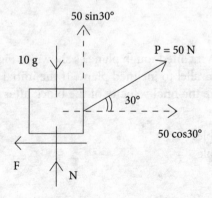

Fig. 13.7 F.B.D. of block

Solution:

The free body diagram of block will be, as shown in Fig. 13.7.
Using the work–energy principle,

Net work done = Change in K.E. of block

$$(50\cos 30° - F)s = \frac{1}{2}m(V_2^2 - V_1^2)$$

$$(50\cos 30° - \mu N)\times 24 = \frac{1}{2}\times 10(V_2^2 - 0) \quad \ldots\ldots (1)$$

$$\sum Y = 0,\ 50\sin 30° + N = 10g$$

$$N = 10.g - 50\sin 30°$$

$$= 73.10\ N$$

Substituting the value of N in equation (1),

$$(50\cos 30° - 0.24\times 73.1) = \frac{1}{2}\times 10(V_2^2)$$

$$V_2 = 2.27\ m/sec$$

If the block traverses by distance 'x' after removal of the pull force,

Using the Work–energy principle,

Net work done = Change in K.E. of block

$$(0 - F)x = \frac{1}{2}m(v_2^2 - v_1^2)$$

$$(0 - \mu N)x = \frac{1}{2}\times 10(0 - 2.27^2)$$

$$(-0.24\times 73.1)x = \frac{1}{2}\times 10(0 - 2.27^2)$$

$$x = 1.47\ m$$

Example: 13.3

A 40 kg block is lying on an inclined rough plane as shown in Fig. 13.8. It is pulled by 600 N force by means of a rope parallel to inclined plane. If the initial velocity of the 40 kg block is 2.4 m/sec then determine the final velocity of the block after traversing 5 m on inclined plane by using:

(i) Work–energy principle
(ii) D' Alembert's principle

Work and Energy

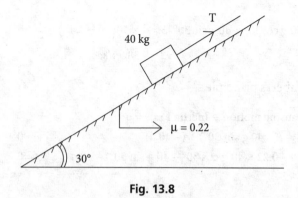

Fig. 13.8

Solution:

Consider free body diagram of block where 40 kg mass is pulled by force 600 N as shown in Fig. 13.8. (a).

Given: $\mu = 0.22$, $u = v_1 = 2.4$ m/sec, $s = 5$ m, $v_2 = ?$

(i) Using the work–energy principle,

W.D. on block = change in K.E.

$$(T - F - 40\ g \sin 30°)s = \frac{1}{2}m(v_2^2 - v_1^2)$$

$$(600 - \mu N - 20\ g)5 = \frac{1}{2} \times 40(v_2^2 - 2.4^2) \qquad \ldots\ldots (1)$$

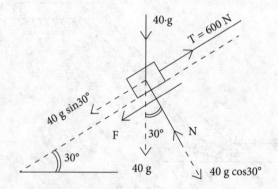

Fig. 13.8 (a) F.B.D. of block

Resolving the forces perpendicular to inclined plane,

$N = 40\ g \cos 30°$

Substituting the value of N in equation (1),

$$\left(600 - 0.22 \times 40 \ g \cos 30° - 40 \ g \sin 30°\right)5 = \frac{1}{2} \times 40\left(v_2^2 - 2.4^2\right)$$

$$v_2 = 9.38 \ m/sec$$

(ii) Using D' Alembert's principle,

Net force causing motion + Inertia force = 0

$(T - F - 40 \ g \sin 30°) + (- 40.a)$ = 0

$(600 - 0.22 \times 40 \ g \cos 30° - 40 \ g \sin 30°) - 40.a$ = 0

$a = 8.23 \ m/sec^2$

$v^2 = u^2 + 2as$

$v^2 = 2.4^2 + 2 \times 8.23 \times 5$

$v = 9.38 \ m/sec$

Example: 13.4

A system of masses is pulled by a force of 720 N as shown in Fig. 13.9. If the co-efficient of friction between all contact surfaces is 0.24, determine the final velocity of the system after traversing 3.6 m starting from the rest. Determine tensions in the string also. Take both pulleys as frictionless. Solve the problem by using the Work–energy principle.

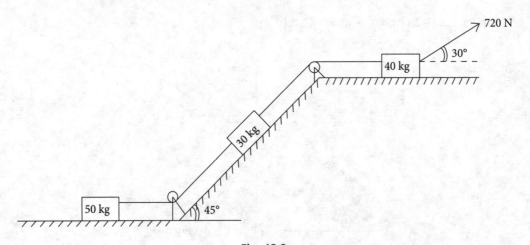

Fig. 13.9

Solution:

If all masses are considered together with connected string the free body diagram will be as shown in Fig. 13.9 (a)

Work and Energy

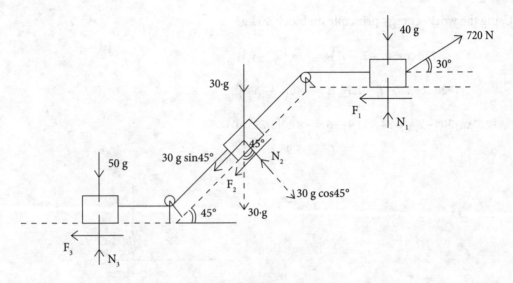

Fig. 13.9 (a) F.B.D. of blocks

Using the Work–energy principle,

Work done on the system of masses = Change in K.E. of blocks

$$(\text{Net forces causing motion}) \times \text{displacement} = \frac{1}{2}m_1(v_2^2 - v_1^2) + \frac{1}{2}m_2(v_2^2 - v_1^2) + \frac{1}{2}m_3(v_2^2 - v_1^2)$$

$$(720\cos 30° - F_1 - F_2 - 30\, g\sin 45° - F_3)s = \frac{1}{2}(v_2^2 - v_1^2)(m_1 + m_2 + m_3)$$

$$(720\cos 30° - \mu N_1 - \mu N_2 - 30\, g\sin 45° - \mu N_3)s = \frac{1}{2}(v_2^2 - v_1^2)(m_1 + m_2 + m_3) \quad \ldots\ldots (1)$$

From Fig. 13.9.(a),

$N_1 + 720\sin 30° = 40 \cdot g$, $N_1 = 32.40$ N $N_2 = 30\, g\cos 45°$, $N_3 = 50 \cdot g$

Given: $\mu = 0.24$, $m_1 = 40$ kg, $m_2 = 30$ kg, $m_3 = 50$ kg, $v_1 = u = 0$, $s = 3.6$ m, $v_2 = ?$

Substituting given values in equation (1),

$$(720\cos 30° - 0.24 \times 32.40 - 0.24 \times 30\, g\cos 45° - 30\, g\sin 45° - 0.24 \times 50\, g) \times 3.6$$

$$= \frac{1}{2}(v_2^2 - 0)(40 + 30 + 50)$$

$$v_2 = 3.79 \text{ m/sec}$$

Let the tension between 40 kg and 30 kg block is T_1 and the same between 30 kg and 50 kg is T_2. As the block 40 kg is moving right side the tension T_1 will be greater than T_2. To determine tensions consider free body diagram of blocks as shown in Fig. 13.9 (b).

Using the work–energy principle on block 40 kg,

$$\left(720\cos 30° - T_1 - F_1\right)s = \frac{1}{2}.m_1\left(v_2^2 - v_1^2\right)$$

$$\left(720\cos 30° - T_1 - \mu_1 N_1\right)s = \frac{1}{2}.40\left(v_2^2 - v_1^2\right)$$

$$\left(720\cos 30° - T_1 - 0.24 \times 32.4\right)3.6 = \frac{1}{2} \times 40\left(3.79^2 - 0\right)$$

$$T_1 = 535.96 \ N$$

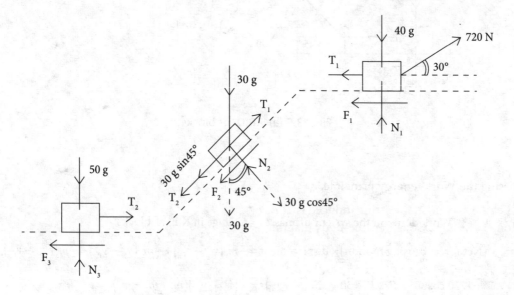

Fig. 13.9 (b) F.B.D. of blocks

Using the work–energy principle on 30 kg block,

$$\left(T_1 - T_2 - 30\ g\sin 45° - F_2\right)s = \frac{1}{2}.m_2\left(v_2^2 - v_1^2\right)$$

$$\left(535.96 - T_2 - 30\ g\ \sin 45° - 0.24 \times 30\ g\ \cos 45°\right)3.6 = \frac{1}{2}.30\left(3.79^2 - 0\right)$$

$$T_2 = 218.06\ N$$

Note: Tension, T_2 may also be determine by using the Work–energy principle on 50 kg block.

Example: 13.5

An ideal pulley is hinged to a triangular plate as shown in the Fig. 13.10. Two masses of 100 kg and 70 kg are suspended with the help of an inextensible string. Determine velocity of blocks after travelling a distance of 4.5 m starting from rest. Determine tension in the string also.

Work and Energy

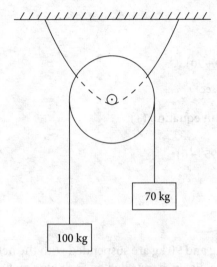

Fig. 13.10

Solution:

Given: $v_1 = u = 0$, $s = 4.5$ m

Let the tension in the string induced is T. As the pulley is ideal, the tension will remain same on each block. Consider free-body diagram of each block as shown in Fig. 13.10 (a)

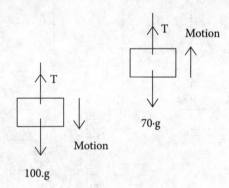

Fig. 13.10 (a)

Using the Work–energy principle on 100 kg and 70 kg separately,

$$(100 \cdot g - T) \times s = \frac{1}{2} \times 100 (v_2^2 - v_1^2)$$

$$(100g - T) \times 4.5 = \frac{1}{2} \times 100 (v_2^2 - 0) \quad \ldots\ldots (1)$$

Similarly,

$$(T - 70g) \times 4.5 = \frac{1}{2} \times 70 (v_2^2 - 0) \quad \ldots\ldots (2)$$

Add equation (1) and (2),

$$(100g - 70g)4.5 = \frac{1}{2} \times (100 + 70).v_2^2$$
$$v = 3.95 \ m/sec$$

Substituting the value of v_2 in equation (1),

$$(100g - T)4.5 = \frac{1}{2} \times 100(3.95^2 - 0)$$
$$T = 807.64 \ N$$

Example: 13.6

Two blocks of masses 120 kg and 90 kg are suspended with the help of two pulleys as shown in Fig. 13.11. Determine the distance travelled by each block when 120 kg block reaches to velocity of 3.6 m/sec starting from rest. Take both pulleys as of same size, frictionless and weightless. Determine tension in the string also.

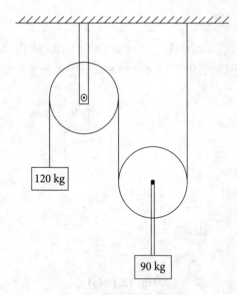

Fig. 13.11

Solution:

Given, $v_1 = u = 0$, $v_2 = 3.6$ m/sec of block 120 kg

Let the tension induced in the string is T and block 120 kg travels distance y. Thus 90 kg block will travel by distance $y/2$ and its velocity v_2 will be half of velocity of block 120 kg, i.e., v_2 of 90 kg block = 1.8 m/sec

Work and Energy

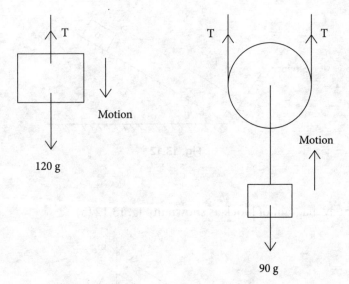

Fig. 13.11 (a) F.B.D. of blocks

Using the Work–energy principle on block 120 kg,

$$(120g - T).y = \frac{1}{2} \times 120 \left(v_2^2 - v_1^2\right)$$

$$(120g - T).y = \frac{1}{2} \times 120 \left(3.6^2 - 0\right) \qquad \ldots (1)$$

Using the Work–energy principle on block 90 kg,

$$(2T - 90g).\frac{y}{2} = \frac{1}{2} \times 90 \left(v_2^2 - v_1^2\right)$$

$$(2T - 90g).\frac{y}{2} = \frac{1}{2} \times 90 \left(1.8^2 - 0\right)$$

$$(2T - 90g).y = 90(1.8)^2 \qquad \ldots (2)$$

Solving equation (1) and (2) we get,

$y = 1.26$ m

$T = 560.06$ N

Thus, the distance travel by 120 kg block is 1.26 m and the distance travel by 90 kg block is 0.63 m.

Example: 13.7

A block of mass m kg is placed on a rough incline at distance 'x' from stopper as shown in Fig. 13.12. If the co-efficient of friction between block and inclined plane is μ then determine the distance travelled by the block when it hits the stopper by velocity V.

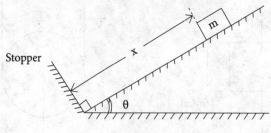

Fig. 13.12

Solution:

Consider free body diagram of block as shown in Fig. 13.12 (a).

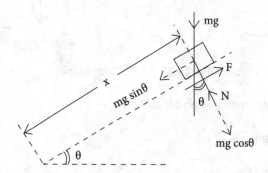

Fig. 13.12 (a) F.B.D. of blocks

Using the Work–energy principle,

W.D. by block = Change in its K.E.

$$(mg\sin\theta - F)x = \frac{1}{2}m(v^2 - 0)$$

$$(mg\sin\theta - \mu N)x = \frac{1}{2}mv^2 \quad \text{..... (1)}$$

$\Sigma Y = 0, \qquad N = mg\cos\theta$

Substituting the value of N in equation (1),

$$(mg\sin\theta - \mu.mg\cos\theta)x = \frac{1}{2}mv^2$$

$$(g\sin\theta - \mu.g\cos\theta)x = \frac{v^2}{2}$$

$$g\sin\theta(1 - \mu\cot\theta).x = \frac{v^2}{2}$$

$$x = \left[\frac{v^2}{2g\sin\theta(1 - \mu\cot\theta)}\right]$$

Note: We can observe that the distance travelled by block is not depending on its mass.

Example: 13.8

Two blocks are connected via inextensible string passes over an ideal pulley as shown in Fig. 13.13. When the system is released from rest, it attains velocity of 3.4 m/sec after travelling some distance. If the coefficient of friction between all contact surfaces is 0.24, determine the distance travelled by blocks and tension in the string.

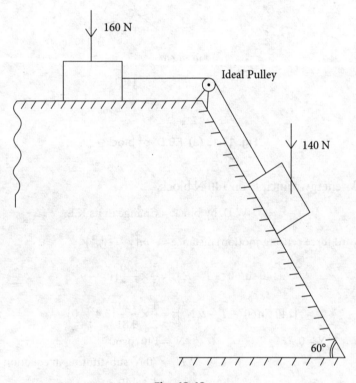

Fig. 13.13

Solution:

Given: $V_1 = u = 0$, $W_1 = 140$ N, $W_2 = 160$ N, $\mu = 0.24$, $V_2 = 3.4$ m/sec

Let the tension in the string is T.

Consider free body diagram of both blocks as given by Fig. 13.13 (a)

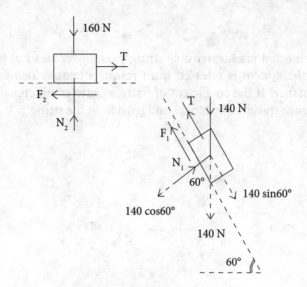

Fig. 13.13 (a) F.B.D. of blocks

Using the work–energy principle on 140 N block,

$$\text{W.D. by block} = \text{Change in its K.E.}$$

$$(\text{Net downword force causing motion}) \cdot \text{distance} = \frac{1}{2}m(v_2^2 - v_1^2)$$

$$(140\sin 60° - T - F_1)s = \frac{1}{2} \times \frac{140}{g}(v_2^2 - v_1^2)$$

$$(140\sin 60° - T - \mu_1 N_1)s = \frac{1}{2} \times \frac{140}{9.81}(3.4^2 - 0) \qquad \ldots\ldots (1)$$

$$\sum Y = 0, \qquad N_1 = 140\cos 60°$$

$$N_1 = 70\ N \text{ substituting in equation (1),}$$

$$(140\sin 60° - T - 0.24 \times 70)s = \frac{1}{2} \times \frac{140}{9.81}(3.4^2) \qquad \ldots\ldots (2)$$

Using the work–energy principle on 160 N block,

$$(T - F_2)s = \frac{1}{2} \times \frac{160}{g}(3.4^2 - 0)$$

$$(T - \mu_2 N_2)s = \frac{1}{2} \times \frac{160}{9.81}(3.4^2) \qquad \ldots\ldots (3)$$

$$\sum Y = 0, \qquad N_2 = 160\ N$$

Substituting in equation (3)

$$(T - 0.24 \times 160)s = \frac{1}{2} \times \frac{160}{9.81}(3.4^2) \qquad \ldots\ldots (4)$$

Work and Energy

Add equation (2) and (4),

  $s = 2.68$ m

Substituting the value of 's' in equation (4),

  $T = 73.58$ N

Example: 13.9

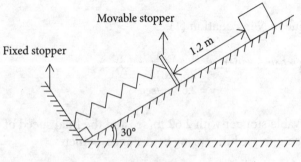

Fig. 13.14

A block of weight 640 N is placed on a 30° incline as shown in Fig. 13.14. If the distance between block and movable stopper is 1.2 m and coefficient of friction between block and inclined plane is 0.24 then determine:

(i) The velocity of block when it hits the movable stopper.
(ii) The compression of spring if stiffness of the spring is 10 N/mm.

Solution:

Given, $m = \dfrac{640}{g}$ kg, $s = 1.2$ m, $\mu = 0.24$, $v_1 = u = 0$

Let the velocity of block becomes v before it hits the movable stopper. Consider free body diagram of block as shown in Fig. 13.14 (a)

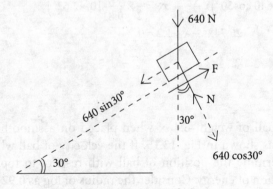

Fig. 13.14 (a) F.B.D. of blocks

Using the work–energy principle,

$$(640\sin 30° - F)s = \frac{1}{2}m(v_2^2 - v_1^2)$$

$$(640\sin 30° - \mu N)s = \frac{1}{2} \times \frac{640}{g}(v^2 - 0) \qquad \ldots (1)$$

$$\sum Y = 0, \quad N = 640\cos 30°$$

Substituting the value of N in equation (1),

$$(640\sin 30° - 0.24 \times 640\cos 30°)1.2 = \frac{1}{2} \times \frac{640}{9.81}(v^2 - 0)$$

$$v = 2.62 \text{ m/sec}$$

Thus block hits movable stopper with 2.62 m/sec and the final speed of the block becomes zero. Let the block compressed the spring by distance 'x' meter.

As stiffness of spring, $k = 10$ N/mm
$$= 10 \times 10^3 \text{ N/m}$$

When the block compresses the spring, the spring does work on the block by amount
= Force exerted by spring × compression

$$= -\frac{1}{2}kx.x$$

$$= -\frac{1}{2} \times 10 \times 10^3.x^2$$

Using the Work–energy principle when the block hits the movable stopper,
Network done by block = Change in K.E.
i.e., Work done by block − Work done by spring = Change in K.E.

$$(640\sin 30° - 0.24 \times 640\cos 30°)x - \frac{10^4}{2}.x^2 = \frac{1}{2} \times \frac{640}{9.81}(0 - 2.62^2)$$

$$26.74x^2 - x + 1.2 = 0$$

$$x = 0.231 \text{ m}$$

Example: 13.10

A smooth spherical ball of weight 6.4 N when placed on a smooth semi-circular log at its top, it slips down as shown in Fig. 13.15. If the velocity of ball when leaving the log is 1.36 m/sec then determine the position of ball with respect to top of the log by using principle of conservation of energy. Consider the radius of log as 0.92 m.

Work and Energy

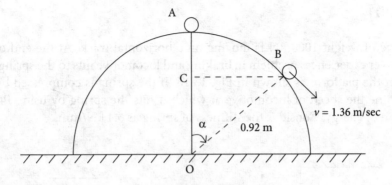

Fig. 13.15

Solution:

Given: $W = 6.4$ N

$V = 1.36$ m/sec

$r = 0.92$ m, $\alpha = ?$

When ball is placed at top 'A',

The Total Energy = $(K.E.)_{ball} + (P.E.)_{ball}$

$$E_1 = 0 + mg \cdot r$$
$$= W \cdot r$$
$$E_1 = 6.4 \times 0.92$$

Let the ball leave the surface of log at position B as shown in Fig. 13.15.

$$= (K.E)_{ball} + (P.E)_{ball}$$
$$= \frac{1}{2}mv^2 + mg(OC)$$
$$= \frac{1}{2}mv^2 + mg \cdot OB\cos\alpha$$
$$E_2 = \frac{1}{2} \times \frac{6.4}{9.81}(1.36)^2 + 6.4 \times 0.92\cos\alpha$$
$$E_2 = 0.603 + 6.4 \times 0.92\cos\alpha \qquad \ldots\ldots (2)$$

Using principle of Conservation of energy,

$$E_1 = E_2$$
$$6.4 \times 0.92 = 0.603 + 6.4 \times 0.92\cos\alpha$$
$$\alpha = 26.16°$$

Example: 13.11

A locomotive of weight 1000 kN is moving on a horizontal track. At the end of journey, suddenly driver experiences problem in braking and locomotive hits to the spring mounted at the end of the platform as shown in Fig. 13.16. If the spring is compressed by 230 mm then determine the speed of locomotive at which it hits the spring by using Principle of Conservation of energy. Consider the stiffness of spring as 6.4 kN/mm.

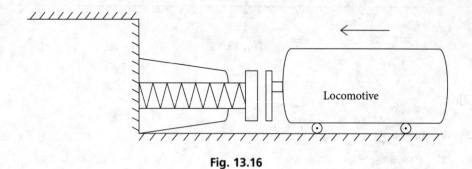

Fig. 13.16

Solution:

Given: $W = 1000$ kN, $x = 230$ mm $= 0.23$ m
$k = 6.4$ kN/mm $= 6.4 \times 10^3 \times 10^3$ N/m

Let locomotive hits the spring by velocity 'v'.

Total energy of the locomotive and spring before collision,

$$E_1 = (K.E)_{locomotive} + (P.E)_{locomotive} + (K.E)_{spring} + (P.E)_{spring}$$

$$E_1 = \frac{1}{2}mv^2 + 0 + 0 + 0$$

$$= \frac{1}{2} \times \frac{1000 \times 10^3}{9.81} v^2 \quad \quad \ldots\ldots (1)$$

Similarly total energy after collision,

$$E_2 = (K.E)_{locomotive} + (P.E)_{locomotive} + (K.E)_{spring} + (P.E)_{spring}$$

$$= 0 + 0 + 0 + \frac{1}{2}.k.x^2$$

$$E_2 = \frac{1}{2} \times 6.4 \times 10^3 \times 10^3 .(0.23)^2 \quad \quad \ldots\ldots (2)$$

Work and Energy

Using Principle of Conservation of energy,

$$E_1 = E_2$$

$$\frac{1}{2} \times \frac{1000 \times 10^3 v^2}{9.81} = \frac{1}{2} \times 6.4 \times 10^6 \times (0.23)^2$$

$$v = 1.82 \ m/sec$$

Example: 13.12

A prismatic bar is suspended from roof as shown in Fig. 13.17. A spring containing stopper is placed between rod and ground. Determine the compression of spring when a 12 kg mass is allowed to fall along the bar to hit the stopper from 0.75 m height by using Principle of Conservation of energy. Consider the stiffness of spring as 1 N/mm.

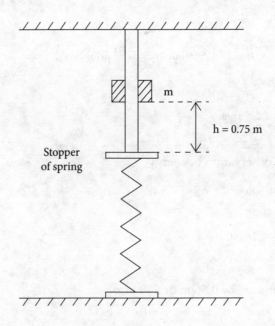

Fig. 13.17

Solution:

When mass, m just hit the stopper, the velocity will be,

$V^2 = u^2 + 2gh$

$\quad = 0 + 2 \times 9.81 \times 0.75$

$V = 3.84 \ m/sec$

Thus at the beginning of hit,

$$E_1 = (K.E)_{mass} + (P.E)_{mass} + (K.E)_{spring} + (P.E)_{spring}$$
$$= \frac{1}{2}mv^2 + 0 + 0 + 0$$
$$= \frac{1}{2} \times 12 \times (3.84)^2$$
$$= 88.47\, Nm$$

Given, $k = 1$ N/mm $= 1 \times 10^3$ N/m

Let the spring is compressed by amount y.

Thus mass and spring displaces by height y after hit.

Total energy after hit,

$$E_2 = (K.E)_{mass} + (P.E)_{mass} + (K.E)_{spring} + (P.E)_{spring}$$
$$= 0 + mg(-y) + 0 + \frac{1}{2}k.y^2$$
$$E_2 = -12 \times 9.81 \times y + \frac{1}{2} \times 1000 \times y^2$$

Using Principle of Conservation of energy,

$$88.47 = -117.72.y + 500.y^2$$
$$500.y^2 - 117.72\, y - 88.47 = 0$$
$$y^2 - 0.235 - 0.18 = 0$$
i.e., $\quad y = 0.56$ m

Thus spring is compressed by 0.56 m or 560 mm

Example: 13.13

A flexible thin wire PQ of uniform material and cross-section is placed over a semi-circular roof ABC of a building as shown in Fig. 13.18. The length of the wire is half of the semi-circular arch. Determine the final velocity of thin wire when its end P passes through C by using Principle of conservation of energy. Consider friction between thin wire and roof as negligible.

Work and Energy

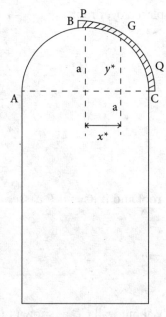

Fig. 13.18

Solution:

Let the radius of arch of semi-circular roof is 'a' and the density, cross-section area of thin wire is ρ and A, respectively.

As the length of thin wire is equal to half of the semi-circular arch, l = perimeter of quarter circle

$$= \left(\frac{\pi.a}{2}\right)$$

Mass of thin wire will be,

$$m = \rho \times V$$
$$= \rho \times A \times l$$
$$= \left(\rho \times A \times \frac{\pi a}{2}\right)$$

The centroid of thin wire will be,

$$x^* = \left(\frac{2a}{\pi}\right) \text{ and } y^* = \left(\frac{2a}{\pi}\right)$$

Thus total energy of wire when placed over roof,

$E_1 = K.E. + P.E.$

$= \dfrac{1}{2}mv^2 + mgh$

here, $v = 0$ and $h = y^* = \left(\dfrac{2a}{\pi}\right)$

$E_1 = 0 + \left(\rho.A\dfrac{\pi a}{2}\right)g\left(\dfrac{2a}{\pi}\right)$

$E_1 = \rho.A.g.a^2$ ..... (1)

When thin wire slips down the roof and if its end P passes through C with velocity V, the total energy at this instant,

$E_2 = K.E. + P.E.$

$= \dfrac{1}{2}mv^2 + mgh$ ..... (2)

Where the wire PQ lies vertically about wall CD as shown in Fig. 13.18 (a)

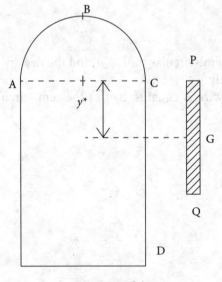

Fig. 13.18 (a)

$h = y^* = -\dfrac{l}{2}$

$h = \dfrac{-1}{2}\left(\dfrac{\pi a}{2}\right)$

$h = \dfrac{-\pi a}{4}$

Work and Energy

Substituting value of m and h in equation (2),

$$E_2 = \frac{1}{2}\left(\frac{\rho.A.\pi a}{2}\right)v^2 + \left(\frac{\rho.A.\pi a}{2}\right)g\left(\frac{-\pi a}{4}\right)$$

$$= \left(\frac{\rho.A\pi av^2}{4}\right) - \left(\frac{\pi^2 \rho A a^2 g}{8}\right) \qquad \text{..... (3)}$$

Using the principle of conservation of energy,

$$E_1 = E_2$$

$$\rho A g a^2 = \frac{\rho A \pi a v^2}{4} - \frac{\pi^2 \rho A a^2 g}{8}$$

$$v^2 = ag\left(\frac{4}{\pi} + \frac{\pi}{2}\right)$$

$$v = 5.28\sqrt{a}$$

Example: 13.14

A bus of weight 50 kN descends an incline of 1 in 110 from rest. The bus descends 16 m in 4 seconds under uniform acceleration. Determine power consumption under uniform acceleration. Further, if bus runs with uniform speed after 4 seconds then also determine the power consumption. Consider total frictional resistance as 800 N.

Solution:

Given: $W = 50$ kN,

$\tan\theta = \dfrac{1}{100}$ since θ is very-2 small angle, thus $\theta = \dfrac{1}{110}$

$\sin\theta = \theta = \dfrac{1}{110}$

$s = 16$ m, $t = 4$ sec, $u = 0$

If bus is accelerated by amount 'a',

Using equation,

$$s = ut + \frac{1}{2}at^2$$

$$16 = 0 + \frac{1}{2}.a.(4)^2$$

$$a = 2 \text{ m/sec}^2$$

Consider free body diagram of bus as shown in Fig. 13.19.

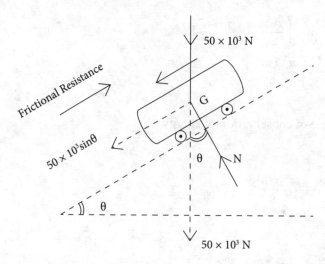

Fig. 13.19 F.B.D. of bus

Net force required to accelerate the bus downward

$$= (ma - mg\sin\theta + \text{frictional resistance}) \quad \ldots\ldots (1)$$

$$= \left(\frac{50\times 10^3}{9.81}\times 2\right) - \left(50\times 10^3 \times \frac{1}{110}\right) + 800$$

$F_{effort} = 10539.13\ N$

Note: In equation (1), $mg\sin\theta$ is taken negative as it will accelerate the bus.

Velocity of bus after 4 seconds,

$V = u + at$

$\quad = 0 + 2 \times 4$

$V = 8$ m/sec

Thus power consumption $= F_{effort} \times V$

$\quad\quad\quad = 10539.13 \times 8$

$\quad\quad P = 84313.08$ Watt or $P = 84.31$ kW

When bus runs with uniform velocity of 8 m/sec, the force required will be given by equation (1), where $a = 0$.

$F_{effort} = 0 - 50\times 10^3 \sin\theta + \text{frictional resistance}$

$\quad\quad = -50\times 10^3 \times \frac{1}{100} + 800$

$\quad\quad = 300\ N$

Work and Energy

Thus power consumption will be $= P = F_{effort} \times V$

$$= 300 \times 8$$
$$= 2400 \text{ Watt}$$
$$P = 2.4 \text{ kW}$$

Example: 13.15

A lorry of weight 200 kN is uniformly accelerated on an inclined 1 in 100 starting from rest. The lorry ascends 20 m distance in 10 seconds during acceleration. Further lorry ascends with uniform velocity. Determine the power consumption:

(i) When lorry is accelerated uniformly.
(ii) When lorry ascends with uniform velocity after 10 sec.

Consider the total resistance to lorry as 2.5 kN.

Solution:

Given: $W = 200$ kN, $u = 0$, $s = 20$ m, $t = 10$ sec,

If angle of incline is θ, $\sin\theta = \left(\dfrac{1}{100}\right)$

Using equation,

$$S = ut + \dfrac{1}{2}at^2$$
$$20 = 0 + \dfrac{1}{2}.a.(10)^2$$
$$a = 0.40 \; m/sec^2$$

The velocity of lorry after 10 sec will be

$V = u + at$

$V = 0 + 0.4 \times 10$

$V = 4$ m/sec

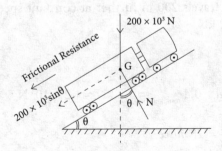

Fig. 13.20

(i) Net effort required to ascend the lorry

$$= [m.a + 200 \times 10^3 \sin\theta + \text{frictional resistance}] \quad \ldots (1)$$

$$= \left(\frac{200 \times 10^3}{9.81} \times 0.4\right) + \left(200 \times 10^3 \times \frac{1}{100}\right) + \left(2.5 \times 10^3\right)$$

$$F_{effort} = 12.65 \times 10^3 \, N$$

Power Consumption when lorry is uniformly accelerated,

$P = F_{effort} \times v$

$= 12.65 \times 10^3 \times 4$

$= 50.62 \times 10^3$ watt $= 50.62$ kW

(ii) When lorry ascends with uniform velocity after 10 seconds, $a = 0$, $v = 4$ m/sec

Using equation (1),

Net effort,

$$F_{effort} = \left(0 + 200 \times 10^3 \times \frac{1}{100} + 2.5 \times 10^3\right)$$

$$= 4.5 \times 10^3 \, N$$

Power Consumption with uniform velocity,

$P = F_{effort} \times v$

$= 4.5 \times 10^3 \times 4$

$= 18 \times 10^3$ watt

$= 18$ kW

Example: 13.16

A loaded van of weight 240 kN is uniformly accelerated to speed of 90 kmph from rest. Determine power consumed, if it is uniformly accelerated for 120 m. Also determine the power consumption if it travels 200 m further at constant speed of 90 kmph. Take total frictional resistance as 1.2 kN.

Solution:

Given: $W = 240 \times 10^3$ N, $u = 0$, $F_{resistance} = 1.2$ kN $= 1.2 \times 10^3$ N

$s = 120$ m and $v = 90 \times \dfrac{5}{18} = 25$ m/sec

(i) Let the van is accelerated by 'a',

Total effort applied by van = Force required to accelerate + Frictional resistance

$$= (m_{van} \cdot a) + F_{resistance}$$

$$F_{Van} = \left(\frac{240 \times 10^3}{9.81} \cdot a\right) + 1.2 \times 10^3 \quad \ldots\ldots (1)$$

Using equation,

$V^2 = u^2 + 2as$

$(25)^2 = 0 + 2 \times a \times 120$

$a = 2.60 \text{ m/sec}^2$

Substituting value of 'a' in equation (1),

$$F_{Van} = \left(\frac{240 \times 10^3}{9.81} \times 2.6\right) + 1.2 \times 10^3 = 64808.56 \text{ N}$$

Power consumed when van is uniformly accelerated,

$= F_{Van} \times v$

$= 64808.56 \times 25$

$= 1620214.07$ watt

$= 1620.21$ kW

(ii) For constant speed, $a = 0$ i.e., inertia force becomes zero

Using equation (1),

Power consumed at constant speed $= F_{Van} \times v$

$= [0 + 1.2 \times 10^3] \times 25$

$= 30,000$ watt

$= 30$ kW

Example: 13.17

A tractor pulls a damaged car of weight 9000 N form rest to a distance of 38 m. Determine work done by tractor if it pulls under:

(i) Constant acceleration when tractor reaches to velocity of 12 m/sec
(ii) Constant velocity

Take the total frictional resistance as 320 N.

Solution:

Given: $W = 9000$ N, $u = 0$, $s = 38$ m

(i) Let the constant acceleration is 'a' and final velocity, $v = 12$ m/sec

$$v^2 = u^2 + 2as$$
$$(12)^2 = 0 + 2 \cdot a \times 38$$
$$a = 1.89 \text{ m/sec}^2$$

Work done by tractor

= Total pulling effort of tractor × distance

= (Force required to accelerate the car + Frictional resistance) × distance

$$= \left[(m_{car} \cdot a) + F_{resistance} \right] \cdot s \quad \ldots\ldots (1)$$

$$= \left[\left(\frac{9000}{9.81} \right) \times 1.89 + 320 \right] \times 38$$

$$W = 78049.91 \, N.m$$

(ii) At constant speed, $a = 0$ i.e., inertia force becomes zero.

Using equation (1),

$$W = [0 + 320] \times 38$$
$$= 12160 \text{ N.m}$$

Example: 13.18

A labourer working at a construction site pulls a corrugated box of weight 600 N with the help of an inextensible string by 10 m as shown in Fig. 13.21. Determine the work done by the person:

(i) If he pulls the box by constant acceleration in 8 seconds
(ii) If he pulls the box by constant velocity

Solution:

Given, $u = 0$, $W = 600$ N, $s = 10$ m
Let the constant acceleration is 'a' and tension in the string is T
$t = 8$ seconds, $T = 600$ N, as pulley is ideal.

(i) Using equation,

$$s = ut + \frac{1}{2} at^2$$
$$10 = 0 + \frac{1}{2} \cdot a \cdot (8)^2$$
$$a = 0.31 \text{ m/sec}^2$$

Work and Energy

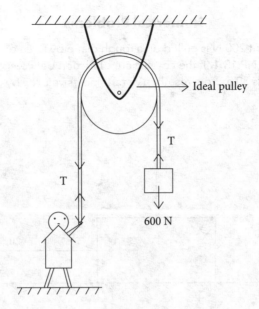

Fig. 13.21

W. D. by person under constant acceleration will be

$\quad$ = [Total effort] × distance

$\quad$ = [Force required to accelerate the box + T] × distance

$\quad$ = $[(m_{box} \times a) + T] \cdot s$ ..... (1)

$$W.D. = \left[\left(\frac{600}{9.81} \times 0.31\right) + 600\right] \times 10$$

$\quad\quad\quad = 6189.60 \, Nm$

(ii) Under constant speed, $a = 0$, i.e., inertia force of box becomes zero
Substituting value of 'a' in equation (1),

$\quad$ W. D. = $[0 + 600] \cdot 10$

$\quad\quad\quad\quad$ = 6000 Nm

Theoretical Problems

T 13.1 Illustrate the work done during a process when a body is pushed or pulled by a force.
T 13.2 Discuss the work done by variable force acting on a body.
T 13.3 Briefly discuss kinetic energy and potential energy of a body.
T 13.4 State whether work is a scalar or vector quantity.
T 13.5 Illustrate the principle of work and energy.
T 13.6 State energy, power and describe the relation between them.
T 13.7 Define principle of conservation of energy with a suitable example.

Numerical Problems

N 13.1 A block of weight 200 N is pulled on a rough plane by force of 80 N at an angle of 20° as shown in Fig.NP 13.1. If the co-efficient of friction between contact surfaces is 0.25 and block travels 30 m from rest then determine its velocity by using the Work–energy principle.

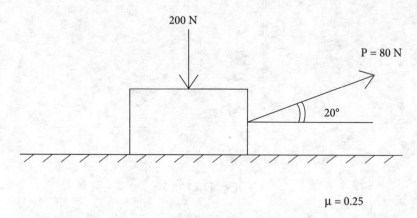

Fig. NP 13.1

N 13.2 A block of weight 500 N is placed on a rough incline plane as shown in Fig. NP 13.2. Determine the tangential force 'P' required to move the block by 9 m from initial velocity of 3 m/sec to final velocity of 5 m/sec.

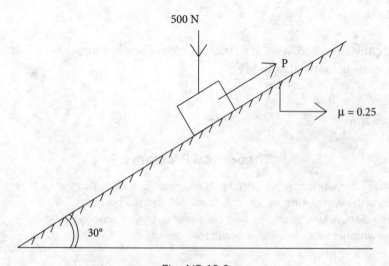

Fig. NP 13.2

N 13.3 Two blocks of mass 90 kg and 60 kg are suspended with the help of an inextensible string as shown in Fig. NP 13.3. If blocks are released from rest, then determine their velocities after travelling a distance of 6.4 m. Determine tension in the string also. Solve the problem by using the Work–energy principle.

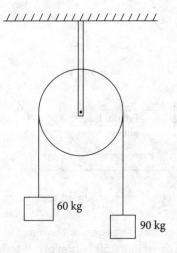

Fig. NP 13.3

N 13.4 Two identical, weightless and frictionless pulleys are used to support two blocks of weight 600 N and 800 N as shown in Fig. NP 13.4. Using the Work–energy principle, determine the tension in the string and velocities of each block after travelling a distance of 5.6 m starting from rest.

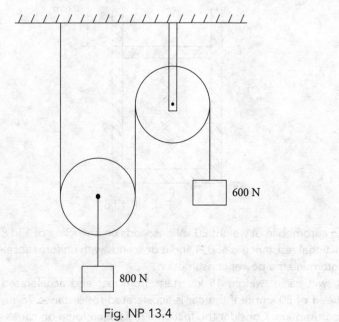

Fig. NP 13.4

N 13.5 Two blocks of weight 400 N and 250 N are pulled from rest by a tangential force 'P' as shown in Fig. NP 13.5. If the system attains a velocity of 4.2 m/sec, determine the tangential force P. Take coefficient of friction for all contact surfaces as 0.22.

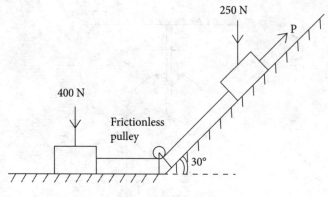

Fig. NP 13.5

N 13.6 Consider a cylinder–piston arrangement where a spring is placed as shown in Fig. NP 13.6. A block of weight 150 N is allowed to fall on piston from height 0.6 m. If the stiffness of spring is 1.20 N/mm, determine the compression of spring by using Principle of Conservation of energy.

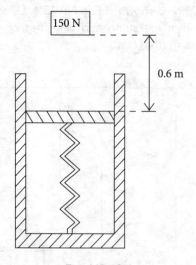

Fig. NP 13.6

N 13.7 An automobile of weight 20 kN descends on an incline of 1 in 85 from rest. If the total frictional resistance is 650 N and it descends with uniform acceleration of 1.38 m/sec², determine the power consumption.

N 13.8 A swift car of weight 15 kN starts from rest and accelerated uniformly to attain a speed of 80 kmph. If the car is accelerated for distance 75 m, determine the power consumption. Consider the frictional resistance force on car as 400 N.

Multiple Choice Questions

1. Work and energy principle is useful where problems are mainly concerned with
   a. acceleration
   b. time
   c. displacement
   d. angular velocity

2. Principle of work and energy relates the parameters
   a. Force, displacement and time
   b. Force, displacement and velocity
   c. Force, displacement and acceleration
   d. Force, time, velocity and mass

3. State the vector quantity out of following quantities
   a. work
   b. energy
   c. velocity
   d. none of these

4. If a particle is displaced by displacement dx under variable force p then the work done is given by
   a. $W = \int p\, dx$
   b. $W = \int x\, dp$
   c. $W = \int p/dx$
   d. none of these

5. State the condition in which work done remains zero
   a. if net force acting on the body is zero
   b. if displacement lies perpendicular to force
   c. if no displacement takes place in the direction of force
   d. all of these

6. A crane lifts up a car, the work done by crane will be
   a. negative
   b. positive
   c. not predictable
   d. none of these

7. What kind of energy is considered in engineering mechanics
   a. potential energy
   b. kinetic energy
   c. both (a) and (b)
   d. none of these

8. Unit of power is
   a. horsepower
   b. joule/sec
   c. watt
   d. all of these

9. In British units, 1 hp is equal to
   a. 100 watt
   b. 746 watt
   c. 735.5 watt
   d. none of these

10. In Metric units, 1 *hp* is equal to
    a. 100 watt
    b. 746 watt
    c. 735.5 watt
    d. none of these

Answers

1. c   2. b   3. c   4. a   5. d   6. a   7. c   8. d   9. b   10. c

Chapter 14

Impulse and Momentum

14.1 Introduction

So far, we have studied the work and energy principle and D' Alembert's principle to analyze the kinetic problems. In this chapter we will study another method i.e., the Principle of Impulse and Momentum to analyze the kinetic problems. This principle is derived from Newton's second law of motion and preferred where large magnitude of force acts for shortest duration. This principle relates the mass, force and motion parameters like velocity, time, etc. Before studying this principle we must understand the meaning of impulse and momentum.

Impulse means when large magnitude of force causes high impact in shortest duration. It is the product of force and shortest duration of collision of bodies. It is a vector quantity and also called impulsive force. For example high impact takes place when a high speed moving car gets collided with side tree or a missile strikes a fighter plane, etc.

Momentum is the product of mass and velocity of a particle. It is a vector quantity and always acts in the direction of moving particle.

14.2 Principle of Impulse and Momentum

Consider a body of mass m is accelerated by force F which changes its velocity from initial velocity u to final velocity v during time t, then according to Newton's second law of motion,

$F = ma$

as $a = \dfrac{dv}{dt}$

$F = m\left(\dfrac{dv}{dt}\right)$

$F.dt = m.dv$

$$\int_0^t F.dt = \int_u^v m.dv$$
$$\int_0^t F.dt = m(v-u)$$

This is knows as impulse–momentum equation. In this equation term $\int_0^t F.dt$ is called impulse of force and term $m(v-u)$ is defined as the change in momentum of the body. Thus impulse of force is nothing but the change of momentum of a body. Hence this principle can be used to determine the impulse force if initial and final velocities of the body are known. Here both terms "impulse of force" and "change in momentum" are vector quantities even they include scalar quantities time and mass with vector quantities force and velocity respectively.

14.3 Principle of Conservation of Momentum

Consider the impulse–momentum equation as stated above:

$$\int_0^t F.dt = m(v-u)$$

If the impulse of force $\int_0^t F.dt$ becomes zero then the above equation becomes

$m(v-u) = 0$

i.e., $mu = mv$

Initial momentum before collision = Final momentum after collision

This situation arises when two bodies are moving independently; collide in such a way that their net resultant force remains zero. This becomes zero due to their mutual action and reaction. Consider an example of firing of a bomb from tank. In this system, the force acting on bomb and tank is equal and opposite but net force remains zero when bomb and tank are taken together. Hence, according to the principle of conservation of momentum, bodies contain constant momentum before and after the collision.

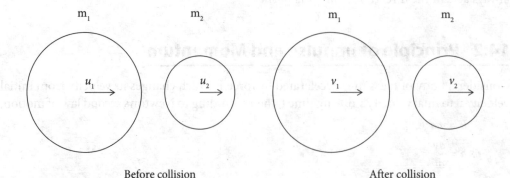

Fig. 14.1

Impulse and Momentum

In a system of masses as shown in Fig. 14.1., if two bodies m_1 and m_2 are moving with velocities u_1 and u_2 before collision and after collision if corresponding velocities become v_1 and v_2, then according to the principle of conservation of momentum,

Initial momentum before collision = Final momentum after collision

$$m_1 u_1 + m_2 v_2 = m_1 v_1 + m_2 v_2$$

As momentum is a vector quantity the above equation can be resolved in rectangular components as given below:

$$m_1(u_1)_x + m_2(u_2)_x = m_1(v_1)_x + m_2(v_2)_x \text{ and}$$
$$m_1(u_1)_y + m_2(u_2)_y = m_1(v_1)_y + m_2(v_2)_y$$

where $(u_1)_x, (u_2)_x, (u_1)_y, (u_2)_y$ are the velocities of the bodies before collision in X and Y direction and $(v_1)_x, (v_2)_x, (v_1)_y, (v_2)_y$ are velocities of the bodies after collision in X and Y direction.

14.4 Collisions of Elastic Bodies

During collision, impulsive force acts on each bodies and causes their compression. The material of bodies plays a crucial role in collision of bodies. Some materials undergo to considerable deformation and come back to the original condition within a shortest time. Such bodies are called elastic bodies and have greater life during entire period of service. Collision of elastic bodies is of following types:

i. Direct central impact
ii. Oblique/indirect central impact
iii. Direct eccentric impact
iv. Oblique/indirect eccentric impact

14.4.1 Direct central impact

Line of impact is the line joining the mass centres O_1 and O_2 of both bodies. Before collision, if velocity of bodies acts parallel to the line of impact then such impact is called direct central impact as shown in Fig. 14.2.

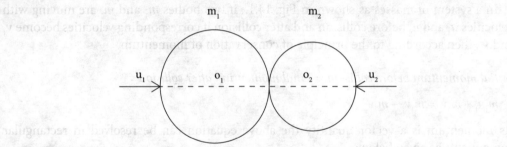

Fig. 14.2 Direct Central Impact

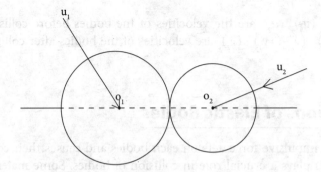

Fig. 14.3 Oblique/Indirect Central Impact

14.4.2 Oblique/Indirect central impact

Before collision, if velocity of bodies lies at an angle with the line of impact then such impact is called oblique/indirect central impact as shown in Fig. 14.3.
Direct and Oblique eccentric impacts are not included here.

14.4.3 Coefficient of restitution

It is a factor which measures how much a body is elastic. It is defined as the ratio of relative velocities after collision and before collision and represented by the symbol e.

Consider Fig. 14.1, where two bodies m_1 and m_2 are moving with the initial velocities u_1 and u_2 ($u_1 > u_2$) in same direction and after collision if their velocities become v_1 and v_2, respectively.

Relative velocity before collision will be $(u_1 - u_2)$

Relative velocity after collision will be $(v_2 - v_1)$

$$e = \frac{(v_2 - v_1)}{(u_1 - u_2)}$$

Impulse and Momentum

The coefficient of restitution depends on the material and size of bodies but does not depend on their mass. Its value lies in the following range:

For perfect non elastic collision, $e = 0$

For semi elastic collision, $\quad 0 < e < 1$

For perfect elastic collision, $\quad e = 1$

Example: 14.1

A pile of mass 1700 kg is hammered into ground by using a 900-kg hammer. The hammer is dropped from a height of 1.2 m. If the hammer rebounds by 0.09 m after collision, then determine:

(i) The velocity of pile after collision
(ii) The coefficient of restitution

Solution:

$m_1 = 1700$ kg, $m_2 = 900$ kg, $h_1 = 1.2$ m, $h_2 = 0.09$ m

Let the initial velocities of pile and hammer is u_1 and u_2 respectively and after collision it becomes v_1 and v_2, respectively. Thus $u_1 = 0$ and at the time of collision, the velocity of hammer will be

$$u_2 = \sqrt{2gh_1}$$
$$= \sqrt{2 \times 9.81 \times 1.2}$$
$$u_2 = 4.85 \ m/sec$$

After collision hammer rebounds to 0.09 m, thus velocity just after collision will be given by

$$v_2 = \sqrt{2gh_2}$$
$$= \sqrt{2 \times 9.81 \times 0.09}$$
$$v_2 = 1.33 \ m/sec$$

i.e., hammer goes in upward direction by 1.33 m/sec just after collision.

Using principle of conservation of momentum,

$$m_1 u_1 + m_2 v_2 = m_1 v_1 + m_2 v_2$$
$$0 + 900 \times 4.85 = 1700 \ v_1 + 900 \ (-1.33)$$
$$v_1 = 3.27 \ m/sec$$

The coefficient of restitution is given by,

$$e = \left(\frac{v_2 - v_1}{u_2 - u_2}\right)$$

$$= \frac{-1.33 - 3.27}{0 - 4.85}$$

$$e = 0.95$$

Example: 14.2

A billiard player wishes to strike two balls kept straight on the table as shown in Fig. 14.4. He strikes straight to one ball with a velocity of 6 m/sec by cue stick of mass 600 gm. If the masses of cue balls are 170 gm and 160 gm, respectively, then determine:

(i) The velocity of the balls and cue stick after strike
(ii) Loss of K.E. during each strike

Take coefficient of restitution as 0.7.

Solution:

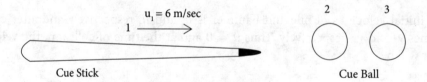

Fig. 14.4

Given:

(i) $m_1 = 600$ gm, $m_2 = 170$ gm, $m_3 = 160$ gm, $e = 0.7$, $u_1 = 6$ m/sec, $u_2 = 0$

Using principle of conservation of momentum between the stick and one ball,

$$m_1 u_1 + m_2 v_2 = m_1 v_1 + m_2 v_2$$

$$0.600 \times 6 + 0 = 0.60 v_1 + 0.170 v_2$$

$$3.53 v_1 + v_2 = 21.18 \quad \ldots\ldots (1)$$

The coefficient of restitution,

$$e = \left(\frac{v_2 - v_1}{u_1 - u_2}\right)$$

$$0.7 = \frac{v_2 - v_1}{6 - 0} \quad \ldots\ldots (2)$$

$$v_2 - v_1 = 4.2$$

Impulse and Momentum

Solving equation (1) and (2), $v_1 = 3.75$ m/sec, $v_2 = 7.95$ m/sec

Velocity of cue stick and first ball after strike will be 3.75 m/sec and 7.95 m/sec, respectively.

Now first ball will strike to second ball with velocity of 7.95 m/sec i.e.,

here $u_2 = v_2 = 7.95$ m/sec and $u_3 = 0$

Let the velocity of first ball after strike become v_2^1

Using principle of conservation of momentum between balls,

$$m_2 u_2 + m_3 u_3 = m_2 v_2^1 + m_3 v_3$$
$$0.170 \times 7.95 + 0 = 0.170 v_2^1 + 0.160 v_3$$
$$1.06 v_2^1 + v_3 = 8.45 \quad \ldots\ldots (3)$$

The coefficient of restitution

$$e = \left(\frac{v_3 - v_2^1}{u_2 - u_3} \right)$$

$$0.7 = \frac{v_3 - v_2^1}{7.95 - 0}$$

$$v_3 - v_2^1 = 5.57 \quad \ldots\ldots (4)$$

Solving equation (3) and (4),

$v_2^1 = 1.39$ m/sec

$v_3 = 6.97$ m/sec

(ii) The loss of K.E. during first strike = (K.E) after strike − (K.E) before strike

$$= \left(\frac{1}{2} m_1 v_1^2 + \frac{1}{2} m_2 v_2^2 \right) - \left(\frac{1}{2} m_1 u_1^2 + \frac{1}{2} m_2 u_2^2 \right)$$

$$= \left[\frac{1}{2} \times 0.6 (3.75)^2 + \frac{1}{2} \times 0.17 (7.95)^2 \right] - \left[\frac{1}{2} \times 0.6 (6)^2 + 0 \right]$$

$$= -1.21 \, Nm$$

The loss of K.E. during second strike,

$$= \left(\frac{1}{2} m_2 (v_2^1)^2 + \frac{1}{2} m_3 v_3^2 \right) - \left(\frac{1}{2} m_2 u_2^2 + \frac{1}{2} m_3 u_3^2 \right)$$

$$= \left[\frac{1}{2} \times 0.17 (1.39)^2 + \frac{1}{2} \times 0.16 (6.97)^2 \right] - \left[\frac{1}{2} \times 0.17 (7.95)^2 + 0 \right]$$

$$= -1.32 \, Nm$$

Example: 14.3

A tennis ball is dropped from a height of 12 m on a hard surface as shown in Fig. 14.5. If the ball rebounds to a height of 8 m, determine:

(i) The coefficient of restitution
(ii) The height to which ball will rebound in the second and third bounces

Solution:

Given, $v_1 = 0$, $h = 12$ m, $h_1 = 8$ m

Let the velocities of hard surface and ball before collision and after collision are u_1, u_2 and v_1, v_2 respectively. Thus, $u_1 = 0$ and $v_1 = 0$

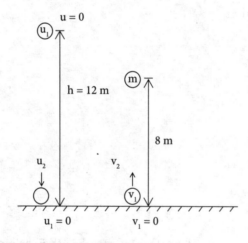

Fig. 14.5

The velocity of ball reaching hard surface will be given by

$v^2 = u^2 + 2gh$
$u_2^2 = 0 + 2 \times 9.81 \times 12$
$u_2 = 15.34$ m/sec

Thus ball collides the hard surface by 15.34 m/sec and then rebounds to height of 8 m, let the ball goes up by velocity v_2 after first bounce,

$0 = v_2^2 - 2 \times 9.81 \times 8$
$v_2 = 12.53$ m/sec

The coefficient of restitution $e = \left(\dfrac{v_2 - v_1}{u_1 - u_2} \right)$

Impulse and Momentum

Where $u_1 = 0$, $v_2 = 15.34$ m/sec

As the ball goes up by 12.53 m/sec, $v_2 = -12.53$ m/sec and $v_1 = 0$

$$e = \left(\frac{-12.53 - 0}{0 - 15.34}\right)$$

$e = 0.82$

As the velocity of hard surface will always zero thus u_1 and $v_1 = 0$

hence, $e = \dfrac{v_2}{u_2}$ ..... (1)

Let the ball rebound to height h_2 and h_3 in second bounce,

$$v_2 = u^2 + 2gh$$
$$0 = u_2^2 - 2gh_1$$
$$u_2 = \sqrt{2gh_1}$$

Similarly $v_2 = \sqrt{2gh_2}$

Using equation (1),

$$e = \frac{\sqrt{2gh_2}}{\sqrt{2gh_1}} = \sqrt{\frac{h_2}{h_1}}$$ ..... (2)

$$0.82 = \sqrt{\frac{h_2}{8}}$$

$h_2 = 5.38$ m

Let the ball rebound to height h_3 in third bounce,

Using equation (2), the height rebound by ball after third bounce,

$$e = \sqrt{\frac{h_3}{h_2}}$$

Substituting values of e and h_2 we get

$h_3 = 3.62$ m

Example: 14.4

A ball of 4 kg is suspended by inextensible string of length 1.8 m. A body of mass 10 kg is thrown horizontally towards 4 kg ball. If the body strikes the ball with velocity of 5 m/sec as shown in Fig. 14.6, then determine:

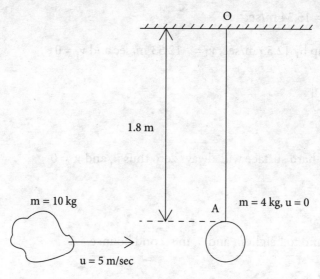

Fig. 14.6

(i) The maximum angle made by the string with the initial vertical position if coefficient of restitution is 0.7.
(ii) The velocity of ball and body after collision.

Solution:

Given: $m_1 = 10$ kg, $u = 5$ m/sec, $m_2 = 4$ kg, $u = 0$

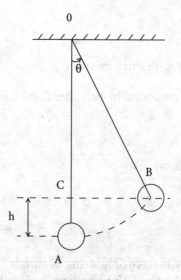

Fig. 14.6 (a)

Impulse and Momentum

Let the ball moves from position *OA* to *OB* and makes an angle θ as shown in Fig. 14.6 (a). If the velocities of body and ball are v_1 and v_2 after collision then using principle of conservation of momentum,

$$m_1 u_1 + m_2 v_2 = m_1 v_1 + m_2 v_2$$
$$10 \times 5 + 0 = 10 v_1 + 4 v_2$$
$$5 v_1 + 2 v_2 = 25 \quad \ldots (1)$$

The coefficient of restitution is given by

$$e = \left(\frac{v_2 - v_1}{u_1 - u_2} \right)$$
$$0.7 = \left(\frac{v_2 - v_1}{5 - 0} \right) \quad \ldots (2)$$
$$v_2 - v_1 = 3.5$$

Solving equation (1) and (2),

$v_2 = 6.07$ m/sec

$v_1 = 2.57$ m/sec

Let the ball goes up by height 'h' as shown in Fig. 14.6 (a), the initial velocity of ball after collision is 2.57 m/sec and finally becomes zero at position *OB*,

$$v^2 = u^2 - 2gh$$
$$0 = (2.57)^2 - 2 \times 9.81 \times h$$
$$h = 0.34 \text{ m}$$

thus, $\cos\theta = \dfrac{OC}{OB}$

$= \dfrac{OA - AC}{OB}$

$= \dfrac{1.8 - h}{1.8}$

$\cos\theta = \dfrac{1.8 - 0.34}{1.8}$

$\theta = 35.8°$

Example: 14.5

A body of mass 32 kg is moving in straight line with a velocity of 24 m/sec. The body split into two pieces of 12 kg and 20 kg which moves at 25° and 40° from straight line as shown in Fig. 14.7. Determine the velocities of both pieces. Take air resistance as negligible.

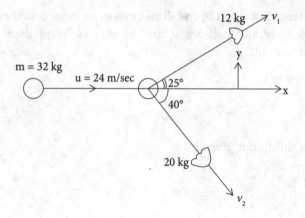

Fig. 14.7

Solution:

Given: $m = 32$ kg, $u = 24$ m/sec, $m_1 = 12$ kg, $m_2 = 20$ kg

Let the pieces fly apart with velocities v_1 and v_2.

Using principle of conservation of momentum in x-direction,

$m.u = m_1 v_1 + m_2 v_2$

$32 \times 24 = 12.v_1 \cos 25° + 20.v_2 \cos 40°$

$v_1 + 1.41\, v_2 = 64$ ..... (1)

Using the principle of conservation of momentum in y-direction,

$0 = 12.v_1 \sin 25° + 20 (-v_2 \sin 40°)$

$v_1 = 2.53\, v_2$ substituting in equation (1),

$v_2 = 16.24$ m/sec

$v_1 = 41.09$ m/sec

Example: 14.6

Due to residual stresses, a body at rest is suddenly burst into two pieces of mass 4 kg and 2 kg respectively. Assuming that the pieces fly apart in opposite direction with a relative velocity of 25 m/sec, determine the speed velocity of each piece. Neglect air resistance.

Solution:

Given: $m_1 = 4$ kg, $m_2 = 2$ kg, $u = 0$

Impulse and Momentum

Let the two pieces fly apart with velocities v_1 and v_2 as shown in Fig. 14.8.

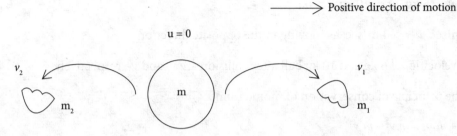

Fig. 14.8

Using the principle of conservation of momentum,

$m.u = m_1v_1 + m_2v_2$

$0 = 4.v_1 + 2(-v_2)$

$v_2 = 2v_1$ ..... (1)

As both pieces fly apart in opposite direction, velocity v_2 will be taken as negative,

The relative velocity = 25

$v_1 - (-v_2) = 25$

$v_1 + v_2 = 25$ ..... (2)

Solving equations (1) and (2),

$v_1 = 8.33$ m/sec

$v_2 = 16.66$ m/sec

Example: 14.7

Two smooth balls of masses 6 kg and 10 kg are moving to each other on a straight smooth surface as shown in Fig. 14.9. If balls are moving with uniform velocities of 8 m/sec and 4 m/sec and coefficient of restitution is 0.6, then determine their velocities after collision.

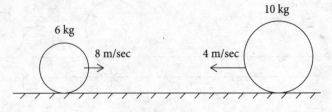

Fig. 14.9

Solution:

Given: $m_1 = 6$ kg, $m_2 = 10$ kg, $e = 0.6$

$u_1 = 8$ m/sec, $u_2 = -4$ m/sec, as moving in the opposite direction.

Let the velocities of 6 kg and 10 kg ball after collision are v_1 and v_2, respectively.

Using the principle of conservation of momentum,

$m_1 u_1 + m_2 u_2 = m_1 v_1 + m_2 v_2$
$(6 \times 8) + (-10 \times 4) = 6 v_1 + 10 v_2$
$3 v_1 + 5 v_2 = 4$ ..... (1)

The coefficient of restitution is given by,

$e = \left(\dfrac{v_2 - v_1}{u_1 - u_2} \right)$

$0.6 = \left(\dfrac{v_2 - v_1}{8 - (-4)} \right)$

$v_2 - v_1 = 7.2$ ..... (2)

From equation (1) and (2),

$v_2 = 3.2$ m/sec and $v_1 = -4$ m/sec, negative sign shows that after collision ball of mass 6 kg moves in opposite direction.

Example: 14.8

Two smooth balls of masses 5 kg and 8 kg are lying on a straight smooth track. The distance between the balls is 42 m. The balls start to move in same direction and just before collision their velocities are 12 m/sec and 5 m/sec respectively. Determine:

(i) The distance and time when collision will take place
(ii) The velocity of each ball after collision if coefficient of restitution is 0.7.

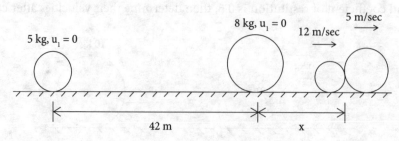

Fig. 14.10

Impulse and Momentum

Solution:

Given: $m_1 = 5$ kg, $m_2 = 8$ kg

$u_1 = u_2 = 0$, $v_1 = 12$ m/sec, $v_2 = 5$ m/sec

$s = 42$ m, $e = 0.7$

(i) Let both balls collide at distance x from 8 kg ball after time 't' seconds, thus distance travelled by 5 kg ball will be $(42 + x)$. If balls are accelerated by a_1 and a_2, respectively,

$$s = ut + \frac{1}{2}at^2$$

$$x = 0 + \frac{1}{2}a_1 t^2 \quad \ldots (1)$$

$$(42 + x) = 0 + \frac{1}{2}a_2 t^2 \quad \ldots (2)$$

$v = u + at$

$v_1 = 0 + a_1 t$ and $v_2 = 0 + a_2 t$

i.e., $a_1 = v_1/t$ and $a_2 = v_2/t$

Substituting a_1 and a_2 in equations (1) and (2),

$$x = \frac{1}{2}v_1 t \text{ and } (42 + x) = \frac{1}{2}v_2 \cdot t$$

$$42 + x = \frac{1}{2} \times 12 \times t = 6t$$

$$x = \frac{1}{2} \times 5t = 2.5t$$

$42 + 2.5t = 6t$

$t = 12$ sec

Thus the distance at collision will be, $x = 2.5\, t$

$= 2.5 \times 12$

$= 30$ m

(ii) If the velocities after collisions are v_1^1 and v_2^1.

The co-efficient of restitution is given by,

$$e = \left(\frac{v_2^1 - v_1^1}{v_1 - v_2}\right)$$

$$0.7 = \frac{v_2^1 - v_1^1}{(12 - 5)}$$

$v_2^1 - v_1^1 = 4.9 \quad \ldots (3)$

Using the principle of conservation of momentum,

$$m_1 v_1 + m_2 v_2 = m_1 v_1^1 + m_2 v_2^1$$

$$(5 \times 12) + (8 \times 5) = 5v_1^1 + 8v_2^1$$

$$5v_1^1 + 8v_2^1 = 100 \quad\quad\quad\quad \ldots (4)$$

Solving equation (3) and (4),

$$v_2^1 = 9.58 \ m/sec$$

$$v_1^1 = 4.68 \ m/sec$$

Example: 14.9

A cricket player takes a catch of a ball falling vertically under gravity from a height of 30 m. If the mass of ball is 163 gm, determine the impulse force and average force borne by the player if he holds and stops the ball in 0.05 seconds.

Solution:

Given, $m = 163$ gm $= 0.163$ kg, $u = 0$, $h = 30$ m, $t = 0.05$ sec

The velocity of ball at the time of catch will be given by

$v^2 = u^2 + 2gh$

$v^2 = 0 + 2 \times 9.81 \times 30$

$v = 24.26$ m/sec

Impulse force = final momentum – initial momentum

$\quad = m(v - u)$

$\quad = 0.163 (24.26 - 0)$

$\quad = 3.95$ N.sec

If the average force is P_{avg},

Impulse force $= P_{avg} \times t$

$\quad 3.95 = P_{avg} \times 0.05$

$\quad P_{avg} = 79$ N

Impulse and Momentum

Example: 14.10

In a Circus, the joker of weight 500 N jumps to the net from a height of 24 m and rebounds further to a height of 4 m. Assume the net is inextensible and performs like a hard surface determine:

(i) Impulse force
(ii) Average force between the joker and the net, if the time interaction of contact is 0.9 seconds.

Solution:

Using equation, the velocity of joker just before hits the net,

$v^2 = u^2 + 2gh$

$0 = u^2 - 2 \times 9.81 \times 24$

$u = 21.69$ m/sec

Let velocity of joker just after rebound is u and final velocity becomes zero at height 4 m.

$v^2 = u^2 + 2gh$

$0 = u^2 - 2 \times 9.81 \times 4$

$u = 8.86$ m/sec

Thus initial momentum $= mv = \dfrac{500}{9.81} \times 21.69$ N sec

final momentum $= mu = \dfrac{500}{9.81} \times 8.86$ N sec

Let the upward and downward velocities are positive and negative respectively,

Impulse force = Final momentum − Initial momentum

$= \dfrac{500}{9.81} \times 8.86 - \left(-\dfrac{500}{9.81} \times 21.69\right)$

$= \dfrac{500}{9.81}(8.86 + 21.69)$

$= 1557.08$ N sec

Let the average force is P_{avg}

Impulse force $= P_{avg} \times 0.9$

$1557.08 = P_{avg} \times 0.9$

$P_{avg} = 1730.09$ N

Example: 14.11

A hockey player hits a ball of 156 gm from rest. The ball leaves the ground at an angle of 15° with a velocity of 20 m/sec. Using impulse momentum equation, determine the average force acting on the ball, if time interaction of contact is 0.04 seconds. Consider frictional forces as negligible.

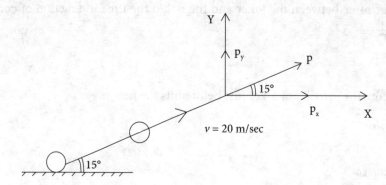

Fig. 14.11

Solution:

Given: $m = 156$ gm $= 0.156$ kg, $u = 0$, $v = 20$ m/sec, $t = 0.04$ sec

Let the average force 'P' acts on the ball. Thus the resolved parts of force P will be P_x and P_y along x and y directions respectively as shown in Fig. 14.11.

Using impulse momentum equations along x-axis,

Impulse force = Final momentum – Initial momentum

$$P_x \times t = m(v_x - u_x)$$

$P_x \times 0.04 = 0.156 (20 \cos 15° - 0)$

$P_x = 75.34$ N

Similarly, using impulse momentum equation along y-axis,

$$P_y \times t = m(v_y - u_y)$$

$P_y \times 0.04 = 0.156 (20 \sin 15° - 0)$

$P_y = 20.19$ N

Thus the total average force,

$$P = \sqrt{P_x^2 + P_y^2}$$
$$= \sqrt{(75 \cdot 3.4)^2 + (20 \cdot 19)^2}$$

$P = 77.99$ N at 15° from x-axis.

Example: 14.12

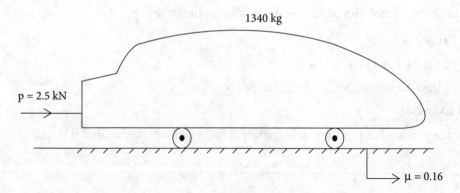

Fig. 14.12

Four persons push a swift car of mass 1340 kg from rest as shown in Fig. 14.12. If the push force 2.5 kN is applied for 22 seconds then determine the final speed of car using impulse–momentum equation. The coefficient of friction between road and tires is 0.16.

If after 22 seconds, the persons left the car then determine the time taken by car to stop.

Solution:

Given: $m = 1340$ kg, $t = 22$ sec, $u = 0$, $\mu = 0.16$

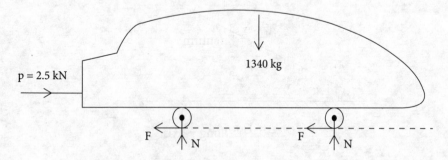

Fig. 14.12 (a) F.B.D. of Car

Let the normal reaction between each tyre and road is N. Thus total reactions of four wheels will be

$4N = 1340.g$

$\quad = 1340 \times 9.81$

$N = 3286.35$ N

Using impulse momentum equation along the motion,

Impulse force = Final Momentum − Initial momentum

$(P - 4F) t = m (v - u)$

$(2.5 \times 10^3 - 4 \times \mu N) t = m (v - u)$

$(2.5 \times 10^3 - 4 \times 0.16 \times 3286.35) 22 = 1340 (v - 0)$

$v = 6.51$ m/sec

When the persons left pushing, only frictional force acts on the car until it stops.

Using impulse momentum equation along the motion,

$(0 - 4F) t = 1340 (0 - 6.51)$

$(0 - 4 \times 0.16 \times 3286.35) t = 1340 (0 - 6.51)$

$t = 4.15$ sec

Example: 14.13

A block of mass 20 kg is placed on an inclined plane as shown in Fig. 14.13. The block starts to slide downward from rest.

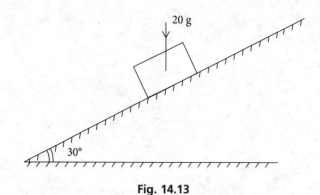

Fig. 14.13

Using impulse–momentum equation, determine the time taken by block to attain a velocity of 9 m/sec. Take coefficient of friction between contact surfaces as 0.22.

Solution:

Given: $m = 20$ kg, $u = 0$, $v = 9$ m/sec, $\mu = 0.22$

Consider free body diagram of block as shown in Fig. 14.13 (a), $N = 20 g \cos 30°$

Impulse and Momentum

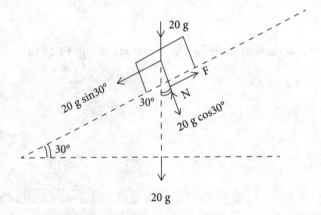

Fig. 14.13 (a) F.B.D. of block

Using impulse–momentum equation,

Impulse force = Final momentum − Initial momentum

Net forwarding force = $m(v - u)$

(Downward force − F) $t = m(v - u)$

$(20 g \sin 30° - \mu N) t = m(v - u)$

$(20 g \sin 30° - 0.22 \times 20 g \cos 30°) t = 20(9 - 0)$

$t = 2.96$ sec

Example: 14.14

Two masses of 30 kg and 20 kg are connected by an inextensible string passing over an ideal pulley as shown in Fig. 14.14. If the coefficient of friction between all contact surfaces is 0.16 then determine the pull required on block 30 kg to attain a velocity of 9.6 m/sec during 6 second. Also determine the tension in the string.

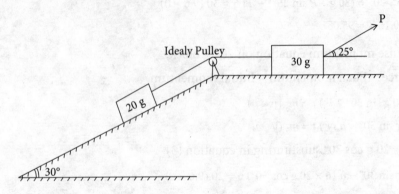

Fig. 14.14

Solution:

Consider free body diagrams of both blocks as shown in Fig. 14.14 (a),

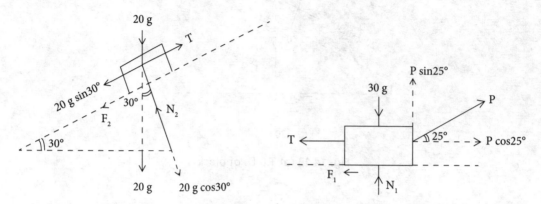

Fig. 14.14 (a) F.B.D. of blocks

Given: $m_1 = 30$ kg, $m_2 = 20$ kg, $u = 0$, $t = 6$ sec, $v = 9.6$ m/sec and $\mu = 0.16$

Let the tension induced in the string is T,

Using impulse momentum for 30-kg block,

Impulse force = Final momentum − Initial momentum

$(P \cos 25° - F_1 - T) t = m_1 (v - u)$

$(P \cos 25° - \mu_1 N_1 - T) t = m_1 (v - u)$ ..... (1)

where, $(N_1 + P \sin 25°) = 30 g$

i.e., $N_1 = (30 g - P \sin 25°)$

Substituting the value of N_1 in equation (1),

$[P \cos 25° - 0.16 (30 g - P \sin 25°) - T] 6 = 30 (9.6 - 0)$

$0.84 P - 0.16 T = 95.08$ ..... (2)

Using impulse momentum equation on 20 kg block,

Impulse force = Final momentum − Initial momentum

$(T - 20 g \sin 30° - F_2) t = m_2 (v - u)$

$(T - 20 g \sin 30° - \mu_2 N_2) t = m_2 (v - u)$ ..... (3)

where, $N_2 = 20 g \cos 30°$ substituting in equation (3)

$(T - 20 g \sin 30° - 0.16 \times 20 g \cos 30°) 6 = 20 (9.6 - 0)$

$T = 157.29$ N

Impulse and Momentum

Substitution value of T in equation (2),

$P = 143.15$ N

Example: 14.15

A rocket of weight 24 N is fired by an army man by using a portable rocket launcher of weight 180 N. If the rocket launcher is recoiled with a velocity of 0.8 m/sec, determine the velocity of rocket during launching.

Solution:

Given:

$m_1 = \dfrac{24}{g}$ kg, $m_2 = \dfrac{180}{g}$ kg

$v_1 = ?$ $v_2 = -0.8$ m/sec

$u_1 = u_2 = 0$

Using the principle of conservation of momentum,

$m_1 u_1 + m_2 u_2 = m_1 v_1 + m_2 v_2$

$0 = \dfrac{24}{g} v_1 + \dfrac{180}{g}(-0.8)$

$v_1 = 6$ m/sec

Example: 14.16

Two blocks 40 kg and 24 kg are connected by inextensible strings and supported as shown in Fig. 14.15. Determine tension in the string and time taken by block 24 kg to attain a velocity of 2 m/sec from rest. Consider pulley as frictionless and weightless. Take coefficient of friction between block 40 kg and plane as 0.25.

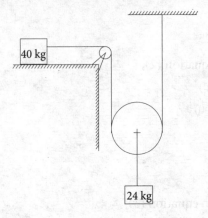

Fig. 14.15

Solution:

Given: $m_1 = 40$ kg, $m_2 = 24$ kg, $\mu = 0.25$, $u = 0$, $v_2 = 2$ m/sec

Let the tension induced in the string is 'T'.

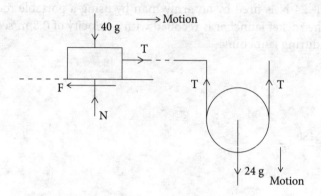

Fig. 14.15 (a) F.B.D. of blocks

Consider free body diagram of blocks as shown in Fig. 14.15 (a). Using impulse momentum equation for 24-kg block,

Impulse momentum equation for 24-kg block,

Impulse force = Final momentum − Initial momentum

$(24g - 2T) \times t = m(v_2 - u_2)$

$(24g - 2T)t = 24(2 - 0)$

$(12g - T)t = 24$ ..... (1)

Similarly, for 40-kg block,

$(T - F)t = m(v_1 - u_1)$ ..... (2)

here, $v_1 = 2.v_2 = 4$ m/sec

and $N = 40g$, $u_1 = 0$, $\mu = 0.25$

Substituting the values in equation (2),

$(T - \mu N)t = (v_1 - u_1)$

$(T - 0.25 \times 40g)t = 40(4 - 0)$

$(T - 10g)t = 160$ ..... (3)

Add equation (1) and (3),

$t = 9.38$ sec

Substituting the value of t in equation (3),

$T = 115.16$ N

Example: 14.17

A bus of mass 18 ton is moving with 80 kmph behind an auto rickshaw of mass 800 kg. The auto rickshaw is moving with a velocity of 60 kmph. Auto rickshaw driver suddenly applies brake. The bus driver also applies the brake but bus hits auto and due to this both vehicles coalesce together and move with a speed of 12 m/sec.
Determine:

(i) The velocity of bus at the time of collision
(ii) Loss of Energy

Solution:

Given,

(i) $m_1 = 800$ kg, $u_1 = 60$ kmph $= \dfrac{60 \times 5}{18} = 16.67$ m/sec

$m_2 = 18000$ kg, $v = 12$ m/sec, $u_2 = ?$

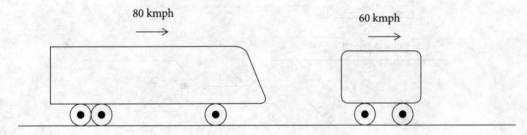

Fig. 14.16

Using the principle of conservation of momentum,

$m_1 u_1 + m_2 u_2 = (m_1 + m_2) v$

$800 \times 16.67 + 18000.u_2 = (800 + 18000) \times 12$

$u_2 = 11.79$ m/sec $= 42.45$ kmph

(ii) Loss of Energy = Final K.E – Initial K.E

$= \dfrac{1}{2}\left[(m_1 + m_2)v^2\right] - \left[\dfrac{1}{2}m_1 u_1^2 + \dfrac{1}{2}m_2 u_2^2\right]$

$= \dfrac{1}{2}\left[(800+18000)(12)^2\right] - \left[\dfrac{1}{2} \times 800 \times (16.67)^2 + \dfrac{1}{2} \times 18000 \times (11.79)^2\right]$

$= -8592.46$ Nm

Negative sign shows that there is a loss of energy.

Example: 14.18

A bus of mass 18 ton is running at 72 kmph on a highway. Another lorry of mass 72 ton is running at 24 kmph on a road as shown by top view in Fig. 14.17. Due to parking of vehicles, the drivers of both vehicles could not see each other and both vehicles collided at point 'C'. If after collision, both vehicles coalesce as single body, determine their net velocity.

Solution:

Given

$m_1 = 18 \, ton = 18 \times 10^3 \, kg, \ m_2 = 72 \, ton = 72 \times 10^3 \, kg,$

$u_1 = 72 \, kmph = \dfrac{72 \times 5}{18}, \quad u_2 = 24 \, kmph = \dfrac{24 \times 5}{18}$

$= 20 \, m/sec \qquad\qquad\quad = 6.67 \, m/sec$

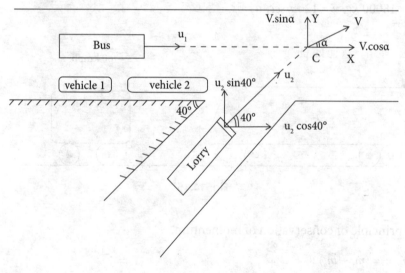

Fig. 14.17

Let the net velocity after collision is 'V' at an angle 'α' from x-axis as shown in Fig. 14.17.

Using the principle of conservation of momentum along x-axis,

$m_1(u_1)_x + m_2(u_2)_x = (m_1 + m_2)V_x$

$(u_1)_x = 20 \, m/sec, \ (u_2)_x = u_2 \cos 40° \text{ and } V_x = V \cos \alpha$

$(18 \times 10^3 \times 20) + (72 \times 10^3 \times 6.67 \cos 40°) = (18 + 72)10^3 \times V \cos \alpha$

$V \cos \alpha = 8.09$ ..... (1)

Impulse and Momentum

Using the principle of conservation of momentum along y-axis,

$m_1(u_1)_y + m_2(u_2)_y = (m_1 + m_2)V_y$

$(u_1)_{y=0}, (u_2)_y = u_2 \sin 40°$ and $V_y = V \sin \alpha$

$0 + (72 \times 10^3 \times 6.67 \sin 40°) = (18 + 72)10^3 \times V \sin \alpha$

$V \sin \alpha = 3.43 \, m/sec$ ..... (2)

from equation (1) and (2),

$V^2(\cos^2 \alpha + \sin^2 \alpha) = (8.09)^2 + (3.43)^2$

$V = 8.79 \, m/sec$

and direction will be given by,

$\dfrac{V \sin \alpha}{V \cos \alpha} = \dfrac{3.43}{8.09}$

$\alpha = 22.93°$

Example: 14.19

In a school function, a boy fires a pellet to a toy suspended by string as shown in Fig. 14.18.

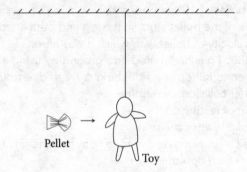

Fig. 14.18

The mass of pellet and toy is 1.08 gm and 100 gm, respectively. If the pellet with initial velocity 260 m/sec gets embedded into the stationary toy then determine:

(i) Their velocity after collision
(ii) Loss in energy

Solution:

Using the principle of conservation of momentum,

$m_1 u_1 + m_2 u_2 = (m_1 + m_2)V$

$1.08 \times 10^{-3} \times 260 + 0 = (1.08 + 100)10^{-3} \times V$

$V = 2.78 \, m/sec$

Loss of energy = (Final energy of pellet and toy) − (Total initial energy of pellet and toy)

$$= \left[\frac{1}{2}(m_1+m_2)V^2\right] - \left[\frac{1}{2}m_1u_1^2 + \frac{1}{2}m_2u_2^2\right]$$

$$= \left[\frac{10^{-3}}{2}(1.08+100)(2.78)^2\right] - \left[\frac{10^{-3}}{2}\times 1.08(260)^2 + 0\right]$$

$$= -36.11 \text{ Nm}$$

Negative sign shows that there is a loss of 36.11 N.m energy.

Theoretical Problems

T 14.1 Discuss the law of conservation of momentum.
T 14.2 Define momentum and impulse.
T 14.3 Explain the principle of conservation of energy. Proof the principle by considering kinetic and potential energy of a falling body under gravity.
T 14.4 Discuss the kinetic energy and potential energy of a body.

Numerical Problems

N 14.1 A hunter fires a bullet of 10 gm horizontally to a log of mass 1.5 kg lying on a rough horizontal surface. If the bullet stuck in the log and both are displaced by 0.75 m then determine, the velocity of bullet with which it was fired.

N 14.2 A pile of mass 1800 kg is hammered into ground by using a hammer of 800 kg. The hammer is allowed to fall on pile from height 1.1 m due to this hammer rebounds to height 0.08 m after collision, determine:
(i) The coefficient of restitution
(ii) The velocity of pile after collision

N 14.3 A hard plastic ball is dropped on a marble floor from height 15 m. If the ball rebounds to a height of 9 m then determine:
(i) The coefficient of restitution
(ii) The height to which ball will rebound in third bounce

N 14.4 A log of mass 2 kg is suspended by thread of length 2.5 m. A stone of mass 0.8 kg is thrown horizontally to the log. If the initial velocity of stone is 4.5 m/sec and coefficient of restitution is 0.72 then determine:
(i) The maximum angle swing by log from vertical position
(ii) The velocity of stone and log just after the collision

N 14.5 Two smooth balls A and B are moving on a smooth plane in the same direction with uniform velocities of 15 m/sec and 6 m/sec respectively. If m_A = 5 kg and m_B = 8 kg, determine the velocities of both balls after collision. Take co-efficient of restitution as 0.70.

N 14.6 A boy catches a stone of mass 50 gm falling under gravity from height 20 m. If boy holds and stops the stone in 0.10 seconds, determine the impulse force and average force bear by boy.

Impulse and Momentum

N 14.7 A football player hits the football at an angle of 18° from the ground with a velocity of 16 m/sec. If time interaction of contact is 0.05 seconds, determine the average force acting on the ball. Take initial velocity and mass of foot-ball as zero and 400 gm respectively. Assume the frictional resistance as negligible.

N 14.8 A block of weight 150 N is kept on a rough inclined plane. If angle of inclination is 30° then determine the velocity of block after traveling 2.7 m from rest. Consider the coefficient of friction between contact surfaces as 0.20.

N 14.9 A loaded truck 30 ton and a car of 1.2 ton are moving in same direction. The uniform velocities of truck and car are 18 m/sec and 12 m/sec when truck hits the car at its rear end. If both vehicles move as a single unit after collision, determine their velocity and the loss of energy.

N 14.10 A hunter fires a bullet of mass 2 gm into a toy horizontally. The mass of toy is 120 gm and after impact both bullet and toy swings by an angle of 25°. Determine the velocity of bullet.

Multiple Choice Questions

1. Principle of impulse and momentum is derived from
    a. Newton's first law of motion
    b. Newton's second law of motion
    c. Newton's third law of motion
    d. none of these

2. State the vector quantity out of following quantities
    a. rate of change of momentum of a body
    b. product of momentum with time
    c. momentum of a body
    d. all of these

3. Impulse of force is equal to
    a. rate of change of momentum of a body
    b. product of momentum with time
    c. momentum of a body
    d. change in momentum of a body

4. Principle of impulse and momentum relates the parameters
    a. Force, displacement and time
    b. Force, mass and velocity
    c. Force, displacement and acceleration
    d. Force, time, velocity and mass

5. The unit of impulse of force is
    a. N/sec
    b. N
    c. N.sec
    d. none of these

6. If two bodies m_1 and m_2 are moving with initial velocities u_1 and u_2 ($u_1 > u_2$) in same direction and their velocities become v_1 and v_2 after direct collision then coefficient of restitution will be given by

   a. $e = \dfrac{(v_2 - v_1)}{(u_2 - u_1)}$
   b. $e = \dfrac{(v_2 - v_1)}{(u_1 - u_2)}$
   c. $e = \dfrac{(v_1 - v_2)}{(u_1 - u_2)}$
   d. none of these

7. Identify the parameter which remains constant during perfect elastic collision
   a. momentum and kinetic energy both
   b. momentum only
   c. kinetic energy only
   d. none of these

8. The value of coefficient of restitution for perfect elastic collision is
   a. e = 0
   b. e = 1
   c. 0 < e < 1
   d. none of these

9. The value of coefficient of restitution for perfect non elastic collision is
   a. e = 0
   b. e = 1
   c. 0 < e < 1
   d. none of these

10. The value of coefficient of restitution for semi elastic collision is
    a. e = 0
    b. e = 1
    c. 0 < e < 1
    d. none of these

Answers
1. b 2. d 3. d 4. d 5. c 6. b 7. a 8. b 9. a 10. c

Chapter 15

Kinematics of Rigid Bodies

15.1 Introduction

So far, the motion of particles has been analyzed under kinematics and kinetics. Earlier, the rectilinear and curvilinear motion of a particle was studied in the chapter 10 and 11, respectively, where a body was idealized as a particle and does not have any rotational motion. However, in some cases we cannot make such assumptions where a body consists considerable rotary motion. In this chapter, we will analyse motion of rigid bodies having rotary and general plane motion for parameters like displacement, time, velocity and acceleration without cause of motion (force).

15.2 Rotational Motion

In this motion all the particles of a rigid body continuously rotate about a fixed axis. Here all the particles have parallel path in the plane of rotation. The examples are moving fan, pump, rotating shaft in a lathe machine and drilling operation, etc.

15.3 Angular Displacement, Angular Velocity and Angular Acceleration

Angular displacement is the displacement of a body on the plane of rotation about a fixed axis. It is a vector quantity and expressed in radians. For example, consider a link rotates clockwise about a fixed axis as shown in Fig. 15.1. If the link turns by angle θ during time t from OX-axis, the angular displacement of all particles on the link will be θ in clockwise direction. In one revolution the angular displacement of link will be equal to 2π radians.

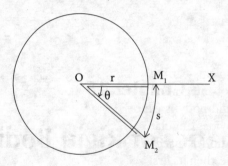

Fig. 15.1

Angular Velocity is the rate by which angular displacement varies with respect to time. It is represented by ω. If a rotating body turns by angle $d\theta$ in time dt then angular velocity will be given by $\omega = \dfrac{d\theta}{dt}$

It is a vector quantity. The unit of angular velocity is rad/sec, however in engineering devices; it is expressed in r.p.m. (revolutions per minute).

If the angular velocity is expressed in r.p.m. then it can be converted into rad/sec as follows:

Consider a shaft rotating with N r.p.m.

Then number of revolutions per second will be $\dfrac{N}{60}$

The angular displacement in one revolution of shaft will be 2π radians

Thus total angular displacement per second will be $\dfrac{2\pi N}{60}$

i.e., $\omega = \dfrac{2\pi N}{60}$ rad/sec

Angular Acceleration is the rate at which angular velocity varies with respect to time. It is represented by α. If a rotating body turns by angular velocity $d\omega$ in time dt then the angular acceleration will be given by

$$\alpha = \dfrac{d\omega}{dt}$$

The angular acceleration can be expressed by more ways

as $\alpha = \dfrac{d\omega}{dt}$

and $\alpha = \dfrac{d^2\theta}{dt^2}$

or, $\alpha = \dfrac{d\omega}{dt} = \dfrac{d\omega}{d\theta} \times \dfrac{d\theta}{dt} = \dfrac{d\omega}{d\theta} \times \omega = \omega \dfrac{d\omega}{d\theta}$

It is also a vector quantity. The unit of angular acceleration is rad/sec^2.

Kinematics of Rigid Bodies

We can conclude that all particles of a body have same angular displacement, velocity and acceleration in a rotary motion. Thus analysis of motion of any particle provides the motion analysis for rigid body.

In rotary motion, the problems can be solved by using two methods, i.e., Method of differentiation and method of integration. The method of differentiation is used where angular displacement of a rigid body is expressed as a function of time then its first and second differential value provides angular velocity and angular acceleration respectively. However, method of integration is used where angular acceleration of a rigid body is expressed as a function of time, then its first and second integral value gives angular velocity and angular displacement respectively.

15.4 Relationship between Linear and Angular Velocity

Refer last section where a link rotates clockwise about a fixed axis as shown in Fig. 15.1. If the length of link is 'r' and during time t, it turns from position M_1 to M_2 by angle θ from OX-axis then it covers linear displacement (arc) 's' on its path.

The linear displacement will be given by,

arc = angle × radius

$$s = \theta \times r$$

if the tangential velocity is v then it will be given by

$$v = \frac{ds}{dt}$$

$$v = \frac{d\theta \times r}{dt}$$

$$v = \omega \times r$$

15.5 Relationship between Linear, Normal and Angular Acceleration

Refer section 11.3 of chapter 11 where tangent acceleration and normal acceleration are denoted by a_t and a_n, respectively.

The linear acceleration or the tangent acceleration is given by

$$a_t = \frac{dv}{dt}$$

since, $v = \omega \times r$

$$a_t = \frac{d(\omega \times r)}{dt}$$

$$a_t = r \times \frac{d\omega}{dt}$$

$$a_t = r \times \alpha$$

Normal or radial acceleration is given by

$a_n = \dfrac{v^2}{r}$ substituting $v = \omega \times r$

$a_n = \dfrac{v^2}{r} = \dfrac{(\omega \times r)^2}{r}$

$a_n = r \times \omega^2$

15.6 Equations of Angular Motion

We know that the angular acceleration is given by

$\alpha = \dfrac{d\omega}{dt}$

$d\omega = \alpha \times dt$

Integrating both side

$\int d\omega = \int \alpha \times dt$

i.e., $\omega = \alpha \times t + C_1$

C_1 is the constant of integration and its value can be obtained when $t = 0$ and angular velocity ω at initial state will be ω_0. Substituting value in the above equation,

$\omega_0 = 0 + C_1$

Thus, $\omega = \alpha \times t + \omega_0$

$\omega = \omega_0 + \alpha t$ ..... (1)

We know that angular velocity is given by

$\omega = \dfrac{d\theta}{dt}$

$d\theta = \omega dt$

$d\theta = (\omega_0 + \alpha t)dt$

$d\theta = \omega_0 dt + \alpha t \, dt$

$\int d\theta = \int \omega_0 dt + \int \alpha t \, dt$

$\theta = \omega_0 t + \dfrac{1}{2}\alpha t^2 + C_2$

C_2 is the constant of integration and its value can be obtained when $t = 0$, the angular displacement will be zero, i.e., $\theta = 0$.

Kinematics of Rigid Bodies

Substituting the value in the above equation, $C_2 = 0$

$$\theta = \omega_0 t + \frac{1}{2}\alpha t^2 \quad \ldots (2)$$

Considering angular acceleration,

$$\alpha = \frac{d\omega}{dt} = \frac{d\omega}{d\theta} \times \frac{d\theta}{dt} = \frac{d\omega}{d\theta} \times \omega = \omega \frac{d\omega}{d\theta}$$

or, $\alpha \times d\theta = \omega d\omega$

$$\int \alpha \times d\theta = \int_{\omega_0}^{\omega} \omega \, d\omega$$

$$\alpha \times \theta = \frac{1}{2}\omega^2 + C_3$$

C_3 is the constant of integration and its value can be obtained when $\theta = 0$; the angular velocity will be ω_0, i.e.,

$$C_3 = -\frac{1}{2}\omega_0^2$$

$$\alpha \times \theta = \frac{1}{2}\omega^2 - \frac{1}{2}\omega_0^2$$

$$\omega^2 = \omega_0^2 + 2\alpha\theta \quad \ldots (3)$$

Equations (1), (2) and (3) are called the equations of angular motion. The equations of linear motion are valid under constant acceleration 'a'; however the equations of angular motion are valid under constant angular acceleration 'α'.

Thus there are three equations of angular motion when a particle moves with constant angular acceleration 'α' about a fixed axis.

1. $\omega = \omega_0 + \alpha t$
2. $\theta = \omega_0 t + \frac{1}{2}\alpha t^2$
3. $\omega^2 = \omega_0^2 + 2\alpha\theta$

Example: 15.1

An electric fan rotating at 320 rpm makes 30 revolutions until stops. Determine:

(i) The angular retardation of fan assuming it to be uniform.
(ii) The time taken by fan until stops
(iii) The r.p.m. of fan after 4 seconds

Solution:

Given, $N_0 = 320$ rpm, $\omega_0 = \dfrac{2\pi \times 320}{60} = 33.51$ rad/sec

$\omega = 0$, $\theta = 30$ revolutions $= 30 \times 2\pi$ radians

(i) Using equation $\omega^2 = \omega_0^2 + 2\alpha\,\theta$

$$0 = (33.51)^2 - 2 \times \alpha \times 30 \times 2\pi$$

$$\alpha = 2.98 \text{ rad/sec}^2$$

(ii) $\omega = \omega_0 + \alpha\,t$

$0 = (33.51) - 2.98 \times t$

$t = 11.24$ seconds

(iii) The angular velocity after 4 seconds,

$\omega = \omega_0 + \alpha\,t$

$= 33.51 - 2.98 \times 4$

$= 21.59$ rad/sec

If the rpm is N after 4 seconds,

$\omega = \dfrac{2\pi N}{60} = 21.59$

$N = 206.17$ rpm

Example: 15.2

A wheel rotating about a fixed axis at 20 rpm is uniformly accelerated for 70 seconds, during which it makes 50 revolutions. Determine: (i) Angular velocity at the end of this interval

(ii) Time required for the speed to reach 100 revolutions per minute.

[UPTU, Co Ist Sem, 2003]

Solution:

Given, $N_0 = 20$ rpm, i.e., $\omega_0 = \dfrac{2\pi \times 20}{60} = 2.09$ rad/sec

$t = 70$ sec, $\theta = 50$ revolutions $= 50 \times 2\pi$ radians

(i) To determine angular velocity, first of all angular acceleration is required to compute. Using the equation,

$$\theta = \omega_0 t + \dfrac{1}{2}\alpha t^2$$

Kinematics of Rigid Bodies

$$50 \times 2\pi = 2.09 \times 70 + \frac{1}{2} \alpha (70)^2$$

$$\alpha = 0.069 \text{ rad/sec}^2$$

The angular velocity at the end of 70 seconds will be,

$$\omega = \omega_0 + \alpha t$$
$$= 2.09 + 0.069 \times 70$$
$$\omega = 6.92 \text{ rad/sec}$$

(ii) Let the wheel is uniformly accelerated up to 100 r.p.m.

$$\omega = \frac{2\pi \times 100}{60} = 10.47 \text{ rad/sec}$$

Using the equation, $\omega = \omega_0 + \alpha t$

$$10.47 = 2.09 + 0.069 \times t$$
$$t = 121.45 \text{ sec}$$

Example: 15.3

A flywheel attains a speed of 300 rpm from rest during 20 seconds. Determine the angular displacement and angular acceleration of flywheel at the end of this interval.

Solution:

Given, $\omega_0 = 0$, $N = 300$ rpm, $t = 20$ sec

$$\omega = \frac{2\pi N}{60} = \frac{2\pi \times 300}{60} = 10\pi$$

Using the equation,

$$\omega = \omega_0 + \alpha t$$
$$10\pi = 0 + \alpha \times 20$$
$$\alpha = 1.57 \text{ rad/sec}^2$$

Angular displacement,

$$\theta = \omega_0 t + \frac{1}{2} \times t^2$$
$$= 0 + \frac{1}{2} \times 1.57 \times (20)^2$$
$$\theta = 314 \text{ radians}$$

Example: 15.4

The angular motion of a work piece in headstock of a lathe machine is defined by equation, $\theta = (t^3 - 2t^2 + 3t)$, where '$\theta$' is in radians and '$t$' is in seconds. Determine the angular position, velocity and angular acceleration at $t = 5$ seconds.

Solution:

The angular position defined by, $\theta = (t^3 - 2t^2 + 3t)$ ..... (1)

The first and second differential of angular displacement will provide angular velocity and angular acceleration, respectively.

Thus angular velocity, $\omega = \dfrac{d\theta}{dt}$

$= \dfrac{d}{dt}(t^3 - 2t^2 + 3t)$

$\omega = (3t^2 - 4t + 3)$ ..... (2)

The angular acceleration, $\alpha = \dfrac{d^2\theta}{dt^2}$

$= \dfrac{d\omega}{dt}$

$= (6t - 4)$ ..... (3)

Substituting $t = 5$ in equation (1), (2) and (3),

$\theta = 90$ rad, $\omega = 58$ rad/sec, $\alpha = 26$ rad/sec²

Example: 15.5

A cord is wrapped around a wheel of radius 0.2 m, which is initially at rest as shown in Fig. 15.2. If a force is applied to the cord and gives it an acceleration, $a = 4\,t$ m/sec², where 't' is in seconds. Determine the angular velocity of the wheel and the angular position of line OP, as a function of time.

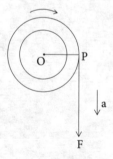

Fig. 15.2

Kinematics of Rigid Bodies

Solution:

Given, $r = 0.2$ m, $\omega_0 = 0$

The acceleration of wheel is given by,

$a = 4t$

where force, F is applied tangentially to the wheel thus 'a' is the tangential acceleration and given as,

$a = a_t = r.\alpha$

Thus, $r\alpha = 4t$

$0.2\alpha = 4t$

$\alpha = 20t$

Thus angular velocity of wheel, $d\omega = \alpha.dt$

$$\int d\omega = \int 20t.\, dt$$

$$\omega = \frac{20.t^2}{2} + C_1$$

when $t = 0$, $\omega = 0$, thus $C_1 = 0$

$\omega = 10t^2$ rad/sec

The angular position of line OP will be,

$d\theta = \omega.dt = 10.t^2.\, dt$

$$\int d\theta = \int 10t^2.\, dt$$

$$\theta = \frac{10.t^3}{3} + C_2$$

when $t = 0$, $\theta = 0$, thus $C_2 = 0$

$$\theta = 10.\frac{t^3}{3} \text{ radians}$$

Example: 15.6

The rotor of an electric motor uniformly accelerates to a speed of 1800 rpm from rest in 5 seconds and then the rotor decelerates uniformly to stop. If the total time, elapsed from start to stop is 12.3 second, determine the number of revolutions made while acceleration and deceleration.

[UPTU, Ist Sem, 2002–03]

Solution:

$N = 1800$ rpm, $\omega = \dfrac{2\pi \times 1800}{60} = 60\pi$ rad/sec

$\omega_0 = 0$, $t = 5$ sec, $T = 12.3$ seconds

Using equation, $\omega = \omega_0 + \alpha t$

$60\pi = 0 + \alpha \times 5$

$\alpha = 12\pi$ rad/sec^2

If angular displacements during accelerations and retardations are θ_a and θ_r respectively,

$\theta_a = \omega_0 t + \dfrac{1}{2}\alpha t^2$

$= 0 + \dfrac{1}{2} \times 12\pi(5)^2$

$\theta_a = 150\pi$ radians

if N_a is the revolutions made by rotor during acceleration,

$\theta_a = 2\pi N_a = 150\pi$

$N_a = 75$ revolutions

As the rotor decelerates after 5 sec, the time taken by rotor during deceleration will be,

$t = T - 5 = 12.3 - 5 = 7.3$ sec

The initial speed of rotor during beginning of deceleration,

$\omega_0 = 60\pi$ rad/sec

and $\omega = 0$

If rotor decelerates by angular acceleration, α_r then using equation,

$\omega = \omega_0 + \alpha t$

$0 = 60\pi - \alpha_r \times 7.3$

$\alpha_r = 25.82$ m/sec^2

Thus, $\omega^2 = \omega_0^2 - 2\alpha_r \theta_r$

$0 = (60\pi)^2 - 2 \times 25.82 \times \theta_r$

$\theta_r = 688.04$ rad/sec^2

If N_r is the revolution made by rotor during deceleration,

$\theta_r = 2\pi N_r = 688.04$

$N_r = 109.5$ revolutions

Kinematics of Rigid Bodies

Example: 15.7

A horizontal bar 1.5 m long and of small cross-section rotates about vertical axis through on end. It accelerates uniformly from 1200 rpm to 1500 rpm in an interval of 5 seconds. What is the linear velocity at the beginning and at the end of the interval?

What are the normal and tangential components of acceleration of the mid-point of the bar after 5 seconds when the acceleration begins?

Solution:

Given, $l = 1.5$ m, $N_0 = 1200$ rpm, $N = 1500$ rpm, $t = 5$ sec

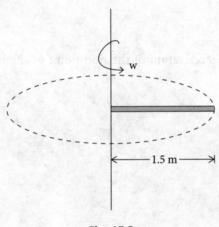

Fig. 15.3

$$\text{Initial angular velocity, } \omega_0 = \frac{2\pi N_0}{60} = \frac{2\pi \times 1200}{60} = 125.66 \text{ rad/sec}$$

$$= \frac{2\pi N_0}{60} = \frac{25 \times 1500}{60} = 157.08 \text{ rad/sec}$$

Using equation, $\omega = \omega_0 + \alpha\, t$

$157.08 = 125.66 + \alpha \times 5$

$\alpha = 6.28$ m/sec²

Thus the linear velocity at the beginning,

$v_0 = r.\, \omega_0$

$\quad = l.\, \omega_0$

$\quad = 1.5 \times 125.66$

$v_0 = 188.49$ m/sec

Linear velocity after 5 sec will be,

$v = r.\omega$

$= l.\omega$

$= 1.5 \times 157.08$

$= 235.62$ m/sec

Normal component of acceleration of the mid-point of the bar after 5 seconds will be given by,

$a_n = r.\omega^2$

$= \dfrac{l}{2}.\omega^2$

$= \dfrac{1.5}{2} \times (157.08)^2$

$a_n = 18.51 \times 10^3$ m/sec^2

Tangential component of acceleration of the mid-point of the bar after 5 seconds will be given by,

$a_t = r.\alpha$

$= \dfrac{l}{2}.\alpha$

$a_t = \dfrac{1.5}{2} \times 6.28$

$= 4.71$ m/sec^2

Example: 15.8

A train starts from rest and moves along a curved track of radius 800 m with uniform acceleration until it attains a velocity of 72 km/hr at the end of the third minute. Determine the tangential, normal and total accelerations in m/sec² of the train at the end of second minute.

[UPTU, IInd Sem, 2000–01]

Solution:

Given, $\omega_0 = 0$, $R = 800$ m, $v = 72$ km/hr, $v = 72 \times 5/18 = 20$ m/sec

$t_1 = 3$ min, $t_2 = 2$ min,

$t_1 = 180$ sec, $t_2 = 120$ sec,

The angular velocity of train after 3 min,

$\omega = \dfrac{V}{R} = \dfrac{20}{800}$

$\omega = 0.025$ rad/sec

Thus angular acceleration up to 3 min will be,

$\omega = \omega_0 + \alpha t$

$0.025 = 0 + \alpha \times 180$

$\alpha = 1.39 \times 10^{-4}$ rad/sec^2

Since, train accelerates uniformly up to 3 mins, α will remain constant up to this duration, i.e., α will remain same at the end second minute.

The tangential acceleration is given by, $a_t = R.\alpha$

a_t at the end of second min. will be $= 800 \times 1.39 \times 10^{-4} = 0.11$ m/sec^2

The normal acceleration is given by,

$a_n = R.\omega^2$

$a_n = 800.\omega^2$ ..... (1)

The angular velocity will vary and after 2 min. it will be,

$\omega = \omega_0 + \alpha t$

$\omega = 0 + 1.39 \times 10^{-4} \times 120$

$= 166.8 \times 10^{-4}$ rad/sec

Substituting value of 'ω' in equation (1),

$a_n = 800 (166.8 \times 10^{-4})^2$

$= 0.22$ m/sec^2

Thus total acceleration, $a = \sqrt{a_t^2 + a_n^2}$

$= \sqrt{(0.11)^2 + (0.22)^2}$

$a = 0.25$ m/sec^2

15.7 General Plane Motion

If a body performs motion in such a way that linear and angular displacement happens together, such a motion is called general plane motion. This kind of motion is not translatory alone or rotary alone but comprises the combination of both translatory and rotary motion of a rigid body.

To illustrate the general plane motion consider a ladder allowed to slip against floor and wall. Let the initial position and final position of ladder is L_iM_i and L_fM_f as shown in Fig. 15.4 and Fig. 15.4 (a), respectively.

First, the ladder is assumed to perform translatory motion as shown by intermediate position L_iM_i to L_fM_i in Fig. 15.4 (b). Further it is assumed that it performs rotary motion, i.e., turn about point M_f on the floor as shown by position L_fM_f in Fig. 15.4 (c). Thus slipping of ladder represents the general plane motion.

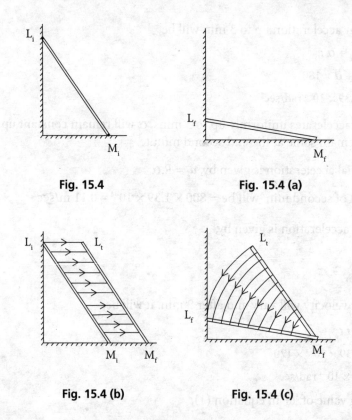

Fig. 15.4

Fig. 15.4 (a)

Fig. 15.4 (b)

Fig. 15.4 (c)

The motion of wheel without slipping is another example of general plane motion. Consider a wheel of radius r turns about its centre O by half revolution from initial position $L_i M_i$ to final position $L_f M_f$ as shown in Fig. 15.5. During half turn, it traverses half of its circumference, i.e., πr.

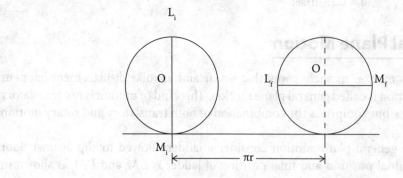

Fig. 15.5

Kinematics of Rigid Bodies

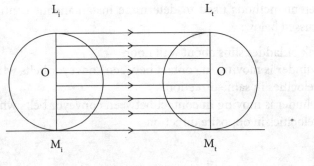

Fig. 15.5 (a)

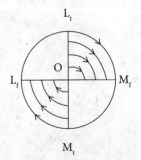

Fig. 15.5 (b)

Let us assume that the wheel first performs translatory motion as shown by intermediate position $L_t M_t$ in Fig. 15.5 (a) and then it performs rotary motion, i.e., turn about centre O as shown by position $L_f M_f$ in Fig. 15.5(b). Finally the wheel has travelled linear displacement (πr) as well as angular displacement (π). Thus motion of wheel without slipping represents the general plane motion.

15.8 Instantaneous Centre

The instantaneous centre is a virtual centre for an instant and used to analyse the general plane motion. It is the simplest method in which general plane motion is converted into rotary motion for an instant about a point, which is known as instantaneous centre. However such point changes continuously with the motion of body. A line joining all instantaneous centres is known as Centrods. The distance of any point on the body (under general plane motion) from instantaneous centre work as radius, r and linear velocity of such point can be determine by using expression, $v = \omega \times r$

There are different methods used to determine instantaneous centre in a general plane motion as discussed below:

i. When a link or ladder slips about wall floor.
ii. When a cylinder is moving in contact between conveyor belts which are running with different velocities in same direction.
iii. When a cylinder is moving in contact between conveyor belts which are running with different velocities in opposite direction.

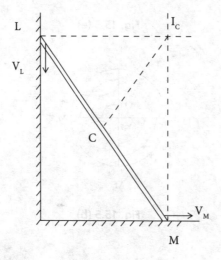

Fig. 15.6

i. When a link or ladder slips about wall and floor

To locate instantaneous centre, consider a body which may be a link or ladder LM placed against smooth wall and floor as shown in Fig. 15.6. Let the body slips with angular velocity ω. As we know the direction of velocities at L and M then draw perpendiculars to them and their point of intersection is known as instantaneous centre I_c. This is the centre for an instant about which all the particles of ladder will rotate, i.e.,

$V_M = \omega \, r$ where $r = MI_c$

Thus, $V_M = \omega \, MI_c$

Similarly, $V_L = \omega \, LI_c$

If velocity at centre C of link is required it will be given by

$V_C = \omega \, CI_c$

Kinematics of Rigid Bodies

ii. **When a cylinder is moving in contact between conveyor belts which are running with different velocities in same direction.**

Consider a cylinder of radius r is rotating with angular velocity ω between two conveyor belts. Let, the belts are running in same direction with different velocities V_A and V_B $(V_A > V_B)$ as shown in Fig. 15.7.

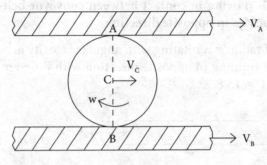

Fig. 15.7

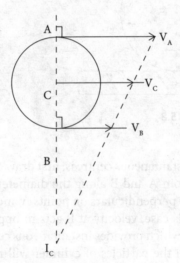

Fig. 15.7 (a)

The procedure to locate the instantaneous centre is, first draw cylinder and marks the contact points of belts, i.e., A and B. Join A and B along the diameter of the cylinder and extend this line considerably as shown in Fig. 15.7.(a). Afterwards draw perpendiculars on points A and B which represent velocities of belts at these points. Further, join extremities of these velocities and extend this line where it cuts the line joining A and B and provides instantaneous centre I_C. Thus, this is the centre for an instant about which all the particles of cylinder will rotate. We can determine the velocity of any point of the wheel about instantaneous centre i.e.

$V_A = \omega AI_C$

Similarly,

$V_B = \omega BI_C$

If velocity at centre C is required then draw perpendicular about C which will cut line joining A and B, it will be given by $V_C = \omega CI_C$

iii. When a cylinder is moving in contact between conveyor belts which are running with different velocities in opposite direction

Consider a cylinder of radius r is rotating with angular velocity ω between two conveyor belts. Let, the belts are running in opposite direction with different velocities V_A and V_B ($V_A > V_B$) as shown in Fig. 15.8.

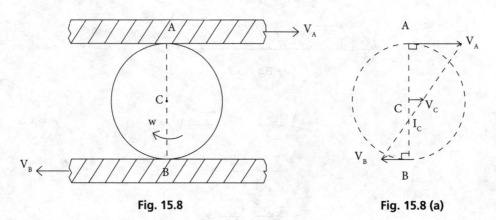

Fig. 15.8   **Fig. 15.8 (a)**

The procedure to locate the instantaneous centre is, first draw cylinder and mark the contact point of belts, i.e., A and B. Join A and B along the diameter of the cylinder as shown in Fig. 15.8.(a). Afterwards draw perpendiculars on points A and B which represent velocities of belts at these points. In this case, velocity at B acts in opposite direction. Further, join extremities of these velocities which provides instantaneous centre I_C. Thus this is the centre for an instant about which all the particles of cylinder will rotate. We can determine the velocity of any point of the wheel about instantaneous centre, i.e.,

$V_A = \omega AI_C$

Similarly,

$V_B = \omega BI_C$

If velocity at centre C is required then draw perpendicular about C which will cut line joining A and B, it will be given by $V_C = \omega CI_C$

Kinematics of Rigid Bodies

Example: 15.9

The straight link AB 40 cm long, has at a given instant, its end B moving along line OX at 0.8 m/sec and the other end 'A' moving along OY, as shown in Fig. 15.9. Find the velocity of the end A and of mid-point 'C' of the link when inclined at 30° with OX.

[UPTU, 2001–02]

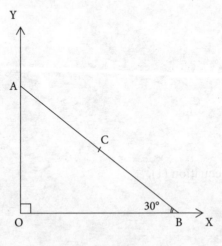

Fig. 15.9

Solution:

Given, AB = 40 cm = 0.40 m, V_B = 0.8 m/sec Consider the free body diagram of link AB as shown in Fig. 15.9 (a).

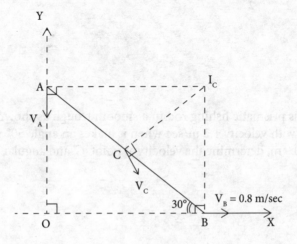

Fig. 15.9 (a)

To determine instantaneous centre, draw perpendiculars on linear velocities V_A and V_B as shown in Fig. 15.9. (a). All points an link AB are having pure rotary motion about instantaneous centre I_C.

If the angular velocity of link AB is 'ω' then linear velocities at points A, B and C will be given by

$$V_A = r.\, w = AI_C \times \omega \qquad \ldots\ldots (1)$$
$$V_B = r.\, w = BI_C \times \omega \qquad \ldots\ldots (2)$$
$$V_C = r.\, \omega = CI_C \times \omega \qquad \ldots\ldots (3)$$

Considering equation (2),

$$0.8 = AB \sin 30° \times \omega$$
$$0.8 = 0.40 \sin 30° \times \omega$$
$$\omega = 4 \text{ rad/sec}$$

Substituting value of ω in equation (1),

$$V_A = AI_C \times \omega$$
$$V_A = AB \cos 30° \times 4$$
$$= 0.4 \cos 30° \times 4$$
$$V_A = 1.39 \text{ m/sec}$$

Considering equation (3),

$$V_C = CI_C \times \omega$$
$$CI_C = \frac{1}{2} \cdot OI_C = \frac{1}{2} \cdot AB = \frac{0.40}{2} = 0.20 \, m$$
$$V_C = 0.2 \times 4$$
$$V_C = 0.8 \text{ m/sec}$$

Example: 15.10

A fisherman puts his prismatic fishing rod in a smooth trough as shown in Fig. 15.10. The end A of rod slides with velocity 1.2 m/sec when it makes an angle 50° with bottom. If the height of trough is 0.9 m, determine the velocity at point 'C' and angular velocity of the rod.

Kinematics of Rigid Bodies

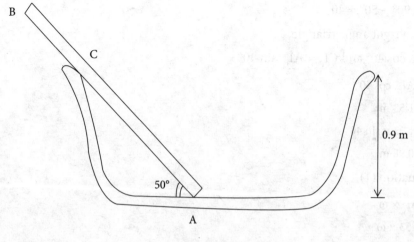

Fig. 15.10

Solution:

Draw perpendiculars on velocities V_A and V_C as shown in Fig. 15.10 (a) which intersect and provide the instantaneous centre I_C. If the rod has angular velocity 'ω'

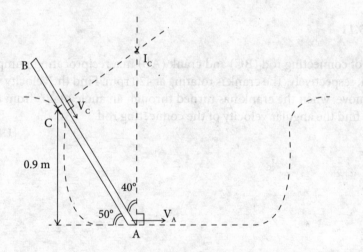

Fig. 15.10 (a)

$V_A = AI_C \times \omega$ ..... (1)

$V_C = CI_C \times \omega$ ..... (2)

As the height of trough is 0.9 m,

$$\sin 50° = \frac{0.9}{AC}$$

$AC = 1.17$ m

$\angle CAI_C = 90° - 50° = 40°$

ΔACI_C is a right angle triangle,

$AC = AI_C \cdot \cos 40°$ and $CI_C = AI_C \cdot \sin 40°$

$\quad 1.17 = AI_C \cdot \cos 40°$

$\quad AI_C = 1.53$ m

$\quad CI_C = 1.53 \times \sin 40°$

$\quad \quad = 0.98$ m

Using equation (1),

$\quad V_A = AI_C \times \omega$

$\quad 1.2 = 1.53 \times \omega$

$\quad \omega = 0.78$ rad/sec

Using equation (2),

$\quad V_C = CI_C \times \omega$

$\quad \quad = 0.98 \times 0.78$

$\quad \quad = 0.76$ m/sec

Example: 15.11

The lengths of connecting rod (BC) and crank (AB) in a reciprocating pump are 1125 mm and 250 mm, respectively. The crank is rotating at 420 rpm. Find the velocity with which the piston will move, when the crank has turned through an angle of 40° from the inner dead centre. Also find the angular velocity of the connecting rod.

[UPTU 2002–03]

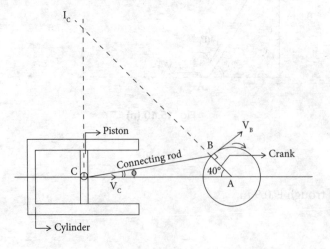

Fig. 15.11

Kinematics of Rigid Bodies

Given, $BC = l = 1125$ mm, $AB = r = 250$ mm

$N = 420$ rpm, $\alpha = 40°$

If the angular velocity of the crank is ω_{ck},

then, $\omega_{ck} = \dfrac{2\pi N}{60}$

$= \dfrac{2\pi \times 420}{60}$

$= 43.98$ rad/sec

Thus linear velocity of the crank at B,

$V_B = r \cdot \omega_{ck} = 250 \times 10^{-3} \times 43.98$

$= 11$ m/sec

If the angular velocity of connecting rod is ω_{cr}

then, $V_c = CI_C \times \omega_{cr}$ ..... (1)

and, $V_B = BI_C \times \omega_{cr}$ ..... (2)

Using sine rule in triangle ABC as shown in Fig. 15.11 (a),

$\dfrac{BC}{\sin 40°} = \dfrac{AB}{\sin \phi}$

$\dfrac{1.125}{\sin 40°} = \dfrac{0.250}{\sin \phi}$

$\phi = 8.21°$

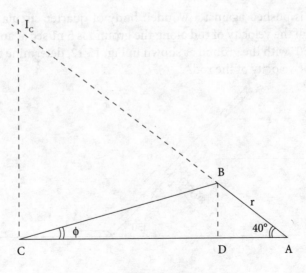

Fig. 15.11 (a)

Consider △ ABC,

$AC = CD + AD$

$= BC \cos \phi + AB \cos 40°$

$= 1.125 \cos 8.21° + 0.25 \cos 40°$

$= 1.30$ m

Consider right angle △ ACI_C,

$\tan 40° = \dfrac{CI_C}{AC} = \dfrac{CI_C}{1.30}$

$CI_C = 1.3 \tan 40° = 1.09$ m

$\cos 40° = \dfrac{AC}{AI_C} = \dfrac{1.30}{AI_C}$

$AI_C = 1.70$ m

Thus, $BI_C = AI_C - AB = 1.70 - 0.25 = 1.45$ m

Substituting value of V_B and BI_C in equation (2),

$11 = 1.45 \times \omega_{cr}$

$\omega_{cr} = 7.59$ rad/sec

Substituting values of CI_C and ω_{cr} in equation (1),

$V_c = 1.09 \times 7.59$

$= 8.27$ m/sec

Example: 15.12

A slender rod AB is pushed against a wooden body of quarter circular shape along the horizontal ground. If the velocity of rod along the ground is 5 m/sec at an instant when rod makes an angle of 50° with the ground as shown in Fig. 15.12, determine the velocity of rod at C and the angular velocity of the rod.

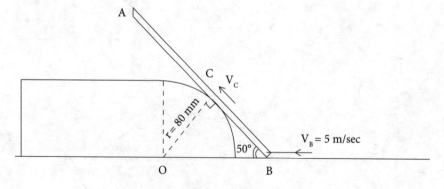

Fig. 15.12

Kinematics of Rigid Bodies

Solution:

Let the angular velocity of rod is 'ω'. Draw perpendiculars on linear velocities to get instantaneous centre, I_C as shown in Fig. 15.12 (a). Thus,

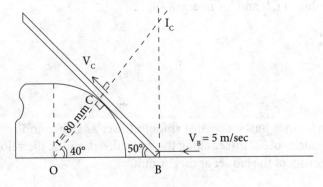

Fig. 15.12 (a)

$$V_B = BI_C \times \omega$$
$$5 = BI_C \times \omega \qquad \ldots\ldots (1)$$
$$V_c = CI_C \times \omega \qquad \ldots\ldots (2)$$

Consider Δ OCB,

$$\sin 50° = \frac{80 \times 10^{-3}}{OB}$$

$OB = 0.104$ m

Consider right angle triangle OBC,

$\angle COB = 90° - 50° = 40°$

Consider right angle triangle OBI_C,

$$\tan 40° = \frac{BI_C}{OB}$$

$$BI_C = OB . \tan 40°$$
$$= 0.104 \tan 40°$$
$$BI_C = 0.087 \text{ m}$$

$$\cos 40° = \frac{OB}{OI_C}$$

$$OI_C = \frac{OB}{\cos 40°} = \frac{0.104}{\cos 40°}$$

$OI_C = 0.136$ m, $CI_C = OI_C - OC = 0.136 - 0.08 = 0.056$ m

Substituting the value of BI_c in equation (1),

$5 = 0.087 \times \omega$

$\omega = 57.47$ rad/sec

Substituting the value of CI_c and 'ω' in equation (2),

$V_c = 0.056 \times 57.47$

$\quad = 3.22$ m/sec

Example: 15.13

A ladder of length 20 m is pushed against smooth tower as shown in Fig. 15.13. If the end 'A' of ladder has velocity of 3.7 m/sec along the ground, determine the velocity of other end 'B' and angular velocity of the ladder at this instant.

Solution:

Draw perpendiculars to V_A and V_B to get instantaneous centre as shown in Fig. 15.13. If the angular velocity of ladder in ω then,

$V_A = AI_c \times \omega$ ..... (1)

$V_B = BI_c \times \omega$ ..... (2)

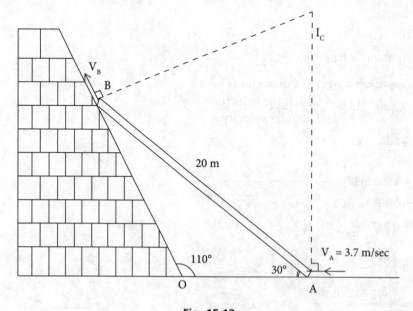

Fig. 15.13

Consider $\triangle\ OAB$, $\angle OBA = 180° - (110° + 30°) = 40°$

Consider $\triangle\ ABI_C$,

$\angle BAI_C = 90° - 30° = 60°$

$\angle ABI_C = 90° - 40° = 50°$

$\angle AI_C B = 180° - (60° + 50°) = 70°$

Using sine rule in $\triangle\ ABI_C$,

$$\frac{AB}{\sin 70°} = \frac{AI_C}{\sin 50°} = \frac{BI_C}{\sin 60°}$$

$$AI_C = \frac{AB \cdot \sin 50°}{\sin 70°} = \frac{20 \times \sin 50°}{\sin 70°} = 16.30 \text{ m}$$

$$BI_C = \frac{20 \times \sin 60°}{\sin 70°} = 18.43 \text{ m}$$

Substituting 'V_A' and AI_C in equation (1),

$3.7 = 16.3 \times \omega$

$\omega = 0.23$ rad/sec

Substituting 'ω' and BI_C in equation (2),

$V_B = 18.43 \times 0.23$

$\quad = 4.24$ m/sec

Example: 15.14

A cylindrical roller of 50 cm in diameter is in contact with two conveyor belts at its top and bottom as shown in Fig. 15.14. If the belts run at the uniform speed of 5 m/sec and 3 m/sec, find the linear velocity and angular velocity of the roller.

[UPTU 2000–01]

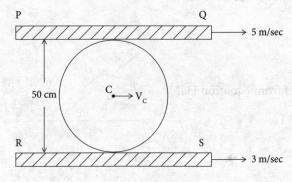

Fig. 15.14

Solution:

Given, $d = 50$ cm $= 0.50$ m

Consider two points A and B on the circumference of the roller which have velocities,

$V_A = V_{PQ} = 5$ m/sec
$V_B = V_{RS} = 3$ m/sec

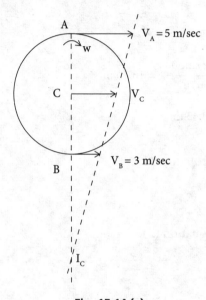

Fig. 15.14 (a)

Draw these parallel velocities according to their magnitudes and join further. Also draw perpendicular to both V_A and V_B as shown in fig 15.14 (a). The intersection of these two lines provides instantaneous centre, I_C. To determine linear velocity of roller draw parallel line through centre 'C' parallel to V_A,

If the angular velocity of roller is 'ω' then,

$V_A = AI_C \times \omega$ ..... (1)
$V_B = BI_C \times \omega$ ..... (2)
$Vc = CI_C \times \omega$ ..... (3)

Subtract equation (2) from equation (1),

$V_A - V_B = (AI_C - BI_C)\, \omega$
$5 - 3 = AB \times \omega$
$2 = 0.50 \times \omega$
$\omega = 4$ rad/sec

Substituting the value of 'ω' in equation (3),

$$V_C = CI_C \times \omega$$
$$= (BC + BI_C)\,\omega$$
$$= (0.25 + BI_C)\,\omega$$
$$V_C = (0.25 \times \omega) + (BI_C \times \omega)$$

Using equation (2),

$$V_C = (0.25 \times 4) + V_B$$
$$= 1.0 + 3.0$$
$$V_C = 4 \text{ m/sec}$$

Example: 15.15

In a cylindrical roller and belt system, two belts run in opposite direction as shown in Fig. 15.15. Determine the linear and angular velocity of roller.

If lower belt RS also runs with same uniform speed of 6 m/sec in the opposite direction then determine the linear and angular velocity of roller.

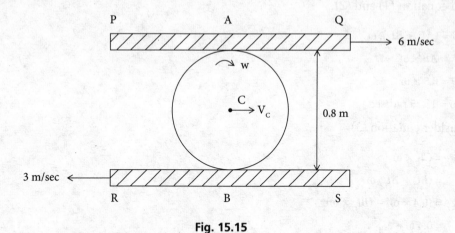

Fig. 15.15

Solution:

Let the linear and angular velocity of roller is V_C and 'ω', respectively. Draw velocities V_A and V_B parallel and opposite to each other representing magnitudes of 6 m/sec and 3 m/sec as shown in Fig. 15.15 (a).

Join the extremities and draw perpendicular to both velocities V_A and V_B. The intersection of these lines provides instantaneous centre, I_C.

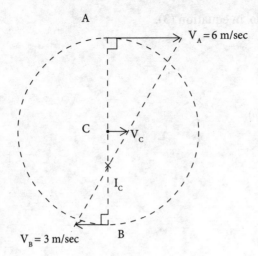

Fig. 15.15 (a)

$V_A = 6 = AI_C \times \omega$ ..... (1)

$V_B = 3 = BI_C \times \omega$ ..... (2)

$V_C = CI_C \times \omega$ ..... (3)

Add equation (1) and (2),

$9 = (AI_C + BI_C) \cdot \omega$

$= AB \times \omega$

$9 = 0.8 \times \omega$

$\omega = 11.25$ rad/sec

Consider equation (3),

$V_C = CI_C \times \omega$

$= (BC - BI_C) \omega$

$V_C = (0.4 \times \omega) - (BI_C \times \omega)$

$= 0.4\, \omega - V_B$

$= 0.4 \times 11.25 - 3$

$V_C = 1.5$ m/sec

If lower belt *RS* runs with uniform speed of 6 m/sec then intersection of perpendicular line and line joining velocities extremities meets at centre of the circle. Thus instantaneous centre lies at centre of the roller as shown in Fig. 15.15(b)

Kinematics of Rigid Bodies

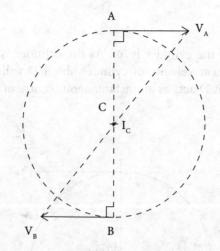

Fig. 15.15 (b)

thus $V_C = 0$ as $CI_C = 0$
and $V_A = V_B = AI_C \times \omega$

$6 = 0.4 \times \omega$

$\omega = 15$ rad/sec

In this situation, the roller will rotate only about its centre.

Example: 15.16

A cylinder of diameter one metre is rolling along a horizontal plane *AB* without slipping. Its centre has uniform velocity of 20 m/sec. Determine the velocity of the points *E* and *F* on the circumference of the cylinder shown in Fig. 15.16.

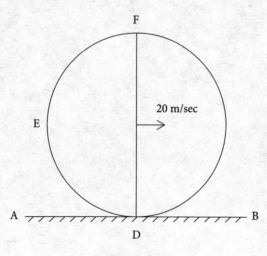

Fig. 15.16

Solution:

Let the angular velocity of the cylinder is 'ω'. As the cylinder is rolling without slipping of horizontal plane, the linear velocity of cylinder about D will be zero and it will turn about this point. Thus point D acts as an instantaneous centre of the cylinder as shown in Fig. 15.16 (a).

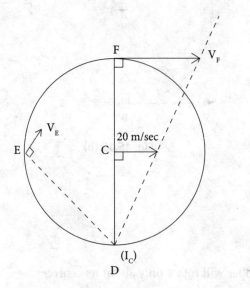

Fig. 15.16 (a)

Join first I_C and velocity of centre and extend further by dash line. Draw perpendicular to diameter FD at point F. The intersection of lines provides linear velocity at F as shown by V_F in Fig. 15.16 (a).

$20 = CI_C \times \omega$ ..... (1)

$V_F = FI_C \times \omega$ ..... (2)

Given, diameter = 1 m

$\therefore FI_C = 1$ m

and $CI_C = 0.5$ m

Consider equation (1),

$20 = 0.5 \times \omega$

$\omega = 40$ rad/sec

Consider equation (2),

$V_F = FI_C \times \omega$

$= 1 \times 40$

$= 40$ m/sec

Kinematics of Rigid Bodies

To determine velocity at E, join point E and D and draw perpendicular to ED which shows V_E,

thus $V_E = ED \times \omega$

$\quad = \sqrt{(EC^2 + CD^2)}\,\omega$

$\quad = \sqrt{(0.5)^2 + (0.5)^2} \times 40$

$V_E = 28.28$ m/sec

Example: 15.17

A compound wheel rolls without slipping between two parallel plates A and B as shown in Fig. 15.17. At a particular instant A moves to the right with a velocity of 1.2 m/sec and B moves to the left with a velocity of 0.6 m/sec. Calculate the velocity of centre of wheel and angular velocity of wheel.

Take $r_1 = 120$ mm and $r_2 = 240$ mm

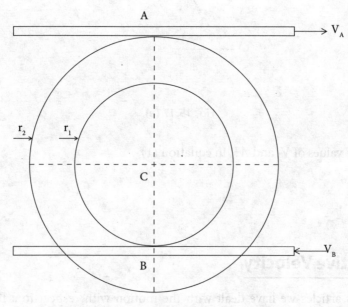

Fig. 15.17

Solution:

Consider suitable scale for radius and velocity and draw figure as shown by Fig. 15.17 (a).
Scale for radius, let 60 mm = 10 mm
Thus $BC = 20$ mm and $AC = 40$ mm.
Similarly for velocity, let 0.3 m/sec = 10 mm,
thus $V_B = 0.6$ m/sec = 20 mm and $V_A = 1.2$ m/sec = 40 mm

Draw 40 mm horizontal line at A and 20 mm at B representing velocity V_A and V_B respectively. Now join V_A and V_B which passes through centre C. Thus centre of wheel acts as instantaneous centre as shown in Fig. 15.17 (a).

Thus, linear velocity of the wheel, $V_C = 0$

Let the wheel rotates by angular velocity of 'ω',

$$V_A = AI_C \times \omega \qquad \qquad \ldots\ldots (1)$$

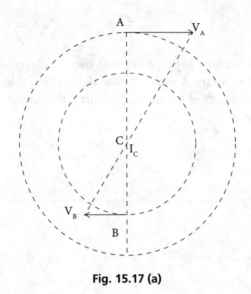

Fig. 15.17 (a)

Substituting the values of V_A and AI_C in equation (1),

$\quad 1.2 = 0.240 \times \omega$

$\quad \omega = 5$ rad/sec

15.9 Relative Velocity

In the previous articles we have dealt with the motion with respect to a fixed point on the earth. Such motion of a particle is called absolute motion. For example, if a stationary observer realizes a bike is moving with a velocity of 70 kmph towards north, it is called the absolute motion of bike. In absolute motion, the motion of earth is neglected. However, if observer is also moving with some velocity then a comparison can be made between these two velocities and such comparison is known as Relative Velocity.

Consider two persons are riding bicycles in same direction with same velocity then an observer sitting on any bicycle does not realize motion with respect to each other. However, if any one of them increases his velocity, the difference of their velocity is realized by the observer; this is called the relative velocity. Such relative velocity appears to be small as both

Kinematics of Rigid Bodies

bicycles are moving in the same direction. However, if they move opposite to each other, then the observer realizes that relative velocity has increased considerably.

Consider two particles A and B are moving under translational motion with respect to a fixed reference frame as shown in Fig. 15.18.

If the translatory velocity of particles A and B is V_A and V_B, respectively, then the relative velocity between particles can be determined as follows:

First, each velocity is resolved in two mutually perpendicular axis, i.e., V_{A_x}, V_{B_x} along X axis and V_{A_y}, V_{B_y} along Y axis. Second, the relative velocity of particle A with respect to particle B can be determined along two mutually perpendicular axis by subtracting their corresponding velocities.

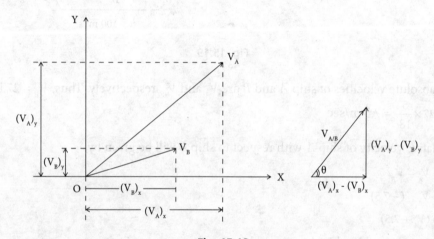

Fig. 15.18

i.e. $(V_{A/B})_x = V_{A_x} - V_{B_x}$

$(V_{A/B})_y = V_{A_y} - V_{B_y}$

The net relative velocity of particle A with respect to particle B is given by

$(V_{A/B}) = \sqrt{(V_{A/B})_x^2 + (V_{A/B})_y^2}$

The direction of relative velocity will be

$\theta = \left[\tan^{-1} \dfrac{(V_{A/B})_y}{(V_{A/B})_x} \right]$

Example: 15.18

Two ships A and B are sailing with constant velocities and approaching to each other as shown in Fig. 15.19. The length of ship A and B are 400 m and 100 m, respectively. If ship B is moving with 27 kmph, determine the velocity of ship A if it takes 40 seconds to cross ship B.

Solution:

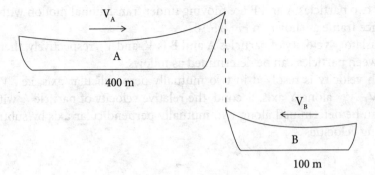

Fig. 15.19

Let the absolute velocities of ship A and B are V_A and V_B, respectively. Thus, $V_B = 27$ kmph, that is, $27 \times \dfrac{5}{18} = 7.5$ m/sec

Thus relative velocity of ship A with respect to ship B will be given by,

$V_{A/B} = V_A - V_B$
$= V_A - (-7.5)$
$= (V_A + 7.5)$

Total distance travel by ship A to cross ship B,

= length of ship A + length of ship B
= (400 + 100) m
= 500 m

Time taken by ship A to cross ship B, $t = 40$ sec
As ships are moving with constant velocities, $a = 0$

$S = ut + \dfrac{1}{2} at^2$
$S = ut + 0$
$500 = (V_{A/B}) \cdot t$
$500 = (V_A + 7.5) \, 40$
$V_A = 5$ m/sec or 18 kmph

Example: 5.19

A heavy truck of length 18 m is moving with constant velocity of 54 kmph parallel to a local train. Truck driver takes 60 seconds to reach from rear end of train to the engine. If truck takes 4 seconds more to cross the train then determine the length and velocity of the train.

Kinematics of Rigid Bodies

Solution:

Consider Fig. 15.20 in which the local train and the truck are moving parallel to each other in the same direction. Let the length of the train is x and its velocity is V_B

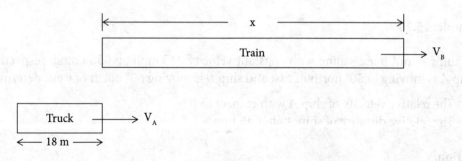

Fig. 15.20

The relative velocity of truck,

$$V_{A/B} = (V_A - V_B)$$

Since truck driver takes 60 sec to travel distance x

$$x = V_{A/B} \times 60$$

$$x = (V_A - V_B) 60 \quad \ldots\ldots (1)$$

To cross the train, truck will travel total distance $(x + 18)$ m and time taken will be $(60 + 4)$ sec thus,

$$(x + 18) = (V_A - V_B) 64 \quad \ldots\ldots (2)$$

Given that,

$$V_A = 54 \text{ kmph}$$
$$= 54 \times \frac{5}{8}$$
$$= 15 \text{ m/sec}$$

Divide equation (1) by equation (2),

$$\left(\frac{x}{18+x}\right) = \frac{(V_A - V_B) 60}{(V_A - V_B) 64}$$

$$x = 270 \text{ m}$$

Substituting the values of x and V_A in equation (1),

$$x = (V_A - V_B) 60$$
$$270 = (15 - V_B) 60$$

$V_B = 10.5$ m/sec
$= 10.5 \times \dfrac{18}{5}$
$= 37.8$ kmph

Example: 15.20

Two ships A and B are sailing with constant velocity 24 kmph and 36 kmph, respectively. If ship A is moving in 30° north of east and ship B is moving 60° north of west, determine:

(i) The relative velocity of ship A with respect to B
(ii) The relative distance of ship A after 45 min

Solution:

(i) Let the absolute velocities of ship A and B are V_A and V_B, respectively.

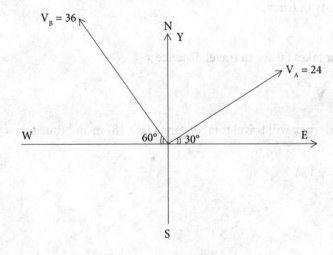

Fig. 15.21

Given, $V_A = 24$ kmph, $V_B = 36$ kmph

Consider Fig. 15.21, where absolute velocities have been resolved along X–Y axis.

Thus, $(V_A)_x = 24 \cos 30°$, $(V_A)_y = 24 \sin 30°$

and $(V_B)_x = -36 \cos 60°$, $(V_B)_y = 36 \sin 60°$

The relative velocity of ship A with respect to ship B,

$(V_{A/B})_x = (V_A)_x - (V_B)_x$

$= 24 \cos 30° - (-36 \cos 60°)$

$= 38.78$ kmph

Kinematics of Rigid Bodies

Similarly, $(V_{A/B})_y = (V_A)_y - (V_B)_y$

$\quad\quad = 24 \sin 30° - 36 \sin 60°$

$\quad\quad = -19.18$ kmph

Thus, relative velocity of ship A,

$V_{A/B} = \sqrt{((V_{A/B})_x)^2 + ((V_{A/B})_y)^2}$

$\quad\quad = \sqrt{(38.72)^2 + (-19.18)^2}$

$\quad\quad = 43.26$ kmph

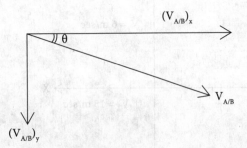

Fig. 15.21 (a)

$\theta = \tan^{-1}\left[\dfrac{(V_{A/B})_y}{(V_{A/B})_x}\right]$

$\quad = \tan^{-1}\left[\dfrac{19.18}{33.78}\right]$

$\quad = 26.32°$

(ii) The relative distance of ship A after 45 min will be,
$x = V_{A/B} \times t$
$t = 45$ min. $= 45/60 = 0.75$ hr
Substituting values of $V_{A/B}$ and t,
$x = 43.26 \times 0.75$
$\quad = 32.45$ km

Example: 15.21

A car is moving with uniform velocity of 54 kmph towards east. Suddenly rain starts falling vertically downward with a velocity of 6 m/sec. Determine the velocity and direction of rain experienced by the driver of the car.

Solution:

Let the velocity of the car and rain is V_A and V_r respectively,

$V_A = 54$ kmph $= 54 \times \dfrac{5}{18} = 15$ m/sec

$V_r = 6$ m/sec

Consider Fig. 15.22,

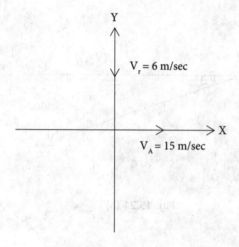

Fig. 15.22

The relative velocity of rain with respect to car along X-axis will be

$(V_{r/A})_x = (V_r)_x - (V_A)_x$

$= 0 - 15$

$= -15$ m/sec

Similarly the relative velocity of rain with respect to car along Y-axis will be

$(V_{r/A})_y = (V_r)_y - (V_A)_y$

$= -6 - 0 = -6$ m/sec (downward)

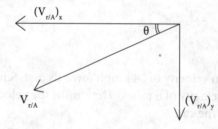

Fig. 15.22 (a)

Kinematics of Rigid Bodies

$$V_{r/A} = \sqrt{(-15)^2 + (-6)^2}$$
$$= 16.16 \text{ m/sec}$$

$$\theta = \tan^{-1}\left[\frac{6}{15}\right]$$
$$= 21.8°$$

Example: 15.22

An observer is standing at a port. At a particular instant, he watches that ship A is 60 m away from him towards north. Ship A starts moving from rest and accelerates towards port by 0.4 m/sec². At the same time, another ship B also crosses the observer moving towards east by uniform velocity of 24 kmph. Determine relative position and relative velocity of ship A with respect to ship B after 10 second from a particular instant.

Solution:

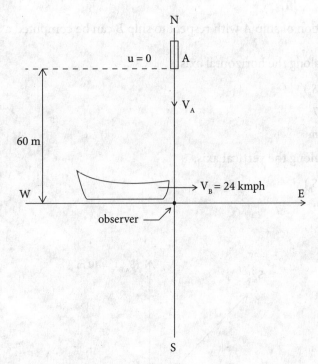

Fig. 15.23

Consider Fig. 15.23 where, for ship A, $u = 0$, $a = 0.4$ m/sec², $t = 10$ sec

For ship B, $V_B = 24$ kmph $= 24 \times \dfrac{5}{18} = 6.67$ m/sec, $t = 10$ sec

First, we must know the position of both ships after 10 seconds.

The distance travelled by ship A during 10 seconds will be

$$= ut + \frac{1}{2}at^2$$
$$= 0 + \frac{1}{2}(0.4)(10)^2$$
$$= 20 \text{ m}$$

Thus, position of ship A from observer after 10 seconds,

$S_A = 60 - 20$

$= 40$ m

The position of ship B after 10 seconds, $a = 0$

$S_B = V_B \times t$

$= 6.67 \times 10$

$= 66.7$ m

Thus relative position of ship A with respect to ship B can be computed as follows:

Relative position along the horizontal axis,

$(S_{A/B})_x = (S_A)_x - (S_B)_x$

$= 0 - 66.67$

$= -66.67$ m

Relative position along the vertical axis,

$(S_{A/B})_y = (S_A)_y - (S_B)_y$

$= 40 - 0$

$= 40$ m

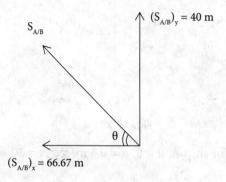

Fig. 15.23 (a)

Kinematics of Rigid Bodies

Thus relative position of ship A to ship B will be,

$$S_{A/B} = \sqrt{(-66.67)^2 + (40)^2}$$
$$= 77.75 \text{ m}$$
$$\theta = \tan^{-1}\left(\frac{40}{66.67}\right)$$
$$= 30.96°$$

The relative velocity of ship A with respect to ship B will be computed as follows:

The absolute velocity of ship A after 10 seconds will be,

$$V_A = u + at$$
$$= 0 + 0.4 \times 10 = 4 \text{ m/sec}$$

$V_A = 4$ m/sec and $V_B = 6.67$ m/sec, remains same as moving with uniform velocity

$$(V_{A/B})_x = (V_A)_x - (V_B)_x$$
$$= 0 - 6.67$$
$$= -6.67 \text{ m/sec}$$
$$(V_{A/B})_y = (V_A)_y - (V_B)_y$$
$$= -4 - 0$$
$$= -4 \text{ m/sec}$$

Relative velocity of ship A with respect to B will be given by,

$$V_{A/B} = \sqrt{((V_{A/B})_x)^2 + ((V_{A/B})_y)^2}$$
$$= \sqrt{(-6.67)^2 + (-4)^2}$$
$$= 7.78 \text{ m/sec}$$
$$\theta = \tan^{-1}\left(\frac{4}{6.67}\right)$$
$$= 30.95°$$

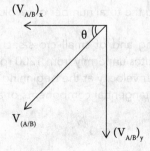

Fig. 15.23 (b)

Theoretical Problems

T 15.1 Illustrate the difference-between curvilinear motion and rotary motion.
T 15.2 Discuss the following terms; angular displacement, angular velocity and angular acceleration of a rotary motion.
T 15.3 Derive the relation between linear velocity and angular velocity.
T 15.4 Prove that the relation between linear acceleration and angular acceleration is given by $a_t = r \times \alpha$
T 15.5 What do you mean by general plane motion? Illustrate with some suitable examples.
T 15.6 Define instantaneous centre and state some examples where it is significantly used.
T 15.7 State the term relative velocity with examples.

Numerical Problems

N 15.1 A flywheel rotating at 50 rpm attains a speed of 400 rpm in 24 seconds. Determine:
(i) Angular acceleration
(ii) Angular displacement

N 15.2 A grinding wheel rotating at 450 rpm makes 36 revolutions until it stops. If the wheel decelerates uniformly, determine:
(i) Time taken by wheel to stop
(ii) The rpm of wheel after 5 seconds

N 15.3 An electric fan is uniformly accelerated to a speed of 240 rpm from rest in 6 seconds. At this instant the power is switched off and fan decelerates uniformly until stop. If the total time for acceleration and retardation in 15 seconds, determine the number of revolutions made by fan during acceleration period and deceleration period.

N 15.4 The angular displacement of a rotating body is given by equation $\theta = (2.5t^3 + 4.2t^2 + 6)$ where 'θ' is expressed in radians and 't' in seconds determine –
(i) Angular velocity when $t = 0$ and $t = 3$ seconds
(ii) Angular acceleration at $t = 3.6$ seconds

N 15.5 A car starts from rest on a curved road of 200 m radius and accelerates at a constant angular acceleration of 0.5 m/sec². Determine the distance and time which the car will travel before the total acceleration attained by it becomes 0.75 m/sec².
[UPTU, IInd Sem, 2013–14]

N 15.6 A flywheel had an initial angular speed of 3000 rev/min in clockwise direction. When a constant turning moment was applied to the wheel, it got subjected to a uniform anticlockwise angular acceleration of 3 rad/sec². Determine the angular velocity of the wheel after 20 seconds, and the total number of revolutions made during this period.
[UPTU, SEM I, 2013–14]

N 15.7 A horizontal bar 1.5 m long and of small cross-section rotates about vertical axis through one end. It accelerates uniformly from 1200 rpm to 1500 rpm in an interval of 5 seconds. What is the linear velocity at the beginning and at the end of the interval? Determine the normal and tangential components of acceleration of midpoint of the bar after 5 seconds.

Kinematics of Rigid Bodies

N 15.8 A cylinder of diameter 1.2 m rolls without slipping on a horizontal plane PQ as shown in Fig. NP 15.1. The centre B has uniform velocity of 36 m/sec. Determine the velocity at points C and D on the circumference of the cylinder.

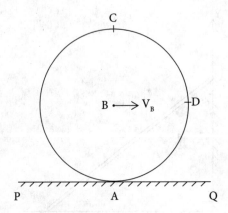

Fig. NP 15.1

N 15.9 Two conveyor belts are running opposite with uniform speed as shown in Fig. NP 15.2. A cylindrical roller of 0.8 m is moving in between conveyor belts. Determine the linear velocity and angular velocity of the roller.

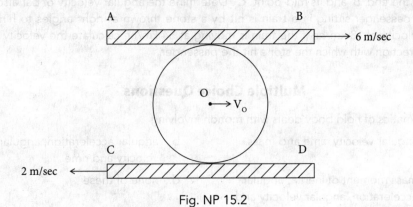

Fig. NP 15.2

N 15.10 A prismatic bar 50 cm long, at an instant is moving with its end 'A' along axis OY with 1.2 m/sec and the other and 'B' is moving along axis OX as shown in Fig. NP 15.3.

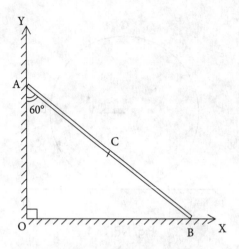

Fig. NP 15.3

If the rod in inclined at 60° with axis OY for a particular instant, determine the velocity of the end 'B' and its mid-point 'C'. Determine the angular velocity of bar also.

N 15.11 A passenger sitting in a train is hit by a stone thrown at right angles to him with a velocity of 18 km/hr. If the train is moving at 54 km/hr, calculate the velocity and the direction with which the stone hit the passenger.

Multiple Choice Questions

1. Kinematics of rigid body deals with motion involving
   a. angular velocity, time and mass
   b. angular acceleration, angular velocity and time
   c. mass moment of inertia, angular acceleration, angular velocity and time
   d. none of these

2. In method of differentiation which parameter is taken as a function of time
   a. angular displacement
   b. angular velocity
   c. angular acceleration
   d. all of these

3. State the vector quantity out of following quantities
   a. angular displacement
   b. angular velocity
   c. angular acceleration
   d. all of these

Kinematics of Rigid Bodies

4. In method of integration which parameter is taken as a function of time
   a. angular displacement
   b. angular velocity
   c. angular acceleration
   d. all of these

5. The unit of angular acceleration is
   a. rad/sec
   b. m/sec²
   c. rad/sec²
   d. none of these

6. Relation between linear velocity and angular velocity is given by
   a. $v = \omega \times r$
   b. $v = r \times \omega^2$
   c. $v = r \times \alpha$
   d. none of these

7. Equation of angular motion is
   a. $\omega = \omega_0 + \alpha t$
   b. $\theta = \omega_0 t + \frac{1}{2} g t^2$
   c. $\omega^2 = \omega_0^2 + 2\alpha s$
   d. all of these

8. General plane motion always contains
   a. translational motion
   b. rotational motion
   c. combination of translational motion and rotational motion
   d. none of these

9. Instantaneous centre is a
   a. real centre
   b. virtual centre
   c. fixed centre
   d. none of these

10. Instantaneous centre method is used to analyse
    a. translational motion
    b. rotational motion
    c. general plane motion
    d. none of these

Answers

1. b 2. d 3. d 4. d 5. c 6. a 7. a 8. c 9. b 10. c

Chapter 16

Kinetics of Rigid Bodies

16.1 Introduction

We analyzed the kinematics of rigid body consisting rotary and general plane motion in the previous chapter. In this chapter, we will analyze the same motion on rigid bodies along with their cause of motion. In addition to this, the principle of Conservation of Energy, the principle of Work and Energy are also used to analyze the problems.

16.2 Kinetics of Rotary Motion

Refer to section 15.2, where we have studied motion for particles of a rigid body rotate continuously about a fixed axis without cause of motion, mass. Here rigid body will be analysed for parameters angular velocity, angular acceleration along with cause of motion and mass.

16.2.1 Moment of momentum

Moment of momentum is also known as angular momentum for a rotating body. Consider a wheel of mass m and radius R is rotating clockwise with angular velocity ω about fixed axis as shown in Fig. 16.1.

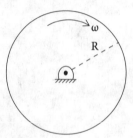

Fig. 16.1

Kinetics of Rigid Bodies

Let a particle of mass dm, at distance r from a fixed axis is moving with velocity v
The momentum of particle from the fixed axis will be, $dT = dm \times v$
Since, $v = \omega r$
$$dT = dm \times \omega r$$
Moment of momentum of particle from the fixed axis will be,
$$dM = dT \times r$$
Moment of momentum of rigid body from the fixed axis will be,

$$\int dM = \int dT \times r$$
$$M = \int dm \times \omega r^2$$
$$M = \omega \int dm\, r^2$$

where the term $\int dm\, r^2$ is the mass moment of inertia of the rigid body and represented by I.

Thus, moment of momentum of rigid body from fixed axis is given by, $M = I\omega$

We know that the linear momentum of a body is given by mv. However, moment of momentum is also known as angular momentum for a rotating body and given by $I\omega$.

16.2.2 Torque and angular momentum

Newton's second law for rotary motion states that the rate of change of angular momentum is directly proportional to the external torque (T) applied on the body and lies in the direction of the torque.

i.e., Torque $\propto$ rate of change of angular momentum

$$T \propto \frac{d}{dt}(I\omega)$$
$$T \propto I\frac{d}{dt}(\omega)$$
$$T \propto I\alpha$$

$T = C.\, I\,\alpha$ where C is a constant value of proportionality and its value remain unity.

Finally, the relation between torque and angular acceleration is given by
$T = I\alpha$

16.3 Kinetic Energy of a Body in Translatory and Rotary Motion

Consider a rigid body of mass m moving in general plane motion with velocity v. If the mass of each particle is dm then kinetic energy (K. E.) of particle during translatory motion will be

$$(K.E. \text{ of particle})_{Translatory} = \frac{1}{2} dm\, v^2$$

The kinetic energy (K. E.) of rigid body during translatory motion will be

$$(K.E. \text{ of body})_{Translatory} = \frac{1}{2} \int dm\, v^2$$

$$(K.E. \text{ of body})_{Translatory} = \frac{1}{2} m v^2$$

When a body rotates about its axis with angular velocity ω then the kinetic energy (K. E.) in rotary motion for a particle will be

$$(K.E. \text{ of particle})_{Rotary} = \frac{1}{2} dm\, v^2$$

Since $v = \omega \times r$

$$(K.E. \text{ of body})_{Rotary} = \frac{1}{2} \int dm (\omega \times r)^2$$

$$= \frac{1}{2} \omega^2 \int dm\, r^2$$

$$(K.E. \text{ of body})_{Rotary} = \frac{1}{2} I \omega^2$$

Thus K.E. of rigid body under general plane motion

$$= (K.E. \text{ of body})_{Translatory} + (K.E. \text{ of body})_{Rotary}$$

$$= \frac{1}{2} m v^2 + \frac{1}{2} I \omega^2$$

$$= \frac{1}{2} m v^2 + \frac{1}{2} (mk^2) \omega^2$$

where k is the radius of gyration of body.

16.4 Principle of Conservation of Energy

According to this principle, when a rigid body has general plane motion under the influence of conservative forces then sum of potential energy and kinetic energy of a rigid body remains constant.

$(P.\,E. + K.\,E.) = \text{constant}$

Thus, $(P.\,E. + K.\,E.)_1 = (P.\,E. + K.\,E.)_2$

16.5 Principle of Work and Energy

According to this principle, the work done by force to displace a rigid body is equal to the change in its kinetic energy.

Thus, Work done = Change in Kinetic Energy of the rigid body

$$W = (K.E.\,of\,body)_{final} - (K.E.\,of\,body)_{initial}$$

$$W = \left(\frac{1}{2}mv^2 + \frac{1}{2}I\omega^2\right)_{final} - \left(\frac{1}{2}mv^2 + \frac{1}{2}I\omega^2\right)_{initial}$$

Example: 16.1

A bracket hinged to the wall supports a disc of mass 16 kg and radius 120 cm as shown in Fig. 16.2. An inextensible thread of negligible mass in wound on the disc by number of turns and supports a pan of mass 100 gm. When 2.5 kg mass is placed on pan determine:

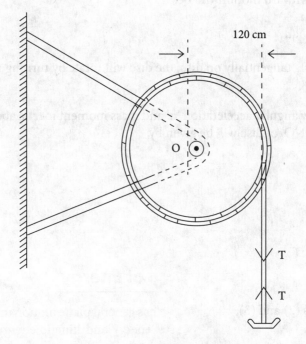

Fig. 16.2

(i) Angular acceleration of the disc
(ii) Acceleration of the pan
(iii) Tension induced in the thread
(iv) Velocity and height travelled by pan after 4 sec when mass is placed on it

Solution:

Given, mass of the disc = m_1 = 16 kg, radius r = 120 cm = 1.2 m, t = 4 sec
mass of pan = 100 gm = 0.10 kg, mass in pan = 2.5 kg, Total mass = m_2 = 0.10 + 2.5 = 2.6 kg
Consider F.B.D of pan as shown in Fig. 16.2 (a).

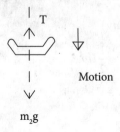

Fig. 16.2 (a) F.B.D. of Pan

Let the tension induced in thread is T. If the pan moves by downward acceleration 'a' then the equation of downward motion will be,

$(m_2 g - T) = m_2 . a$
$(2.6 g - T) = 2.6 \times a$ ..... (1)

Since tension 'T' acts tangentially on disc, the disc will rotate by turning moment about 'O'
$= T \times r$ ..... (2)

If the disc rotates by angular acceleration 'α' and mass moment inertia about polar axis is I_{zz} then moment about 'O' of disc will be given by,

$= I_{zz} \times \alpha$

$= \dfrac{m_1 r^2}{2} \times \alpha$

$= \dfrac{m_1 r (r . \alpha)}{2}$

$= \dfrac{m_1 r . a}{2}$ ..... (3)

Equating equations (2) and (3),

$T \cdot r = \dfrac{m_1 r . a}{2}$

$T = \left(\dfrac{m_1 a}{2}\right)$

$= \left(\dfrac{16 \times a}{2}\right)$

$T = 8.a$ ..... (4)

Kinetics of Rigid Bodies

Substituting the value of 'T' in equation (1),
$$2.6 g - 3a = 2.6 \times a$$
$$a = 2.41 \text{ m/sec}^2$$
Since $a = r.\alpha$
$$2.41 = 1.2 \times \alpha$$
$$\alpha = 2.01 \text{ rad/sec}^2$$
Substituting the value of 'a' in equation (4),
$$T = 8 \times 2.41$$
$$T = 19.28 \text{ N}$$
$$t = 4 \text{ sec}, u = 0$$

Using equation,
$$v = u + at$$
$$= 0 + 2.41 \times 4$$
$$v = 9.64 \text{ m/sec}$$

Height travelled by pan will be,
$$h = ut + \frac{1}{2}at^2$$
$$= 0 + \frac{1}{2} \times 2.41 \times (4)^2$$
$$= 19.28 \, m$$

Example: 16.2

A prismatic log of 10 kg mass and radius 100 mm is wound by number of turns using inextensible thread of negligible mass. The thread is tied to hook and the log is released as shown in Fig. 16. 3.

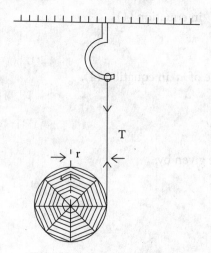

Fig. 16.3

If the log rolls down without slip then determine:

(i) Tension 'T', induced in the thread
(ii) Height fallen by log after 2 sec from release
(iii) Velocity attained by log after 2 sec

Solution:

When log is released, it is pulled downward by gravitational force. During downward motion it bears general plane motion, let the log has downward acceleration 'a' and angular acceleration α.

Using equation of downward motion for log,
$$10 \times g - T = 10 \cdot a \quad \ldots\ldots (1)$$

Turning moment about centre of log due to tension in thread,
$$= T \times r \quad \ldots\ldots (2)$$

The turning moment of log due to angular acceleration, α
$$= I_{zz} \times \alpha \quad \ldots\ldots (3)$$

Equating equation (2) and (3)
$$T \times r = I_{zz} \times \alpha$$
$$T.r = \frac{mr^2}{2} \times \alpha$$
$$T = \frac{m(r.\alpha)}{2}$$
$$= \frac{10 \times a}{2}$$
$$T = 5a \quad \ldots\ldots (4)$$

Substituting the value of 'T' in equation (1),
$$10g - 5a = 10a$$
$$a = 6.54 \text{ m/sec}^2$$

(i) Substituting the value of 'a' in equation (4),
$$T = 5 \times 6.54$$
$$T = 32.7 \text{ N}$$

(ii) $u = 0, t = 2$ sec

Height fallen by log will be given by,
$$h = ut + \frac{1}{2}at^2$$
$$= 0 + \frac{1}{2} \times 6.54 \times (2)^2$$
$$= 13.08 \, m$$

Kinetics of Rigid Bodies

(iii) $v = u + at$
    $= 0 + 6.54 \times 2$
    $v = 13.08$ m/sec

Example: 16.3

A right circular cylinder of mass 'm' and radius 'r' is suspended from a thread that is wound around its circumference. If the cylinder is allowed to fall freely, determine the acceleration of its mass centre G and the tension induced in the cord.

[U.P.T.U. 2003–04]

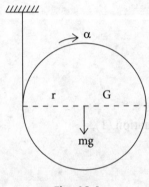

Fig. 16.4

Solution:

When the cylinder is allowed to fall freely, it is pulled downward due to gravitational force and tension which is acting tangentially on it causes rotation during downward motion. Thus it bears general plane motion. Let the acceleration of mass centre of cylinder is 'a' and angular acceleration is α.

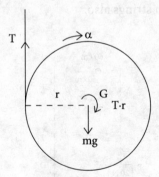

Fig. 16.4 (a)

If the tension in the thread is T then using equation of downward motion for cylinder,
$mg - T = m.a$ ..... (1)

The turning moment about centre G due to tension in thread,
$$= T \times r \quad \ldots\ldots (2)$$

The turning moment of cylinder due to angular acceleration α,
$$= I_{zz} \times \alpha \quad \ldots\ldots (3)$$

Equating equation (2) and (3),

$$T.r = I_{zz}.\alpha$$

$$T.r = \frac{m.r^2}{2} \cdot \alpha$$

$$T = \frac{mr\alpha}{2}$$

$$T = \frac{m(r\alpha)}{2}$$

$$T = \frac{ma}{2}$$

Substituting the value of T in equation (1),

$$mg - \frac{ma}{2} = ma$$

$$mg = \frac{3ma}{2}$$

$$a = \frac{2}{3} \cdot g$$

Example: 16.4

Two weights, each of 20 N, are suspended from a two-step pulley as shown in Fig. 16.5. Find the accelerations of weight 'A' and 'B'. The weight of pulley is 200 N and its radius of gyration is 200 mm. Determine tension in strings also.

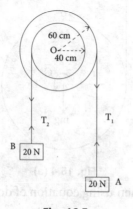

Fig. 16.5

Kinetics of Rigid Bodies

Solution:

However both weights are of equal magnitude but weight A will produce large turning moment about 'O' as it is acting on large radius. Thus block 'A' will move downward and block 'B' will move upward. If the pulley rotates by an angular acceleration 'α' and acceleration of weights A and B are a_1 and a_2, respectively.

Using the equation of motions,

$$20 - T_1 = \frac{20}{g} \cdot a_1 \qquad \ldots (1)$$

$$T_2 - 20 = \frac{20}{g} \cdot a_2 \qquad \ldots (2)$$

Turning moment on pulley due to tension will be,
$$\Sigma M = (T_1 \cdot r_1 - T_2 \cdot r_2)$$
$$= (T_1 \times 0.6 - T_2 \times 0.4) \qquad \ldots (3)$$

The turning moment of pulley for angular acceleration, α
$$\Sigma M = I \times \alpha$$
$$(0.6\, T_1 - 0.4\, T_2) = mk^2 \times \alpha$$

Given mass of pulley $m = \frac{200}{g}$ and radius of gyration $k = 0.2$ m

$$(0.6 T_1 - 0.4 T_2) = \frac{200}{g} \times (0.2)^2 \times \alpha \qquad \ldots (4)$$

Since $a = r \cdot \alpha$
$a_1 = 0.6\alpha$ and $a_2 = 0.4\alpha$

Equations (1) and (2) may be written as,

$$20 - T_1 = \frac{20}{g} \times 0.6\alpha \text{ and } T_2 - 20 = \frac{20}{g} \times 0.4\alpha$$

$$T_1 = \left(20 - \frac{20}{g} \times 0.6\alpha\right) \text{ and } T_2 = \left(20 + \frac{20}{g} \times 0.4\alpha\right)$$

Substituting the values of T_1 and T_2 in equation (4),

$$0.6\left(20 - \frac{20}{g} \times 0.6\alpha\right) - 0.4\left(20 + \frac{20}{g} \times 0.4\alpha\right) = \frac{200}{g} \times (0.2)^2 \times \alpha$$

$$\alpha = 2.13\, rad/sec^2$$

acceleration of block A, $a_1 = 0.6\alpha = 0.6 \times 2.13 = 1.28$ m/sec^2
acceleration of block B, $a_2 = 0.4\alpha = 0.4 \times 2.13 = 0.85$ m/sec^2

$$T_1 = \left(20 - \frac{20}{g} \times 0.6\alpha\right) = \left[20 - \left(\frac{20 \times 0.6 \times 2.13}{9.81}\right)\right] = 17.39N$$

$$T_2 = \left(20 + \frac{20}{g} \times 0.4\alpha\right) = \left[20 + \left(\frac{20 \times 0.4 \times 2.13}{9.81}\right)\right] = 21.73N$$

Example: 16.5

In the previous example, determine the distance travelled by block A and B when block B reaches to the velocity of 2 m/sec.

Solution:

From previous example,
$a_1 = 1.28$ m/sec², $a_2 = 0.85$ m/sec², $u = 0$
Given, $v_B = 2$ m/sec
Let block A and B travel distance S_A and S_B, respectively,

Using equation,
$v^2 = u^2 + 2as$
$v_B^2 = u^2 + 2a_2 S_B$
$(2)^2 = 0 + 2 \times 0.85 \times S_B$
$S_B = 2.35$ m

Let pulley rotates by an angular velocity 'ω',

$$\omega = \frac{v_B}{0.4} = \frac{v_A}{0.6}$$

$$v_A = \frac{0.6 \times v_B}{0.4}$$
$$= \frac{0.6 \times 2}{0.4}$$
$$= 3 \, m/sec$$

Using equation,
$v_A^2 = u^2 + 2a_1 S_A$
$(3)^2 = 0 + 2 \times 1.28 \times S_A$
$S_A = 3.51$ m

Example: 16.6

A system of cylinder mass M, radius r is welded to a prismatic bar of mass 'm', length 'l' as shown in Fig. 16.6. If the system is released from the horizontal position, determine its angular acceleration about hinge O.

Kinetics of Rigid Bodies

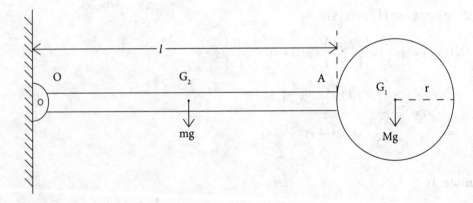

Fig. 16.6

Solution:

Let the system has mass moment of inertia I and angular acceleration α about hinge O. The turning moment of the system about O will be,

$$\sum M = Mg(l+r) + mg \times \frac{l}{2} \qquad \ldots(1)$$

The turning moment of the system due to angular acceleration will be

$$\sum M = I \times \alpha \qquad \ldots(2)$$
$$\text{where } I = I_{cylinder} + I_{bar} \qquad \ldots(3)$$

Using the parallel axis theorem,

$$I_{cylinder} \text{ from end 'O'} = I_{G_1} + M(l+r)^2$$
$$= \left(\frac{Mr^2}{2}\right) + M(l+r)^2$$

$$I_{bar} \text{ from end 'O'} = \frac{ml^2}{3}$$

From equation (3),

$$I = \left(\frac{M.r^2}{2}\right) + M(l+r)^2 + \frac{m.l^2}{3}$$

Substituting the value of I in equation (2),

$$\sum M = \left[\left(\frac{M.r^2}{2}\right) + M(l+r)^2 + \frac{ml^2}{3}\right].\alpha \qquad \ldots(4)$$

Equating equation (1) and (4),

$$\left[Mg(l+r)+mg\frac{l}{2}\right]=\left[\left(\frac{Mr^2}{2}\right)+M(l+r)^2+\frac{ml^2}{3}\right].\alpha$$

$$\alpha = \frac{\left[Mg(l+r)+mg\dfrac{l}{2}\right]}{\left[\left(\dfrac{Mr^2}{2}\right)+M(l+r)^2+\dfrac{ml^2}{3}\right]}$$

Example: 16.7

A sphere of radius 10 cm and mass 25 kg is attached to a thin rod of length 1 m and mass 3 kg. It is free to rotate about the axis shown in Fig. 16.7. Determine the angular acceleration of the system about O.

Solution:

If the moment of inertia and angular acceleration of the system about hinge 'O' is I and α respectively, the turning moment about 'O' will be,

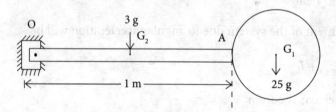

Fig. 16.7

$$\sum M = 25g(1+0.1)+3g.\frac{1}{2}$$
$$= 284.49\, Nm \qquad\qquad \ldots (1)$$

The turning moment due to angular acceleration,
$$\Sigma M = I \times \alpha \qquad\qquad \ldots (2)$$

Where,
$$I = I_{sphere} + I_{rod}$$

$$=\left[I_{G_1}+M(l+r)^2\right]_{sphere}+\left[\frac{ml^2}{3}\right]_{rod}$$

$$=\left[\frac{2}{5}\times 25(0.1)^2+25(1+0.1)^2\right]_{sphere}+\left[\frac{3\times 1^2}{3}\right]_{rod}$$

$$= 31.35\ kg.m^2$$

Substituting value of I in equation (2) and equating with equation (1),
31.35 × α = 284.49
α = 9.07 rad/sec²

Example: 16.8

In a rolling mill, the solid cylindrical shaft is released from rest on a rough ramp to store in underground stockyard. If the mass and radius of shaft is 'm' and 'r', respectively, then determine:

(i) Acceleration of the mass centre
(ii) Distance travelled by shaft after 't' sec along the ramp

Assume that the shaft does not slip during descending and the angle of inclination of ramp is θ

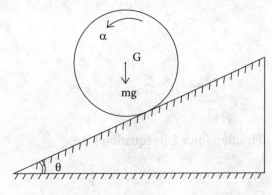

Fig. 16.8

Solution:

(i) Consider free body diagram of shaft as shown in Fig. 16.8 (a) where the shaft bears general plane motion due to frictional force F and weight component $mg.\sin\theta$. If angular acceleration is 'α' and acceleration of the mass centre is 'a', then resolving the forces for motion along the ramp and perpendicular to the ramp,

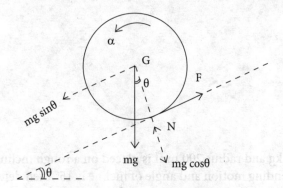

Fig. 16.8 (a) F.B.D. of Cylindrical shaft

i.e., $mg\sin\theta - F = m \cdot a$ ..... (1)
and $N = mg\cos\theta$ ..... (2)

Taking moment of the force system on shaft about G,
$\Sigma M = F \times r$ ..... (3)

Turning moment about centre G producing angular acceleration, α
$\Sigma M = I \times \alpha$ ..... (4)

Equating equation (3) and (4),
$F \times r = I \times \alpha$ ..... (5)

For cylindrical shaft, $I = \dfrac{mr^2}{2}$

$$F \times r = \dfrac{mr^2}{2} \times \alpha$$
$$F = \dfrac{m(r\alpha)}{2}$$
$$F = \dfrac{m.a}{2}$$

Substituting the value of friction force F in equation (1),

$$mg\sin\theta - \dfrac{ma}{2} = ma$$
$$mg\sin\theta = \dfrac{3}{2}ma$$
$$a = \dfrac{2}{3}g\sin\theta$$

(ii) As $u = 0$, using the equation

$$s = ut + \dfrac{1}{2}at^2$$
$$= 0 + \dfrac{1}{2} \times \dfrac{2}{3}g\sin\theta.t^2$$
$$s = \dfrac{gt^2}{3}\sin\theta$$

Example: 16.9

A sphere of mass 10 kg and radius 200 mm is placed on a rough incline. If the sphere does not slip during descending motion and angle of incline is 15° then determine:

Kinetics of Rigid Bodies

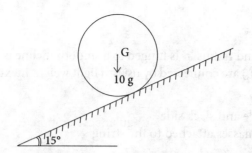

Fig. 16.9

(i) Acceleration of the mass centre
(ii) Distance travelled by sphere on incline after 3 sec from rest

Solution:

(i) Given, $m = 10$ kg, $r = 200$ mm, $u = 0$, $t = 3$ sec, $\theta = 15°$

Consider previous example, where cylinder was descending. However, in this problem sphere is used. Equations (1) to (5) will remain same of last example. Here the moment of inertia of sphere is, $I = \dfrac{2}{5}mr^2$

Consider equation (5),

$$F \times r = I \times \alpha$$
$$F \times r = \dfrac{2}{5}mr^2 \times \alpha$$
$$F = \dfrac{2}{5}m(r\alpha)$$
$$F = \dfrac{2}{5}m.a$$

Substituting the value of frictional force F in equation (1),

$$mg\sin\theta - \dfrac{2}{5}ma = m.a$$
$$mg\sin\theta = \dfrac{7}{5}ma$$
$$\text{i.e., } a = \dfrac{5}{7}.g\sin\theta$$

(ii) Using equation,

$$s = ut + \dfrac{1}{2}at^2$$
$$= 0 + \dfrac{1}{2} \times \dfrac{5}{7}g\sin 30° \times (3)^2$$
$$= 15.77 \, m$$

Example: 16.10

A rough disk of mass 'm' and radius 'r' is hinged to a smooth incline plane as shown in Fig. 16.10. Two masses ($m_1 > m_2$) are connected by using a light weight inextensible string passes over disk, Determine:

(i) Tensions on tight side and slack side
(ii) Acceleration of the masses attached to the string

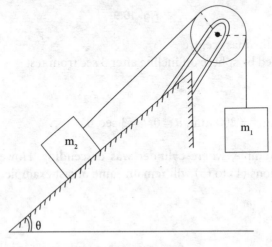

Fig. 16.10

Solution:

Let the masses are accelerated by acceleration 'a'. The free body diagram of both masses are shown in Fig. 16.10 (a).

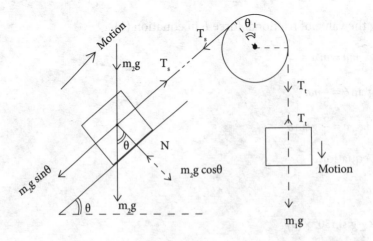

Fig. 16.10 (a)

Kinetics of Rigid Bodies

Using the equation of motion for m_1 and m_2 masses,
$$m_1 g - T_t = m_1 a \qquad \ldots (1)$$
$$T_s - m_2 g \sin\theta = m_2 a \qquad \ldots (2)$$

If the disc rotates by angular acceleration α, then turning moment about hinge will be
$$\Sigma M = I \times \alpha \qquad \ldots (3)$$

The turning moment due to tension will be,
$$\Sigma M = (T_t - T_s) \cdot r \qquad \ldots (4)$$

Equating equations (3) and (4),
$$I \times \alpha = (T_t - T_s) \cdot r$$
$$\frac{mr^2}{2} \times \alpha = (T_t - T_s) \cdot r$$
$$(T_t - T_s) = \frac{m(r\alpha)}{2}$$
$$(T_t - T_s) = \frac{m \cdot a}{2} \qquad \ldots (5)$$

Add equations (1) and (2),
$$m_1 g - T_t + T_s - m_2 g \sin\theta = (m_1 + m_2) a$$
$$(m_1 - m_2 \sin\theta) g - (T_t - T_s) = (m_1 + m_2) a$$
$$(m_1 - m_2 \sin\theta) g - \frac{ma}{2} = (m_1 + m_2) \cdot a$$
$$(m_1 - m_2 \sin\theta) g = \left(m_1 + m_2 + \frac{m}{2}\right) \cdot a$$
$$a = \frac{2g(m_1 - m_2 \sin\theta)}{(2m_1 + 2m_2 + m)}$$

Substituting the value of 'a' in equations (1) and (2),
$$T_t = m_1 g \cdot \left[\frac{m + 2m_2(1 + \sin\theta)}{(m + 2m_1 + 2m_2)}\right]$$
$$T_s = m_2 g \cdot \left[\frac{m\sin\theta + 2m_1(1 + \sin\theta)}{(m + 2m_1 + 2m_2)}\right]$$

Example: 16.11

In the previous example, if $m_1 = 12$ kg, $m_2 = 15$ kg and $m = 9$ kg and $\theta = 30°$. Determine:
(i) Tensions on tight side and slack side.
(ii) Acceleration of both masses

(iii) Angular acceleration of disc if radius of disc is 120 mm
(iv) Coefficient of kinetic friction of disc

Solution:

(i) Tension on tight side will be,

$$T_t = m_1 g \left[\frac{m + 2m_2(1 + \sin\theta)}{(m + 2m_1 + 2m_2)} \right]$$

$$= 12 \times 9.81 \left[\frac{9 + (2 \times 15)(1 + \sin 30°)}{9 + (2 \times 12) + (2 \times 15)} \right]$$

$$T_t = 100.90 \ N$$

Tension on slack side will be,

$$T_s = m_2 g \left[\frac{m \sin\theta + 2m_1(1 + \sin\theta)}{(m + 2m_1 + 2m_2)} \right]$$

$$= 15 \times 9.81 \left[\frac{9 \sin 30° + 2 \times 12(1 + \sin 30°)}{9 + (2 \times 12) + (2 \times 15)} \right]$$

$$= 94.60 \ N$$

(ii) Using equation (5),

$$(T_t - T_s) = \frac{m.a}{2}$$

$$(100.90 - 94.60) = \left(\frac{9 \times a}{2} \right)$$

$$a = 1.4 \ m/sec^2$$

(iii) r = 120 mm = 0.12 m and moment of inertia of disc, $I = \frac{m.r^2}{2}$

From equation (3) and (4),

$$I \times \alpha = (T_t - T_s) \times r$$

$$\left(\frac{m.r^2}{2} \right) \times \alpha = (100.90 - 94.60)r$$

$$\left(\frac{9 \times 0.120 \times \alpha}{2} \right) = 6.3$$

$$\alpha = 11.67 \ rad/sec^2$$

Kinetics of Rigid Bodies

(iv) If coefficient of kinetic friction is μ then we know that,

$$\frac{T_t}{T_s} = e^{\mu \theta^1}$$

where the angle of contact between thread and pulley will be (90 + θ) as shown in Fig. 16.10(a)

i.e. $\theta^1 = 90 + 30° = 120°$

$$= \left(\frac{120° \times \pi}{180}\right)$$

$$\left(\frac{100.90}{94.60}\right) = e^{\mu\left(\frac{120°\pi}{180}\right)}$$

$$\mu = 0.03$$

Example: 16.12

A thin uniform bar of mass 'm' and length 'L' is suspended from two vertical inextensible strings of equal length so that the bar is in equilibrium in horizontal position as shown in Fig. 16.11. If the right hand string BD is cut, find the tension in the left hand string AC at that instant, and also calculate the acceleration of the bar. [UPTU 2001–01]

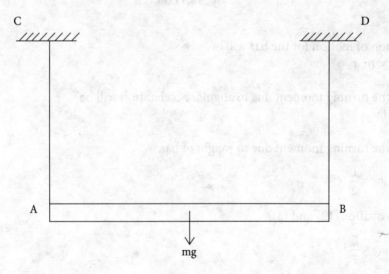

Fig. 16.11

Solution:

The bar AB freely rotates about end A when string BD is cut. For a particular instant, its motion is as shown in Fig. 16.11(a). Let the bar rotates by an angular acceleration 'α' and the acceleration of the bar is 'a'.

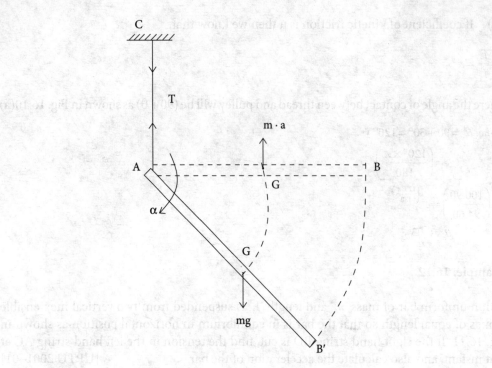

Fig. 16.11 (a)

The equation of motion for the bar will be,
$$mg - T = m.a \qquad \qquad \dots (1)$$

About A, the turning moment due to angular acceleration will be
$$\Sigma M_A = I \times \alpha \qquad \qquad \dots (2)$$

About A, the turning moment due to weight of bar,
$$\Sigma M = mg \times \frac{L}{2} \qquad \qquad \dots (3)$$

Equating equations (2) and (3),
$$I \times \alpha = mg \times \frac{L}{2}$$

where I of rod about A, $I = \dfrac{mL^2}{3}$

$$\frac{mL^2}{3} \times \alpha = mg \times \frac{L}{2}$$

$$\alpha = \left(\frac{3g}{2L}\right)$$

Kinetics of Rigid Bodies

The linear acceleration of the mass centre of the bar will be normal and given by,

$$a = r \times \alpha$$
$$= \left(\frac{L}{2} \times \alpha\right)$$

Substituting the value of α,

$$a = \frac{L}{2} \times \frac{3g}{2L}$$
$$a = \left(\frac{3}{4} \cdot g\right)$$

Substituting the value of 'a' in equation (1),

$$mg - T = m\left(\frac{3}{4} \cdot g\right)$$
$$T = \left(\frac{mg}{4}\right)$$

Example: 16.13

A slender prismatic rod AB of mass 'm' and length 'l' is held in equilibrium as shown in Fig. 16.12. If string BC snaps suddenly, determine:

(i) Angular acceleration of rod
(ii) Linear acceleration of rod
(iii) Reaction of hinge 'O' if the mass of the rod is 30 kg

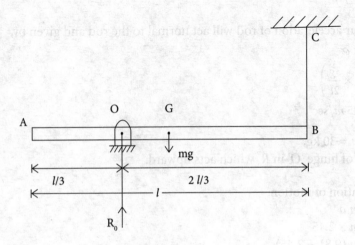

Fig. 16.12

Solution:

Let the rod has angular acceleration α about 'O'.

(i) Thus turning moment of rod about 'O' will be
$$\Sigma M = I \times \alpha \quad \ldots\ldots (1)$$

The moment of inertia of rod about O will be

$$I = I_G + m(OG)^2 \qquad OG = \left(\frac{l}{2} - \frac{l}{3}\right)$$

$$= \frac{ml^2}{12} + m\left(\frac{l}{6}\right)^2 \qquad = \left(\frac{l}{6}\right)$$

$$= ml^2\left(\frac{1}{12} + \frac{1}{36}\right)$$

$$= \frac{ml^2}{9}$$

The turning moment of rod due to weight about 'O' will be

$$\Sigma M = mg \times OG = mg \times \frac{l}{6} \quad \ldots\ldots (2)$$

Equating equations (1) and (2),

$$I \times \alpha = mg \times \frac{l}{6}$$

$$\frac{ml^2}{9} \times \alpha = mg \times \frac{l}{6}$$

$$\alpha = \left(\frac{3g}{2l}\right)$$

(ii) The linear acceleration of rod will act normal to the rod and given by,

$$a = r \times \alpha$$

$$= \left(\frac{l}{6} \times \frac{3g}{2l}\right)$$

$$a = 2.45 \, m/sec^2$$

(iii) Given, $m = 30$ kg,
If the reaction of hinge 'O' in R_0 which acts upward.

Using the equation of motion,
$$mg - R_0 = m.a$$
$$mg - R_0 = m \times 2.45$$
$$R_0 = m(9.81 - 2.45)$$
$$= 30(9.81 - 2.45)$$
$$= 220.80 \, N$$

Example: 16.14

A stainless steel hoop of mass 1.2 kg and radius 150 mm rolls on a horizontal plane as shown in Fig. 16.13.

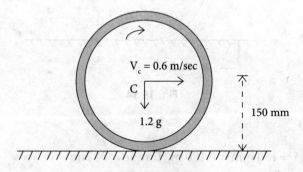

Fig. 16.13

If hoop rolls without slipping with a translatory velocity of 0.6 m/sec then, determine:
(i) angular velocity of hoop (ii) kinetic energy of hoop

Solution:

Given, $m = 1.2$ kg, $r = 150$ mm $= 0.15$ m, $V_c = 0.6$ m/sec

(i) The angular velocity of hoop will be, $\omega = \dfrac{v_c}{r} = \dfrac{0.6}{0.15} = 4\ rad/sec$

(ii) The kinetic energy of hoop will be,

$$= \frac{1}{2}mv_c^2 + \frac{1}{2}I\omega^2 \qquad \ldots\ldots (1)$$

The moment of inertia of hoop = mr^2

$$= \frac{1}{2}mv_c^2 + \frac{1}{2}mr^2\omega^2$$
$$= \frac{1}{2} \times 1.2(0.6)^2 + \frac{1}{2} \times 1.2(0.15)^2(4)^2$$
$$= 0.43\ Nm$$

Example: 16.15

A cylindrical Cast iron shaft of mass 100 kg and radius 180 mm is kept on a rough surface as shown in Fig. 16.14. Determine the frictional force and acceleration of the mass centre of shaft when a torque of 74 Nm is applied clockwise and shaft rotates without slipping.

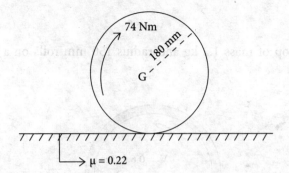

Fig. 16.14

Solution:

Given, $m = 100$ kg, r = 180 mm = 0.180 m, Torque = 74 Nm

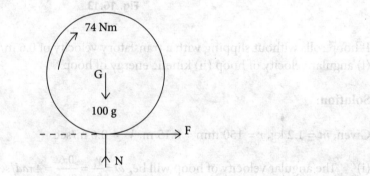

Fig. 16.14 (a) F.B.D. of shaft

Consider free body diagram of shaft,
$$N = 100 \times g$$
thus $F = \mu.N$
$$= 0.22 \times 100 \times 9.81$$
$$F = 215.82 \ N$$

$Torque = \sum M = I \times \alpha$

$$74 = \left(\frac{m \cdot r^2}{2}\right) \cdot \alpha$$

$$74 = \frac{100 \times (0.18)^2}{2} \times \alpha$$

$$\alpha = 45.68 \ rad/sec^2$$

The acceleration of the mass centre of shaft = $a = r.\alpha$
$$= 0.18 \times 45.68$$
$$= 8.22 \ m/sec^2$$

Example: 16.16

A road roller of weight 11772 N and radius 0.5 m is pulled with a force of 2000 N on a rough ground. If the roller starts from rest and rolls without slipping, find the distance travelled by the centre of the roller at which roller acquires a velocity of 4 m/sec.

[UPTU, 2003–04]

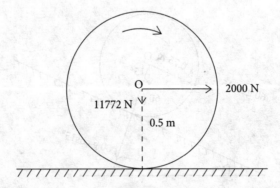

Fig. 16.15

Solution:

Given, $W = 11772$ N, $r = 0.5$ m, $F = 2000$ N, $u = 0$, $v = 4$ m/sec
Initial kinetic energy of roller = 0
Final kinetic energy when pulled = (K.E.)$_{\text{linear motion}}$ + (K.E.)$_{\text{rotary motion}}$

$$= \frac{1}{2}mv^2 + \frac{1}{2}I \cdot \omega^2 \qquad \ldots (1)$$

where, $m = \dfrac{11772}{9.81}$ kg, $I = \dfrac{m \cdot r^2}{2} = \dfrac{11772}{2 \times 9.81}(0.5)^2$ kgm^2

and $\omega = \dfrac{v}{r} = \dfrac{4}{0.5} = 8$ rad/sec

Substituting all values in equation (1),

$$\text{Final K.E.} = \left[\frac{1}{2} \times \frac{11772}{9.81}(4)^2\right] + \left[\frac{1 \times 11772}{2 \times 2 \times 9.81}(0.5)^2 \times (8)^2\right]$$
$$= 14400 \, N.m$$

If roller is pulled by distance 'S', then work done on roller = $F \times S = 2000 \times S$
Using the work–energy principle,

Work done on roller = Change in K.E.
$2000 \times S = 14400$
$S = 7.2$ m

Example: 16.17

A prismatic circular bar AISI 304 is placed at rest on 15° rough inclined plane. If bar rolls down without slip, determine the distance travelled by it when attain a velocity of 4 m/sec by using work energy principle. Take mass and radius of bar as 60 kg and 0.75 m respectively.

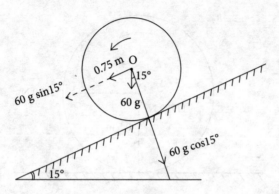

Fig. 16.16

Solution:

Given, $m = 60$ kg, $r = 0.75$ m, $u = 0$, $v = 4$ m/sec, $S = ?$
The bar bears general plane motion during rolling motion.

Initial K.E. of bar $= 0$

Final K.E. of bar $= (K.E.)_{linear\ motion} + (K.E.)_{rotary\ motion}$

$$= \frac{1}{2}mv^2 + \frac{1}{2}I\omega^2$$

where, $I = \dfrac{m \cdot r^2}{2} = \dfrac{60 \times (0.75)^2}{2}$ kgm²

and $\omega = \dfrac{v}{r} = \dfrac{4}{0.75} = 5.33$ rad/sec

Final K.E. $= \dfrac{1}{2} \times 60 \times (4)^2 + \dfrac{1}{2} \times \dfrac{60(0.75)^2}{2} \times (5.33)^2$

$= 719.7$ Nm

Using the work–energy principle,
   W. D. by bar = Change in K.E. of bar
   $F \times S$ = Final K.E. − Initial K.E.
   $(60 g \sin 15°) S = 719.7 - 0$
   $S = 4.72$ m

Kinetics of Rigid Bodies

Example: 16.18

Determine the height fallen by the log in example 16.2 by using work–energy principle if log attains a velocity of 13.08 m/sec.

Solution:

Let the log fall down by height 'h'.
Work done by log = mgh
$$= 10 \times 9.81 \times h \qquad \ldots\ldots (1)$$

Initial K.E. of log = 0
Final K.E. of log = $(K.E.)_{linear\ motion} + (K.E.)_{rotary\ motion}$
$$= \frac{1}{2}mv^2 + \frac{1}{2}I\omega^2 \qquad \ldots\ldots (2)$$

Where $v = 13.08$ m/sec, $r = 100$ mm $= 0.1$ m

$$\omega = \frac{v}{r} = \frac{13.08}{0.1} = 130.8\ rad/sec$$

$$I = \frac{m \cdot r^2}{2} = \frac{10(0.1)^2}{2} = 0.05\ kgm^2$$

Substitute all values in equation (2),

$$Final\ K.E. = \left[\frac{1}{2} \times 10 \times (13.08)^2 + \frac{1}{2} \times 0.05 \times (130.8)^2\right]$$
$$= 1283.15\ Nm$$

Using the work–energy principle,
Work done = Change in K.E.
$10 \times 9.81 \times h = 1283.15 - 0$
$h = 13.08$ m

Example: 16.19

In example 16.4, determine distance travelled by both block A and B when block B reaches to velocity of 2 m/sec by using work energy principle.

Solution:

Initial K.E. of blocks = 0
Final K.E. of blocks = K.E. of blocks + K.E. of pulley

$$= \left[\frac{1}{2}m_A v_A^2 + \frac{1}{2}m_B v_B^2\right] + \left[\frac{1}{2}I\omega^2\right] \qquad \ldots\ldots (1)$$

where $m_A = m_B = \dfrac{W}{g} = \dfrac{20}{g}$ kg

$v_B = 2$ m/sec

The angular velocity of pulley,

$$\omega = \dfrac{v_B}{0.4} = \dfrac{2}{0.4} = 5 \text{ rad/sec}$$

as, $\omega = \dfrac{v_A}{0.6}$

$v_A = 0.6 \times \omega = 0.6 \times 5 = 3$ m/sec

Work done by both blocks will be,
= W.D. by block A − W.D. on block B
= $(20 \times S_A - 20 \times S_B)$ ..... (2)

Using the work–energy principle,

W.D. = Change in K.E.

$$20(S_A - S_B) = \left[\dfrac{1}{2} \times \dfrac{20}{g}(3)^2 + \dfrac{1}{2} \times \dfrac{20}{g}(2)^2 + \dfrac{1}{2} \times \dfrac{200}{g}(0.2)^2 \times (5)^2 \right]$$

$S_A - S_B = 1.17$ ..... (3)

Let the pulley turn by angle '$d\theta$' when blocks travel for distance S_A and S_B.

thus $d\theta = \dfrac{S_A}{0.6} = \dfrac{S_B}{0.4}$

$S_A = 1.5 S_B$ ..... (4)

Substituting the value of S_A in equation (3),
$1.5 S_B - S_B = 1.17$
$S_B = 2.35$ m

Substituting the value of S_B in equation (4),
$S_A = 3.52$ m

Example: 16.20

In the example 16.2, determine the angular velocity and velocity of log by using principle of conservation of energy when it falls down by 13.08 m.

Solution:

Using the principle of conservation of energy,

$[\text{P.E.} + \text{K.E}]_{\text{initial position}} = [\text{P.E.} + \text{K.E}]_{\text{final position}}$

Kinetics of Rigid Bodies

Since the log is in equilibrium at initial position, the total energy is zero. Thus

$$0 = \left[mgh + \frac{1}{2}mv^2 + \frac{1}{2}I\omega^2 \right]_{\text{Final position}} \quad \ldots\ldots (1)$$

At final position, $h = -13.08$ m when initial position of log is taken as reference line.

and $v = r\omega = 0.1 \times \omega$

$$I = I_G = \frac{m \cdot r^2}{2}$$

Substituting the values in equation (1),

$$0 = mg(-13.08) + \frac{m}{2}(0.1 \times \omega)^2 + \frac{1}{2}\left(\frac{m \cdot r^2}{2}\right)\omega^2$$

$$0 = (-9.81 \times 13.08) + \frac{(0.1)^2}{2}\omega^2 + \frac{(0.1)^2}{4} \cdot \omega^2$$

$$\omega = 130.8 \text{ rad/sec}$$

$$v = r\omega = 0.1 \times 130.8 = 13.08 \text{ m/sec}$$

Example: 16.21

In the example 16.12, determine the linear velocity and angular velocity of the centre of gravity of the bar by using the principle of conservation of energy when string BD is cut and the bar reaches to its vertical position from initial position. Take length of the bar as 5 m.

Solution:

Let the length of the bar is 'l'. When the string BD is cut, bar AB reaches position AB. If the velocity and angular velocity of centre of gravity of bar becomes 'V' and 'ω', respectively.

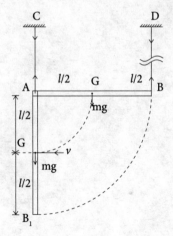

Fig. 16.17

Using the principle of conservation of energy,

[P.E. + K.E.]$_{\text{initial position}}$ = [P.E. + K.E.]$_{\text{final position}}$

Since the bar AB is in equilibrium at horizontal condition, the total energy at initial condition will be zero. Thus,

$$0 = \left[mgh + \frac{1}{2}mv^2 + \frac{1}{2}I\omega^2 \right]_{\text{Final position}} \quad \ldots\ldots (1)$$

At final condition, $h = -\frac{l}{2}$ when AB is taken as reference line.

and $v = r\omega = \frac{l}{2} \cdot \omega$

$I = I_G = \left(\frac{ml^2}{12}\right)$

Thus equation (1) becomes,

$$0 = mg\left(-\frac{l}{2}\right) + \frac{1}{2}m\left(\frac{l \cdot \omega}{2}\right)^2 + \frac{1}{2}\left(\frac{ml^2}{12}\right) \cdot \omega^2$$

$$\frac{mgl}{2} = \frac{m}{2} \cdot \frac{l^2\omega^2}{4} + \frac{m}{2} \cdot \frac{l^2\omega^2}{12}$$

$$g = l\left(\frac{\omega^2}{4} + \frac{\omega^2}{12}\right)$$

$$g = l \cdot \frac{\omega^2}{3}$$

given, $l = 5$ m

$$9.81 = \frac{5 \times \omega^2}{3}$$

$\omega = 2.43$ rad/sec

and $v = r \cdot \omega$

$= \frac{l}{2} \times \omega$

$v = \frac{5}{2} \times 2.43$

$= 6.08$ m/sec

Theoretical Problems

T 16.1 Discuss the moment of momentum.
T 16.2 Prove that the relation between torque and angular acceleration is given by $T = I\alpha$.
T 16.3 Derive the expression for kinetic energy of a body under general plane motion.
T 16.4 Illustrate the principle of conservation of energy for a body moving under general plane motion.
T 16.5 State the principle of work and energy for a body moving under general plane motion.

Numerical Problems

N 16.1 A right circular cylinder is wrapped by inextensible string around the circumference. If the mass of cylinder is 4 kg and radius is 0.10 m; determine tension induced in the cord and acceleration of its mass centre G, when allowed to fall freely.

N 16.2 A solid shaft carries a flywheel of mass 20 kg and radius of gyration 0.30 m. If the flywheel is rotating at 900 rpm and stops after 600 revolutions due to frictional resistance. Determine:
(i) The angular deceleration
(ii) Time taken by flywheel to stop
(iii) The frictional torque due to frictional resistance

N 16.3 Consider a step pulley supporting two equal weights of 50 N with the help of string as shown in Fig. NP 16.1. If the weight and radius of gyration of pulley are 600 N and 0.45 m, respectively, determine:

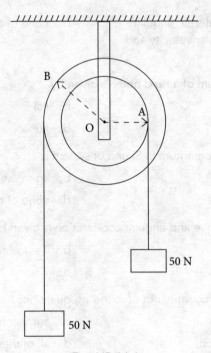

Fig. NP 16.1

(i) The accelerations of both blocks
(ii) Tension in both strings
(iii) Take OA = 50 cm and OB = 75 cm

N 16.4 A flywheel of mass 200 kg and radius of gyration 0.80 m decelerates from 600 rpm to 250 rpm in 1.5 minutes. Determine:
(i) The loss in kinetic energy
(ii) The retarding torque on flywheel

N 16.5 A solid cylinder is released from rest on an inclined plane at an angle θ from horizontal. The mass of the cylinder is 'M' and radius is 'R'. Determine the velocity of cylinder after it has rolled down the incline through a distance 'S'. Solve by using the principle of conservation of energy.

N 16.6 A metal hoop of weight 30 N of radius 0.20 m is placed on an incline, inclined at 30°. If the coefficient of kinetic friction is 0.20. Determine:
(i) The angular acceleration of the hoop
(ii) The velocity of hoop after moving a distance of 1.8 m down the incline

N 16.7 A cylinder of mass 50 kg and radius 0.80 m rolls down on a 20° incline. Using work energy principle, determine the distance travelled by cylinder when it reaches a velocity of 3.6 m/sec.

Multiple Choice Questions

1. Kinetics of rigid body deals with motion involving
   a. angular velocity, time
   b. angular acceleration, angular velocity and time
   c. mass moment of inertia, angular acceleration, angular velocity and time
   d. none of these

2. The angular momentum of a rigid body is given by
   a. $I\omega^2$
   b. $I\omega$
   c. mv
   d. $I\alpha$

3. The unit of angular momentum in SI unit of system is
   a. kg-m/sec
   b. kg-m²/sec
   c. N-m²/sec
   d. none of these

4. Relation between torque and angular acceleration is given by
   a. $T = I\alpha$
   b. $T = I\omega^2$
   c. $T = I\omega$
   d. $T = mv$

5. Determine the vector quantity out of following quantities
   a. momentum
   b. angular momentum
   c. angular acceleration
   d. all of these

Kinetics of Rigid Bodies

6. Identify which principle can be used for the analysis of kinetic motion of a rigid body
   a. principle of conservation of energy
   b. principle of conservation of work and energy
   c. D'Alembert's principle
   d. all of these

Answers

1. c 2. b 3. b 4. a 5. d 6. d

Chapter 17

Virtual Work

17.1 Introduction

We have used free body diagrams and equilibrium equations to determine unknown forces in the previous chapters 2–6. These analyses were made on a certain part of the system of bodies. However, when system becomes complex with large number of members, then it can be analyzed as a single body by another approach, called the principle of virtual work.

17.2 Principle of Virtual Work

Before studying this principle we must understand the term Virtual, means imaginary or not real which forms the basis of this principle. The virtual work arises when a force displaces the body by virtual infinitesimal displacement. The virtual terms for linear displacements in horizontal and vertical axis are designated by δ_x and δ_y respectively. Similarly virtual angular displacement is designed by δ_θ.

This principle states that when a body is maintained in equilibrium by a system of forces, then the algebraic sum of their individual virtual work (total virtual work) will be equal to zero where the virtual displacement (linear or angular) must be consistent with geometrical conditions.

This principle is widely used in a system of large connected members; however, it can be applied on a particle or free body diagram also.

During analysis, there are some forces which do not produce virtual work under certain conditions:

(i) The forces acting along the length of the members, i.e., axial forces (tension or compression).
(ii) Weight of the body when body moves along horizontal surface.

Virtual Work

(iii) Normal reaction of the body when body moves along horizontal surface.
(iv) Reaction of a frictionless hinge when body rotates about hinge.
(v) Frictional force acting on a wheel when it is rolling without slipping.

17.3 Work Done by Forces

Refer to section 13.2, if a force p pushes the body by displacement 'd' along horizontal smooth surface as shown in Fig. 13.1, the work done is given by

$W = p \times d$

Similarly, when force p pushes the body by virtual displacement δ_x along horizontal smooth surface as shown in Fig. 17.1, then virtual work will be given by

$W = p \times \delta_x$

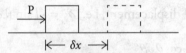

Fig. 17.1

Here, both force and virtual displacement are acting in the same direction thus the virtual work will be taken as positive. However if force and virtual displacement are acting in the opposite direction then the virtual work will be taken as negative.

Consider another example where a ladder AB of length l and weight W is placed between rough floor and smooth wall as shown in Fig. 17.2.

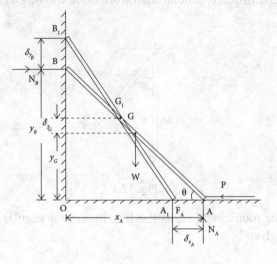

Fig. 17.2

Let the ladder is making an angle θ with the floor and the initial distances of its salient points from the reference point O are as follows:

(i) The end points A and B are at distances x_A and y_B, respectively.
(ii) The weight W is acting at height y_G from its centre of gravity, G.

Let the force p pushes ladder from position AB to A_1B_1 by virtual displacement δ_{x_A} and causes corresponding virtual displacements δ_{y_B} and δ_{y_G} at points B and G, respectively. The normal reactions of floor and wall are N_A and N_B, respectively, which do not produce virtual work. The frictional force F_A of floor is acting tangentially against the virtual displacement δ_{x_A}.

Using the principle of virtual work,

$$p \times (\delta_{x_A}) + W \times (-\delta_{y_G}) + F_A \times (-\delta_{x_A}) = 0$$

From Fig. 17.2,
$x_A = l \cos\theta$, thus virtual displacement, i.e., $\delta_{x_A} = -l \sin\theta . \delta\theta$

$y_G = \dfrac{l}{2}\sin\theta$ i.e. $\delta_{y_G} = \dfrac{l}{2}\cos\theta . \delta\theta$

Here negative sign of virtual displacements is not considered in the principle of virtual work equation as this shows relation between distances and angle of inclination θ, i.e., if θ is increased then distance x_A decreases; however, y_B increases.

17.4 Work Done by Moments

Consider a link OA is hinged about O as shown in Fig. 17.3. If a torque or moment M acting at link turns by angular displacement $d\theta$ then work done will be given by
   $W = M \times d\theta$

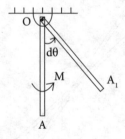

Fig. 17.3

Similarly, if torque or moment M turns the link by virtual angular displacement $\delta\theta$, the virtual work will be given by

Work done by moment = $M \times \delta\theta$

Example: 17.1

A block of weight 10 kN is pushed by a force 'P' as shown in Fig. 17.4. If the co-efficient of friction between the contact surfaces is 0.3, using principle of virtual work, determine the value of force 'P' by which block impends right side.

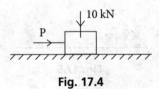

Fig. 17.4

Solution:

Given, $\mu = 0.3$.
Let the block is pushed by virtual distance δx as shown in Fig. 17.4 (a).

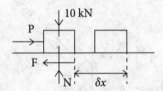

Fig. 17.4 (a)

Using the principle of virtual work,
$P.\delta x - F\delta x = 0$
$\delta x(P - F) = 0$
$P - \mu.N = 0$ ..... (1)
$\Sigma Y = 0$,
$N = 10$ kN

Substituting the value of N in equation (1),
$P - 0.3 \times 10 = 0$
$P = 3$ kN

Here virtual work of weight and normal reaction will be zero due to zero virtual distance.

Example: 17.2

A block of weight 20 kN is placed on a smooth inclined plane as shown in Fig. 17.5. Determine force P by which the block can be held in equilibrium by the principle of virtual work.

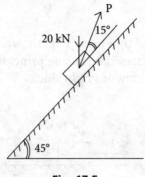

Fig. 17.5

Solution:

As the inclined plane is smooth, no frictional force acts. Consider the concurrent coplanar force system where let the block move up by virtual distance 'δx' along the inclined plane.

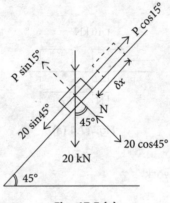

Fig. 17.5 (a)

Using the principle of virtual work,
P cos15°.δx − 20 sin45°. δx = 0

$$P = \frac{20 \sin 45°}{\cos 15°}$$

$P = 14.64$ kN

Here virtual work for forces P sin15°, 20 cos45° and normal reaction 'N' will be zero due to zero virtual distance.

Example: 17.3

In the previous question if inclined plane is rough such that the co-efficient of friction between block and inclined plane is 0.28 then determine maximum force 'P' by which block can be held in equilibrium.

Virtual Work

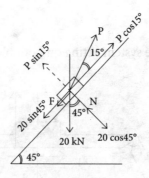

Fig. 17.6

Solution:

Given, $\mu = 0.28$, when maximum force 'P' will be applied, the block will tend to move upward and frictional force will act downward as shown in Fig. 17.6.

Let the block is displaced up by virtual distance δ_x. Using the principle of virtual work,
$P \cos 15°.\delta x - 20 \sin 45°.\delta x - F.\delta x = 0$
$(P \cos 15° - 20 \sin 45° - \mu.N).\delta x = 0$ ..... (1)

Resolving forces perpendicular to the inclined plane,
$\Sigma Y = 0$,
$N + P \sin 15° = 20 \cos 45°$
$N = (20 \cos 45° - P \sin 15°)$

Substituting values of N in equation (1),
$[P \cos 15° - 20 \sin 45° - 0.28(20 \cos 45° - P \sin 15°)].\delta x = 0$
$P = 17.43$ kN

Example: 17.4

In example 4.14, determine force P to impend block A right side using principle of virtual work. Take weight of blocks A and B 200 kN and 500 kN, respectively.

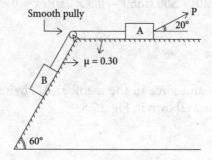

Fig. 17.7

Solution:

Given, μ = 0.30. Consider the force system acting on both blocks as shown in Fig. 17.7. (a).

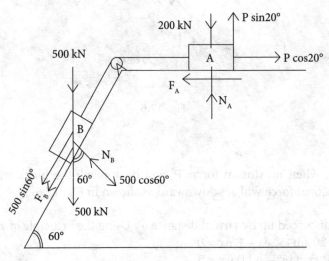

Fig. 17.7 (a)

Let the block 'A' moves to right side by virtual distance 'δx'. Thus the block B will move by same virtual distance on inclined plane towards pulley.

Using the principle of virtual work,
$P \cos 20° \times \delta x - F_A \times \delta x - 500 \sin 60° \times \delta x - F_B \times \delta x = 0$
$(P \cos 20° - \mu_A N_A - 500 \sin 60° - \mu_B N_B) \delta x = 0$ ..... (1)
$\Sigma Y = 0$ for both blocks,
$P \sin 20° + N_A = 200$, i.e., $N_A = (200 - P \sin 20°)$
and $N_B = 500 \cdot \cos 60°$
$= 250$ kN

Substituting the values of N_A and N_B in equation (1),
$[P \cos 20° - 0.3(200 - P \sin 20) - 500 \cdot \sin 60° - 0.3 \times 250] \delta x = 0$
$P = 544.96$ kN

Example: 17.5

Determine support reactions and force in the member AC by using principle of virtual work for given triangular truss as shown in Fig. 17.8.

Virtual Work

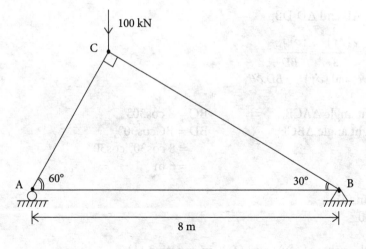

Fig. 17.8

Solution:

Consider Fig. 17.8. (a) where reaction of roller bearing is displaced truss by virtual distance $(\delta y)_A$.

Using the principle of virtual work,

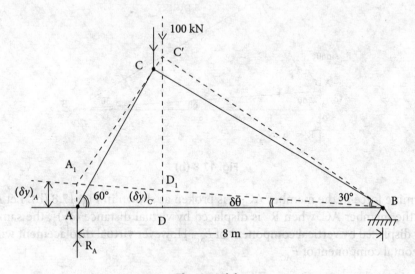

Fig. 17.8 (a)

$R_A.(\delta y)_A + 100[-(\delta y)_{C'}] = 0$
$R_A.(\delta y)_A - 100\,(\delta y)_{C'} = 0$ ..... (1)

Consider ΔA_1AB and ΔD_1DB,

$$\tan \delta\theta = \delta\theta = \frac{(\delta y)_A}{8} = \frac{(\delta y)_{C'}}{BD}$$

$(\delta y)_A = 8.\delta\theta$ and $(\delta y)_{C'} = BD.\delta\theta$ ..... (2)

Consider right angle ΔACB, $BC = 8 \cos 30°$
    In right angle ΔBCD, $BD = BC\cos 30°$
                                                   $= 8 \cos 30°.\cos 30°$
                                                   $= 6$ m

From equation (2),
   $(\delta y)_C = 6.\delta\theta$

Substituting the value of $(\delta y)_A$ and $(\delta y)_C$ in equation (1),
   $R_A.8\delta\theta - 100 \times 6\delta\theta = 0$
   $R_A = 75$ kN
   $\Sigma Y = 0$,
   $R_A + R_B = 100$
   $R_B = 25$ kN

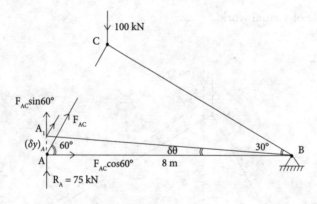

Fig. 17.8 (b)

To determine force in the member AC, it is broken as shown in Fig. 17.8 (b). Let F_{AC} is the force in the member AC, when R_A is displaced by virtual distance $(\delta y)_A$, the same will be virtually displaced by vertical component of F_{AC}. However virtual displacement will be zero for horizontal component of F_{AC}.

Using the principle of virtual work,
   $R_A.(\delta y)_A + F_{AC}.\sin 60°.(\delta y)_A = 0$

$$F_{AC} = \frac{-75}{\sin 60°}$$

   $F_{AC} = -86.60$ kN (Compression)

Virtual Work

Example: 17.6

Determine force in the member AD by using the principle of virtual work in a cantilever truss as shown in Fig. 17.9.

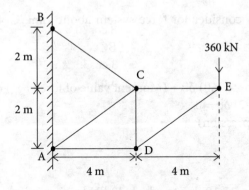

Fig. 17.9

Solution:

Let the force 360 kN is virtually displaced by $(\delta y)_E$ from E to E' and force F_{AD} virtually displaced by $(\delta x)_D$ from D to D' as shown in Fig. 17.9(a).

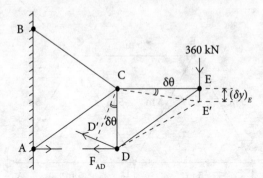

Fig. 17.9 (a)

Using the principle of virtual work,

$$360 \times (\delta y)_E + F_{AD} \times (\delta x)_D = 0 \qquad \ldots\ldots (1)$$

where,

$$\delta\theta = \frac{(\delta y)_E}{CE} = \frac{(\delta x)_D}{CD}$$

$$\delta\theta = \frac{(\delta y)_E}{4} = \frac{(\delta x)_D}{2}$$

$$(\delta y)_E = 4\delta\theta \text{ and } (\delta x)_D = 2\delta\theta$$

Substituting values in equation (1),
$$360 \times 4\delta\theta + F_{AD} \times 2\delta\theta = 0$$
$$F_{AD} = -720 \text{ kN (Compression)}$$

Alternative Method:
If virtual turning '$\delta\theta$' is consider for force system about joint 'C' then using principle of virtual work,
$$\Sigma M_C \times \delta\theta = 0$$

i.e., (Moment value of force 360) $\delta\theta$ + (Moment value of force F_{AD}) $\delta\theta$ = 0
$$(360 \times 4) \delta\theta + (F_{AD} \times 2) \delta\theta = 0$$
$$F_{AD} = -720 \text{ kN (Compression)}$$

Example: 17.7

Determine force P in Fig. 17.10 due to which 40 kN tension is induced in the member EC. Solve by using virtual work. Also determine force in member BC.

Solution:

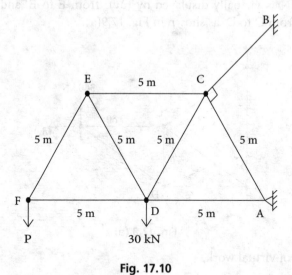

Fig. 17.10

Given F_{EC} = + 40 kN

Let the members FD and EC are turned about joint 'D' by virtual angle $\delta\theta$ as shown in Fig. 17.10 (a).

Virtual Work

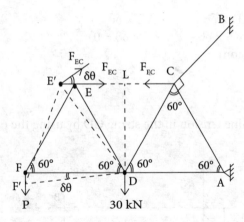

Fig. 17.10 (a)

Using principle of virtual work,
$\Sigma M_D \times \delta\theta = 0$
$(P \times FD \times \delta\theta) - (F_{EC} \times LD \times \delta\theta) = 0$
$(P \times 5 \times \delta\theta) - 40 \times 5 \sin 60° \times \delta\theta) = 0$
$P = 34.64$ kN

Note:
Here the moment of 'P' and the virtual angle '$\delta\theta$' both have anticlockwise turning, thus taken as positive; however, the moment of F_{EC} and the virtual angle both are in opposite direction, thus taken as negative.

To determine force in member BC, let the truss is virtually turned about A as shown in Fig. 17.10(b).

Using the principle of virtual work,
$\Sigma M_A \times \delta\theta = 0$

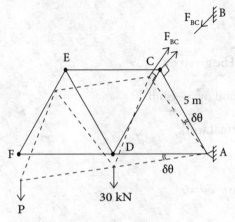

Fig. 17.10 (b)

P × AF × δθ + 30 × AD × δθ − F_{BC} × AC × δθ = 0
34.64 × 10 × δθ + 30 × 5 × δθ − F_{BC} × 5 × δθ = 0
F_{BC} = 99.28 kN, (Tension)

Example: 17.8

In example 2.28, determine tension in the string CD by using the principle of virtual work.

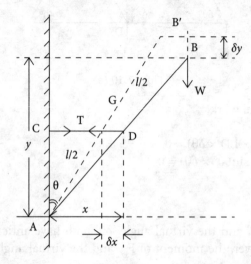

Fig. 17.11

Solution:

Let the tension produces virtual distance δx as shown in Fig. 17.11 by shifting rod AB to AB′.

$x = \dfrac{l}{2} \cdot \sin\theta$ and $y = l\cos\theta$

Thus virtual distance will be given by,

$\delta x = \dfrac{l}{2} \cdot \cos\theta \cdot \delta\theta$ and $\delta y = -l\sin\theta \cdot \delta\theta$

Using the principle of virtual work,

$T(+\delta x) + W(-\delta y) = 0$

$T\left(\dfrac{l}{2}\cos\theta\delta\theta\right) - W \cdot l \sin\theta \cdot \delta\theta = 0$

$T = 2W \tan\theta$

Virtual Work

Example: 17.9

In example 3.17, determine force P by using the principle of virtual work.

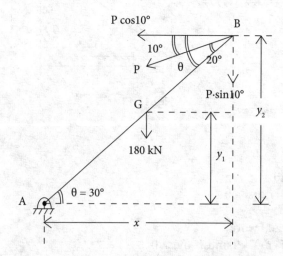

Fig. 17.12

Solution:

Let horizontal component of force P turn the rod AB by virtual angle $\delta\theta$ about hinge A.

AB = 10 m
$x = 10\cos\theta$, $\quad \delta x = -10.\sin\theta.\delta\theta$
$y_1 = 5.\sin\theta$, $\quad \delta y_1 = 5.\cos\theta.\delta\theta$
$y_2 = 10.\sin\theta$, $\quad \delta y_2 = 10.\cos\theta.\delta\theta$

Using the principle of virtual work,
$P\cos10°(+\delta x) + 180(-\delta y_1) + P\sin10°(-\delta y_2) = 0$
$P\cos10°(10.\sin\theta.\delta\theta) - 180(5\cos\theta\delta\theta) - P\sin10°(10.\cos\theta.\delta\theta) = 0$
$10P(\sin\theta\cos10° - \cos\theta.\sin10°) = 180 \times 5\cos\theta$
$10P \times \sin(\theta - 10°) = 180 \times 5\cos\theta$

Substitute $\theta = 30°$

$$P = \frac{180 \times 5\cos30°}{\sin(30° - 10°)}$$
$= 227.89\ kN$

Example: 17.10

In example 2.29, determine tension in the string BD by using the principle of virtual work.

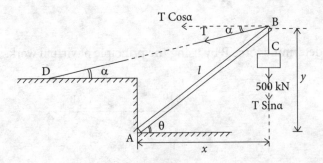

Fig. 17.13

Solution:

$\theta = 45°, \alpha = 30°$
$x = l.\cos\theta, y = l.\sin\theta$
$\delta x = -l.\sin\theta\delta\theta, \delta y = l.\cos\theta.\delta\theta$

Using principle of virtual work,
$T\cos\alpha.(+\delta x) + 500(-\delta y) + T\sin\alpha.(-\delta y) = 0$
$T\cos\alpha.(l\sin\theta.\delta\theta) + 500(-l\cos\theta.\delta\theta) + T\sin\alpha.(-l\cos\theta\delta\theta) = 0$
$T\cos\alpha \sin\theta - 500\cos\theta - T\sin\alpha \cos\theta = 0$
$T[\cos\alpha\sin\theta - \sin\alpha\cos\theta] = 500.\cos\theta$
$T.\sin(\theta - \alpha) = 500.\cos\theta$
$T.\sin(45° - 30°) = 500.\cos 45°$
$T = 1366.03$ kN

Example: 17.11

A uniform ladder of weight 'W' and length 'l' is placed on a rough plane and against a smooth wall. If the ladder impends to slip when it makes an angle of 'θ' with the plane, determine the maximum frictional force between the ladder and the floor.

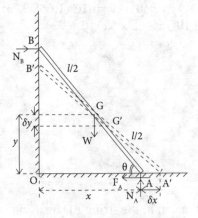

Fig. 17.14

Virtual Work

Solution:

First the ladder is placed at the angle 'θ' with the rough plane. When ladder tends to slip, the frictional force F_A acts towards wall side. Consider force system acting on ladder as shown in Fig. 17.14. The forces acting at points A and G are taken at distance x and y, respectively, from point 'O' for virtual distances.

If the ladder is slipped by virtual distances δx and δy,

Using principle of virtual work,
$$-F_A . \delta x + W . \delta y = 0 \qquad \ldots (1)$$

Here virtual work by N_A and N_B reactions will be zero as virtual distances for these reactions are zero.

Here, $x = l \cos\theta$ i.e. $\delta x = -l \sin\theta . \delta\theta$

and $y = \dfrac{l}{2} \sin\theta$ i.e. $\delta y = \dfrac{l}{2} \cos\theta . \delta\theta$

Substitute the virtual distance in equation (1),

$$-F_A \times l \sin\theta \delta\theta + W \times \frac{l}{2} \cos\theta . \delta\theta = 0$$

$$F_A = \frac{W}{2\tan\theta}$$

Example: 17.12

A ladder of length 'l' rests against a smooth wall, with the angle of inclination of 45°. If the coefficient of friction between ladder and the floor is 0.5, determine the maximum distance along the ladder upto which a person whose weight is 1.5 times the weight of the ladder may ascend before the ladder begins to slip.

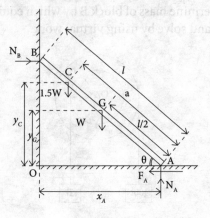

Fig. 17.15

Solution:

First draw Fig. 17.15 where $\theta = 45°$ and $\mu = 0.5$

Using the principle of virtual work,
$$-F_A \times (\delta x)_A + W.(\delta y)_G + 1.5W(\delta y)_c = 0 \quad \ldots (1)$$
where, $x_A = l\cos\theta$ i.e., $(\delta x)_A = -l\sin\theta.\delta\theta$

$y_c = a.\sin\theta$ i.e. $(\delta y)_c = a\cos\theta.\delta\theta$

$y_G = \dfrac{l}{2}.\sin\theta$ i.e. $(\delta y)_G = \dfrac{l}{2}\cos\theta.\delta\theta$

Substituting virtual distance in equation (1),

$$-\mu N_A (l\sin\theta\delta\theta) + W\left(\dfrac{l}{2}\cos\theta\delta\theta\right) + 1.5W(a\cos\theta\delta\theta) = 0$$

$$-0.5 N_A l + \dfrac{Wl}{2} + 1.5W.a = 0 \quad \ldots (2)$$

Using the equilibrium condition,
$\Sigma Y = 0$,
$N_A = 1.5W + W$
$\quad = 2.5W$

From equation (2),

$$-0.5 \times 2.5W \cdot l + \dfrac{W}{2} \cdot l + 1.5W.a = 0$$

$$a = 0.5l$$

i.e., man can go along the ladder upto half of its length.

Example: 17.13

A compound pulley of radius $r_1 = 300$ mm and $r_2 = 200$ mm is used to support 400 N block as shown in Fig. 17.16. Determine mass of block B by which equilibrium can be maintained. Take pulley as frictionless and solve by using virtual work.

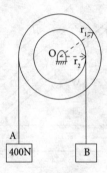

Fig. 17.16

Virtual Work

Solution:

Let pulley rotates by virtual angle $\delta\theta$ in anti–clockwise direction then the blocks A and B will virtually displaced by δA and δB, respectively. If the mass of block B is m

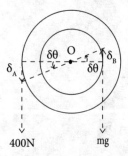

400N   mg

Fig. 17.16 (a)

$$\delta\theta = \frac{\delta_A}{r_1} = \frac{\delta_B}{r_2}$$

$$\delta\theta = \frac{\delta_A}{300} = \frac{\delta_B}{200}$$

Thus, $\delta A = 300.\delta\theta$ and $\delta B = 200.\delta\theta$
Using the principle of virtual work,
  $400 \times \delta A - m \times g \times \delta B = 0$
  $400 \times 300\delta\theta - m \times 9.81 \times 200.\delta\theta = 0$
  $m = 61.16$ kg

Example: 17.14

Three identical frictionless pulleys are supporting blocks as shown in Fig. 17.17. Using the principle of virtual work, determine the mass 'm' which can hold the system in equilibrium. Assume the mass of pulley as negligible.

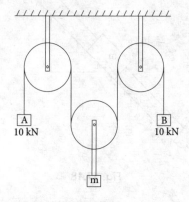

Fig. 17.17

Solution:

Let the blocks 'A' and 'B' moves up by virtual distance δy_A and δy_B, respectively. Thus the mass 'm' will move down by virtual distance,

$$\delta = \left(\frac{\delta y_A + \delta y_B}{2} \right) \quad \quad \ldots (1)$$

As pulleys are identical,

$\delta y_A = \delta y_B = \delta y$

From equation (1), $\delta = \left(\frac{\delta y + \delta y}{2} \right) = \delta y$

Using principle of virtual work,
 m.g × δ – 10 × 10³ × δy – 10 × 10³ × δy = 0
 m × 9.81 × δy – 10⁴ × 2 × δy = 0
 m = 2038.74 kg

Example: 17.15

Four masses are attached by inextensible string as shown in Fig. 17.18. Using the principle of virtual work, determine the minimum mass m_4 to impend downward motion. Consider all pulleys as ideal pulleys and coefficient of friction between surface and masses as 0.22.

Solution:

Consider the force system acting on masses as shown in Fig. 17.18. (a)

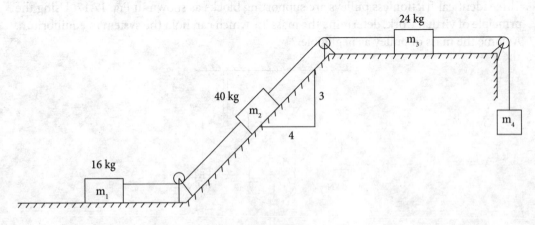

Fig. 17.18

Virtual Work

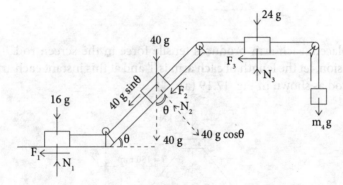

Fig. 17.18 (a)

Let mass m_4 moves downward by virtual distance 'δ', then all masses m_1, m_2 and m_3 will move same virtual distance 'δ' towards right side.

Using the principle of virtual work,
$$m_4 g \cdot \delta - F_1 \times \delta - F_2 \times \delta - 40 g \sin\theta \times \delta - F_3 \times \delta = 0$$
$$\delta[m_4 g - \mu_1 N_1 - \mu_2 N_2 - 40 g \sin\theta - \mu_3 N_3] = 0 \qquad \ldots\ldots (1)$$
where $\mu_1 = \mu_2 = \mu_3 = 0.22$
$N_1 = 16 g$, $N_2 = 40 g \cos\theta$, $N_3 = 24 g$

$\tan\theta = \dfrac{3}{4}$, i.e., $\theta = 36.37°$

Substituting the values in equation (1),
$m_4 \times g - 0.22 \times 16 g - 0.22 \times 40 g \cos 36.87° - 40 g \sin 36.87° - 0.22 \times 24 \times g = 0$
$m_4 = 39.84$ kg

Example: 17.16

A person used a scissor screw jack to change the punctured wheel of his car. Determine the force induced in the screw rod by using virtual work when a load of 3.2 kN has raised to the position as shown in Fig. 17.19. Take weight of the arms as negligible and the length of each arm as 180 mm.

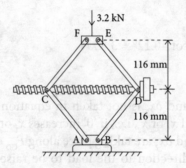

Fig. 17.19

Solution:

When load is placed, each arm produces tensile force in the screen rod. Thus rod CD is subjected to tension, let the length of each arm is 'l' and at this instant each arm makes angle 'θ' from screw rod as shown in Fig. 17.19 (a).

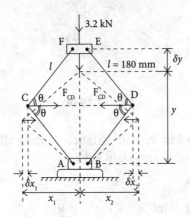

Fig. 17.19 (a)

Let the load is raised by virtual height by δy as shown in Fig. 17.19(a).
$x_1 = x_2 = l.\cos\theta$, i.e., $\delta x_1 = \delta x_2 = -l\sin\theta\delta\theta$
$y = 2l\sin\theta$ i.e., $\delta y = 2l\cos\theta.\delta\theta$

For given position, using principle of virtual work,
$F_{CD} \times \delta x_1 + F_{CD} \times \delta x_2 - 3.2 \times \delta y = 0$
$2F_{CD}.\delta x_1 - 3.2.\delta y = 0$ ..... (1)

Substituting the values in equation (1),
$2F_{CD} \times (l\sin\theta\delta\theta) - 3.2 (2l\cos\theta.\delta\theta) = 0$
$F_{CD} = 3.2 \cot\theta$ ..... (2)
where,
$$\sin\theta = \frac{116}{180}$$
$\theta = 40.12°$

From equation (2) $F_{CD} = 3.2 \cot 40.12° = 3.8$ kN

Note:

(i) The negative sign δx_1 and δx_2 is not taken in equation (1) because this sign shows relation between 'θ' and x_1 or x_2, i.e., if 'θ' increases x_1 or x_2 decreases.
(ii) Virtual distances δx_1 and δx_2 are taking place along F_{CD} direction, thus taken positive but δy is in opposite direction to the load to be raised, thus taken as negative in equation (1).

Example: 17.17

In a smith's shop, a worker uses a fireplace tong to hold a hot metal piece. Determine force 'P' by which its jaws can apply 500 N force on metal piece.

Take, HF = 200 mm, FE = 150 mm, EC = 150 mm and BC = 100 mm.

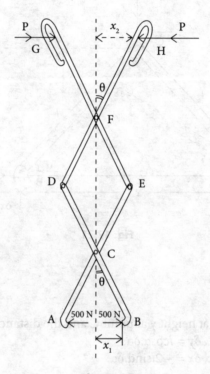

Fig. 17.20

Solution:

When the worker exerts force 'P' on the tong, its jaws at A and B exert force towards centreline. However metal piece exerts same force on jaws as shown in Fig. 17.20. Let the horizontal distance of B and H are x_1 and x_2, respectively from the vertical line.

$x_1 = BC.\sin\theta = 100.\sin\theta$,   i.e., $\delta x_1 = 100.\cos\theta.\delta\theta$
and $x_2 = FH.\sin\theta = 200.\sin\theta$,   i.e., $\delta x_2 = 200.\cos\theta.\delta\theta$

Using the principle of virtual work on one member of tong,
$-500 \times \delta x_1 + P \times \delta x_2 = 0$
$-500 \times 100.\cos\theta.\delta\theta + P \times 200.\cos\theta.\delta\theta = 0$
P = 250 N

Example: 17.18

Two weightless links AC and BC of equal length l are supporting weight 'W' as shown in Fig. 17.21. Determine force P to be applied at roller bearing so that the system may be held in equilibrium.

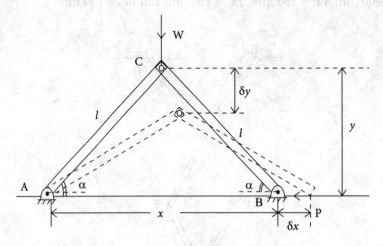

Fig. 17.21

Solution:

Let the weight 'W is acting at height y at point C and the distance between A and B is x.
Thus, $y = l\sin\alpha$ i.e., $\delta y = l\cos\alpha.\delta\alpha$
$x = 2l\cos\alpha$ i.e., $\delta x = -2l\sin\alpha.\delta\alpha$

Using the principle of virtual work,
$$P.(-\delta x) + W.\delta y = 0$$
$$-P.(2l\sin\alpha.\delta\alpha) + W(l\cos\alpha.\delta\alpha) = 0$$
$$P = \frac{W}{2}\cdot\cot\alpha$$

Example: 17.19

Two links of equal length 'l' are hinged and arranged vertically as shown in Fig. 17.22. They are connected at their lower ends by un-stretched spring of length 's'. When a vertical force 'p' is applied, determine the spring constant 'k' to maintain equilibrium at the position shown.

[MTU 2013–14]

Virtual Work

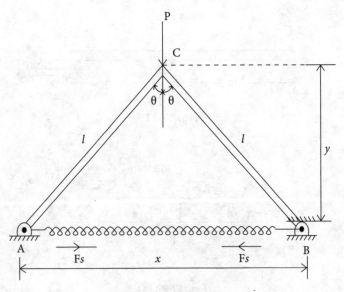

Fig. 17.22

Solution:

Initial length of spring = s
Let the spring is stretched when both links make angle θ from vertical axis.
Final length of spring after stretching will be given by = $2l\sin\theta$
Stretched amount along AB of spring = $(2l\sin\theta - s)$
i.e., $x = (2l\sin\theta - s)$
And $y = l\cos\theta$
If force 'p' moves down by virtual distance 'δ_y' and force in spring is F_s,
$\delta x = 2l\cos\theta \cdot \delta\theta$ and $\delta y = -l\sin\theta \cdot \delta\theta$

Using principle of virtual work,
$P(+\delta y) + F_s(-\delta x) = 0$
$P(l\sin\theta \cdot \delta\theta) + k(2l\sin\theta - s)(2l\cos\theta \delta\theta) = 0$
$P\sin\theta = k(2l\sin\theta - s) \times 2\cos\theta$

$$k = \frac{P.\tan\theta}{2(2l\sin\theta - s)}$$

Example: 17.20

A simple supported beam of span 10 m is carrying a point load at its mid points as shown in Fig. 17.23. Using virtual work, determine reaction of hinge and roller bearing.

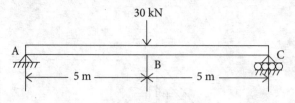

Fig. 17.23

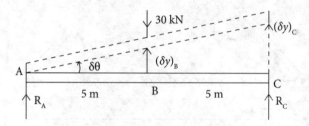

Fig. 17.23 (a)

Solution:

Let the beam is kept fixed about A and virtually turned by angle '$\delta\theta$' which causes virtual displacement $(\delta y)_C$ for reaction R_C and $(\delta y)_B$ for point load 30 kN at mid-point as shown in Fig. 17.23(a).
$\Sigma Y = 0$,
$R_A + R_C = 30$ kN ..... (1)

Using principle of virtual work,
$R_C.(+\delta y)_C + 30.(-\delta y)_B = 0$
i.e., $R_C.(+\delta y)_C - 30.(\delta y)_B = 0$ ..... (2)
$(\delta y)_B$ is taken as negative as it is opposite to the direction of 30 kN.

From right angle triangles,

$$\tan \delta\theta = \delta\theta = \frac{(\delta y)_C}{10} = \frac{(\delta y)_B}{5}$$

$(\delta y)_C = 10.\delta\theta$ and $(\delta y)_B = 5.\delta\theta$

Substituting the values in equation (2),
$R_C.(10\delta\theta) - 30(5.\delta\theta) = 0$
$R_C = 15$ kN

Using equation (1),
$R_A = 15$ kN

Virtual Work

Example: 17.21

Using virtual work, determine support reactions for simple supported beam as shown in Fig. 17.24.

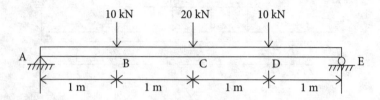

Fig. 17.24

Solution:

Let the beam is turned by virtual angle, $\delta\theta$ about A so that the virtual distance at B, C, D and E are $(\delta y)_B$, $(\delta y)_C$, $(\delta y)_D$ and $(\delta y)_E$, respectively, as shown in Fig. 17.24 (a).

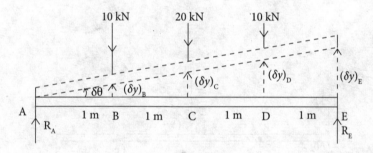

Fig. 17.24 (a)

Using the principle of virtual work,
$R_E \cdot (+\delta y)_E + 10 \cdot (-\delta y)_D + 20 \cdot (-\delta y)_C + 10(-\delta y)_B = 0$
$R_E \cdot (\delta y)_E - 10 \cdot (\delta y)_D - 20 \cdot (\delta y)_C - 10(\delta y)_B = 0$ ..... (1)

From right angle triangles,

$$\tan\delta\theta = \delta\theta = \frac{(\delta y)_E}{4} = \frac{(\delta y)_D}{3} = \frac{(\delta y)_C}{2} = \frac{(\delta y)_B}{1}$$

$(\delta y)_E = 4.\delta\theta$, $(\delta y)_D = 3.\delta\theta$, $(\delta y)_C = 2.\delta\theta$, $(\delta y)_B = \delta\theta$

Substituting virtual distances in equation (1),
$R_E (4.\delta\theta) - 10(3.\delta\theta) - 20(2.\delta\theta) - 10(\delta\theta) = 0$
$R_E = 20$ kN
Since, $R_A + R_E = 40$ kN
thus $R_A = 40 - 20 = 20$ kN

Example: 17.22

Determine support reactions for overhanging beam as shown in Fig. 17.25.

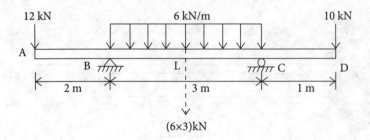

Fig. 17.25

Solution:

Let the beam is turned about hinge B by virtual angle $\delta\theta$ in anti–clockwise direction as shown in Fig. 17.25 (a). Weight of UDL will be 18 kN which will act at its mid length, i.e., at point L.

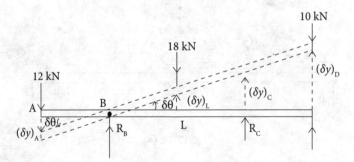

Fig. 17.25 (a)

Using the principle of virtual work,
$$R_C \cdot (+\delta y)_C + 10 \cdot (-\delta y)_D + 18 \cdot (-\delta y)_L + 12 \cdot (+\delta y)_A = 0 \qquad \ldots (1)$$

From right angle triangles,
$$\delta\theta = \frac{(\delta y)_D}{4} = \frac{(\delta y)_C}{3} = \frac{(\delta y)_L}{1.5} = \frac{(\delta y)_A}{2}$$
$$(\delta y)_D = 4.\delta\theta,\ (\delta y)_C = 3.\delta\theta,\ (\delta y)_L = 1.5\delta\theta,\ (\delta y)_A = 2\delta\theta$$

Substituting the values in equation (1),
$R_C(+3.\delta\theta) - 10(4.\delta\theta) - 18(1.5\delta\theta) - 12(2.\delta\theta) = 0$
$R_C = 14.33$ kN
Since, $R_B + R_C = 12 + 18 + 10 = 40$
$R_B = 40 - 14.33 = 25.66$ kN

Virtual Work

Example: 17.23

A simple supported beam supports a UDL and a moment as shown in Fig. 17.26. Using the principle of virtual work, determine support reactions.

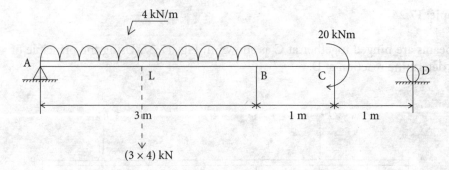

Fig. 17.26

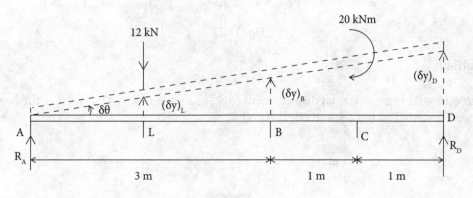

Fig. 17.26 (a)

Solution:

The beam is turned for virtual angle $\delta\theta$ about A as shown in Fig. 17.26 (a). The weight of UDL is 12 kN and acts at mid-point 'L'. The moment 20 kNm is acting clockwise thus virtual angle for this will be $(-\delta\theta)$ as beam is turned in anticlockwise direction.

Using principle of virtual work,
$$R_D \cdot (+\delta y)_D + 20 \cdot (-\delta\theta) + 0 \cdot (\delta y)_B + 12(-\delta y)_L = 0$$
$$R_D \cdot (\delta y)_D - 20 \cdot \delta\theta - 12 \delta y_L \qquad \ldots (1)$$

From right angle triangles,

$$\delta\theta = \frac{(\delta y)_D}{5} = \frac{(\delta y)_B}{3} = \frac{(\delta y)_L}{1.5}$$
$$(\delta y)_D = 5 \cdot \delta\theta, \ (\delta y)_B = 3 \cdot \delta\theta, \ (\delta y)_L = 1.5 \delta\theta$$

Substituting virtual distances in equation (1),
$R_D (5.\delta\theta) - 20\delta\theta - 12.(1.5\delta\theta) = 0$
$R_D = 7.6$ kN
Since, $R_A + R_D = 12$ kN, $R_A = 12 - 7.6 = 4.4$ kN

Example: 17.24

Two beams are hinged together at C as shown in Fig. 17.27. Using the principle of virtual work, determine reaction at D.

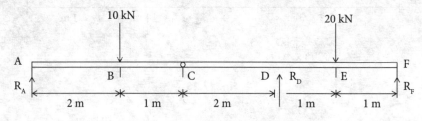

Fig. 17.27

Solution:

Let reaction R_D is given virtual displacement $(\delta y)_D$ as shown in Fig. 17.27(a). Let the beams turned for virtual angles $\delta\theta_1$ and $\delta\theta_2$ at F and A, respectively.

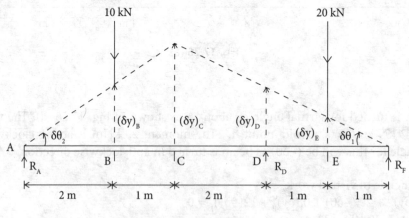

Fig. 17.27 (a)

Using principles of virtual work,

$$\left[R_D.(+\delta y)_D\right] + \left[20.(-\delta y)_E\right] + \left[10.(-\delta y)_B\right] = 0 \quad \ldots\ldots (1)$$

Virtual Work

From right-angled triangles of beam CDF,

$$\delta\theta_1 = \frac{(\delta y)_C}{4} = \frac{(\delta y)_D}{2} = \frac{(\delta y)_E}{1}$$

i.e. $(\delta y)_C = 4.\delta\theta_1$, $(\delta y)_D = 2.\delta\theta_1$, $(\delta y)_E = \delta\theta_1$

From right-angled triangles of beam AC,

$$\delta\theta_2 = \frac{(\delta y)_C}{3} = \frac{(\delta y)_B}{2}$$

$(\delta y)_C = 3.\delta\theta_2$, $(\delta y)_B = 2.\delta\theta_2$

since $(\delta y)_C$ is common virtual distance for both beams,
thus, $(\delta y)_C = 3.\delta\theta_2 = 4.\delta\theta_1$

i.e. $\delta\theta_1 = \frac{3}{4} \cdot \delta\theta_2$

$(\delta y)_B = 2.\delta\theta_2$, $(\delta y)_D = 2.\delta\theta_1 = 2\left(\frac{3}{4} \cdot \delta\theta_2\right)$

$$= \frac{3}{2} \cdot \delta\theta_2$$

and $(\delta y)_E = \delta\theta_1 = \frac{3}{4} \cdot \delta\theta_2$

Substituting the values of virtual distances in equation (1),

$$R_D\left(\frac{3}{2} \cdot \delta\theta_2\right) - 20\left(\frac{3}{4} \cdot \delta\theta_2\right) - 10(2\delta\theta_2) = 0$$

$$R_D = 23.33 \ kN$$

Example: 17.25

An overhanging beam is supporting point load, UDL and moment as shown in Fig. 17.28. Using the principle of virtual work, determine reaction of roller bearing.

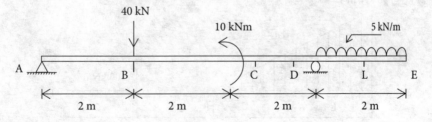

Fig. 17.28

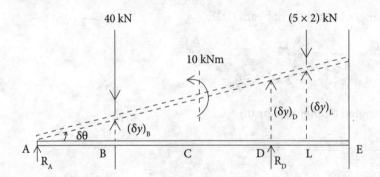

Fig. 17.28 (a)

Solution:

Using the principles of virtual work,

$$[R_D(+\delta y)_D] + [(5 \times 2)(-\delta y)_L] + 10.\delta\theta + [40(-\delta y)_B] = 0$$
$$R_D \cdot (\delta y)_D - 10 \cdot (\delta y)_L + 10.\delta\theta - 40(\delta y)_B = 0 \quad\quad\quad \text{..... (1)}$$

From the right–angled triangles,

$$\delta\theta = \frac{(\delta y)_L}{7} = \frac{(\delta y)_D}{6} = \frac{(\delta y)_B}{2}$$
$$(\delta y)_L = 7.\delta\theta, \ (\delta y)_D = 6.\delta\theta \ \text{and} \ (\delta y)_B = 2.\delta\theta$$

Substituting the values of virtual distances in equation (1),
$R_D (6.\delta\theta) - 10(7.\delta\theta) + 10.\delta\theta - 40(2.\delta\theta) = 0$
$R_D = 23.33$ kN

Example: 17.26

Determine reaction of the hinge 'A' by using the principle of virtual work for simple supported beam as shown in Fig. 17.29.

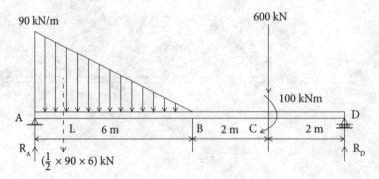

Fig. 17.29

Virtual Work

Solution:

The load of UVL (Uniform Varying Load) will be,

$$= \frac{1}{2} \times 6 \times 90$$
$$= 270 \ kN$$

This will act at point 'L'. Thus $AL = \frac{1}{3} \times 6 = 2m$

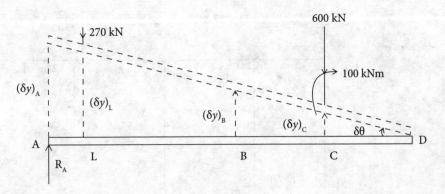

Fig. 17.29 (a)

Using the principle of virtual work,
$R_A \cdot (+\delta y)_A + 270 \cdot (-\delta y)_L + 0 \cdot (\delta y)_B + 600(-\delta y)_C + 100 \cdot \delta\theta = 0$
$R_A \cdot (\delta y)_A - 270 \cdot (\delta y)_L - 600(\delta y)_C + 100 \cdot \delta\theta = 0$ ..... (1)

From the right–angled triangles,

$$\delta\theta = \frac{(\delta y)_A}{10} = \frac{(\delta y)_L}{8} = \frac{(\delta y)_C}{2}$$

$(\delta y)_A = 10 \cdot \delta\theta, \ (\delta y)_L = 8 \cdot \delta\theta, \ (\delta y)_C = 2 \cdot \delta\theta$

Substituting the values in equation (1),
$R_A (10 \cdot \delta\theta) - 270(8 \cdot \delta\theta) - 600(2 \cdot \delta\theta) + 100 \cdot \delta\theta = 0$
$R_A = 326 \ kN$

Example: 17.27

Using virtual work, determine reaction of roller bearing for given overhanging beam as shown in Fig. 17.30.

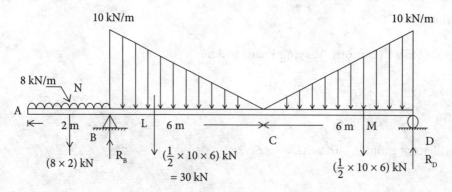

Fig. 17.30

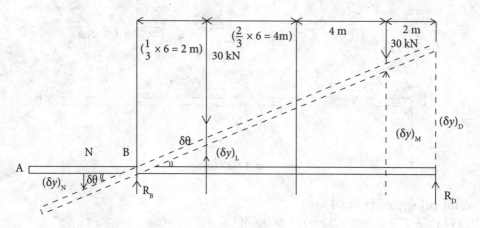

Fig. 17.30 (a)

Using the principle of virtual work,

$$\left[R_D(+\delta y)_D\right]+\left[30(-\delta y)_M\right]+\left[30(-\delta y)_L\right]+\left[(8\times 2)(+\delta y)_N\right]=0$$
$$R_D\cdot(\delta y)_D-30(\delta y)_M-30(\delta y)_L+16(\delta y)_N=0 \quad\ldots\ldots (1)$$

From the right–angled triangles,

$$\delta\theta=\frac{(\delta y)_N}{1}=\frac{(\delta y)_L}{2}=\frac{(\delta y)_M}{10}=\frac{(\delta y)_D}{12}$$

from equation (1),
$R_D\cdot(12.\delta\theta)-30(10.\delta\theta)-30(2.\delta\theta)+16.(\delta\theta)=0$
$R_D = 28.67$ kN

Virtual Work

Example: 17.28

Determine reactions of both roller–bearing for given compound beam arrangement as shown in Fig. 17.31 by using the principles of virtual work.

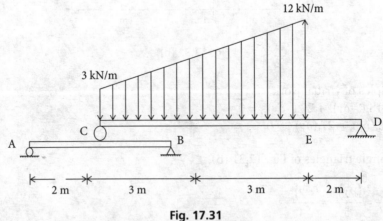

Fig. 17.31

Solution:

Consider free body diagram of beam CD as shown in Fig. 17.31(a). The given load is a combination of UDL and UVL intensities of 3 kN/m and 9 kN/m, respectively.

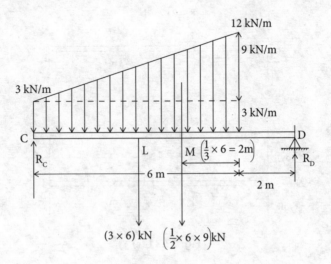

Fig. 17.31 (a)

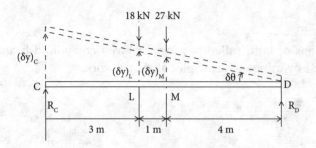

Fig. 17.31 (b)

Using principles of virtual work,
$R_C.(+\delta y)_C + 18.(-\delta y)_L + 27.(-\delta y)_M = 0$
$R_C.(\delta y)_C - 18.(\delta y)_L - 27.(\delta y)_M = 0$ ..... (1)

From right angle triangles of Fig. 17.31 (b),

$$\delta\theta = \frac{(\delta y)_C}{8} = \frac{(\delta y)_L}{5} = \frac{(\delta y)_M}{4}$$

Substituting the virtual distance in equation (1),
$R_c.(8.\delta\theta) - 18(5.\delta\theta) - 27(4.\delta\theta) = 0$
$R_c = 24.75$ kN

Consider free body diagram of beam AB as shown in Fig. 17.31 (c)

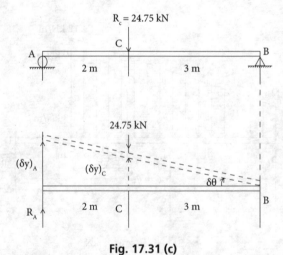

Fig. 17.31 (c)

Using the principle of virtual work,
$R_A.(+\delta y)_A + 24.75(-\delta y)_C = 0$
$R_A.(\delta y)_A - 24.75.(\delta y)_C = 0$ ..... (2)

Virtual Work

From right-angled triangles,

$$\delta\theta = \frac{(\delta y)_A}{5} = \frac{(\delta y)_C}{3}$$

$(\delta y)_A = 5.\delta\theta, \ (\delta y)_C = 3.\delta\theta$

Substituting the virtual distance in equation (2),
$R_A.(5.\delta\theta) - 24.75(3.\delta\theta) = 0$
$R_A = 14.85$ kN

Example: 17.29

Using the principle of virtual work, determine reaction of roller bearing of an overhanging beam as shown in Fig. 17.32.

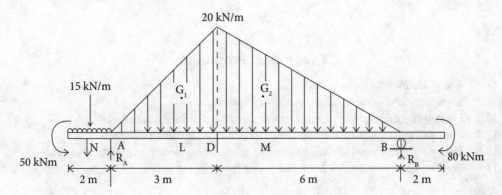

Fig. 17.32

Solution:

The weight and their position of C.G. are shown in Fig. 17.32 (a).

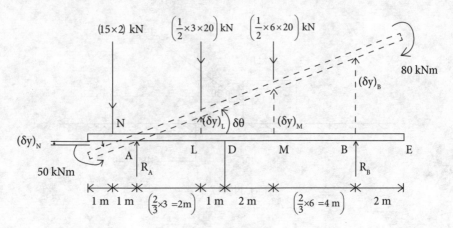

Fig. 17.32 (a)

Using principle of virtual work,

$$R_B(+\delta y)_B + \left[\left(\frac{1}{2} \times 6 \times 20\right)(-\delta y)_M\right] + \left[\left(\frac{1}{2} \times 3 \times 20\right)(-\delta y)_L\right]$$
$$+ \left[(15 \times 2)(+\delta y)_N\right] - 80.\delta\theta + 50.\delta\theta = 0$$
$$R_B.(\delta y)_B - 60.(\delta y)_M - 30.(\delta y)_L + 30.(\delta y)_N - 30.\delta\theta = 0$$

From right angle triangles,

$$\delta\theta = \frac{(\delta y)_B}{9} = \frac{(\delta y)_M}{5} = \frac{(\delta y)_L}{2} = \frac{(\delta y)_N}{1}$$
$$(\delta y)_B = 9.\delta\theta, \ (\delta y)_M = 5.\delta\theta, \ (\delta y)_L = 2.\delta\theta, \ (\delta y)_N = \delta\theta$$
$$R_B(9.\delta\theta) - 60(5.\delta\theta) - 30(2.\delta\theta) + 30(\delta\theta) - 30.\delta\theta = 0$$
$$R_B = 40 \ kN$$

Theoretical Problems

T 17.1 Briefly define the term virtual work.
T 17.2 State the difference between work and virtual work.
T 17.3 Explain the principle of virtual work with some suitable examples.
T 17.4 Briefly discuss where the principle of virtual work is most preferred.
T 17.5 Discuss the forces which are neglected while using principle of virtual work.
T 17.6 Briefly explain the virtual work produced by moment acting on the body.

Numerical Problems

N 17.1 A pull force 'P' is applied on a block 'A' at an angle 30° as shown in Fig. NP 17.1. Determine the maximum value of force 'P' by which block 'A' tends to move right side. Take coefficient of friction between contact surfaces as 0.26 and weight of block as 24 kN.

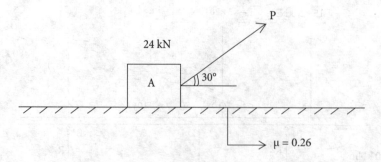

Fig. NP 17.1

Virtual Work

N 17.2 Two weights A and B are connected by a string passing over an ideal pulley as shown in Fig. NP 17.2. If the weights A and B are 60 kN and 400 kN, respectively, then determine the pull to the applied at A tangential to the incline in such a way that it causes A to impend downward. Take the coefficient of friction between all contact surfaces as 0.20.

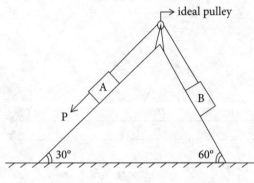

Fig. NP 17.2

N 17.3 A ladder of weight 250 N is placed between smooth wall and smooth surface. Determine the force to keep the ladder in equilibrium by using principle of virtual work. Take angle of inclination of ladder with wall as 45°.

N 17.4 Two identical smooth pulleys are used to support weight 'W' by an effort 600 N as shown in Fig. NP 17.3. Determine the maximum weight W can be maintained in equilibrium by 600 N effort. Assume weight of pulleys as negligible.

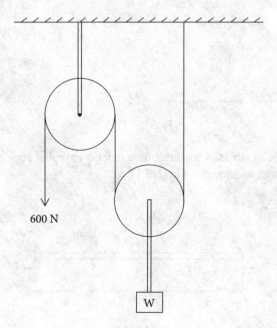

Fig. NP 17.3

N 17.5 Determine force in the member BD by using principle of virtual work for truss as shown in Fig. NP 17.4.

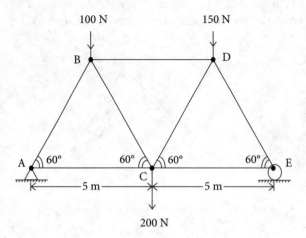

Fig. NP 17.4

N 17.6 Two weights W_A = 200 N and W_B = 300 N are kept in equilibrium by weight W suspended as shown in Fig. NP 17. 5. If coefficient of friction for all surfaces is 0.26, determine weight 'W' by which blocks A and B impend right side. Consider pulley as frictionless.

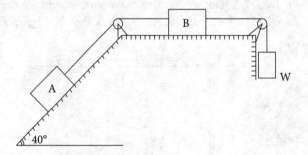

Fig. NP 17.5

N 17.7 Using principle of virtual work determine reaction of roller support bearing of a simple supported beam as shown in Fig. NP 17.6.

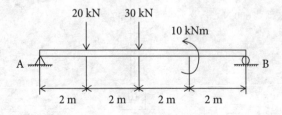

Fig. NP 17.6

N 17.8 Determine reaction of supports of overhanging beam as shown in Fig. NP 17.7.

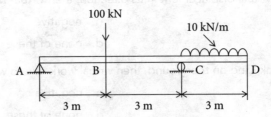

Fig. NP 17.7

N 17.9 Using the principle of virtual work, determine reaction of roller support bearing for simple supported beam as shown in Fig. NP 17.8.

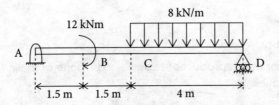

Fig. NP 17.8

N 17.10 Using the principle of virtual work, determine reaction of hinge for overhanging beam as shown in Fig. NP 17.9.

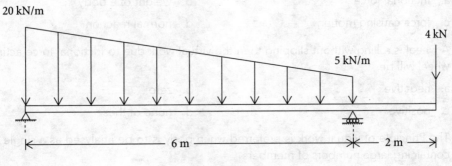

Fig. NP 17.9

Multiple Choice Questions

1. If a force causes virtual displacement in its direction, then virtual work will be
   a. zero
   b. negative
   c. positive
   d. none of these

2. A boy pulls a trunk lying on the ground, then virtual work due to weight of trunk will be
   a. zero
   b. negative
   c. positive
   d. none of these

3. Virtual work due to a force is given by the multiplication of force and
   a. virtual displacement along the line of action of force
   b. virtual displacement perpendicular to the line of action of force
   c. displacement along the line of action of force
   d. none of these

4. Virtual work is a
   a. vector quantity
   b. scalar quantity
   c. may be vector as it is product of two vector quantities
   d. none of these

5. A clockwise moment M is acting on a beam. If a force turns the beam by anticlockwise virtual angle $\delta\theta$ then the virtual work due to moment will be
   a. zero
   b. $-M.\delta\theta$
   c. $M.\delta\theta$
   d. none of these

6. Which force always produces negative virtual work
   a. frictional force
   b. weight of a body
   c. force causing motion
   d. normal reaction

7. A wheel is rolling without slipping then the virtual work due to frictional force acting on wheel will be
   a. negative
   b. zero
   c. positive
   d. none of these

8. The Principle of virtual work is preferred when body is to be analysed as a single body containing large numbers of members
   a. true
   b. false
   c. not applicable
   d. none of these

Answers

1. c   2. a   3. a   4. b   5. b   6. a   7. a   8. a

Index

A

Acceleration, 3, 4, 8, 324, 498–506, 508, 509, 511, 514–520, 522, 523, 529–540, 560–562, 567–570, 573–575, 577, 578, 580–584, 586, 591, 597, 598, 619, 621, 623–625, 661–673, 708, 709, 711, 712, 714–732
  Normal component, 540, 544, 545, 547, 548, 553, 555, 672
  Tangential component, 672
Angle of projection, 540, 544, 545, 547, 548, 553, 555
Angular Acceleration, 661–668, 670, 673, 708, 709, 711–722, 725–730
Angular Displacement, 661–664, 667, 668, 673, 675, 742, 744
angular momentum, 708, 709
Angular Motion, 664, 665, 668
Angular Velocity, 661–668, 672, 673, 676–678, 680–689, 692–694, 708, 710, 718, 731, 736, 737
Average velocity, 499, 510, 516, 517
Axis of Symmetry, 330, 331, 375, 376, 379, 430, 435

B

Band Brakes, 229
Beams, 103, 105, 143, 387, 446–449, 770, 771
  Types of Beams, 105, 446

Bending Moment, 446, 448–452, 455, 458, 460, 462, 464, 466, 468–470, 472, 474, 477, 478, 480, 482, 484, 486, 488, 490, 492
  Diagram, 42, 45, 49, 50, 53, 55, 79, 103, 116, 126, 129, 131, 152, 186, 188, 190, 191, 193, 195, 197, 199, 202, 203, 205, 217, 229, 230, 301, 449, 454, 455, 458–460, 462, 464, 466, 468–470, 472, 474, 475, 477–484, 486–488, 490, 492, 532, 562, 564, 571, 577, 579, 581, 586, 600–603, 605, 608, 609, 611, 620, 650, 654, 679, 721, 724, 732, 742, 775, 776

C

Center of Gravity, 101
  of Cone and Hemisphere, 375
  of Hemisphere, 379
Centroid, 101, 106, 323, 325–337, 339–343, 345, 348, 351–355, 357–362, 365, 366, 369, 370, 372–374, 387, 401, 407, 415, 423, 448, 472, 617
  of Line, 323, 668, 669
  Quarter Circle, 337, 342, 358, 360, 367, 372, 395, 397, 398, 399, 617
  Semicircle, 222, 337, 339, 355, 360, 361, 362, 420
  of Triangle, 10, 16, 100, 106, 273, 394, 448, 472
Centroid of Section, 330
  of Composite Sections and Bodies, 351, 400

Coefficient, 150, 156, 157, 162, 164, 166, 168, 169, 174, 175, 178, 185, 189, 194, 196, 204, 213, 215, 216, 219, 221–223, 225–227, 230, 231, 233, 236, 238, 563, 573, 577, 581, 609, 611, 634–638, 640, 641, 643, 644, 649–651, 653, 726, 727, 757, 760
   of Friction, 149–151, 153, 155, 156, 158–160, 162, 164, 166, 168, 169, 174, 175, 178, 185, 188, 189, 191, 194, 196, 201, 204, 211, 213, 216, 219, 221–223, 225–227, 231, 233, 236, 238, 568, 570, 573, 577, 581, 597, 602, 607, 609, 611, 649–651, 653, 722, 745, 746, 757, 760
   of kinetic friction, 230, 563, 726, 727
   of Restitution, 634–638, 640, 641, 643–645
Collisions, 633, 645
   of Elastic Bodies, 633
   Impact, 523, 631, 633, 634
Conservation, 591, 595, 612–616, 619, 632, 633, 635–637, 641–644, 646, 653, 655–657, 708, 710, 736–738
   of Energy, 591, 595, 612–616, 619, 655, 708, 710, 736–738
   of Momentum, 632, 633, 635–637, 641–644, 646, 653, 655–657, 708, 709
Continuum, 2
Contraflexure, 452, 488, 493
Cross, 17, 19, 20, 25, 26, 84, 117, 132, 446, 448, 513, 695–697,
   or Vector Product, 17, 19
Curvilinear Motion, 529, 531, 534, 537, 661

D

D'Alembert's Principle, 560–564, 566, 567, 569, 570, 572–574, 576, 578, 579, 582, 584, 586
Displacement, 4, 30, 498–505, 514, 515, 529, 560, 561, 591–593, 597, 661–664, 667, 668, 673, 675, 742–744, 750, 766, 770
Dot, 17, 18, 19, 24
   or Scalar Product, 17

E

Effort, 97, 196, 208, 209, 211, 229, 622–625
Efficiency, 211–216
   of screw jack, 208, 211, 213, 214

Energy, 3, 591, 593–605, 607, 608, 610, 612–616, 618, 619, 631, 655, 657, 658, 708–711, 731, 733–738
   kinetic energy, 593–596, 709–711, 731, 733
   potential energy, 593, 595, 596, 710
Equations of motion, 500, 501, 509, 521

F

Force, 1, 3–14, 20, 21, 22, 26, 27, 29, 30, 34–36, 38, 40, 41–47, 49, 50, 52, 53, 55–57, 59, 63, 64, 66, 70, 74–77, 82, 83, 85, 86–88, 96, 97, 99–102, 107, 109–120, 122–124, 128–130, 134, 149–165, 168–175, 179, 185, 186, 188, 189, 192–195, 197, 200, 201, 204, 208, 209, 211–214, 217, 218, 223, 229, 230, 233, 238, 248–251, 261, 283, 289, 296, 299, 301, 304, 306, 308, 310, 323, 388, 446, 448, 449, 450–452, 454, 458–460, 462, 464, 466–470, 472, 475–477, 479–484, 487, 490, 492, 560–564, 567–569, 572, 591–593, 597–602, 620, 623–625, 631–633, 646–652, 654, 661, 668, 669, 711–715, 721–723, 731, 733, 742–748, 750–753, 755–757, 760–765
   Collinear Coplanar Force, 5
   Concurrent Coplanar Force, 5, 35, 40, 45, 46, 209, 248, 249, 746
   Concurrent Non-Coplanar Force, 7
   Coplanar force, 4–6, 35, 40, 45, 46, 96, 102, 107, 111, 122, 123, 185, 209, 248, 301, 746
   Non-Coplanar Force, 4, 7, 8, 101
   Non-Parallel Non-Coplanar Force, 7
   Non-Parallel Coplanar Force, 6
   Parallel Coplanar Force, 6, 111, 122
   Parallel Non-Coplanar Force, 7, 8
   Parallelogram Law of force, 9, 16, 37, 39
   Polygon Law, 11, 12, 37
   Triangle Law, 10, 11, 16, 17, 37, 44
Free Body Diagram, 42, 49, 186, 188, 190, 191, 193, 195, 197, 199, 202, 203, 205, 217
Friction, 129, 149–160, 162, 164, 166, 168, 169, 173–175, 178, 185, 188, 189, 191, 194, 196, 201, 204, 207, 208, 211–213, 215, 216, 219, 221–223, 225–227, 229–234, 236, 238, 563, 568, 569, 570, 573, 577, 581, 597, 602, 607, 609, 611, 616, 649–651, 653, 722, 726, 727, 745, 746, 757, 760

Index

angle, 3, 12, 17, 19, 24, 25, 34, 36, 42, 44, 45, 63, 71, 72, 81, 83, 86, 99, 115, 116, 118, 127, 132, 151–153, 155, 178, 185, 188, 189, 191, 192, 196, 201, 204, 207, 211, 214, 217, 218, 226, 227, 230, 233, 245, 337–340, 370, 539, 540, 542, 544–548, 553, 555, 571, 580, 592, 599, 621, 634, 640, 641, 648, 656, 661–663, 680, 682, 684, 685, 721, 722, 727, 736, 744, 750, 752, 753, 755–757, 759, 762, 765, 766–769, 776, 778

Belt Friction, 216, 232
Coefficient of Friction, 156, 162, 164, 166, 168, 169, 174, 175, 178, 185, 189, 194, 196, 204, 213, 216, 219, 221–223, 225–227, 231, 233, 236, 238, 573, 577, 581, 609, 611, 649–651, 653, 757, 760
Ladder Friction, 185
Screw Friction, 206, 207
Wedge Friction, 196

G

Graphical method, 9, 248, 514
Gravitational Law of Attraction, 8

H

Hogging moment, 449, 450, 455

I

Impulse, 631, 632, 646–652, 654
Instantaneous, 499, 500, 530–532, 595, 675–678, 680, 681, 685, 686, 688–690, 692, 694
Centre, 52, 59, 63, 100, 120, 122, 323–325, 330, 337, 339, 351, 364–366, 368, 375–377, 379, 430, 431, 434, 471, 498, 534, 579, 674,–678, 680–682, 685, 686, 688–694, 714–716, 721–723, 729, 731–733, 737, 744
velocity, 3, 4, 498–506, 508–510, 512, 514–518, 520–522, 524, 525, 529, 530–538, 540–549, 551–555, 560, 561, 563, 585, 587, 591, 593, 595–598, 600, 602, 604, 606, 607, 609, 611, 612, 614–616, 618, 620–624, 631–644, 646–648, 650, 651, 653, 655–657, 661–668, 671–673, 675–689, 691–701, 703, 708–711, 714, 718, 731, 733–737
Inflexion, 452

K

Kinematics, 498, 529, 661, 708
Kinetic energy, 593–596, 709–711, 731, 733
Kinetics, 498, 560, 708

L

Laws of Motion, 8, 560
Line of impact, 633, 634
Linear, 387, 452, 517, 663, 665, 671–673, 675, 680, 683, 685, 687–689, 692, 694, 709, 729, 730, 733, 737, 742
acceleration, 3, 4, 8, 324, 498–506, 508, 509, 511, 514–520, 522, 523, 529–540, 560–562, 567–570, 573–575, 577, 578, 580–584, 586, 591, 597, 598, 619, 621, 623–625, 661–673, 708, 709, 711, 712, 714–732
Velocity, 3, 4, 498–506, 508–510, 512, 514–518, 520–522, 524, 525, 529, 530–538, 540–549, 551–555, 560, 561, 563, 585, 587, 591, 593, 595–598, 600, 602, 604, 606, 607, 609, 611, 612, 614–616, 618, 620–624, 631–644, 646–648, 650, 651, 653, 655–657, 661–668, 671–673, 675–689, 691–701, 703, 708–711, 714, 718, 731, 733–737
Lead, 207, 213–216
angle, 3, 12, 17, 19, 24, 25, 34, 36, 42, 44, 45, 63, 71, 72, 81, 83, 86, 99, 115, 116, 118, 127, 132, 151–153, 155, 178, 185, 188, 189, 191, 192, 196, 201, 204, 207, 211, 214, 217, 218, 226, 227, 230, 233, 245, 337–340, 370, 539, 540, 542, 544–548, 553, 555, 571, 580, 592, 599, 621, 634, 640, 641, 648, 656, 661–663, 680, 682, 684, 685, 721, 722, 727, 736, 744, 750, 752, 753, 755–757, 759, 762, 765, 766–769, 776, 778
Load Intensity, 450

M

Mass Moment of Inertia, 387, 389, 390, 430, 431, 435, 437, 709, 719
Method of Joint, 248, 249, 302
Method of Section, 248, 249, 301, 302, 306, 308, 311, 313
Moment, 4, 13, 20, 96–104, 106, 107, 109–115, 128, 132, 213–215, 230–233, 248, 304, 307, 330, 387–398, 400, 401, 403, 405, 406, 408–411, 413–418, 420, 422, 423, 425–427, 429–431, 434–437, 446, 449–452, 454, 455, 460, 462, 468–470, 472, 474, 477–480, 482–484, 486, 488, 490, 492, 579, 708, 709, 712, 714, 716, 717, 719, 720, 722, 723, 725, 726, 728, 730, 731, 744, 752, 753, 769, 771
  Parallel Axis Theorem, 389, 391, 393, 397, 398, 431, 719
  Perpendicular Axis Theorem, 391, 396, 433
Moment of Inertia, 387–398, 400, 401, 403, 405, 406, 408–411, 413–418, 420, 422, 423, 425–427, 429–431, 434–437, 709, 719, 720, 723, 726, 730, 731
  of Circle, 366, 396
  of Composite Sections, 351, 400
  Moment of momentum, 708, 709
  of Quarter Circle and Semicircle, 337
  of Rectangle, 106, 367, 392, 393, 448
  of Semicircle, 355, 360
  of Triangle, 10, 16, 100, 106, 273, 394, 448, 472
Momentum, 4, 631–633, 635–637, 641–644, 646–657, 708, 709
  angular momentum, 708, 709
  Moment of momentum, 708, 709
Motion, 1, 2, 4, 8, 42, 103, 104, 149, 150, 154, 165, 168, 194–196, 227, 387, 498, 500–502, 504–506, 508, 509, 512, 514, 516, 517, 520, 521, 529–531, 534, 536, 537, 539–541, 546, 548–552, 560–562, 564, 567, 568, 572, 580, 597–599, 602, 631, 650, 661, 663–665, 668, 673–675, 676, 680, 694, 695, 708–710, 712, 714, 715, 721, 722, 725, 727, 728, 730, 733, 734, 760
  Rectilinear, 498, 500, 502, 504–506, 516, 520, 529, 661
  Translatory motion, 673, 675, 709, 710

N

Newton, 3, 8, 560, 561, 631, 709
  Newton's First law, 8, 560
  Newton's Second law, 8, 560, 561, 631, 709
  Newton's Third law, 8, 560
  Plane motion, 500, 661, 673–676, 708–710, 714, 715, 721, 734

O

Oblique eccentric impact, 634

P

Perfect Truss, 245–249
Particle, 2, 8, 323, 498–502, 504, 505, 509, 514, 515, 521, 529–537, 539–542, 560, 591–594, 631, 661, 663, 665, 694, 695, 709, 710, 742
Pitch, 207, 211, 213, 215, 216
Plane Motion, 500, 661, 673–676, 708–710, 714, 715, 721, 734
Polygon Law of forces, 37
potential energy, 593, 595, 596, 710
Projectile, 539–542, 546, 548–552
  motion, 1, 2, 4, 8, 42, 103, 104, 149, 150, 154, 165, 168, 194–196, 227, 387, 498, 500–502, 504–506, 508, 509, 512, 514, 516, 517, 520, 521, 529–531, 534, 536, 537, 539–541, 546, 548–552, 560–562, 564, 567, 568, 572, 580, 597–599, 602, 631, 650, 661, 663–665, 668, 673–675, 676, 680, 694, 695, 708–710, 712, 714, 715, 721, 722, 725, 727, 728, 730, 733, 734, 760
Power, 3, 4, 149, 226, 229, 594, 595, 619–623

R

Radius of Gyration, 388, 389, 710, 716, 717
Range, 169, 170, 220, 221, 540, 542, 544, 545, 546, 547, 549, 553, 554, 565, 566, 635
Reactions, 42, 51, 52, 54, 56, 65, 67, 103, 104, 130, 134, 135, 140, 141, 142, 143, 185, 200, 251, 253, 254, 261, 273, 277, 299, 303, 304, 307, 312, 448, 449, 450, 453, 472, 498, 649, 744, 748, 757, 767, 768, 769, 775

Index

Rectilinear, 498, 500, 502, 504, 505, 506, 516, 520, 529, 661
  Coordinates, 21, 323, 324, 325, 327, 329, 330, 331, 335, 351, 529, 540
  Motion, 1, 2, 4, 8, 42, 103, 104, 149, 150, 154, 165, 168, 194–196, 227, 387, 498, 500–502, 504–506, 508, 509, 512, 514, 516, 517, 520, 521, 529, 530, 531, 534, 536, 537, 539, 540, 541, 546, 548–552, 560–562, 564, 567, 568, 572, 580, 597–599, 602, 631, 650, 661, 663–665, 668, 673–676, 680, 694, 695, 708–710, 712, 714, 715, 721, 722, 725, 727, 728, 730, 733, 734, 760
Relative Velocity, 634, 642, 643, 694, 695, 696, 697, 698, 699, 700, 701, 703
Rigid Body, 1, 2, 4, 9, 10, 11, 13, 34, 96, 98, 102, 661, 663, 673, 708, 709, 710, 711
Rotary motion, 529, 580, 661, 663, 673, 675, 680, 708 709, 710, 733
Rotational Motion, 500, 661

S

Sagging moment, 449, 450, 455
Shear Force, 446, 448–452, 454, 458–460, 462, 464, 466–469, 470, 472, 475–477, 479, 481–483, 484, 487, 490, 492
  diagram, 42, 45, 49, 50, 53, 55, 79, 103, 116, 126, 129, 131, 152, 186, 188, 190, 191, 193, 195, 197, 199, 202, 203, 205, 217, 229, 230, 301, 449, 454, 455, 458–460, 462, 464, 466, 468–470, 472, 474, 475, 477–479, 480–484, 486–488, 490, 492 532, 562, 564, 571, 577, 579, 581, 586, 600–603, 605, 608, 609, 611, 620, 650, 654, 679, 721, 724, 732, 742, 775, 776
Sign Convention, 449, 476, 481
Scalar, 4, 15, 17, 499, 591–593, 632
  product, 15, 17, 18, 19, 20, 24, 25, 26, 97, 98, 219, 324, 389, 631
  quantities, 2, 4, 17, 185, 576, 592, 632
Screw Jack, 207–209, 211–216, 761
Slack sides, 216

T

tight side, 216, 217, 225, 724–726
Time of flight, 540–543, 546, 547, 554, 555
Translatory motion, 673, 675, 709, 710
Triangle Law of forces, 37
Torque, 216, 222–226, 229, 230, 232, 233, 235–238, 709, 731, 732, 744
Trusses, 245
  Redundant Truss, 247
  Three Dimensional, 20, 245
  Two Dimensional, 34, 37, 44, 96, 102

U

Uniform acceleration, 500, 508, 509, 619, 672
Uniform velocity, 499, 522, 620–622, 691, 699, 701, 703
Units, 2–4, 20, 595
  Base units, 2–4
  Fundamental units, 2, 3
  SI units, 3

V

Variable acceleration, 500, 501, 504
Varignon's theorem, 101, 112, 120–122, 124, 323, 324
Vector, 4, 12–23, 29, 30, 37, 97, 99, 103, 251, 498, 499, 529–532, 592, 631–633, 661, 662
  Fixed Vector, 13, 14
  Free Vector, 14, 99, 103
  Multiplication of vectors, 12, 16, 17
  Null Vector, 13, 15
  Product, 15, 17–20, 24–26, 97, 98, 219, 324, 389, 631
  Sliding Vector, 13
  Unit Vector, 13, 15, 19, 21–23, 30
Vectors Addition, 16, 17
Vectors Equality, 15
Vectors Subtraction, 17
Vector Quantities, 4, 592, 632

Velocity, 3, 4, 498–506, 508–510, 512, 514–518, 520–522, 524, 525, 529–538, 540–549, 551–555, 560, 561, 563, 585, 587, 591, 593, 595–598, 600, 602, 604, 606, 607, 609, 611, 612, 614–616, 618, 620–624, 631–644, 646–648, 650, 651, 653, 655–657, 661–668, 671–673, 675–689, 691–701, 703, 708–711, 714, 718, 731, 733–737
of projection, 521, 540, 542, 544, 545, 547, 548, 553, 555

Virtual Work, 742–778

W

Work, 3, 5, 7, 19, 30, 150, 207, 351, 560, 591–595, 597–605, 607, 608, 610, 612, 623, 624, 631, 668, 675, 708, 711, 733–736, 742–778

Work Done, 30, 560, 591–594, 597, 612, 623, 624, 711, 733, 735, 736, 743, 744

Work–Energy principle, 593, 597–605, 607, 608, 610, 612, 733–736